ADVANCES IN BIOMATERIALS SCIENCE AND BIOMEDICAL APPLICATIONS

Edited by **Rosario Pignatello**

Advances in Biomaterials Science and Biomedical Applications
http://dx.doi.org/10.5772/56420
Edited by Rosario Pignatello

CBS, Edition 2016

Contributors

Chowdhury, Xiaohong Wang, Irene Tereshko, Valery Tereshko, Patrick Frayssinet, Ben Ayed Foued, Shaojun Yuan, Gordon Xiong, Ariel Roguin, Swee Hin Teoh, Cleo Choong, Tiago Pereira, Andrea Gartner, Paulo Armada-Da-Silva, Cátia Pereira, Miguel França, Diana Morais, Miguel Rodrigues, Ascenção Lopes, José Domingos, Ana Lúcia Luís, Ana Colette Maurício, Irina Amorim, Raquel Gomes, Xiongbiao Chen, Mituso Niinomi, Ylenia Zambito, Masaru Murata, Young-Kyun Kim, Kyung-Wook Kim, Jeong Keun Lee, In-Woong Um, Stefano Geuna, Frank Xue Jiang, Yan-Ru Lou, Carmen Escobedo-Lucea, Arto Urtti, Marjo Yliperttula, Juan Valerio Cauich-Rodríguez, Juliana Carvalho, Mhamdi Lotfi, M. Nejib, M. Naceur, Lucie Germain, Jean-Michel Bourget, Maxime Guillemette, Teodor Veres, François A. Auger, Ruggero Bettini, Susan Scholes, Thomas Joyce

Published by InTech

Janeza Trdine 9, 51000 Rijeka, Croatia

Publishing Process Manager Oliver Kurelic

Technical Editor InTech DTP team

Cover InTech Design team

Additional hard copies can be obtained from orders@intechopen.com

Advances in Biomaterials Science and Biomedical Applications, Edited by Rosario Pignatello
p. cm.
ISBN 978-953-51-1051-4

INTECH

open science | open minds

Contents

Preface

A recent editorial production from InTech resulted in the publication of three volumes focused on biomaterials. In those books, also edited by myself, the fundamental and applicative aspects of biomaterials, in the wide connotation of the word, have been reviewed and supported by the experimental work of many scientists, who from many years have dedicated their research to this fascinating world, composed of many different skills, techniques and competencies.

When I was invited by the Publisher to coordinate a further editorial task on Biomaterials, I was glad to help in collecting new contributions in this area of research and science. The scientific production in the field is, in fact, rapidly growing and updating, mainly on the fronts of new and original applications of already known or novel compounds and polymers. As proof, we easily received a high number of articles to be selected for composition of this new volume.

The chosen title gives a clear suggestion to the need of focusing all the basic studies, for instance the physico-chemical characterization of biomaterials, towards their potential applications in biomedicine and drug delivery, or in any other relevant area of diagnosis, therapy, surgical manipulation, and rehabilitation. Traditional, or 'known', biomaterials can now be handled to meet specific medical needs, based on the large experience of their chemical, physical and biological properties. Conversely, newly produced materials can be directly designed and tailored to such requirements, so that novel and somewhat unexpected areas of application are continuously disclosed.

These considerations have been the basis of this editorial product. The contribution presented consists of review articles, original researches and experimental reports from eminent international experts of the multidisciplinary world, which is required for an effective development and utility of biomaterials. 21 chapters have been organized to explore different aspects of biomaterials science. From advanced means for the characterization and toxicological assessment of new materials, passing through some 'classical' applications in nanotechnology and tissue engineering, toward novel specific uses of these products, the volume is intended to give readers a view of the wide range of disciplines and methodologies that have been exploited to develop biomaterials with the physical and biological features needed for specific clinical and medical purposes.

I hope that you reading these interesting chapters will prompt your interest research towards the exciting field of biomaterials science and applications

Rosario Pignatello
Universita degli Studi di Catania, Italy

Section 1

Characterization of Novel Biomaterials

Chapter 1

Biomedical Applications of Materials Processed in Glow Discharge Plasma

V. Tereshko, A. Gorchakov, I. Tereshko,
V. Abidzina and V. Red'ko

Additional information is available at the end of the chapter

http://dx.doi.org/10.5772/55548

1. Introduction

There is exhaustive literature about interactions of charged particles with solid surfaces [l, 2]. For a long period only high energies were assumed to cause any significant modifications. However, low-energy ion bombardments (up to 5 keV) of metal and alloy samples were shown to be very efficient too: the increase of dislocation density (up to 10 mm in depth from the irradiated surface) was detected [3–7]. In fact, a bulk long-range modification of materials in the glow discharge plasma (GDP) took place. The above results were obtained by the use of transmission electron microscopy for well annealed samples with initially small dislocation density (armco-Fe, Ni3Fe, *etc.*) [4, 6]. For materials with initially increased dislocation density (unannealed copper, M2 high-speed steel, titanium alloys) reorganization of dislocation structure is the most considerable: either intensive formation of the dislocation fragments or grinding of the fragments with corresponding increase in their disorientation is observed. These reorganizations also take place well below the irradiated surface. When the ion energy decreases by 1 keV, the modified layer became even deeper [7].

The above results can only be explained by taking the nonlinear nature of atom interactions into account. The ion bombardment is assumed to induce nonlinear oscillations in crystal lattices leading to self-organization of the latter. Modelling shows the formation of new collective atom states. The observed phenomena include the redistribution of energy, clusterization, structure formation when the atoms stabilizes in new non-equilibrium positions, localized structures, auto-oscillations, and travelling waves and pulses [3–7].

The next step was to look at the influence of low-energy GDP on liquids. Water that occupies up 70 percent of the Earth's surface and is the main component of all living things was taken for investigation. Water molecules are able to create molecular associates using Van der Waals forces as well as labile hydrogen interactions [8–11]. Owing to hydrogen bonds molecules of water are capable to form not only random associates (one having no ordered structure) but clusters, i.e. associates having some ordered structure [9–11]. The network of hydrogen bonds and the high order of intermolecular cooperativity facilitate long-range propagation of molecular excitations [12, 13]. This allows, in principle, to consider water and water-based solutions as systems sensitive to weak external forces. Indeed, the study of luminescence at long time scale shows that the structural equilibrium in water is not stable: it changes after dissolution of small portions of added substances and after exposition of aqueous samples to UV and mild X-ray irradiation [14].

The results obtained by Lobyshev, *et al* opened up the new avenues to water and aqueous solutions as non-equilibrium systems capable of self-organization [14]. The key property of self-organization is, however, nonlinearity to which, in models of water, hasn't paid the required attention yet. The present paper is aimed to cover this flaw. Basic models of nonlinear chains that can be related to water structure were investigated. We observed self-organization processes resulting in the displacement of atoms and their stabilization in new positions, which can be viewed as the formation of water clusters.

In experiments, we exposed crop seeds, baking yeast and water to GDP. The results were very promising: the seed sprouts showed greater growth and the yeast showed greater metabolic activity compared to the control samples. The results on volunteers with different diseases, who either drunk the processed water or was injected intravenously with the processed physiological solution, were encouraging too. The diagnostics of volunteers' blood immune cells (lymphocytes and leukocytes) showed significant normalization of their state toward homeostasis.

Next part of this paper is devoted to the study of properties of implants processed in GDP. The modern medicine is characterized by active introduction of high technologies to clinical practice. It requires sufficient biocompatibility of implanted mechanical, electromechanical and electronic devices with natural tissues. The properties of materials are crucial, since insufficient biocompatibility can lead to the negative reactions to the implant from the side of surrounding tissues causing inflammatory processes, dysfunction of the endothelium, disturbance of homeostasis, destruction and the necrosis of bone tissue and so forth [15, 16]. The formation of hydrophilic coatings and the modification of chemical composition and topography of the implant surface make it possible to reduce the frequency of the development of negative processes. The bone, fibrous and endothelial tissues are uniquely structured, and the attempts to design the next generations of implants are focused on the development of unique nanotopography of the surface of implants based on the imitation of nature. Our and other studies showed the effectiveness of vacuum-plasma technology for improving biocompatibility and durability (mechanical and chemical) of implanted materials [17–19]. New avenues in the application of above technology to the titanium implants and their influence to surrounding tissues are explored in this paper.

http://dx.doi.org/10.5772/55548

2. Modelling atomic and molecular chains

Molecular dynamics were used to develop the model. To describe the atomic and molecular interactions, Morse (1) and Born-Mayer (2) potentials were chosen [2].

Morse potential takes the form

$$U(r) = J\left\langle \exp\left[-2\alpha\left(r - r_0\right)\right] - 2\exp\left[-\alpha\left(r - r_0\right)\right]\right\rangle \quad (1)$$

where J and α are the parameters of dissociation energy and anharmonicity respectively; $\Delta r = (r - r_0)$ is the displacement from an equilibrium.

Born-Mayer potential takes the form

$$U(r) = T \cdot e^{\frac{-r}{a}} \quad (2)$$

where T, a, and r are the energy constant, the shielding and atomic lengths respectively.

We assume the existence of multiple equilibria corresponding to thermodynamic as well non-thermodynamic branches. Expanding the potentials in a Taylor series (up to the fifth order term), find the interaction force

$$F = -\frac{dU(r)}{dr} = -K\Delta r + A\Delta r^2 - B\Delta r^3 + C\Delta r^4 - D\Delta r^5 \quad (3)$$

For the Morse potential

$$K = 2\alpha^2 J,\ A = 3\alpha^3 J,\ B = 2.3\alpha^4 J,$$
$$C = 1.25\alpha^5 J,\ D = 1.1\alpha^6 J \quad (4)$$

where K, A, B, C, D are the coefficients of elasticity, quadratic cubic, fourth and fifth orders nonlinearities respectively.

For the Born-Mayer potential

$$K = \frac{T}{a^2},\quad A = \frac{T}{2a^2},\quad B = \frac{T}{6a^4},\quad C = \frac{T}{24a^5},\quad D = \frac{T}{120a^6}. \quad (5)$$

The coefficient values are presented in Table 1.

Coefficient	Born-Mayer potential	Morse potential
K, N/m	9,341•10^{4}	1,140•10^{4}
A, N/m^2	1,951•10^{15}	8,244•10^{14}
B, N/m^3	2,716•10^{25}	3,046•10^{25}
C, N/m^4	2,836•10^{35}	7,980•10^{35}
D, N/m^5	2,370•10^{445}	3,385•10^{446}

Table 1. Coefficients for Morse and Born-Mayer potentials.

There are many models that describe water molecules [13]. The molecular structure of water is presented in Figure 1. The covalent and hydrogen bonds are marked by the grey springs and the bold lines respectively. For simplicity, in our simulations we consider a chain, i.e. 1D lattice, of water molecules (see the marked area of Figure 1).

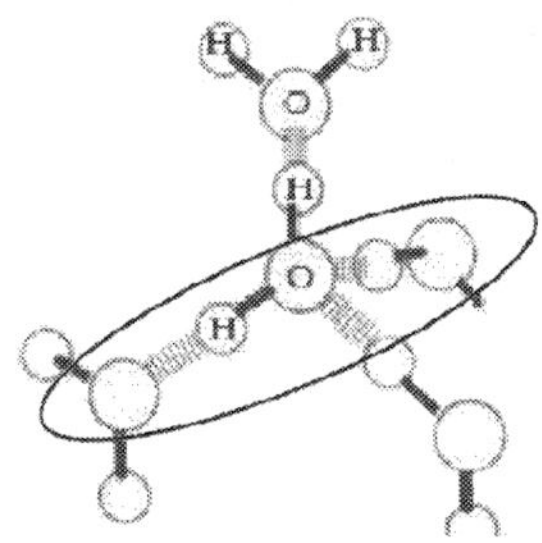

Figure 1. Molecular structure of water in a solid phase. The ellipse marks a piece of 1D chain used in simulations.

Considering only single component of $r = (x, y, z)$, say x, and viewing the atom as interacting nonlinear oscillators, the system equations take the form:

$$
\begin{aligned}
& m\frac{d^2x_1}{dt^2} = -K'x_1 + Ax_1^2 - Bx_1^3 + Cx_1^4 - Dx_1^5 + K(x_2 - x_1) - \\
& -A(x_2 - x_1)^2 + B(x_2 - x_1)^3 - C(x_2 - x_1)^4 + D(x_2 - x_1)^5 - \beta'\frac{dx_1}{dt}, \\
& m\frac{d^2x_i}{dt^2} = -K(x_i - x_{i-1}) + A(x_i - x_{i-1})^2 - B(x_i - x_{i-1})^3 + C(x_i - x_{i-1})^4 - \\
& -D(x_i - x_{i-1})^5 + K(x_{i+1} - x_i) - A(x_{i+1} - x_i)^2 + B(x_{i+1} - x_i)^3 - C(x_{i+1} - x_i)^4 + \\
& +D(x_{i+1} - x_i)^5 - \beta\frac{dx_i}{dt}, \\
& m\frac{d^2x_n}{dt^2} = -K(x_n - x_{n-1}) + A(x_n - x_{n-1})^2 - B(x_n - x_{n-1})^3 + C(x_n - x_{n-1})^4 - \\
& -D(x_n - x_{n-1})^5 - K'x_n + Ax_n^2 - Bx_n^3 + Cx_n^4 - Dx_n^5 - \beta'\frac{dx_n}{dt},
\end{aligned}
\tag{6}
$$

where x_i, $i = 1,\ldots, n$ is displacement of i-th oscillator from the its equilibrium position, K′ is the coefficient of elasticity on the chain borders, and ß and ß′ are the damping factors inside the chain and on its borders respectively. The system (6) was solved by the Runge–Kutta method.

Relaxation processes of atoms after stopping the external influence were under investigation. Sources that gave impulses to atoms of the chains were both direct ion impact on the first atom of the chain (single impact) and random impacts on randomly chosen atoms of the chain (plasma treatment). In practice the atom bonds are important to keep unbroken, so all types of influences were low-energy ones.

2.1. Hydrogen atom chain

We carried out the simulations for chain consisting of 50 hydrogen atoms (Figure 2). Morse potential was chosen; N_i defines the number of i-th atoms. For single impact the first atom of the chain was displaced with velocity $V = 500$ m/s, which corresponds to 10^{-3} eV of the exposed energy. In case of plasma treatment the following atoms were exposed to low-energy impacts: atom N_1 ($V = 538$ m/s), atom N_{10} ($V = 1682$ m/s) and atom N_{30} ($V = 1237$ m/s).

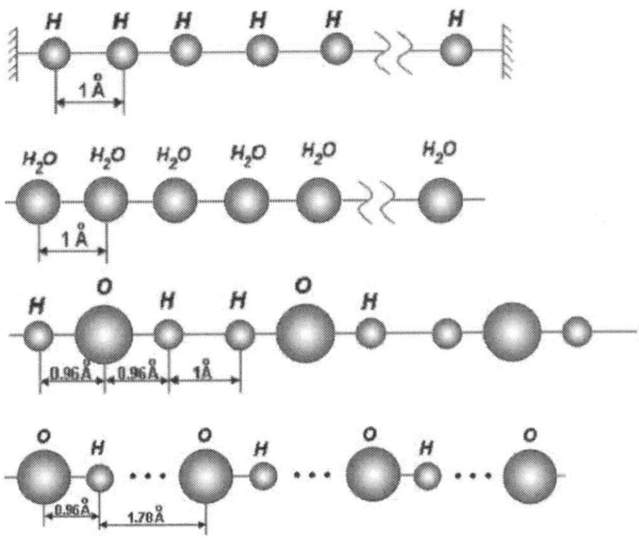

Figure 2. Chain of hydrogen atoms.

Figure 3 illustrates the atom displacements after the plasma treatment. The atoms are stabilized in the new positions that can be described as (nano)clusters (atoms N_{1-29} and N_{30-50}). After atom relaxation the simulations were continued fourth times longer, and the persistent stabilization was always observed.

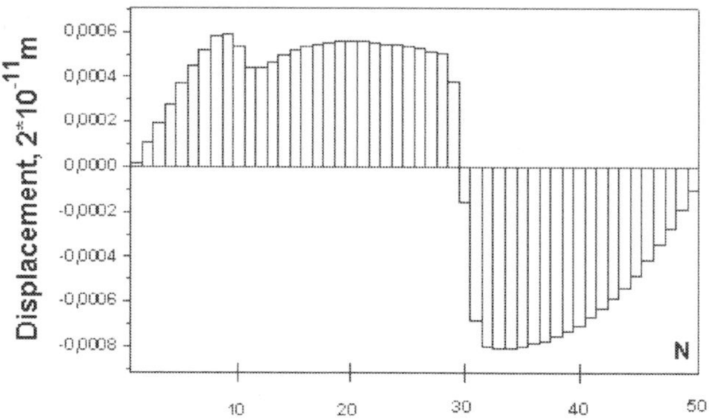

Figure 3. Displacement of 50 atoms of the excited nonlinear chain at the time of stabilization. *N* defines the atom number in the chain.

2.2. H–O–H molecule chain

We investigated the chain of H–O–H molecules shown in Figure 4. From two to eight molecules (6–24 atoms) were used. The equilibrium distances between H and O atoms inside the molecule are 0.96 Å, and the equilibrium distance between the molecules is 1 Å, which corresponds to … Single impact were assumed, and the velocity of the first atom was varied from 100 to 1600 m/s, which corresponds to 10^{-5}–10^{-2} eV of the exposed energy. Again, Morse potential was used.

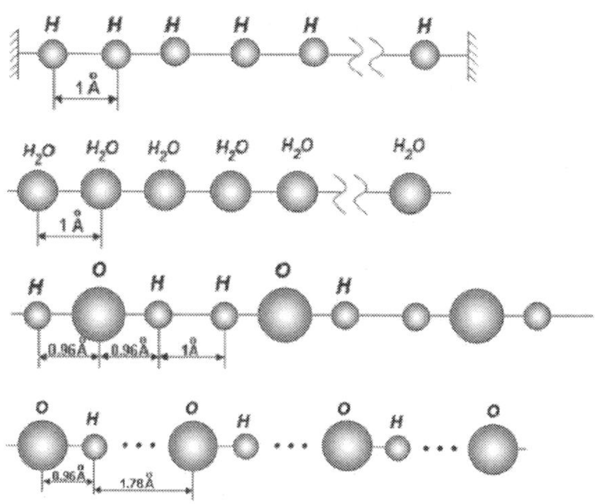

Figure 4. Atom chain of water consisting of hydrogen and oxygen atoms.

http://dx.doi.org/10.5772/55548

Figure 5 illustrates the atom displacements versus the velocity that the first atom received from external impact. In all cases significant shrinkage, or collapse, of the chains is observed. One can see that the length of the collapsed chain depends on the above velocity, the minimal length being detected at some low impact energies. For example, for the chain consisting of eight H–O–H molecules the minimal length of the chain was observed at $V = 1200$ m/s (see Fig. 5b).

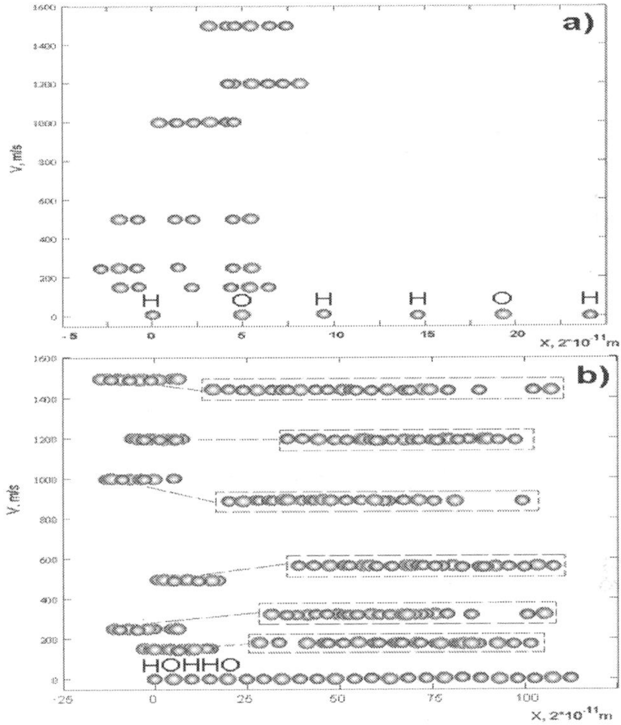

Figure 5. Atom displacements in the chain of H–O–H molecules versus the velocity received by the first atom: a) chain consisting of two molecules, b) chain consisting of eight molecules. In dashed areas the collapsed chains are shown enlarged.

2.3. 1D water molecule chain

Finally, we investigated the chain shown in Figure 6. It corresponds to the 1D cut of water molecule (see the area marked by the ellipse in Figure 1). The solid and dotted lines correspond to the covalent and hydrogen bonds respectively. As one can see the covalent bonds yields the equilibrium between O and H atoms at 0.96 Å whereas the hydrogen bonds yields the equilibrium at about twice longer distance (1.78 Å).

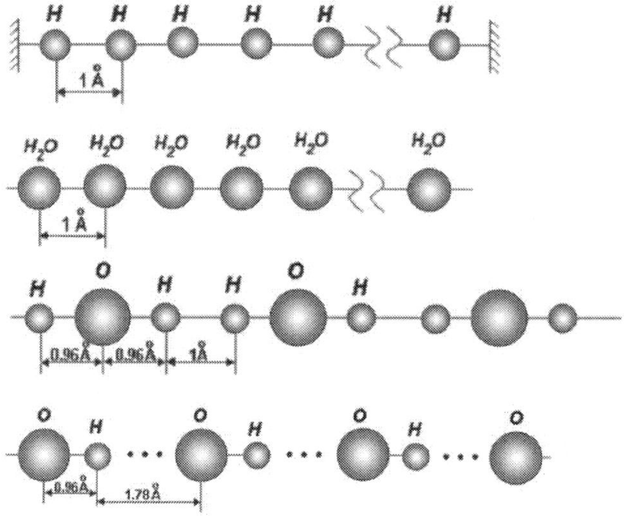

Figure 6. chain of water consisting of hydrogen and oxygen atoms. The solid and dotted lines mark the covalent and hydrogen bonds respectively.

In simulations we considered simple chain consisting of two 1D water molecules. The initial velocity of the first atom was taken at $V = 500$ m/s. The Morse and Born-Mayer potentials were used for this investigation.

Figure 7 represents the initial and final stabilized conditions of atom chains (after direct low-energy ion impact to the first atom of the chain) calculated with Born-Mayer (Figure 7a) and Morse (Figure 7b) potentials

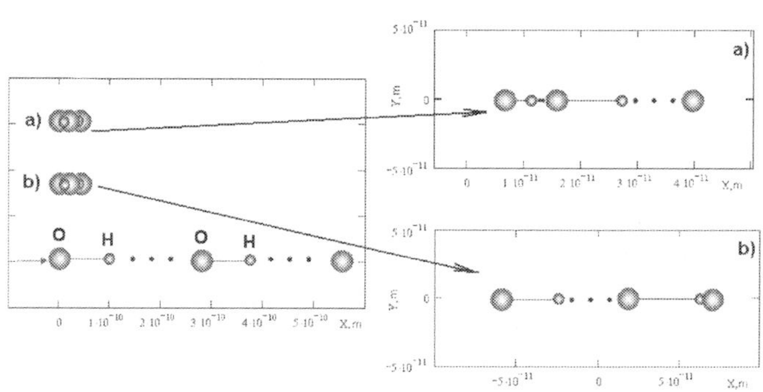

Figure 7. Initial and final positions of atoms of 1D water molecule after direct low-energy impact to the first atom (oxygen): a) Born-Mayer potential, b) Morse potential. The right figures show the enlarged final chains.

3. Biomedical applications

Biological objects are known for their high sensitivity to weak external fields. The evidence that electromagnetic fields can have "non-thermal" biological effects is now overwhelming. When the production of heat shock proteins is triggered electromagnetically it needs 100 million times less energy than when triggered by heat [20]. Low-frequency weak magnetic fields may lead to the resonant change of the rate of biochemical reactions although the impact energy is by ten orders of magnitude less than k_BT where k_B is the Boltzmann constant and T is the temperature of the medium [21].

The therapeutic ability of the low intensity electromagnetic radiations is actively discussed [22]. The low-power millimeter wave irradiation and magnetic-resonance therapy are used in practical medicine already, which differ significantly from the drug treatment by the fact that they do not clog organism with the undesirable chemical compounds, i.e. xenobiotics. In this chapter we discuss the biomedical application of vacuum-plasma technologies.

3.1. Activating and therapeutic properties of water processed in GDP

To understand the above extreme sensitivity of living objects, investigations in influences of weak fields on water appear to be essential. Indeed, water plays a major role in biological processes. A man consumes about 2 l of drinking water a day. Water is the main component of human, animal, plant and generally every living being body. A new-born child body contains 97% of water, decreasing to 70–75% with aging. In particular, human brain consists of about 85% of water.

So, we performed experiments with water, crop seeds and baking yeast *S. cerevisiae*. The crop seeds and yeast were processed directly in GDP. Also, the untreated crop seeds and yeast were poured with the water processed in GDP. In all cases practically the same biotrophic effects were observed. Namely, the seed sprouts showed the growth in 3–4 times higher than the control samples. Both the processed yeast and the unprocessed one that immersed in the processed (by GDP) water showed greater metabolic activity compared to the control samples.

The obtained results allow suggesting that the discovered phenomena can be used for direct correction of pathological states. Therefore we processed water and physiological solution. The samples were exposed to low-energy ion irradiation in GDP of residual gases. The ion energy depends on the voltage in the plasma generator. The latter was kept at 1.2 keV while the current in the plasma generator was maintained at 70 mA. The temperature in the chamber was controlled during the irradiation process and did not exceed 298 K (25° C). The irradiation time was 60 minutes.

In test experiments, volunteers with different diseases either drunk the processed water or they were injected intravenously with the similarly processed physiological solution. The course of treatment included 3–5 sessions of 0.5 l physiological solution transfusion. The preliminary results appeared to be very promising. We were most interested in the therapeutic treatment of the global inflammatory processes such as cardio-vascular diseases and pancreatic (insular) diabetes complicated by the acute and chronic forms of atherosclerosis. Also, different types of oncology, say, leukemia, etc., were under investigation.

The blood immune cells were taken for diagnostics. The immune system is known as one of the leading homeostatic systems in the organisms. It may serve as a mirror that reflects practically all adaptations and pathological rearrangements. The immunocompetent cells, lymphocytes and leukocytes, have a set of properties that may be used as an indicator of the organism state. In addition, the structural organization of blood lymphocytes and leukocytes makes possible a most efficient use of microspectral analysis and different fluorescent probes for their studies [23].

We used the dual-wavelength microfluorimetry analysis. The selected cell populations (mono- and polynuclears of blood immunocytes) were mapped as clusters of points on the phase plane in coordinates of the red and green luminescence intensities, i.e. on the wavelengths I_{530} (abscissa) and I_{640} (ordinate). Figure 8 presents the above phase plane for an oncologic out-patient before and after the monthly course treatment. The black and grey pluses represent lymphocyte and leukocyte cells respectively. The white ellipses mark the distribution of fluorescent signals of lymphocytes (lower ellipse) and leukocytes (upper ellipse) in norm. As seen, the treatment results into significant normalization toward the homeostasis.

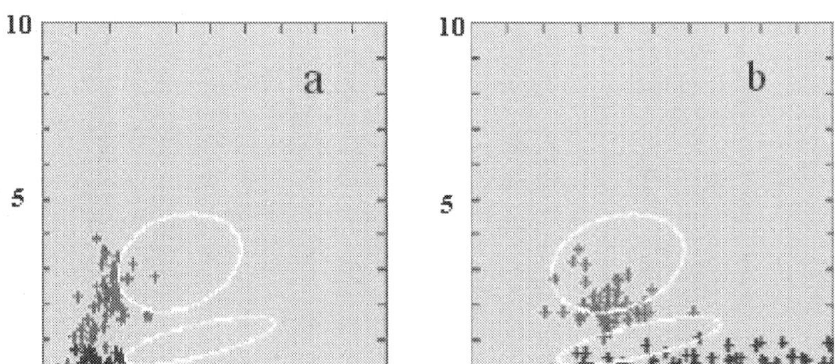

Figure 8. Dual-wavelength microfluorimetry analysis (abscissa and ordinate represent the luminescent intensity I_{530} and I_{640} respectively) of blood immunocytes of oncologic patient (second stage breast cancer). The state before (a) and after (b) the water treatment course (see the text). The black and grey pluses represent lymphocytes and leukocytes respectively. The white ellipses mark the distribution of fluorescent signals of lymphocytes (lower ellipse) and leukocytes (upper ellipse) in norm.

3.2. Biocompatibility of titanium alloys and stainless steel processed in GDP

Stainless steel, titanium and its alloys are among the most utilized biomaterials and are still the materials of choice for many structural implantable device applications [24, 25]. We processed both the titanium and stainless steel samples and investigated changing in their properties caused by GDP.

Current titanium implants face long-term failure problems due to poor bonding to juxtaposed bone, severe stress shielding and generation of debris that may lead to bone cell death and perhaps eventual necrotic bone [26–28]. Improving the bioactivity of titanium implants, especially with respect to cells, is a major concern in the near and intermediate future. Surface properties such as wettability, chemical composition and topography govern the biocompatibility of titanium. Conventionally processed titanium currently used in the orthopedic and dental applications exhibits a micro-rough surface and is smooth at the nanoscale. Surface smoothness on the nanoscale has been shown to favor fibrous tissue encapsulation [27–29]. An approach to design the next-generation of implants has recently focused on creating unique nanotopography (or roughness) on the implant surface, considering that natural bone consists of nanostructured materials like collagen and hydroxyapatite. Some researchers have achieved nano-roughness in titanium substrates by compacting small (nanometer) constituent particles and/or fibers [30]. However, nanometer metal particles can be expensive and unsafe to fabricate. For this reason, alternative methods of titanium surface treatment are desirable.

For the investigation of biocompatibility of implanted materials the tests *in vitro* with the cultures of different cells (fibroblasts, lymphocytes, macrophages, epithelial cells, *etc.*) are used. The influence of material is typically evaluated according to such indicators as adhesion, change in the morphological properties, inhibition of an increase in the cellular population, oppression of metabolic activity and others.

The adhesion of cells, as is known, plays exceptionally important role in the biological processes, such as formation of tissues and organs during embryogenesis, reparative processes, immune and inflammatory reactions, *etc.* Capability for movement is the characteristic property of fibroblasts, cells of immune system and cells, which participate in the inflammation. Moreover, in immunocytes and leukocytes it consists not only in the free recirculation in the blood stream or lymph but also in the penetration into vascular walls and active migration into the surrounding tissues. Adhesion and flattening of cells to the base layer always precede their locomotion. The degree of flattening is important preparatory step to the cell amoeboid mobility. We concentrate our attention on the above components in experiments with titanium alloys.

Titanium samples were cut into pieces (1 cm × 0.5 cm) and placed in a specially constructed plasma generator. They were exposed to glow discharge plasma by ions of the residual gases of the vacuum. The ion energy depended on the voltage in the plasmatron and did not exceed 1–10 keV. Irradiated fluence was 10^{17} ion•cm^{-2}. The temperature of the specimens was controlled during the irradiation process and did not exceed 343 K while the irradiation time varied from 5 to 60 minutes. Rutherford Backscattering Spectrometry (RBS) was used to study the changes after the irradiation. Cell adhesion to titanium samples was tested with L929 mouse connective tissue (fibroblasts-like cells). L929 cells were cultured in Dulbecco's modified Eagle's medium with 10% fetal bovine serum. Initial cell density was $5 \bullet 10^5$ cells/ml. The samples were placed into the sterile disposable 9 cm diameter tissue culture Petri dishes. 2 ml growth medium with cells were distributed into each Petri then incubated in the 5% CO_2 at 37º C for 2 hours. After that period, cultures were prepared for scanning electron microscopy (SEM).

RBS data for the irradiated sample show the presence of iron on the surface that occurred from high-carbon steel cathode as a result of secondary emission process (Figure. 9). Percentage of iron and thickness of the layer were calculated using RUMP, the program for simulation and analysis based on RBS and Elastic Recoil Detection techniques.

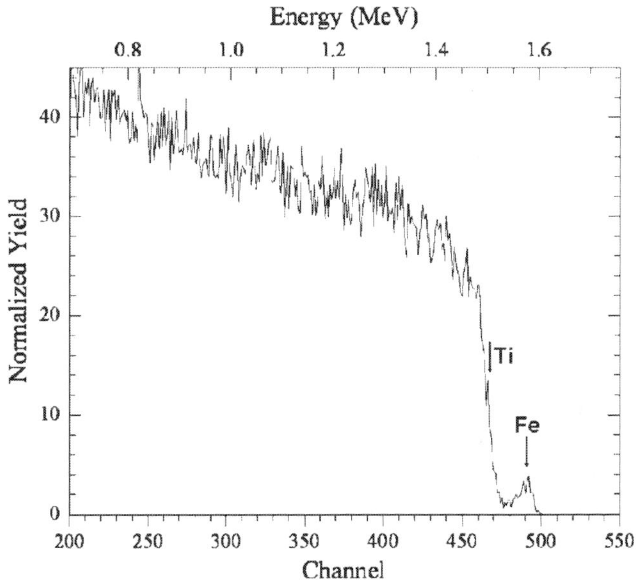

Figure 9. RBS spectrum of the titanium sample irradiated for 5 min at 10 kV.

http://dx.doi.org/10.5772/55548

The obtained data for different voltage and time of the irradiation are presented in Table 2.

Voltage, kV	Time of irradiation, min	Fe:Ti atomic ratio	Density of flattened cells per μm²	Percentage of flattened cells	Increase factor in amount of all cells in comparison with control sample
0.4	60	0.0277:1	534±20	50.2±2.0	1.78
1.2	30	0.0560:1	413±9	43.7±0.9	1.63
10	5	0.0549:1	381±15	42.5±1.7	1.53
Control	0	0:1	26±8	4.4±1.5	1

Table 2. Data obtained from the experiments with titanium samples exposed to GDP.

Calculated data indicate an increase in the density of flattened cells as well as in the cell amount in comparison with the control sample. According to Table 2 one conclude that best adhesion (column 4) and most prolific cell attachment (column 5) correspond to the samples that were exposed to GDP for maximum time at minimum voltage. For this sample we observed less percentage of iron and thickness of the iron layer in comparison with others that were exposed to higher voltage plasma irradiation.

Figure 10 demonstrates SEM images of control and irradiated samples. In comparison with the control sample, analysis of cell attachment for the irradiated samples shows high confluence (attachment ratio) and better spreading.

We also performed experiments on the adhesion of immune-competent cells of human blood to the stainless steel samples. Figure 11 represents the microphotography of the healthy person lymphocytes and leukocytes adhered to the irradiated and non-irradiated plates. As can be seen from photographs, cells, which are located on the different samples, are essentially different. The morphology of leukocytes and lymphocytes, which were adhered to the irradiated material, indicates the expressed amoeboid mobility.

In the majority of the cases endoprosthetics is conducted not in the healthiest people. This fact is very important and it must be considered. Figure 12 displays the results of similar study of the blood nucleus of person who suffers from second stage hypertonia, coronary artery disease and atherosclerosis. From the above data one can conclude that the nature of adhesion of cells to the base layer depends on both the physico-chemical state of this base layer and the state of organism, the owner of cells.

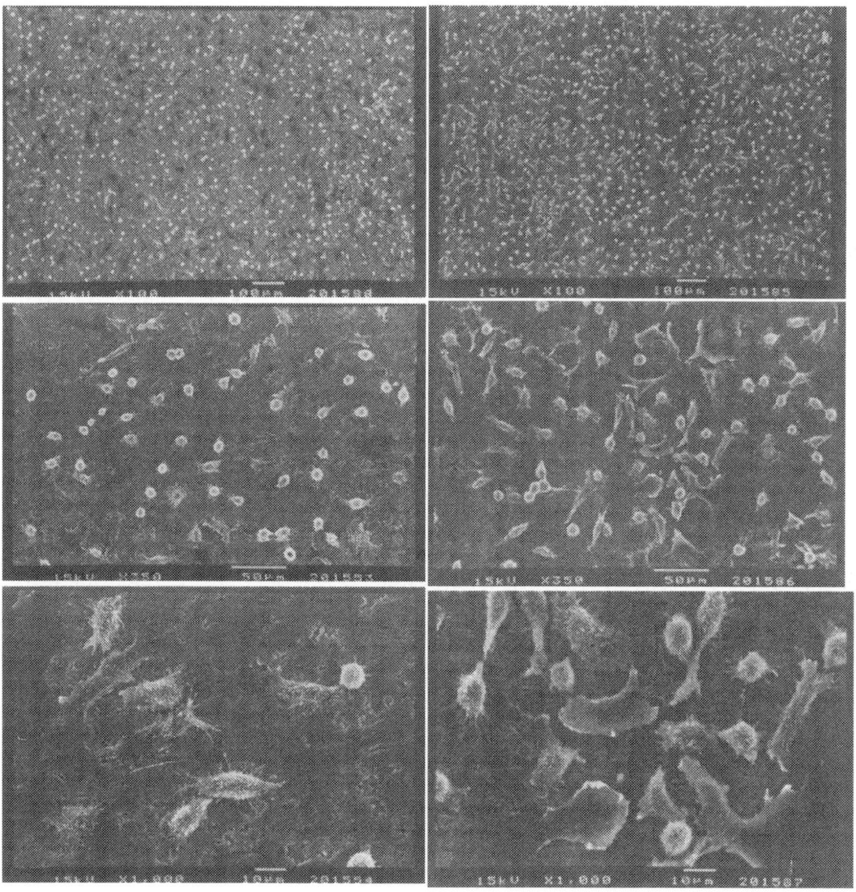

(a) (b)

Figure 10. SEM images of cell attachment on (a) the control sample and (b) the titanium sample that was irradiated for 5 min at 10 kV.

http://dx.doi.org/10.5772/55548

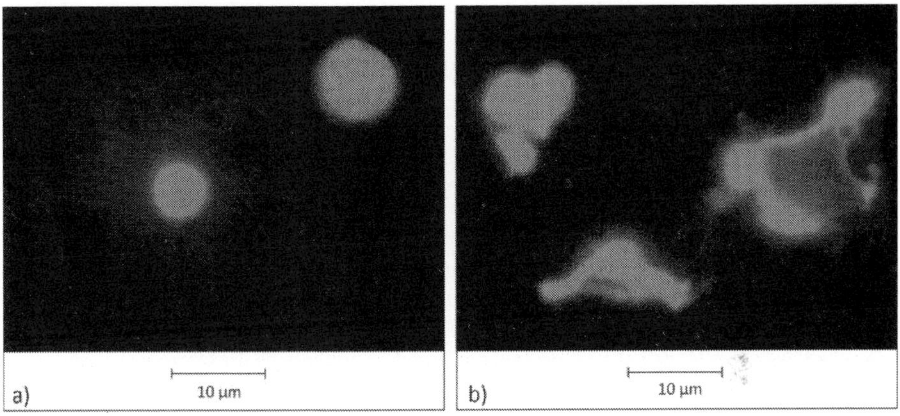

Figure 11. Luminescent microscopy (1000×) of lymphocytes and granulocytes of the blood of healthy donor adhered to (a) non-irradiated and (b) irradiated in GDP surface of the stainless steel samples. The cell nucleus fluorochromization is performed by propidium iodide (λ_{fl} = 615 nm).

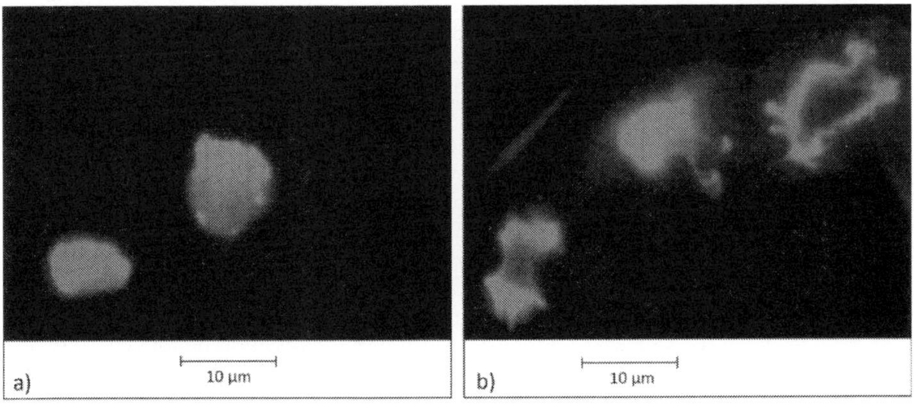

Figure 12. Luminescent microscopy (1000×) *of* lymphocytes and granulocytes of the blood of donor suffering from second stage hypertonia, coronary artery disease and atherosclerosis. The cell nuclei are adhered to (a) non-irradiated and (b) irradiated in GDP surface of the stainless steel samples. The fluorochromization is performed by propidium iodide (λ_{fl} = 615 nm).

4. Discussion and conclusions

Studying the homogeneous chains, like the hydrogen atom chain, exposed to low energies we observed clusterization. It is important to stress that this is truly self-organization phenomenon induced by an external excitation. The chains utilize the excitation energy to initialize nonlinear oscillations and redistribute the energy throughout the chain, which leads to the pattern formation. In the case of multiple impacts on randomly chosen atoms (so-called plasma processing) the atom displacements are by an order higher than in the case of single impact. Thus, the plasma treatment leads to more active self-organization processes and atom rearrangements.

In the cases of inhomogeneous chains containing H and O atoms another type of structures is developed. The shrinkage of chains is so significant that we can say about the collapsed structures. This collapse is observed irrespective of the choice of the atom interaction potentials, whereas the collapsed chain patterns are found to depend on the latter.

To conclude, the performed simulations demonstrated that the system nonlinearity is, in fact, the main reason for the development of self-organization processes leading to significant modifications even in case of low-energy impacts.

In experiments with water and biological objects processed in GDP significant biotrophic effects were detected. The crop seeds and yeast processed directly or indirectly (being immersed in the water processed in GDP) showed markedly greater metabolic activity compared to the control samples. Using the water and the physiological solutions processed in GDP we observed significant therapeutic effects in the test treatments of cardiovascular, oncologic and other diseases. The obtained results suggest the use of discovered phenomena for direct corrections of pathological states by shifting a body state towards its homeostasis. Understanding the mechanisms of the latter will be our next priority.

Next part of this study is devoted to experiments with the titanium alloys and stainless steel exposed to GDP. The experiments with titanium samples reveal an increase in the density of flattened (to the sample surface) cells as well as in the cell amount in comparison with the control sample. These are nothing but preparatory step to the cell amoeboid mobility. Indeed, adhesion and flattening of cells to the base layer always precede their locomotion. According to the results, best adhesion and most prolific cell attachment correspond to the samples that were exposed to GDP for maximum time at minimum voltage. Similar results were obtained in the experiments with stainless steel samples: the morphology of leukocytes and lymphocytes, which were adhered to the irradiated material, indicated the expressed amoeboid mobility. The results with the blood nucleus of person who suffers from several diseases revealed some deviations in the morphology of adhered cells compared to the healthy blood. Thus, the nature of adhesion of cells to the base layer depends on both the physico-chemical state of this base layer and the state of organism, the owner of cells. This circumstance determines even more stringent requirements for the material of the implants.

Author details

V. Tereshko[1*], A. Gorchakov[2], I. Tereshko[3], V. Abidzina[3] and V. Red'ko[4]

*Address all correspondence to: valery.tereshko@uws.ac.uk

1 School of Computing, University of the West of Scotland, Paisley, UK

2 MISTEM, Mogilev, Belarus

3 Department of Physics, Belarusian-Russian University, Mogilev, Belarus

4 Department of Physical Methods of Control, Belarusian-Russian University, Mogilev, Belarus

References

[1] Ziegler, J. F, Biersack, J. P, & Littmark, U. The Stopping and Range of Ions in Solids. New York: Pergamon; (1985).

[2] Eckstein, W. Computer Simulation of Ion-Solids Interaction. Berlin: Springer; (1991).

[3] Tereshko, I. V, Khodyrev, V. I, Tereshko, V. M, Lipsky, E. A, Goncharenya, A. V, & Ofori-sey, S. Self-organizing processes in metals by low-energy ion beams. Nucl. Instr. and Meth. B (1993)., 80, 115-119.

[4] Tereshko, I. V, Khodyrev, V. I, Lipsky, E. A, Goncharenya, A. V, & Tereshko, A. M. Materials modification by low-energy ion irradiation. Nucl. Instr. and Meth. B (1997)., 128, 861-864.

[5] Tereshko, I. V, Glushchenko, V. V, & Tereshko, A. M. Computer simulation of the defect structure formation in crystal lattices by low-energy ion irradiation. Comput. Mater. Sci. (2002)., 24, 139-143.

[6] Tereshko, I, Abidzina, V, Tereshko, A, & Elkin, I. Nanostructural evolution of steel and titanium alloys exposed to glow discharge plasma. Nucl. Instr. and Meth. B (2007)., 261, 678-681.

[7] Tereshko, I. V, Abidzina, V. V, Elkin, I. E, Tereshko, A. M, Glushchenko, V. V, & Stoye, S. Formation of nanostructures in metals by low-energy ion irradiation. Surf. & Coat. Tech. (2007)., 201, 8552-8556.

[8] Stillinger, F. N. Water revisited. Science (1980)., 209(4455), 451-457.

[9] Liu, K, Cruzan, J. D, & Saykally, R. J. Water clusters. Science (1996). , 271(5251), 929-933.

[10] Keutsch, F. N, & Saykally, R. J. Water clusters: untangling the mysteries of the liquid, one molecule at a time. PNAS (2001). , 98(19), 10533-10540.

[11] Galamba, N. Cabral BJC. The changing hydrogen-bond network of water from the bulk to the surface of a cluster: a Born-Oppenheimer molecular dynamics study. J. Am. Chem. Soc. (2008). , 130, 17955-17960.

[12] Luck, W. A. The importance of cooperativity for the properties of liquid water. J. Mol. Struct. (1998).

[13] Shelton, D. P. Collective molecular rotation in water and other simple liquids. Chem. Phys. Lett. (2000).

[14] Lobyshev, V. I, Shikhlinskaya, R. E, & Ryzhikov, B. D. Experimental evidence for intrinsic luminescence of water. J. Mol. Liquids (1999).

[15] Park, J. B, & Lakes, R. S. Biomaterials: An Introduction. New York: Plenum; (1992).

[16] Ratner, B. D, Hoffman, A. S, Schoen, F. J, & Lemons, J. E. editors. Biomaterials Science: Introduction to Materials in Medicine. New York: Academic; (1996).

[17] Abidzina, V, Deliloglu-gurhan, I, Ozdal-kurt, F, Sen, B. H, Tereshko, I, Elkin, I, Budak, S, Muntele, C, & Ila, D. Cell adhesion study of the titanium alloys exposed to glow discharge. Nucl. Instr. and Meth. B. (2007). , 261, 624-626.

[18] Mandl, S, & Rauschenbach, B. Improving the biocompatibility of medical implants with plasma immersion ion implantation. Surf. Coat. Technol. (2002).

[19] Lopez-heredia, M. A, Legeay, G, Gaillard, C, & Layrolle, P. Radio frequency plasma treatments on titanium for enhancement of bioactivity. Acta biomater. (2008). , 4, 1953-1962.

[20] Blank, M, & Goodman, R. Stimulation of stress response by low frequency electromagnetic fields: possibility of direct interaction with DNA. IEEE Trans. Plasma Sci. (2000). , 28, 168-172.

[21] Binhi, V. N, & Savin, A. V. Effects of weak magnetic fields on biological systems: physical aspects. Physics- Uspekhi (2003). , 46(3), 259-291.

[22] Betskii, O. V, Devyatkov, N. D, & Kislov, V. V. Low intensity millimeter waves in medicine and biology. Crit. Rev. Biomed. Eng. (2000).

[23] Gorchakov, A. M, & Karnaukhov, V. N. Melenets YuV, and Gorchakova FT. Identification of pathological conditions by luminescence analysis of immunocompetent blood cells. Biophysics (1999). , 44(3), 550-555.

[24] Hermawan, H, & Ramdan, D. Djuansjah JRP. Metals for biomedical applications. In: Fazel R, editor. Biomedical Engineering- From Theory to Applications. Rijeka: InTech; (2011). , 411-430.

[25] Williams, D. F. Titanium for medical applications. In: Brunette DM, Tengvall P, Textor M, Thomsen P, editors. Titanium in Medicine. Berlin: Springer; (2001). , 13-24.

[26] Buser, D, Nydegger, T, Oxland, T, Cochran, D. L, Schenk, R. K, Hirt, H. P, Snétivy, D, & Nolte, L. P. Interface shear strength of titanium implants with a sandblasted and acid-etched surface: a biomechanical study in the maxilla of miniature pigs. J. Biomed. Mater. Res. (1999). , 45(2), 75-83.

[27] Kaplan, F. S, Hayes, W. C, Keaveny, T. M, Boskey, A, & Einhorn, T. A. Biomaterials. In: Simon SP, editor. Orthopedic Basic Science. Columbus: American Academy of Orthopedic Surgeons; (1994). , 460-478.

[28] Kaplan, F. S, Hayes, W. C, Keaveny, T. M, Boskey, A, Einhorn, T. A, & Iannotti, J. P. Form and function of bone. In: Simon SP, editor. Orthopedic Basic Science. Columbus: American Academy of Orthopedic Surgeons; (1994). , 127-185.

[29] Boyan, B. D, Dean, D. D, Lohmann, C. H, Cochran, D. L, Sylvia, V. L, & Schwartz, Z. The titanium-bone cell interface in vitro: the role of the surface in promoting osteointegration. In: Brunette DM, Tengvall P, Textor M, Thomsen P, editors. Titanium in Medicine. Berlin: Springer; (2001). , 561-586.

[30] Webster, T. J, & Ejiofor, J. U. Increased osteoblast adhesion on nanophase metals: Ti, Ti6Al4V, and CoCrMo. Biomaterials (2004). , 25, 4731-4739.

Chapter 2

Mechanical Properties of Biomaterials Based on Calcium Phosphates and Bioinert Oxides for Applications in Biomedicine

Siwar Sakka, Jamel Bouaziz and Foued Ben Ayed

Additional information is available at the end of the chapter

http://dx.doi.org/10.5772/53088

1. Introduction

Calcium phosphates (CaP) have been sought as biomaterials for reconstruction of bone defect in maxillofacial, dental and orthopaedic applications [1-31]. Calcium phosphates have been used clinically to repair bone defects for many years. Calcium phosphates such as hydroxyapatite ($Ca_{10}(PO_4)_6(OH)_2$, HAp), fluorapatite ($Ca_{10}(PO_4)_6F_2$, FAp), tricalcium phosphate ($Ca_3(PO_4)_2$, TCP), TCP-HAp composites and TCP-FAp composites are used for medical and dental applications [3, 10-29]. In general, this concept is determined by advantageous balances of more stable (frequent by hydroxyapatite or fluorapatite) and more resorbable (typically tricalcium phosphate) phases of calcium phosphates, while the optimum ratios depend on the particular applications. The complete list of known calcium phosphates, including their major properties (such, the chemical formula, solubility data) is given in Table 1. The detailed information about calcium phosphates, their synthesis, structure, chemistry, other properties and biomedical applications have been comprehensively reviewed recently in reference [24].

Calcium phosphate-based biomaterials and bioceramics are now used in a number of different applications throughout the body, covering all areas of the skeleton. Applications include dental implants, percutaneous devices and use in periodontal treatment, treatment of bone defects, fracture treatment, total joint replacement (bone augmentation), orthopedics, cranio-maxillofacial reconstruction, otolaryngology and spinal surgery [32-35]. Depending upon whether a bioresorbable or a bioactive material is desired, different calcium orthophosphates might be used.

In the past, many implantations failed because of infection or a lack of knowledge about the toxicity of the selected materials. In this frame, the use of calcium phosphates is logical due to their similarity to the mineral phase of bone and teeth [36-40]. However, according to available literature, the first attempt to use calcium phosphates as an artificial material to repair surgically-created defects in rabbits was performed in 1920 [41]. More than fifty years later, the first dental application of a calcium phosphate (erroneously described as TCP) in surgically-created periodontal defects [42] and the use of dense HAp cylinders for immediate tooth root replacement were reported [43]. Since Levitt et al. described a method of preparing an apatite bioceramics from FAp and suggested its possible use in medical applications in 1969[44]. According to the available databases, the first paper with the term "bioceramics" in the abstract was published in 1971 [45], while those with that term in the title were published in 1972 [46-47]. However, application of ceramic materials as prostheses had been known before [48-49]. Further historical details might be found in literature [50]. Commercialization of the dental and surgical applications of Hap-based bioceramics occurred in the 1980's, largely through the pioneering efforts by Jarcho [51], de Groot [52] and Aoki [53]. Due to that, HAp has become a bioceramic of reference in the field of calcium phosphates for biomedical applications. Preparation and biomedical applications of apatites derived from sea corals (coralline HAp) [54–56] and bovine bone were reported at the same time [57]. Since 1990, several other calcium phosphate cements have been developed [58-62], injectable cements have been formulated [63], and growth factors have been delivered via these cements [64]. The tetracalcium phosphate [TTCP: $Ca_4(PO_4)_2O$] and dicalcium phosphate anhydrous [DCPA: $CaHPO_4$] system was approved in 1996 by the Food and Drug Administration (FDA) for repairing craniofacial defects in humans, thus becoming the first TTCP–DCPA system for clinical use [65]. However, due to its brittleness and weakness, the use of TTCP–DCPA system was limited to the reconstruction of non-stress-bearing bone [66-67]. To expand the use of TTCP–DCPA system to a wide range of load-bearing maxillofacial and orthopedic repairs, recent studies have developed natural biopolymers that are elastomeric, biocompatible and resorbable [68]. Calcium phosphates in a number of forms and compositions are currently either in use or under consideration in many areas of dentistry and orthopedics. For example, bulk materials, available in dense and porous forms, are used for alveolar ridge augmentation, immediate tooth replacement and maxillofacial reconstruction [35, 69]. Other examples include orbital implants (Bio-Eye) [70-71], increment of the hearing ossicles, spine fusion and repair of bone defects [72-73]. In order to permit growth of new bone onto bone defects, a suitable bioresorbable material should fill the defects. Otherwise, in-growth of fibrous tissue might prevent bone formation within the defects [69-73]. Today, a large number of different calcium phosphate bioceramics for the treatment of various defects are available on the market.

The performance of living tissues is the result of millions of years of evolution, while the performance of acceptable artificial substitutions those man has designed to repair damaged hard tissues are only a few decades old. Archaeological findings exhibited in museums showed that materials used to replace missing human bones and teeth have included animal or human (from corpses) bones and teeth, shells, corals, ivory (elephant tusk), wood, as well as some metals (gold or silver). For instance, the Etruscans learned to substitute missing teeth with bridges made from artificial teeth carved from the bones of oxen, while in ancient Phoenicia loose teeth were bound together with gold wires for tying artificial ones to neighboring teeth.

Compound	Acronym	Formula	Ca/P	pK_s [a]
Monocalcium phosphate monohydrate	MCPM	$Ca(H_2PO_4)_2.H_2O$	0.5	1.14
Monocalcium phosphate anhyrous	MCPA	$Ca(H_2PO_4)_2$	0.5	1.14
Dicalcium phosphate dihydrate	DCPD	$CaHPO_4.2H_2O$	1	6.59
Dicalcium phosphate anhydrous	DCPA	$CaHPO_4$	1	6.90
Aporphous calcium phosphates	ACP	$Ca_xH_y(PO_4)_z.nH_2O$ n = 3-4.5; 12-20%H_2O	1.2-2.2	b
Octacalcium phosphate	OCP	$Ca_8(HPO_4)_2(PO_4)_4.5H_2O$	1.33	96.6
α- Tricalcium phosphate	α-TCP	α- $Ca_3(PO_4)_2$	1.5	25.5
β- Tricalcium phosphate	β-TCP	β- $Ca_3(PO_4)_2$	1.5	28.9
Calcium-deficient Hydroxyapatite	CDHAp	$Ca_{10-x}(HPO_4)_x(PO_4)_{6-x}(OH)_{2-x}$ $(0 < x < 1)$	1.5-1.67	85
Hydroxyapatite	HAp	$Ca_{10}(PO_4)_6(OH)_2$	1.67	116.8
Fluorapatite	FAp	$Ca_{10}(PO_4)_6F_2$	1.67	120
Oxapatite	OAp	$Ca_{10}(PO_4)_6O$	1.67	69
Tetracalcium phosphate	TTCP	$Ca_4(PO_4)_2O$	2	38-44

[(a)]: Solubility at 25°C ($pK_s = -logK_s$);

[(b)] : Cannot be measured precisely.

Table 1. Calcium phosphates and their major properties [3, 24]

Calcium phosphates are established materials for the augmentation of bone defects. They are available as allogenic, sintered materials. Unfortunately, these calcium phosphates exhibit relatively poor tensile and shear properties [74]. In practice, the strength of the calcium phosphate cements is lower than that of bone, teeth, or sintered calcium phosphate bioceramics [75] and, together with their inherent brittleness, restricts their use to non-load bearing defects [76] or pure compression loading [74]. Typical applications are the treatment of maxillo-facial defects or deformities [77] and cranio facial repair [78] or augmentation of spine and tibial plateau [74]. A successful improvement of the mechanical properties would significantly extend the applicability of calcium phosphates [79] and can be achieved by forming composite materials [80]. Second phase additives to the calcium phosphate composites have been either fibrous reinforcements or bioinert oxides that interpenetrate the porous matrix.

Hydroxyapatite and other calcium phosphates bioceramics are important for hard tissue repair because of their similarity to the minerals in natural bone, and their excellent biocompatibility and bioactivity [81-86]. When implanted in an osseous site, bone bioactive materials such as HAp and other CaP implants and coatings provide an ideal environment for cellular reaction and colonization by osteoblasts. This leads to a tissue response termed osteoconduction in which bone grows on and bonds to the implant, promoting a functional interface [81, 84, 87]. Extensive efforts have significantly improved the properties and performance of HAp and other CaP based implants [88-92]. Calcium phosphate cements can be molded or injected to form a scaffold in situ, which can be resorbed and replaced by new bone [93, 65-67]. Chemically, the vast majority of calcium phosphate bioceramics is based on HAp, β-TCP, α-TCP and/or biphasic calcium phosphate (BCP), which is an intimate mixture of either β-TCP - HAp [94-100] or α-TCP - HAp [101-111]. The preparation technique of these calcium phosphates has been extensively reviewed in literature [1, 4, 37, 102-104]. When compared to both β- and α-TCP, HAp is a more stable phase under physiological conditions, as it has a lower solubility (Table 1) [37, 109-110]. Therefore, the BCP concept is determined by the optimum balance of a more stable phase of HAp and a more soluble TCP. Due to a higher biodegradability of the β - or α -TCP component, the reactivity of BCP increases with the TCP-HAp the increase in ratio. Thus, in vivo bioresorbability of BCP can be controlled through the phase composition [95]. As implants made of calcined HAp are found in bone defects for many years after implantation, bioceramics made of more soluble calcium phosphates is preferable for the biomedical purposes [94-110]. HAp has been clinically used to repair bone defects for many years [3]. However, Hap has poor mechanical properties [3]. Their use at high load bearing conditions has been restricted due to their brittleness, poor fatigue resistance and strength.

The main reason behind the use of β-TCP as bone substitute materials is their chemical similarity to the mineral component of mammalian bone and teeth [1-3]. The application of tricalcium phosphate as a bone substitute has received considerable attention, because it is remarkably biocompatible with living bodies when replacing hard tissues and because it has biodegradable properties [1-29]. Consequently, β-TCP has been used as bone graft substitutes in many surgical fields such as orthopedic and dental surgeries [3, 11-12, 16-17]. This use leads to an ultimate physicochemical bond between the implants and bone-termed osteointegration. Even so, the major limitation to the use of β-TCP as load-bearing biomaterial is their mechanical properties which make it brittle, with poor fatigue resistance [3, 10, 21-29]. Moreover, the mechanical properties of tricalcium phosphate are generally inadequate for many load-carrying applications (3 MPa – 5 MPa) [3, 10, 20-29]. Its poor mechanical behaviour is even more evident when used to make highly porous ceramics and scaffolds. Hence, metal oxides ceramics, such as alumina (Al_2O_3), titania (TiO_2) and some oxides (e.g. ZrO_2, SiO_2) have been widely studied due to their bioinertness, excellent tribological properties, high wear resistance, fracture toughness and strength as well as relatively low friction [19, 21-22, 29-31]. However, bioinert ceramic oxides having high strength are used to enhance the densification and the mechanical properties of β-TCP. In this chapter, we will try to improve the strength of β-TCP by introducing a bioinert oxide like alumina. This is because there are few articles reporting the toughening effects of an inert oxide (like alumina (Al_2O_3)) on the mechanical properties of β-TCP [22, 27, 29]. Alumina has a high strength and is bio-inert with human tissues [19, 22, 27, 29]. In order to improve the biocompatibility of alumina and the strength of tricalcium phosphate effectively, and in order to search for an approach to produce high performances of alumina-tricalcium phosphate composites, β-TCP is introduced with different percentages in the alumina matrix. The aim of our study is to elaborate and characterize the TCP-Al_2O_3 composites for biomedical applications.

This chapter proposes to study the sintering of the alumina-tricalcium phosphate composites at various temperatures (1400°C, 1450°C, 1500°C, 1550°C and 1600°C) and with different percentages of β-TCP (10 wt%, 20 wt%, 40 wt% and 50 wt%). The characterization of biomaterials will be realized by using dilatometry analysis, differential thermal analysis (DTA), X-ray diffraction (XRD), magic angle spinning nuclear magnetic resonance (MAS NMR), scanning electron microscopy analysis (SEM) and by using the mechanical properties, such as rupture strength (σ_r) of these biomaterials.

2. Materials and methods

The synthesized tricalcium phosphate and alumina (Riedel-de Haën) were mixed in order to prepare biomaterial composites. The β-TCP powder was synthesized by solid-state reaction from calcium carbonate ($CaCO_3$) and calcium phosphate dibasic anhydrous ($CaHPO_4$) [27]. Stoichiometric amounts of high purity powders such as $CaHPO_4$ (Fluka, purity ≥ 99%) and $CaCO_3$ (Fluka, purity ≥ 98.5%), were sintered at 1000°C for one hour to obtain the β-TCP according to the following reaction:

$$2\,CaHPO_4(s) + CaCO_3(s) \rightarrow Ca_3(PO_4)_2(s) + H_2O(g) + CO_2(g) \tag{1}$$

The β-TCP and the alumina powders were mixed in an agate mortar. The powder mixtures were milled in ethanol for 24 hours. After milling, the mixtures were dried in a rotary vacuum evaporator and passed through a 70-mesh screen. After drying the powder mixtures at 80°C for 24 hours, they were molded in a cylinder having a diameter of 20 mm and a thickness of 6 mm, and pressed under 150 MPa. The green compacts were sintered at various temperatures for different lengths of time in a vertical furnace (Pyrox 2408). The heating rate is 10°C min^{-1}. The size of the particles of the powder was measured by means of a Micromeritics Sedigraph 5000. The specific surface area (SSA) was measured using the BET method and using N_2 as an adsorption gas (ASAP 2010) [112]. The primary particle size (D_{BET}) was calculated by assuming the primary particles to be spherical:

$$D_{BET} = 6 \,/\, S\rho \tag{2}$$

where ϱ is the theoretical density and S is the surface specific area.

The microstructure of the sintered compacts was investigated using the scanning electron microscope (Philips XL 30) on the fractured surfaces of the samples. The grains' mean size was measured directly using SEM micrographs. The powder was analyzed by using Xray diffraction (XRD). The Xray patterns were recorded using the Seifert XRD 3000 TT diffractometer. The Xray radiance was produced by using CuK_α radiation (λ = 1.54056 Å). The crystalline phases were identified with the powder diffraction files (PDF) of the International Center for Diffraction Data (ICDD). Linear shrinkage was determined using dilatometry (Setaram TMA 92 dilatometer). The heating and cooling rates were 10°C min^{-1} and 20°C min^{-1}, respectively. Differential thermal analysis (DTA) was carried out using about 30 mg of powder (DTATG, Setaram Model). The heating rate was 10°C min^{-1}. The ^{31}P and ^{27}Al magic angle spinning nuclear magnetic resonance (^{31}P MAS NMR) spectra were run on a Brucker 300WB spectrometer. The ^{31}P and ^{27}Al observational frequency were 121.49 MHz and 78.2 MHz, respectively. The ^{31}P MAS-NMR chemical shifts were referenced in parts per million (ppm) referenced to 85 wt% H_3PO_4. The ^{27}Al MAS-NMR chemical shifts were referenced to a static signal obtained from an aqueous aluminum chloride solution.

The Brazilian test was used to measure the rupture strength of biomaterials [113-114]. The rupture strength (σ_r) values were measured using the Brazilian test according to the equation:

$$\sigma_r = \frac{2 \cdot P}{\Pi \cdot D \cdot t} \tag{3}$$

where P is the maximum applied load, D the diameter, t the thickness of the disc and σ_r the rupture strength (or mechanical strength).

3. Results and discussion

3.1. Characterization of different powders

The X-ray diffraction analysis of β-TCP powder and α-alumina powder are presented in Figure 1. As it can be noticed from this figure, the X-ray diffraction pattern of tricalcium phosphate powder reveals only peaks of β-TCP (ICDD data file no. 70-2065) without any other phase (Figure 1a). Consequently, the XRD pattern obtained from the alumina powder illustrates α phase peaks relative to ICDD data file no. 43-1484 (Figure 1b).

The ^{31}P MAS-NMR solid spectrum of the tricalcium phosphate powder is presented in Figure 2a. We observe the presence of several peaks of tetrahedral P sites (at 0.36 ppm, 1.46 ppm and 4.83 ppm), while there are other peaks (at -7.43 ppm, -9.09 ppm and -10.35 ppm) which reveal a low quantity of calcium pyrophosphate which was formed during the preparation of the β-TCP.

The ^{27}Al MAS-NMR solid spectrum of the alumina powder is presented in Figure 2b. We notice the presence of two peaks which are characteristic of aluminum: one peak at 7.36 ppm

http://dx.doi.org/10.5772/53088

corresponding to octahedral Al sites (Al^{VI}) and the other at 37.36 ppm which corresponds to pentahedral Al sites (Al^{V}). The results obtained for ^{31}P MAS-NMR and ^{27}Al MAS-NMR are similar to those previously reported by different authors [14, 22, 25-28, 31].

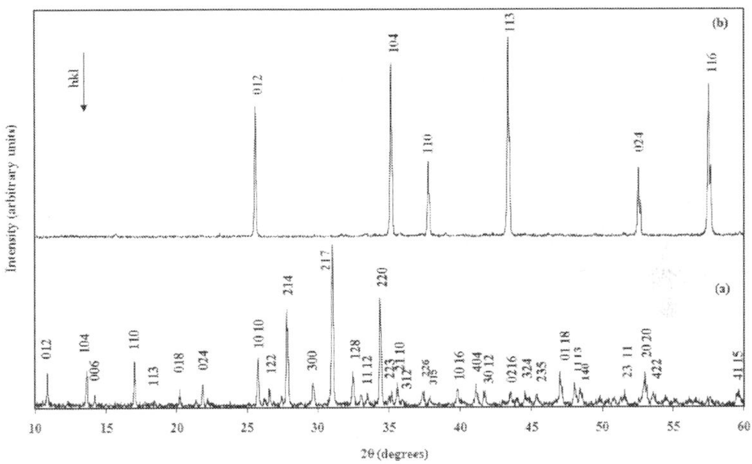

Figure 1. The XRD patterns of: (a) β-TCP powder and (b) α-Al_2O_3 powder

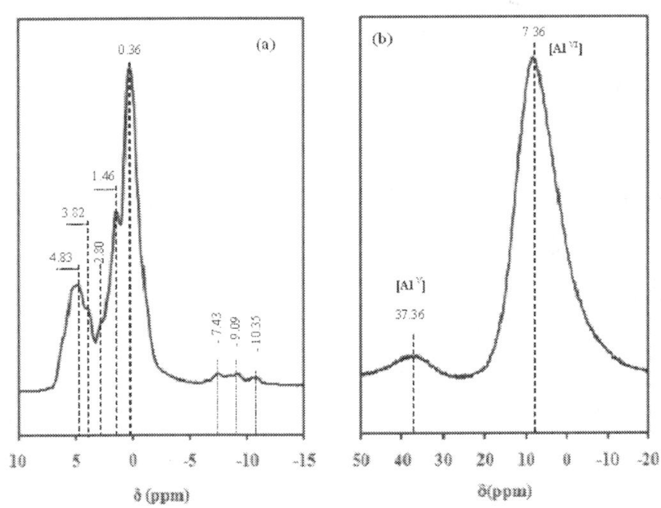

Figure 2. The ^{31}P MAS-NMR spectra of: (a) β-TCP and the ^{27}Al MAS-NMR spectra of: (b) α-Al_2O_3

The experimental characteristics of the different powders used in this study are illustrated in Table 2. Table 2 summarizes the SSA, the DTA measurements, the sintering temperature and the theoretical density of the different powders. The powder particles are assumed to be spherical; the size of the particles can be calculated using Eq. (2). The results from the average grain size obtained by the SSA (D_{BET}) and from the average grain size obtained by granulometric repartition (D_{50}) are presented in Table 2. Compared with those of the β-TCP powder, the grains of the alumina powder have a dense morphology. These (D_{BET}) values obtained by the SSA do not correspond to those obtained from the granulometric repartition (Table 2). The discrepancy may be due to the presence of agglomerates which are formed during the preparation of the β-TCP powder at 1000°C.

Compounds	SSA (m^2/g) ± 1.0	D_{BET} (μm) ± 0.2	D_{50} (μm)[a] ± 0.2	DTA measurements (endothermic peak)	T(°C)[b]	d[c]
TCP	0.70	2.79	6	1100°C-1260°C (β → α) 1470°C (α → α')	1000 - 1300	3.070 (β) 2.860 (α)
Alumina	2.87	0.53	3	-	1400 - 1600	3.98 (α)

[a]: mean diameter,

[b]: sintering temperature domain,

[c]: theoretical density

Table 2. Characteristics of the powders used in the study

Differential thermal analysis studies of the different powders used in this study detected a potential phase change during the sintering process. The DTA thermogram of β-TCP, α-Al_2O_3 and different Al_2O_3 - TCP composites are presented in Figure 3. The DTA curve of alumina reported no process relative to the sintering temperature (Figure 3a). Figure 3b shows the DTA curve of β-TCP. The DTA thermogram of β-TCP shows two endothermic peaks, relative to the allotropic transformations of tricalcium phosphate (Figure 3b). The peak between 1100°C – 1260°C is related to the first allotropic transformation of TCP (β to α), while the last peak at 1470°C is related to the second allotropic transformation of TCP (α to α'). As a matter of fact, this result is similar to the result previously reported by Destainville et al. and Ben Ayed et al. [9, 14]. Figure 3c shows the DTA curve of Al_2O_3-50 wt% TCP composites. This DTA curve is practically similar to the one shown in Figure 3b. Indeed, the DTA thermogram of the composites also shows two endothermic peaks. Figure 3 (d), (e) and (f) illustrate the DTA curves of Al_2O_3–40 wt% TCP composites, Al_2O_3–20 wt% TCP composites and Al_2O_3–10 wt% TCP composites, respectively. The DTA thermograms of each composites show only one endothermic peak between 1100°C and 1260°C, which are relative to the allotropic transformation of TCP (β to α). In these curves, we notice that the endothermic peak relative to a second allotropic transformation of TCP (α to α') has practically disappeared when the percentage of the alumina increases in the Al_2O_3 - TCP composites (Figure 3(d), (e) and (f)).

http://dx.doi.org/10.5772/53088

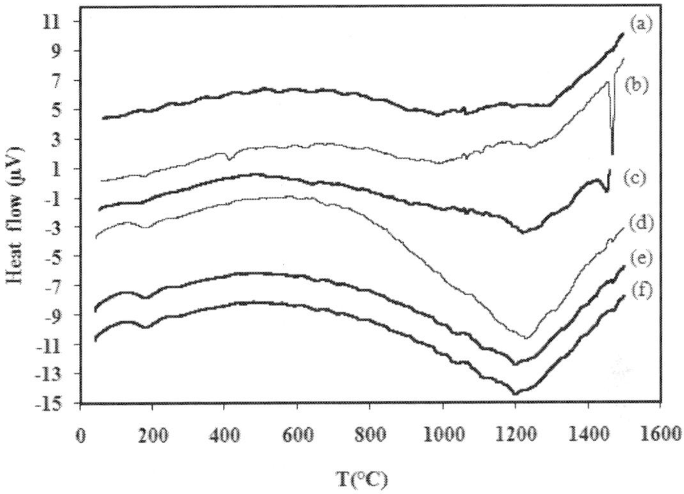

Figure 3. DTA curves of (a) α-Al_2O_3, (b) β-TCP, (c) Al_2O_3 – 50 wt% TCP composites, (d) Al_2O_3 – 40 wt% TCP composites, (e) Al_2O_3 – 20 wt% TCP composites and (f) Al_2O_3 – 10 wt% TCP composites

Figure 4 shows the dilatometric measurements of the different powders used in this study (β-TCP, α-Al_2O_3 and Al_2O_3 - TCP composites). A large sintering domain was observed for the three powders (β-TCP, alumina and composites). The sintering temperature of the initial powder began at about 900°C and at about 1400°C for the β-TCP and alumina (Figure 4a-b and Table 2). The sintering temperature of Al_2O_3-50 wt% TCP composites began at 1100°C (Figure 4c). It is to be noted that the presence of 50 wt% TCP in the alumina matrix decreases the sintering temperature of the alumina by around 300°C (Figure 4c). This variation of the sinterability is relative to the difference between the physicochemical compositions of these powders and the mixture of their different composites.

3.2. The mechanical properties of alumina–tricalcium phosphate composites

The influence of the sintering temperature on the rupture strength of Al_2O_3-TCP composites is shown in Figure 5. The mechanical resistance of Al_2O_3-TCP composites is studied at various temperatures (1400°C, 1450°C, 1500°C, 1550°C, 1600°C) for one hour with different percentages of β-TCP (50 wt%, 40 wt%, 20 wt% and 10 wt%). Thus, Figure 5 illustrates the rupture strength of the Al_2O_3-TCP composites relative to the percentages of the alumina and the sintering temperature. Consequently, the rupture strength of Al_2O_3 mixed with 10 wt% β-TCP reached its maximum value when sintered at 1600°C for one hour; it then decreased with the increase of this percentage. This is how the rupture strength of the Al_2O_3-10 wt% TCP composites reached 13.5 MPa.

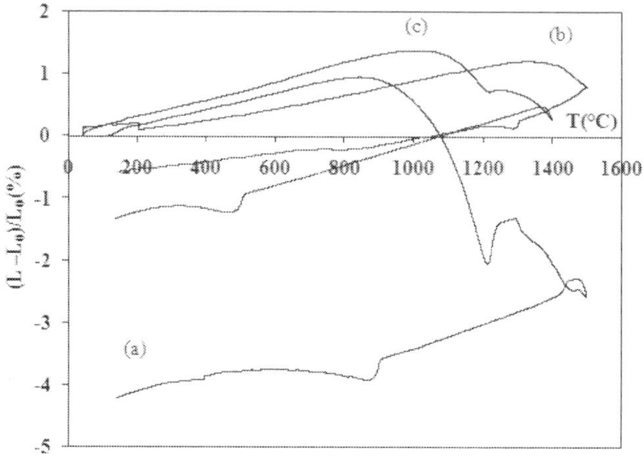

Figure 4. Linear shrinkage versus temperature of: (a) β-TCP, (b) α-Al_2O_3 and (c) Al_2O_3 – 50 wt% TCP composites

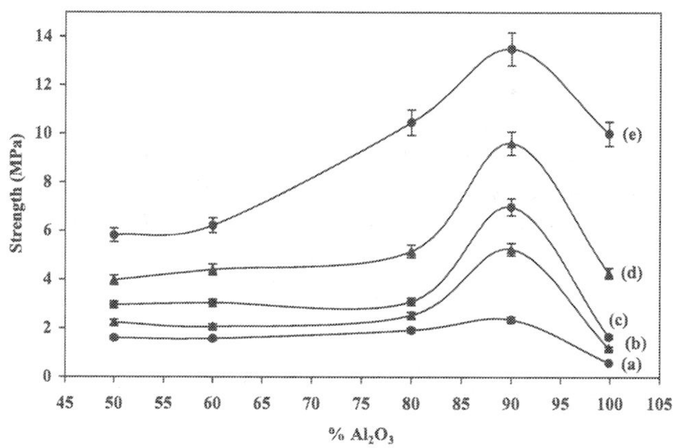

Figure 5. The rupture strength of the TCP-Al_2O_3 composites sintered for 1 hour at various temperatures: (a) 1400°C, (b) 1450°C, (c) 1500°C, (d) 1550°C and (e) 1600°C

Figure 6 shows the evolution of the rupture strength of the Al_2O_3 - 10 wt% TCP composites sintered at various temperatures (1500°C, 1550°C and 1600°C) for different sintering times (0 min, 30 min, 60 min and 90 min). The optimum value of the rupture strength was indeed obtained after an hour-long sintering process at 1600°C. Thus, the mechanical resistance of the samples reached 13.5 MPa.

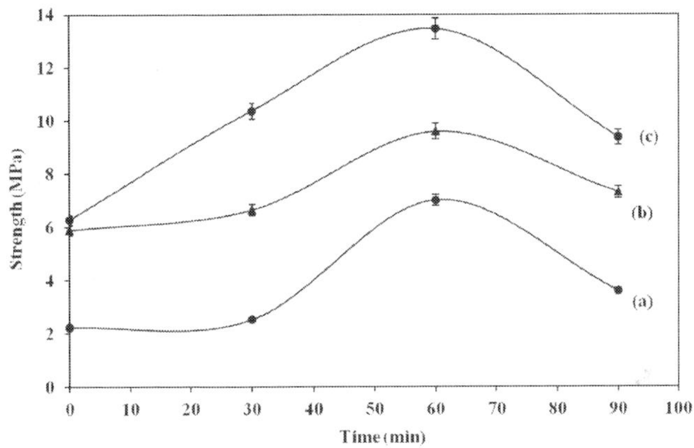

Figure 6. The rupture strength of the Al_2O_3–10 wt% TCP composites sintered for different lengths of time at various temperatures: (a) 1500°C, (b) 1550°C and (c) 1600°C

In this study, we showed that the presence of different amounts of alumina in the β-TCP improves the mechanical properties of Al_2O_3-TCP composites. In fact, the mechanical properties of Al_2O_3-10 wt% TCP composites reached the optimum value by being sintered at 1600°C for one hour. Thus, the rupture strength of these composites reached 13.5 MPa. Table 3 displays several examples of the mechanical properties of the calcium phosphates and the bone tissues. In comparison, we notice that the properties of Al_2O_3-10 wt% TCP composites are close to those of pure β-TCP, pure Fap and TCP-33.16 wt% Fap composites, which have a rupture strength of 5.3 MPa, 14 MPa and 13.7 MPa, respectively [15, 21, 25]. However, the mechanical properties of our composites are more closely comparable to those of the pure Fap and the TCP–33.16 wt% Fap composites (Table 3). Generally, the values found for the mechanical strength of our composites are not identical to those in Table 3, because the authors have used different mechanical modes other than the Brazilian test. In addition, many factors influence the mechanical properties of the samples such as: the use of particular initial powders as well as the conditions of the treatment process.

The sintering of materials is a complex process, involving the evolution of the microstructure through the action of several different transport mechanisms such as: surface diffusion, evaporation- condensation, grain boundary diffusion [3]. However, producing dense TCP–Al_2O_3 composites with a fine uniform microstructure through the sintering process does not seem to be a routine process because the β-TCP has a lower sinterability and a lower sintering temperature than those of pure alumina.

Materials	σ_r (a) (MPa)	σ_c (b) (MPa)	σ_f (c) (MPa)	References
β-TCP	4-6	-	92	[21,27]
Fap	10-14	-	-	[15]
Hap	-	5.35	-	[115]
TCP - 75 wt% Al_2O_3	8.60	-	-	[27]
TCP – 26.52 wt% Fap	9.60	-	-	[21]
TCP – 26.52 wt% Fap- 5 wt% Al_2O_3	13.60	-	-	[22]
TCP – 33.16 wt% Fap	13.70	-	-	[25]
Hap – TCP (40:60)	-	4.89	-	[115]
Al_2O_3 - 26.5 wt% Fap	21.7	-	-	[31]
Cortical bone	-	130-180	50-150	[1, 2]
Cancellous bone	-	2-12	-	[1, 2]

(a): Rupture strength (Brazilian test),

(b): compressive strength

(c): Flexural strength.

Table 3. Literature examples of the mechanical properties of calcium phosphates bioceramics and bone tissues

3.3. Characterization of alumina-tricalcium phosphate composites after the sintering process

Figure 7 shows the ^{31}P MAS-NMR spectra of the Al_2O_3-TCP composites obtained after the sintering process for 1 hour at 1550°C with different percentages of β-TCP (50 wt%, 40 wt%, 20 wt% and 10 wt%). The addition of 50 wt% Al_2O_3 to the TCP matrix shows the presence of several peaks which are assigned to the tetrahedral environment of P sites (1.03 ppm; 1.93 ppm; 3.50 ppm and 4.84 ppm) (Figure 7a). In fact, the increasing of the percentage of alumina in the TCP matrix decreases the number of the tetrahedral phosphorus site peaks which are reduced to a large single peak with 90 wt% alumina (Figure 7 b-d). Moreover, the tetrahedral environment of the phosphorus in tricalcium phosphate is not changed after the sintering process with different percentages of alumina. But the effect of the addition of alumina to the β-TCP matrix provokes the structural rearrangement of the coordination of phosphorus in β-TCP. Similar results were previously reported in literature [14, 26-27, 31].

The ^{27}Al MAS-NMR spectra of Al_2O_3-TCP composites sintered for 1 hour at 1550°C with different percentages of β-TCP (50 wt%, 40 wt%, 20 wt% and 10 wt%) are shown in Figure 8. The chemical shifts at 35 ppm and at 7.3 ppm indicate the presence of octahedral Al sites (Al^{VI}) and pentahedral Al sites (Al^{V}), respectively. The peak of pentahedral Al sites increases with the increase of the percentage of alumina in the Al_2O_3-TCP composites (Figure 8 b-8d). We notice especially the appearance of another octahedral Al peak at 18.6 ppm for alumina sintered with 40 wt% and 20 wt% of β-TCP (Figure 8 b-d). The aluminum in the alumina is primarily in one pentahedral Al site (35 ppm) and in one octahedral Al site (7.3 ppm) (Figure 8e). For the alumina sintered with different percentages of β-TCP (50 wt%, 40 wt%, 20 wt% and 10 wt%), the spectra show two octahedral aluminum environments: AlO_6 at about 7.3 ppm and at 18.6 ppm (Fig 8b-d). Indeed, the intensity of the octahedral signal at 18.6 ppm increases with the percentage of alumina. The estimated concentrations of AlO_5 and AlO_6 are reported in Table 4. During the sintering process, the aluminum in the Al_2O_3-TCP composites provokes the structural rearrangement of the coordination of aluminum. Similar results were provisionally reported by different authors [22, 27, 31]. Indeed, these authors show that the coordination of the aluminum in octahedral sites was forced to change into another coordination in pentahedral sites [31]. The same authors point out that the structural rearrangement of the coordination of aluminum was probably produced by the formation of calcium aluminates which was produced after the sintering process and the reaction between calcium phosphates and alumina [31]. In conclusion, the ^{31}P magic angle scanning nuclear magnetic resonance analysis of different composites reveals the presence of tetrahedral P sites, while the ^{27}Al magic angle scanning nuclear magnetic resonance analysis shows the presence of both octahedral and pentahedral Al sites.

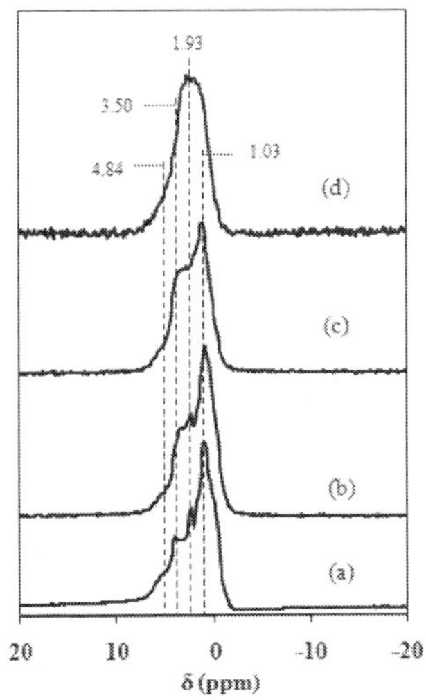

Figure 7. The ^{31}P MAS-NMR spectra of the Al_2O_3-TCP composites sintered for 1 hour at 1550°C with different percentages of β-TCP: (a) 50 wt%, (b) 40 wt%, (c) 20 wt% and (d) 10 wt%

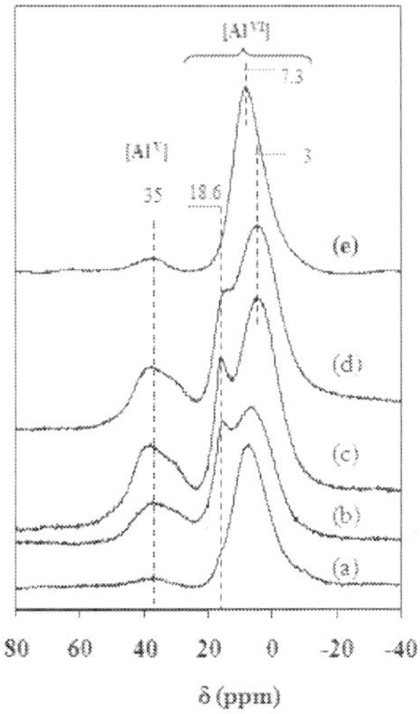

Figure 8. The ^{27}Al MAS-NMR spectra of the Al_2O_3-TCP composites sintered for 1 hour at 1550°C with different percentages of β-TCP: (a) 50 wt%, (b) 40 wt%, (c) 20 wt%, (d) 10 wt% and (e) 0 wt%

Compounds	AlO_5 (%)	AlO_6 (Type 1) (%)		AlO_6 (Type 2) (%)
δ (ppm)	30-40	7		18
TCP - 50 wt % Al_2O_3	0.50		99.50	
TCP - 60 wt % Al_2O_3	15.38	65.37		19.25
TCP - 80 wt % Al_2O_3	19.71	67.81		12.48
TCP- 90 wt % Al_2O_3	14.85	73.71		11.44
Al_2O_3	1.40		98.60	

Table 4. Properties of pentahedral and octahedral Al sites in the Al_2O_3-TCP composites sintered with different percentages of β-TCP at various temperatures for 1 hour

Figure 9 presents XRD patterns of Al_2O_3-TCP composites sintered at 1550°C for 1 hour with different percentages of β-TCP. Besides, the spectra show the characteristic peaks of β-TCP (ICDD data file no. 70-2065) and α- Al_2O_3 (ICDD data file no. 43-1484). This analysis shows that the peak of alumina is predominant in the elaboration of any composite.

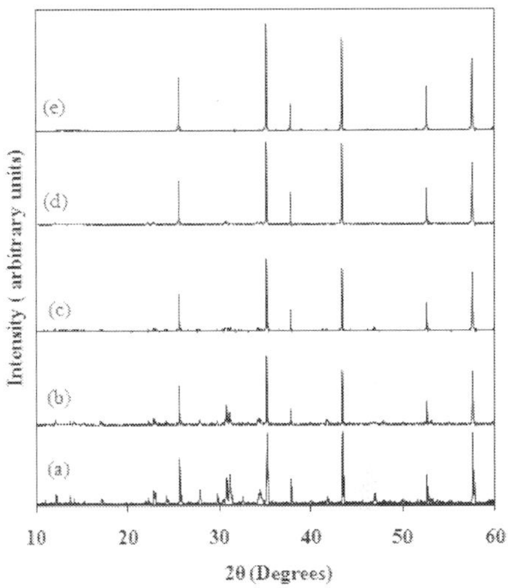

Figure 9. The XRD patterns of the Al_2O_3-TCP composites sintered at 1550°C for 1 hour with different percentages of β-TCP: (a) 50 wt%, (b) 40 wt%, (c) 20 wt%, (d) 10 wt% and (e) 0 wt%

The SEM technique helps to investigate the texture and porosity of any biomaterial. Figure 10 shows the fracture surface of the Al_2O_3-TCP composites sintered at 1550°C for 1 hour with different percentages of β-TCP. These micrographs show the coalescence between β-TCP grains produced with all the percentages of added alumina (Figure 10 a-d). The samples sintered with 50 wt%, 40 wt% and 20 wt% β-TCP present cracks and an important intragranular porosity (Figure 10 a-c). This result is a proof of the fragility of the composites elaborated with the different percentages of alumina as shown in Figure 10a-c. In fact, the microstructure of the composites shows different cracks relative to the allotropic transformation of TCP (β to α) (Figure 10 a-c). But the intensity of the cracks in the composites decreases with the increase in the percentage of β-TCP (Figure 10 a-d). Thus, the absence of micro-cracking and the reduction of the sizes of the pores in the Al_2O_3-10 wt% TCP composites explain the increase in the rupture strength of the samples (Figure 10 d_1-d_2). Indeed, the composites present excellent mechanical properties and a good aptitude for sinterability (Figure 10 d_1-d_2). The SEM micrographs of the alumina sintered without β-TCP shows an intergranular porosity (Figure 10e).

The effects of the sintering temperature on the microstructure of the Al_2O_3-10 wt% TCP composites are presented in Figure 11. The SEM micrographs show the coalescence between the grains with the increase of the sintering temperature. At 1500°C, the samples present an important intergranular porosity (Figure 11a). The microstructure of Al_2O_3-10 wt% TCP composites sintered at 1550°C shows a continuous phase relative to β-TCP phases and small-sized grains relative to the alumina phases (Figure 11b). Furthermore, at 1600°C, the boundaries between grains are evident in the micrographs (marked with arrows in Figure

11c), confirming the best mechanical resistance of Al_2O_3-10 wt% TCP composites in this temperature and with the addition of 10 wt% β-TCP (Figure 11c). In fact, the continuous phases in addition to the formed spherical pores prove that a liquid phase has appeared at 1600°C relative probably to the allotropic transformation of the β-TCP. This similar result was observed by Bouslama and colleagues [25].

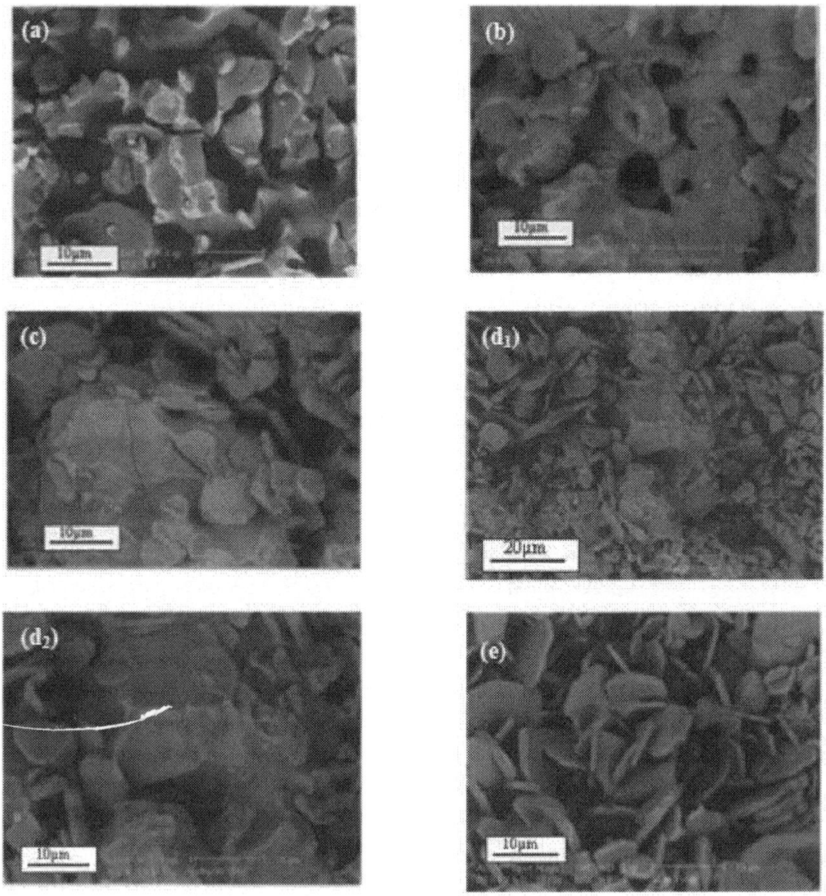

Figure 10. The SEM micrographs of the Al_2O_3 - TCP composites sintered at 1550°C for 1 hour with different percentages of β-TCP: (a) 50 wt%, (b) 40 wt%, (c) 20 wt%, (d_1-d_2) 10 wt% and (e) 0 wt%.

http://dx.doi.org/10.5772/53088

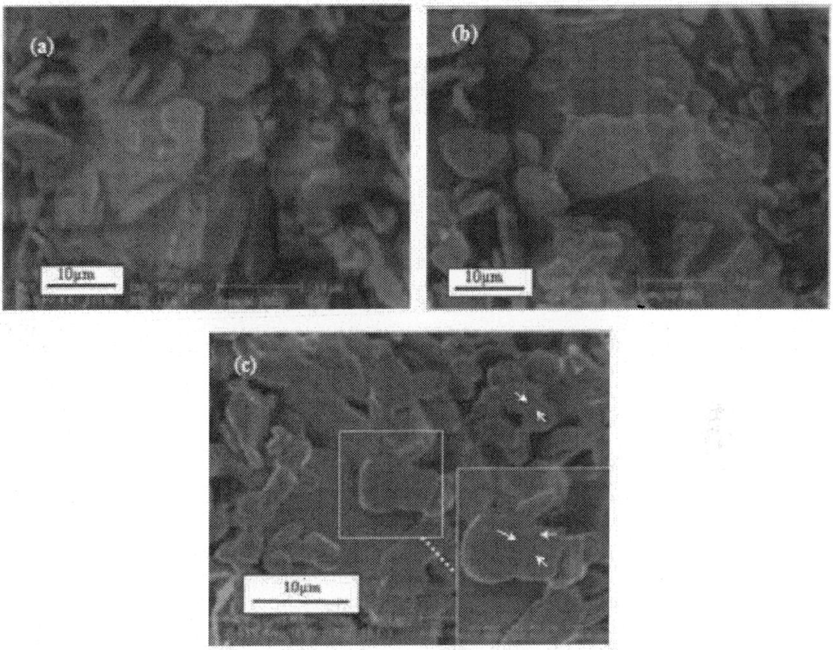

Figure 11. The SEM micrographs of the Al_2O_3-10 wt% TCP composites sintered for 1 hour at: (a) 1500°C, (b) 1550°C and (c) 1600°C.

The results of the microstructural investigations of Al_2O_3-10 wt% TCP composites sintered at 1600°C for different lengths of time (0 min, 30 min, 60 min and 90 min) are shown in Figure 12. These micrographs reveal the influence of different lengths of time on the microstructural developments during the sintering process at 1600°C. The microstructure of the samples leads to the formation of important cracks of different sizes with composites sintered for 0 min, 30 min and 90 min (Figure 12 ba-b and 12d). The continuous phases are relative to the β-TCP phase while the grains of a small size are relative to the alumina phase (Figure 12 ba-b and 12d). In Figure 12c, we notice the coalescence between the grains after the sintering process for 60 min confirming the best mechanical resistance in these conditions.

Furthermore, the sintering behavior of the Al_2O_3-TCP composites has been studied relative to the β-TCP content. It has been shown that alumina should be used in order to prevent the β-α transition of the tricalcium phosphate during the sintering process. At any rate, the results obtained in the present work would be valuable in the performance of Al_2O_3 - TCP composites resembling bone tissue engineering (Table 3). In fact, our preliminary tests indicated that the rupture strength of Al_2O_3-TCP composites is from 2 to 14 MPa. The optimum value of the Al_2O_3 - 10 wt% TCP composites sintered at 1600°C for one hour reached 13.5

MPa. This is true for the values of calcium phosphates fabricated by conventional techniques [116] and is close to a cancellous bone (2-12 MPa) [1-2, 117].

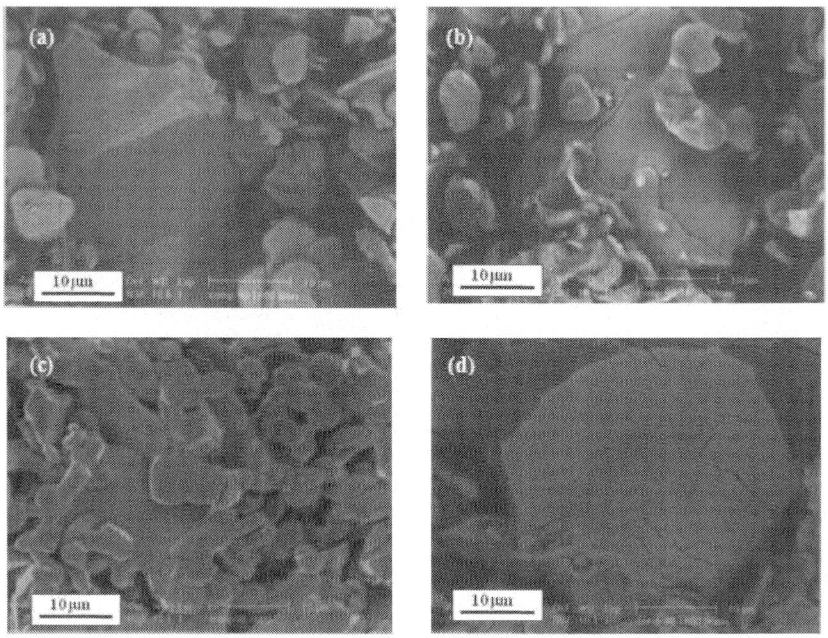

Figure 12. SEM micrographs of the Al_2O_3 - 10 wt% TCP composites sintered at 1600°C for: (a) 0 min, (b) 30 min, (c) 60 min and (d) 90 min.

According to the authors' best knowledge, the highest values of the mechanical characteristics for different samples are detailed in Table 3. As it can be seen from this table, the various techniques used to prepare dense sintered bioceramics affect the final density, as well as the composition of the phase and, consequently, the final mechanical properties of samples. The Al_2O_3 - 10 wt% TCP composites show high rupture strength, which is in concordance with the other results [10, 15, 21, 25-27, 29]. The rupture strength obtained for TCP-Al_2O_3 composites is from 2 to 14 MPa, within the values reported in the literature (Table 3). Moreover, the wide variation in the reported rupture strength of the composites is due to the synthesis route of the β-TCP powder, the size of its particle as well as to its density; it is also due to the application of different processing parameters.

At first, the objective of this work was to characterize the mechanical properties of alumina – TCP composites produced after the sintering process. A sintering stage appears to be of great importance to produce biomaterials with the required properties. Several processes occur during the sintering process of tricalcium phosphate and bioinert oxide. Firstly, the TCP powders are synthesized by solid reaction. Secondly, alumina – TCP powders are sintered for production of dense bioceramics with subsequent shrinkage of the samples. Thirdly, the mechanical properties of alumina–TCP composites are accompanied by a concurrent increase in grain size and a formation of cracks in the alumina sintered with different percentages of TCP (20 wt%, 40 wt% and 50 wt%). Besides, sintering causes the toughening and the increase of the mechanical strength of alumina–10 wt% TCP composites. An extensive study on the effect of the sintering temperature and time on the properties of alumina–TCP composites revealed a correlation between these parameters and density, porosity, grain size, chemical composition and strength of different composites. The degree of densification and mechanical properties of alumina–TCP composites appeared to depend on the sintering temperature and the duration of sintering. Alumina–TCP powders can be pressed and sintered up to theoretical density at 1400°C–1600°C. Processing them with higher percentage of TCP (20 wt%, 40 wt% and 50 wt%) may lead to exaggerated grain growth and formation of cracks because of the formation of α-TCP at higher temperatures. Indeed, the allotropic transformation of TCP is a function of the sintering temperature. The presence of cracks in the alumina–TCP composites is reported to inhibit the mechanical properties. A definite correlation between mechanical strength and grain size in sintered alumina–TCP composites was found: the strength started to decrease at lower sintering temperature and with higher percentage of TCP (20 wt%, 40 wt% and 50 wt%). The sintering process of alumina – 10 wt% TCP composites makes it possible to decrease the grain size and achieve higher densities. This leads to finer microstructures, higher thermal stability of alumina–10 wt% TCP composites and subsequently better mechanical properties of the prepared bioceramics composites. The mechanical properties of alumina – TCP composites is from 2 to 14 MPa. Generally, the mechanical properties of samples increase with the decrease in grain size. In fact, the mechanical strength of alumina–10 wt% TCP composites reaches a maximum value with the decrease in the size of the grains of composites. The optimum measured value of the strength of the alumina–10 wt% TCP composites was 13.5 MPa. This value is compared to those of cancellous bone. Similar values for porous HAp are in the ranges of 2-10 MPa [118]. Generally, variations of mechanical properties of samples are caused by a statistical nature of the strength distribution, influence of remaining microporosity, grain size, presence of impurities and Ca/P ratio [118].

In conclusion, an interfacial reaction between β-TCP and alumina has been studied in the nanocomposites of Al_2O_3-TCP. It was found that the alumina did not completely react with the β-TCP and did not form calcium aluminates. Moreover, it has been shown that the alumina prevents the formation of cracks in the microstructure of composites containing 10 wt % of β-TCP. The mechanical characteristics should be taken into consideration in order to better assess the relationship between the processing conditions, the microstructural design as well as the mechanical response.

4. Conclusions

The biomaterials of alumina-tricalcium phosphate composites have been characterized by using MAS NMR, XRD and SEM analysis after the sintering process. The effect of β-TCP additive on the alumina matrix was observed in different thermal analyses: dilatometry analysis and DTA analysis. The mechanical properties have been investigated by the Brazilian test. This investigation has allowed us to define the sintering temperature and the percentage of added alumina for which β-TCP should have an optimal densification and better mechanical properties. This study has also allowed us to summarize the effect of the sintering temperature and the length of sintering time on the mechanical properties of the Al_2O_3-TCP composites. The produced Al_2O_3-TCP composites with different percentages of β-TCP (50 wt%; 40 wt%; 20 wt% and 10 wt%) exhibited much better mechanical properties than the reported values of β-TCP without alumina. The Al_2O_3-TCP composites showed a higher rupture strength at 1600°C, which certainly increased with the alumina content and reached the optimum value with 90 wt%. However, no cracks were observed in the microstructure of the composites which contained this percentage of alumina. This is due to the allotropic transformation of the tricalcium phosphate. The partial or reversal transformation of tricalcium phosphate (β to α or α to α') during the cooling period could induce a residual stress within the dense bioceramics, marking it much more brittle. Accordingly, the optimum performance of alumina-tricalcium phosphate composites achieved 13.5 MPa. Furthermore, the best mechanical properties of the composites were obtained after the sintering process at 1600°C for 1 hour. With different weight rations of tricalcium phosphate: alumina (50:50, 40:60 and 20:80), the performance of the composites was hindered by the formation of both cracks and intragranular porosity.

Acknowledgements

The authors thank Mr Ahmed BAHLOUL for his assistance in this work.

Author details

Siwar Sakka, Jamel Bouaziz and Foued Ben Ayed*

*Address all correspondence to: benayedfoued@yahoo.fr

Laboratory of Industrial Chemistry, National School of Engineering, Sfax University, Sfax, Tunisia

References

[1] Hench L L. Bioceramics: From Concept to clinic. J. Am. Ceram. Soc. 1991; 74 (7) 1487.

[2] Hench L L. An Introduction to Bioceramics. J. Wilson (ed.). Vol. 1. World Scientific, Singapore; 1993.

[3] Elliott J C. Structure and Chemistry of the Apatite and Other Calcium Orthophosphates. Amsterdam : Elsevier Science B.V.; 1994.

[4] Landi E, Tampieri A, Celotti G, Sprio S. Densification behaviour and mechanisms of synthetic hydroxyapatites. J. Eur. Ceram. Soc. 2000; 20, 2377.

[5] Ben Ayed F, Bouaziz J, Bouzouita K. Pressureless sintering of fluorapatite under oxygen atmosphere. J. Eur. Ceram. Soc. 2000; 20 (8) 1069.

[6] Ben Ayed F, Bouaziz J, Khattech I, Bouzouita K. Produit de solubilité apparent de la fluorapatite frittée. Ann. Chim. Sci. Mater. 2001; 26 (6) 75.

[7] Ben Ayed F, Bouaziz J, Bouzouita K. Calcination and sintering of fluorapatite under argon atmosphere. J. Alloys Compd. 2001; 322 (1-2) 238.

[8] Varma H K, Sureshbabu S. Oriented growth of surface grains in sintered b tricalcium phosphate bioceramics. Materials letters 2001; 49, 83.

[9] Destainville A, Champion E, Bernache – Assolant D. Synthesis. Characterization and thermal behavior of apatitic tricalcium phosphate. Mater. Chem. Phys. 2003; 80, 269.

[10] Wang C X, Zhou X, Wang M. Influence of sintering temperatures on hardness and Young's modulus of tricalcium phosphate bioceramic by nanoindentation technique. Materials Characterization 2004; 52, 301.

[11] Hoell S, Suttmoeller J, Stoll V, Fuchs S, Gosheger G. The high tibial osteotomy, open versus closed wedge, a comparison of methods in 108 patients. Arch.Trauma Surg. 2005; 125, 638–43.

[12] Gaasbeek R D, Toonen H G, Van Heerwaarden R J, Buma P. Mechanism of bone incorporation of β-TCP bone substitute in open wedge tibial osteotomy in patients. Biomaterials 2005; 26, 6713–6719.

[13] Jensen S S, Broggini N, Hjorting-Hansen E, Schenk R, Buser D. Bone healing and graft resorption of autograft, anorganic bovine bone and beta-tricalcium phosphate. A histologic and histomorphometric study in the mandibles of minipigs.. Clin. Oral. Implants Res. 2006; 17, 237–243.

[14] Ben Ayed F, Chaari K, Bouaziz J, Bouzouita K. Frittage du phosphate tricalcique. C. R. Physique, 2006; 7 (7) 825.

[15] Ben Ayed F, Bouaziz J, Bouzouita K. Résistance mécanique de la fluorapatite. Ann. Chim. Sci. Mater. 2006; 31 (4) 393.

[16] Gutierres M, Dias A G, Lopes M A, Hussain N S, Cabral A T, Almeid L. Opening wedge high tibial osteotomy using 3D biomodelling Bone like macroporous structures: case report. J. Mater. Sci. Mater. Med. 2007; 7 (18) 2377–2382.

[17] DeSilva G L, Fritzler A, DeSilva S P. Antibiotic-impregnated cement spacer for bone defects of the forearm and hand. Tech. Hand Up Extrem Surg. 2007; 11, 163–7

[18] Ben Ayed F, Bouaziz J. Élaboration et caractérisation d'un biomatériau à base de phosphates de calcium. C. R. Physique, 2007; 8 (1) 101-108.

[19] Ben Ayed F, Bouaziz J. Sintering of tricalcium phosphate–fluorapatite composites by addition of alumina. Ceramics Int. 2008; 34 (8) 1885-1892.

[20] Ben Ayed F, Bouaziz J. Sintering of tricalcium phosphate–fluorapatite composites with zirconia. J. Eur. Ceram. Soc., 2008; 28 (10) 1995-2002.

[21] Bouslama N, Ben Ayed F, Bouaziz J. Sintering and mechanical properties of tricalcium phosphate–fluorapatite composites. Ceramics Int. 2009; 35, 1909-1917.

[22] Bouslama N, Ben Ayed F, Bouaziz J. Mechanical properties of tricalcium phosphate-fluorapatite-alumina composites. Physics Procedia 2009; 2, 1441-1448.

[23] Chaari K, Ben Ayed F, Bouaziz J, Bouzouita K. Elaboration and characterization of fluorapatite ceramic with controlled porosity. Materials Chemistry and Physics 2009; 113, 219-226.

[24] Guha A K, Singh S, Kumarresan R, Nayar S, Sinha A. Mesenchymal cell response to nanosized biphasic calcium phosphate composites. Coll. Surf. B biointerface 2009; 73, 146-51.

[25] Bouslama N, Ben Ayed F, Bouaziz J. Effect of fluorapatite additive on densification and mechanical properties of tricalcium phosphate. J. Mechanical Behaviour of Biomedical Materials 2010; 3, 2-13.

[26] Ben Ayed F. Biomaterials - Physics and Chemistry, Chapter 18: Elaboration and characterisation of calcium phosphate biomaterial for biomedical application, ISBN 978-953-307-418-4. Croatia: In Tech; 2011. p 357 – 374.

[27] Sakka S, Ben Ayed F, Bouaziz J. Mechanical properties of tricalcium phosphate–alumina composites. IOP Conf. Series : Materials Science and Engineering, 2012. 28, 012028

[28] Sellami I, Ben Ayed F, Bouaziz J. Effect of fluorapatite additive on the mechanical properties of tricalcium phosphate-zirconia composites. IOP Conf. Series : Materials Science and Engineering, 2012. 28, 012029.

[29] Ben Ayed F. Current microscopy contributions to advances in science and technology, Chapter: The effect of the sintering process on the microstructure and the mechanical properties of biomaterials. published by Formatex Research Center Spain; to be published in 2012.

[30] Levin I, Brandon D. Metastable Alumina Polymorphs: Crystal Structures and Transition. Sequences. J. Am. Ceram. Soc. 1998, 81.

[31] Guidera A, Chaari K, Bouaziz J. Elaboration and Characterization of Alumina-Fluorapatite Composites. J. Biomat. Nano. 2011; 2, 103-113.

[32] Doremus RH. Bioceramics. J Mater Sci 1992; 27, 285–97.

[33] Vallet-Regı´ M. Ceramics for medical applications. J Chem Soc Dalton Trans 2001, 97–108.

[34] Rahaman MN, Yao A, Bal BS, Garino JP, Ries MD. Ceramics for prosthetic hip and knee joint replacement. J Am Ceram Soc 2007; 90, 1965–88.

[35] Best SM, Porter AE, Thian ES, Huang J. Bioceramics: past, present and for the future. J Eur Ceram Soc 2008; 28, 1319–27.

[36] Lowenstam HA, Weiner S. On biomineralization. Oxford University Press, 1989; pp 324.

[37] LeGeros RZ. Calcium phosphates in oral biology and medicine. Basel: Karger; 1991, 201.

[38] Weiner S, Wagner HD. Material bone: structure-mechanical function relations. Ann Rev Mater Sci 1998; 28, 271–98.

[39] Weiner S, Traub W, Wagner HD. Lamellar bone: structure-function relations. J Struct Biol 1999; 126, 241–55.

[40] Weiner S, Dove PM. An overview of biomineralization processes and the problem of the vital effect. In: Dove PM, de Yoreo JJ, Weiner S, editors. Biomineralization, series: reviews in mineralogy and geochemistry, vol. 54. Washington, D.C., USA: Mineralogical Society of America; 2003. p 1–29.

[41] Albee FH. Studies in bone growth – triple calcium phosphate as stimulus to osteogenesis. Ann Surg 1920; 71, 32–9.

[42] Nery EB, Lynch KL, Hirthe WM, Mueller KH. Bioceramic implants in surgically produced infrabony defects. J Periodontol 1975; 46, 328–47.

[43] Denissen HW, de Groot K. Immediate dental root implants from synthetic dense calcium hydroxylapatite. J Prosthet Dent 1979; 42, 551–6.

[44] Levitt GE, Crayton PH, Monroe EA, Condrate RA. Forming methods for apatite prosthesis. J Biomed Mater Res 1969; 3, 683–5.

[45] Blakeslee KC, Condrate Sr RA. Vibrational spectra of hydrothermally prepared hydroxyapatites. J Am Ceram Soc 1971; 54, 559–63.

[46] Garrington GE, Lightbody PM. Bioceramics and dentistry. J Biomed Mater Res 1972; 6, 333–43.

[47] Cini L, Sandrolini S, Paltrinieri M, Pizzoferrato A, Trentani C. Materiali bioceramici in funzione sostitutiva. Nota preventiva. (Bioceramic materials for replacement purposes. Preliminary note.). La Chirurgia Degli Organi Di Movimento 1972; 60, 423–30.

[48] Hulbert SF, Young FA, Mathews RS, Klawitter JJ, Talbert CD, Stelling FH. Potential of ceramic materials as permanently implantable skeletal prostheses. J Biomed Mater Res 1970; 4, 433–56.

[49] Hench LL, Splinter RJ, Allen WC, Greenlee TK. Bonding mechanisms at the interface of ceramic prosthetic materials. J Biomed Mater Res 1971; 2, 117–41.

[50] Hulbert SF, Hench LL, Forces D, Bowman L. History of bioceramics. In: Vincenzini P, editor. Ceramics in surgery. Amsterdam, The Netherlands: Elsevier, 1983; p 3–29.

[51] Jarcho M. Calcium phosphate ceramics as hard tissue prosthetics. Clin Orthop Relat Res 1981; 157, 259–78.

[52] de Groot K. Bioceramics consisting of calcium phosphate salts. Biomaterials 1980; 1 (1) 47–50.

[53] Aoki H, Kato KM, Ogiso M, Tabata T. Studies on the application of apatite to dental materials. J Dent Eng 1977; 18, 86–9.

[54] Roy DM, Linnehan SK. Hydroxyapatite formed from coral skeletal carbonate by hydrothermal exchange. Nature 1974; 247, 220–2.

[55] Holmes RE. Bone regeneration within a coralline hydroxyapatite implant. Plast Reconstr Surg 1979; 63, 626–33.

[56] Elsinger EC, Leal L. Coralline hydroxyapatite bone graft substitutes. J Foot Ankle Surg 1996; 35, 396–9.

[57] LeGeros RZ, LeGeros JP. Calcium phosphate bioceramics: past, present, future. Key Eng Mater 2003; 240–242, 3–10.

[58] Durucan C, Brown PW. Low temperature formation of calcium-deficient hydroxyapatite-PLA/PLGA composites. J Biomed Mater Res 2000; 51A, 717–25.

[59] Ginebra MP, Rilliard A, Ferna´ndez E, Elvira C, Roma´n JS, Planell JA. Mechanical and rheological improvement of a calcium phosphate cement by the addition of a polymeric drug. J Biomed Mater Res 2001; 57, 113–8.

[60] Yokoyama A, Yamamoto S, Kawasaki T, Kohgo T, Nakasu M. Development of calcium phosphate cement using chitosan and citric acid for bone substitute materials. Biomaterials 2002; 23, 1091–101.

[61] Barralet JE, Gaunt T, Wright AJ, Gibson IR, Knowles JC. Effect of porosity reduction by compaction on compressive strength and microstructure of calcium phosphate cement. J Biomed Mater Res 2002; 63B, 1–9.

[62] Bohner M, Gbureck U, Barralet JE. Technological issues for the development of more efficient calcium phosphate bone cements: a critical assessment. Biomaterials 2005; 26, 6423–9.

[63] Bohner M, Baroud G. Injectability of calcium phosphate pastes. Biomaterials 2005; 26, 1553–63.

[64] Link DP, van den Dolder J, van den Beucken JJ, Wolke JG, Mikos AG, Jansen JA. Bone response and mechanical strength of rabbit femoral defects filled with injectable CaP cements containing TGF-b1 loaded gelatin microspheres. Biomaterials 2008; 29, 675–82.

[65] Friedman CD, Costantino PD, Takagi S, Chow LC. Bone source hydroxyapatite cement: a novel biomaterial for craniofacial skeletal tissue engineering and reconstruction. J Biomed Mater Res 1998; 43B, 428–32.

[66] Shindo ML, Costantino PD, Friedman CD, Chow LC. Facial skeletal augmentation using hydroxyapatite cement. Arch Otolaryngol Head Neck Surg 1993; 119, 185–90.

[67] Chow LC. Calcium phosphate cements: chemistry, properties, and applications. Mat Res Soc Symp Proc 2000; 599, 27–37.

[68] Muzzarelli RAA, Biagini G, Bellardini M, Simonelli L, Castaldini C, Fraatto G. Osteoconduction exerted by methylpyrolidinone chitosan in dental surgery. Biomaterials 1993; 14, 39–43.

[69] Hing KA, Best SM, Bonfield W. Characterization of porous hydroxyapatite. J Mater Sci Mater Med 1999; 10, 135–45.

[70] Jordan DR, Gilberg S, Bawazeer A. Coralline hydroxyapatite orbital implant (Bio-Eye): experience with 158 patients. Ophthal Plast Reconstr Surg 2004; 20, 69–74.

[71] Yoon JS, Lew H, Kim SJ, Lee SY. Exposure rate of hydroxyapatite orbital implants a 15-year experience of 802 cases. Ophthalmology 2008; 115, 566–72.

[72] Schnettler R, Stahl JP, Alt V, Pavlidis T, Dingeldein E, Wenisch S. Calcium phosphate-based bone substitutes. Eur J Trauma 2004; 4, 219–29.

[73] Zyman ZZ, Glushko V, Dedukh N, Malyshkina S, Ashukina N. Porous calcium phosphate ceramic granules and their behaviour in differently loaded areas of skeleton. J Mater Sci Mater Med 2008; 19, 2197–205.

[74] Larsson S, Hannink G. Injectable boneegraft substitutes: current products, their characteristics and indications, and new developments. Injury 2011; 42, 30e4.

[75] Dorozhkin SV. Calcium orthophosphate cements for biomedical application. J Mater Sci 2008; 43, 3028e57.

[76] Dorozhkin SV. Calcium orthophosphate-based biocomposites and hybrid biomaterials. J Mater Sci 2009; 44, 2343e87.

[77] Bohner M, Gbureck U, Barralet JE. Technological issues for the development of more efficient calcium phosphate bone cements: a critical assessment. Biomaterials 2005; 26, 6423e9.

[78] Von Gonten AS, Kelly JR, Antonucci JM. Load-bearing behavior of a simulated craniofacial structure fabricated from a hydroxyapatite cement and bioresorbable fiber mesh. J Mater Sci Mater Med 2000; 11, 95e100.

[79] Canal C, Ginebra MP. Fibre-reinforced calcium phosphate cements: a review. J Mech Behav Biomed Mater 2011; 4, 1658e71.

[80] Callister Jr WD, Rethwisch DG. Materials science and engineering: an introduction. Hoboken: John Wiley Sons Inc; 2009.

[81] LeGeros RZ. Biodegradation and bioresorption of calcium phosphate ceramics. Clin Mater 1993;14, 65–88.

[82] Suchanek W, Yoshimura M. Processing and properties of hydroxyapatitebased biomaterials for use as hard tissue replacement implants. J Mater Res 1998; 13, 94–117.

[83] Hing KA, Best SM, Bonfield W. Characterization of porous hydroxyapatite. J Mater Sci Mater Med 1999; 10, 135–45.

[84] Ducheyne P, Qiu Q. Bioactive ceramics: the effect of surface reactivity on bone formation and bone cell function. Biomaterials 1999; 20, 2287–303.

[85] Pilliar RM, Filiaggi MJ, Wells JD, Grynpas MD, Kandel RA. Porous calcium polyphosphate scaffolds for bone substitute applications – in vitro characterization. Biomaterials 2001; 22, 963–72.

[86] Chu TMG, Orton DG, Hollister SJ, Feinberg SE, Halloran JW. Mechanical and in vivo performance of hydroxyapatite implants with controlled architectures. Biomaterials 2002; 23, 1283–93.

[87] Hench LL, Polak JM. Third-generation biomedical materials. Science 2002; 295, 1014–7.

[88] Tamai N, Myoui A, Tomita T, Nakase T, Tanaka J, Ochi T, et al. Novel hydroxyapatite ceramics with an interconnective porous structure exhibit superior osteoconduction in vivo. J Biomed Mater Res 2002; 59, 110–7.

[89] Simon JL, Roy TD, Parsons JR, Rekow ED, Thompson VP, Kemnitzer J, et al. Engineered cellular response to scaffold architecture in a rabbit trephine defect. J Biomed Mater Res 2003; 66A, 275–82.

[90] Deville S, Saiz E, Nalla RK, Tomsia AP. Freezing as a path to build complex composites. Science 2006; 311, 515–8.

[91] Miranda P, Pajares A, Saiz E, Tomsia AP, Guiberteau F. Fracture modes under uniaxial compression in hydroxyapatite scaffolds fabricated by robocasting. J Biomed Mater Res 2007; 83A, 646–55.

[92] Miranda P, Pajares A, Saiz E, Tomsia AP, Guiberteau F. Mechanical properties of calcium phosphate scaffolds fabricated by robocasting. J Biomed Mater Res 2008; 85A, 218–27.

[93] Brown WE, Chow LC. A new calcium phosphate water setting cement. In: Brown PW, editor. Cements research progress. Westerville, OH: Am Ceram Soc; 1986. p 352–79.

[94] Daculsi G, Weiss P, Bouler JM, Gauthier O, Millot F, Aguado E. Biphasic calcium phosphate/hydrosoluble polymer composites: a new concept for bone and dental substitution biomaterials. Bone 1999; 25 (Suppl. 2), 59S–61S.

[95] Alam I, Asahina I, Ohmamiuda K, Enomoto S. Comparative study of biphasic calcium phosphate ceramics impregnated with rhBMP-2 as bone substitutes. J Biomed Mater Res 2001; 54, 129–38.

[96] Daculsi G, Laboux O, Malard O, Weiss P. Current state of the art of biphasic calcium phosphate bioceramics. J Mater Sci Mater Med 2003; 14, 195–200.

[97] Daculsi G. Biphasic calcium phosphate granules concept for injectable and mouldable bone substitute. Adv Sci Technol 2006; 49, 9–13.

[98] LeGeros RZ, Lin S, Rohanizadeh R, Mijares D, LeGeros JP. Biphasic calcium phosphate bioceramics: preparation, properties and applications. J Mater Sci Mater Med 2003;14, 201–9.

[99] Lecomte A, Gautier H, Bouler JM, Gouyette A, Pegon Y, Daculsi G, et al. Biphasic calcium phosphate: a comparative study of interconnected porosity in two ceramics. J Biomed Mater Res B (Appl Biomater) 2008; 84B, 1–6.

[100] Tancret F, Bouler JM, Chamousset J, Minois LM. Modelling the mechanical properties of microporous and macroporous biphasic calcium phosphate bioceramics. J Eur Ceram Soc 2006; 26, 3647–56.

[101] Langstaff SD, Sayer M, Smith TJN, Pugh SM. Resorbable bioceramics based on stabilized calcium phosphates. Part II: evaluation of biological response. Biomaterials 2001; 22, 135–50.

[102] Sayer M, Stratilatov AD, Reid JW, Calderin L, Stott MJ, Yin X, et al. Structure and composition of silicon-stabilized tricalcium phosphate. Biomaterials 2003; 24, 369–82.

[103] Yin X, Stott MJ, Rubio A. a- and b-tricalcium phosphate: a density functional study. Phys Rev B 2003; 68, 205205.

[104] Yin X, Stott MJ. Theoretical insights into bone grafting Si-stabilized a-tricalcium phosphate. J Chem Phys 2005; 122, 024709.

[105] Reid JW, Pietak AM, Sayer M, Dunfield D, Smith TJN. Phase formation and evolution in the silicon substituted tricalcium phosphate/apatite system. Biomaterials 2005; 26, 2887–97.

[106] Yin X, Stott MJ. Surface and adsorption properties of a-tricalcium phosphate. J Chem Phys 2006;124, 124701.

[107] Ruan JM, Zou JP, Zhou JN, Hu JZ. Porous hydroxyapatite – tricalcium phosphate bioceramics. Powder Metall 2006; 49, 66–9.

[108] Reid JW, Tuck L, Sayer M, Fargo K, Hendry JA. Synthesis and characterization of single-phase silicon substituted a-tricalcium phosphate. Biomaterials 2006; 27, 2916–25.

[109] da Silva RV, Bertran CA, Kawachi EY, Camilli JA. Repair of cranial bone defects with calcium phosphate ceramic implant or autogenous bone graft. J Craniofac Surg 2007; 18, 281–6.

[110] O'Neill WC. The fallacy of the calcium – phosphorus product. Kidney Int 2007; 72, 792–6.

[111] Sanchez-Sa′ lcedo S, Arcos D, Vallet-Regı′ M. Upgrading calcium phosphate scaffolds for tissue engineering applications. Key Eng Mater 2008; 377, 19–42.

[112] Brunauer S, Emmet P H, Teller. Adsorption of Gases in Multimolecular Layers. J. Amer. Chem. Soc. J. 1938; 60, 310.

[113] ISRM. Suggested methods for determining tensile strength of rock materials, Int. J. Rock Mech. Min. Sci. Geomech. Abstr. 1978; 15, 99.

[114] ASTM C496, Standard test method for splitting tensile strength of cylindrical concrete specimens Annual Book of ASTM, Standards, vol. 0.042, ASTM, Philadelphia, 1984; p 336.

[115] Balcik C, Tokdemir T, Senkoylo A, Koc N, Timucin M, Akin S, Korkusuz P, Korkusuz F. Early weight bearing of porous HA/TCP (60/40) ceramics in vivo: a longitudinal study in a segmental bone defect model of rabbit. Acta Biomaterialia 2007; 3, 985-996.

[116] Deville S, Saiz E, Nalla RK, Tomsia A P. Freezing as a path to build complex composites. Science. 2006; 311 (5760) 515-518.

[117] Murugan R, Ramakrishna S. Development of nanocomposites for bone grafting. Compos Sci Technol. 2005; 65 (15-16) 2385-2406.

[118] Suchanek WL, Yoshimura M. Processing and properties of hydroxyapatite based biomaterials for use as hard tissue replacement implants. J Mater Res 1998; 13, 94–117.

Chapter 3

Degradation of Polyurethanes for Cardiovascular Applications

Juan V. Cauich-Rodríguez, Lerma H. Chan-Chan,
Fernando Hernandez-Sánchez and
José M. Cervantes-Uc

Additional information is available at the end of the chapter

http://dx.doi.org/10.5772/53681

1. Introduction

Polyurethanes are a family of polymers used in a variety of biomedical applications but mainly in the cardiovascular field due to their good physicochemical and mechanical properties in addition to a good biocompatibility. Traditionally, segmented polyurethanes (SPUs), have been used in cardiovascular applications (Kuan et al., 2011) as permanent devices such as pacemaker leads and ventricular assisting devices; however, due to their great chemistry versatility, SPUs can also be tailored to render biodegradable systems for the tissue engineering of vascular grafts and heart valves. Therefore many research work have been focused on varying the chemical composition to enhance biostability or more recently to control the biodegradability of polyurethanes depending on specific applications in the cardiovascular field (Bernacca et al., 2002; Stachelek et al., 2006; Thomas et al., 2009; Wang et al., 2009; Hong et al., 2010; Arjun et al., 2012; Styan et al., 2012). In this way, polyurethanes for biomedical applications can be classified in two main types, according their relative stability in the human body as either biostables or biodegradables. In this chapter, general aspects of polyurethanes chemistry are presented first and then, the various types of degradations that can affect these polymers both *in vivo* and simulated *in vitro* conditions. Emphasis is also made on the mechanism of degradation under various conditions and the techniques used for following the changes in their properties.

2. Chemistry of polyurethanes

2.1. Synthesis of polyurethanes

Polyurethanes (PU's) properties depend both on the method of preparation and the monomers used. In general, PU's can be prepared in one shot process or more commonly by a two step method, especially for the case of segmented polyurethanes (SPU's). These materials are thermoplastic block copolymers of the (AB)n type consisting of alternating sections of hard segments, composed of a diisocyanate and a low molecular weight diol chain extender, and soft segments, generally composed of various types of polyols, also called macrodiols. In the two steps method for SPU synthesis, a prepolymer is first obtained and then chain extended as illustrated in Figure 1. In the first step, an excess of the diisocyanate reacts with the soft segment polyol to form the prepolymer. Here, the characteristic urethane linkages are formed through the reaction between the isocyanate groups and the hydroxyl-terminated end groups of the polyol. In the second step, the low molecular weight chain extender is used to link the prepolymer segments yielding a high molecular weight polymer. During this stage, additional urethane functional groups are formed when using a diol chain extender whereas ureas are produced when a diamine is used.

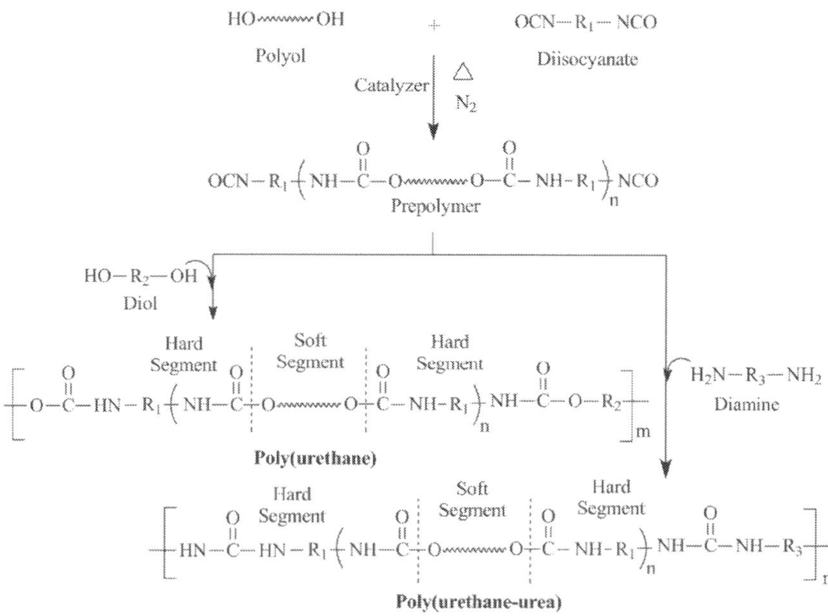

Figure 1. Standard two-step reaction to prepare segmented poly (urethane)s and poly(urethane-urea)s

The properties of polyurethanes as those shown in Figure 1 depend on the various types of monomers that are used during their manufacturing (see Table 1). Historically, biostable polyurethanes were first developed by using polyether type of polyol and different aromatic diisocyanates. Further developments in this area were focused on the substitution of the polyether macrodiol by novel hydrocarbon, polycarbonate or siloxane macrodiols (Gunatillake et al., 2003) or a combination of these which in general are responsible for the flexibility of the SPUs (Król, 2007). In addition to the polyol chemical composition, their molecular weight and concentration have an important effect in the polyurethane behavior. They can be incorporated in various concentrations but up to 50-75% of the polyol is common.

Commercial and experimental polyurethanes have been synthesized by the combination of the aforementioned monomers. Poly(tetramethylene oxide) (PTMO) is the most common polyether in conventional medical formulations (Silvestri et al., 2011). Thus, for example, the Pellethane® 2363 80A and Elasthane™ 80A are poly(ether-urethane)s obtained by the reaction of PTMO, MDI and BD monomers; Tecoflex® by Thermedics is also a poly(ether-urethane) synthesised by the reaction of PTMO, HMDI and BD monomers while Biomer® is a poly(ether-urea-urethane) synthesized from PTMO, MDI and ethylenediamine. Bionate®, Myo Lynk™ and Chronoflex are polyurethanes prepared with polycarbonate diol. These commercial polyurethanes are typical examples of biostables polymers.

The use of vegetable raw materials containing hydroxyl groups such as starch, castor oil, vegetable oil, natural rubber, cellulose, etc, makes possible to obtain biodegradable polyurethanes (Krol, 2007; Aranguren et al. 2012). However, ester polyol commonly used to synthesize biodegradable polyurethanes are polycaprolactone, polylactic acid and adipate polyols. Polyethyleneglycol is a polyether which has been copolymerized with poly lactic acid and/or polycaprolactone because its higher hydrofilicity can accelerate the biodegradation when this is required (Guan et al., 2005b; Wang et al., 2011b).

The most frequently used diisocyanates in the synthesis of biodegradable polyurethanes for biomedical applications are aliphatic or cycloaliphatic as MDI and TDI which can release carcinogenic and mutagenic aromatic diamines (Heijkants *et. al.*, 2005). Aliphatic diisocyanates are less reactive than the aromatic counterparts but have a greater resistance to hydrolysis compared to aromatic diisocyanates, although this resistance frequently results in lower mechanical properties (Gogolewski, 1989).

In general, there are two types of compounds that are generally used as chain extenders, diols or diamines, which can either be aliphatic or aromatic, depending on the required properties in the synthesized polyurethanes. New chain extenders, including amino acids have been also used during polyurethane synthesis as isocyanates can react vigorously with amine, alcohol, and carboxylic acids (Thomson, 2005). These novel chain extenders have been used to synthesize biodegradable polyurethanes (Skarja et al., 1998; Marcos-Fernández et al., 2006; Sarkar et al., 2007).

Monomeric component	Type	Chemical compound	Type of polyurethanes
Polyol (macrodiols)	Polyethers	Poly(ethylene oxide) (PEO)	Biostables, Biodegradables (Sarkar et al., 2009; Lu et al., 2012)
		Poly(propylen oxide) (PPO)	Biostables, Biodegradables (Francolini et al., 2011)
		Poly(tetramethylene oxide) (PTMO)	Biostables (Silvestri et al., 2011; Jiang et al., 2012)
		Poly(caprolactone) (PCL)	Biodegradables (Sarkar et al., 2009; Lu et al., 2012)
		Poly(lactic acid) (PLA)	Biodegradable (Wang et al., 2011a)
		Poly hydroxyalkanoates (PHA)	Biodegradables (Li et al., 2009; Liu et al., 2009)
		Poly(ethylene adipate) (PEA)	Biodegradables (Macocinschi et al., 2009)
	Others	Poly(carbonate) (PCU)	Biostables (Spirkova et al., 2011)
		Polybutadiene (PBD)	Biostables (Thomas et al., 2009)
		Poly(dimethylsiloxane) (PDMS)	Biostables (Park et al., 1999; Madhavan et al., 2006)
Diisocyanate	Aromatic	Methylene diphenyl diisocyanate (MDI)	Biostables (Gunatillake et al., 1992; Styan et al., 2012)
		2,4-toluene diisocyanate (TDI)	Biostables (Labow et al., 1996; Basak et al., 2012)
	Aliphatic	4,4′-methylene bis(cyclohexyl isocyanate) (HMDI)	Biostables, Biodegradables (Thomas et al., 2009; Chan-Chan et al., 2010)
		1,6- hexamethylene diisocyanate (HDI)	Biostables, Biodegradables (Wang et al., 2011a; Baudis et al., 2012)
		1,4-butane diisocyanate (BDI)	Biostables, Biodegradables (Heijkants et al., 2005; Hong et al., 2010)
		Isophorone diisocyanate (IPDI)	Biostables, Biodegradables (Jiang et al., 2007; Ding et al., 2012; He et al., 2012)
		L-lysine ethyl ester diisocyanate (LDI)	Biodegradable (Abraham et al., 2006; Guelcher et al., 2008; Han et al., 2009; Wang et al., 2011b)
Chain Extender	Diols	Ethylene glycol (EG)	Biostables, Biodegradables (Król, 2007)
		Diethylenglycol	
		1,4-butanediol (BD)	
		1,6-hexanediol (HD)	
	Diamines	Aliphatic diamines	
		Aromatic diamines	
	Others	Amino acids	Biodegradables (Kartvelishvili et al., 1997; Skarja et al., 1998; Marcos-Fernández et al., 2006; Sarkar et al., 2007; Chan-Chan et al., 2012)

Table 1. Common monomers used in the synthesis of biostable and biodegradable polyurethanes

During polyurethane synthesis, several side reactions may occur leading to branching, crosslinking, or changes in the stoichiometry of reactants. For example, undesirable branching and crosslinking may occur at elevated temperatures between isocyanates and urethanes (Allophanate formation) and isocyanates and ureas (Biuret reactions). Furthermore, the

http://dx.doi.org/10.5772/53681

presence of water causes isocyanate groups to form unstable carbamic acids, which subsequently decompose to amines with the liberation of CO_2 gas (see Figure 2). These newly formed amines react with isocyanates to form ureas, thus changing reactant stoichiometry and leading to lower molecular weight polymers. Additives are sometime used for improving specific properties of the polyurethane, for example Vitamin E and Santowhite®, two hindered phenolic antioxidants, prevents oxidative chain scission and crosslinking of poly(ether urethane) by capturing oxygen radicals (Schubert et al., 1997; Christenson et al., 2006). Di-tert-butylphenol and bisphosphonates have been incorporated to promote bromoalkylation of urethane nitrogens in prepolimerized polyurethanes to inhibit the oxidation or calcification (Alferiev et al., 2001; Stachelek et al., 2007). Other compounds as fluorocarbon or polydimethylsiloxane end groups have been attached to the surface of polyurethanes in order to enhance their biostability and hemocompatibility (Ward et al., 2007; Xie et al., 2009; Jiang et al., 2012).

In general, monomer type and stoichiometry, type and concentration of catalyzer, temperature and moisture, and the use of additives are important in parameters for controlling the properties of these polymers.

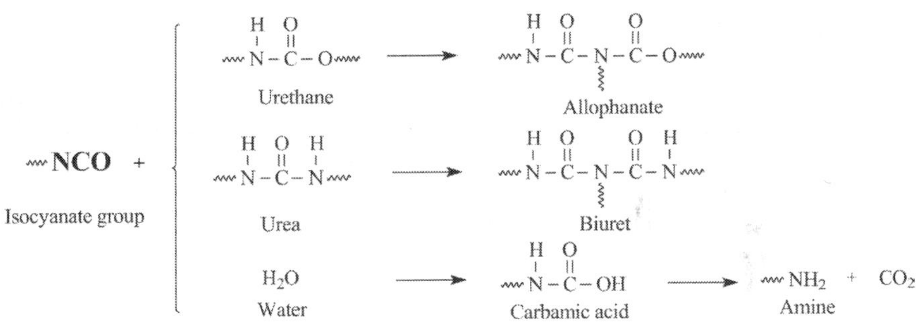

Figure 2. Secondary reactions involved during polyurethane synthesis

3. *In vivo* degradation

SPU's traditionally has been used in medical devices due to their excellent mechanical properties and an acceptable hemocompatibility. However, in the long term they suffer from poor biostability (Santerre et al., 2005). The main reason of this behavior is that living tissues are a very aggressive environment and even when the degradation of these polymers can be simulated by *in vitro* experiments, after *in vivo* usage they can be severally degraded. The *in vivo* failure of polymeric cardiovascular devices has been attributed to a combination of hydrolysis, oxidation, environmental stress cracking and calcification. However, depending on the composition of the polymer one of these predominate over the other.

Polymer degradation in the biological environment results from the synergistic effects of the enzymes present in biological fluids, oxidizing agents and mechanical loads. For example, α-2-macroglobulin, cholesteryl esterase, A2 fosfolipase, K protease and B Cathepsin are enzymes that are known to degrade polyurethanes (Zhao et al., 1993; Dumitriu, 2002). Even when some enzymes require very specific biological substrates, some of them seem to recognise and act over non biological substrates such as polymers (Santerre et al., 2005). White blood cells play also an important role in the *in vivo* degradation. Some experiments conducted using implanted metallic cages have shown that neutrophiles, monocytes, monocyte derived macrophages (MDM) attach to polymer surfaces, leading to the presence of multinucleated giant cells and foreign body reaction. It is generally accepted that one of the immediate immune responses by the body is the release of reactive oxygen species (ROS). In addition, neutrophils and monocytes release hypochlorous acid (HClO) and lysosomal hydrolases as part of their reaction to foreign surfaces. It has been also reported that activated MDM release ROS leading to the formation of hydrogen peroxide (Christenson et al., 2006; McBane et al., 2007). In addition, during the inflammatory reaction macrophages are able to lower the local pH up to 4. This condition can be simulated by following the ISO 10993 section 5.

Suntherland et al. (Sutherland et al., 1993) suggested that poly(ether urethanes) (PEU) cannot be significantly degraded by preformed products of phagocytic cells (such as cationic proteins and proteases) or by activated oxygen species such as superoxide and hydrogen peroxide. In view of the chemically stable nature of PEU, they hypothesized that the *in vivo* degradation of these materials might involve attack by chlorine-based and/or nitric oxide (NO)-derived oxidants, major oxidative products of activated phagocytes. Therefore, they exposed Pellethane to polymorphonuclear neutrophils (PMN) isolated from heparinized venous blood drawn from normal adult donors. The results reported support the idea that PMN-generated chlorine compounds are likely responsible for initial damage to PEU after brief implantation and in addition to macrophage-derived NO and/or peroxynitrite (ONOO-).

Van Minen et al. (van Minnen et al., 2008) studied the *in vivo* (26 weeks of subcutaneous implantation in rats and 2.5 years in rabbits) degradation of porous aliphatic SPU based on butanediisocyanate, DL-lactide-co-caprolactone soft segments and extended with BD-BDI-BD block urethane. After 1 week macrophages were observed along with giant cells and after 4 week phagocytosis was observed. The number of these cells was reduced with time but after 3 years fragments of the polymer remained. Furthermore, few macrophages were observed in the lymph nodes suggesting their local degradation.

Adhikari et al. (Adhikari et al., 2008) studied the *in vivo* degradation of two-part injectable biodegradable polyurethane prepolymer systems (prepolymer A and B) consisting of lactic acid and glycolic acid based polyester star polyols, pentaerythritol (PE) and ethyl lysine diisocyanate (ELDI) using sheep femoral cortical defect model. No adverse acute or chronic inflammatory tissue response was noted in the interface tissues of the precured polymer implants. By 6 weeks, there was direct apposition of new bone to the polymers. New bone and fibrovascular tissue was also observed within the porous spaces of the precured polymers by 6 weeks, and fluorochrome analysis suggested that this bone had started to be laid down at between 4 and 5 weeks. The polymer without β-tricalcium phosphate (TCP) showed histological evidence of some degradation by 6 weeks with progressive increase in polymer loss by 12 and 24 weeks. The polymer with β-TCP showed no evidence of degradation at 6 weeks and only minimal loss at 12 weeks. By 12 weeks, there had been considerable degradation of the polymers and at week 24, polymer was completely degraded.

The *in vivo* degradation of segmented poly(urethane urea)s (SPUUs) with hard segments derived only from methyl 2,6-diisocyantohexanoate (LDI) and PCL, PTMC (polytrimethylene carbonate), P(TMC-co-CL), P(CL-co-DLLA), or P(TMC-co-DLLA) as soft segment was conducted by Asplund et al. (Asplund et al., 2008). The *in vivo* study of SPUU-PCL using male Sprague-Dawley rats displayed the typical foreign body response seen with most inert polymeric implant materials. The reaction at 1 week thus displayed an infiltration of ED1 positive macrophages closest to the implant surface, an outside layer of fibroblasts and some collagen formation. At 6 weeks, the foreign body capsule had matured, displaying lower numbers of interfacial macrophages and an increased amount of collagen in the fibrotic capsule. The thickness of the foreign body capsule was similar to the controls. These observations seemed also to be reflected in the number of ED1 positive macrophages, as well as in the total number of cells throughout the reactive capsule.

Hafeman et al. (Hafeman et al., 2008) synthesized polyurethane scaffolds by one-shot reactive liquid molding of hexamethylene diisocyanate trimer (HDIt) or lysine triisocyanate (LTI) and a polyol as hardener. Trifunctional polyester polyols of 900-Da and 1,800-Da molecular weight were prepared from a glycerol starter and 60% ε-caprolactone, 30% glycolide, and 10% D,L-lactide monomers, and stannous octoate catalyst. Tissue response was evaluated by subcutaneous implantation in male Sprague-Dawley rats for up to 21 days. During this time, initial infiltration of plasma progressed to the formation of dense granulation tissue. All of the implants showed progressive invasion of granulation tissue with little evidence of an overt inflammatory response or cytotoxicity. Fibroplasia and angiogenesis appeared to be equivalent among the different formulations. Extracellular matrix with dense collagen fibers progressively replaced the characteristic, early cellular response. The LTI scaffolds exhibited a greater extent of degradation at 21 days, although the incorporation of PEG into the HDIt scaffold accelerated its degradation significantly. Degradation rates were much higher *in vivo*. With time, each of the materials showed signs of fragmentation and engulfment by a transient, giant cell, foreign body response. After the remnant material was resorbed, giant cells were no longer evident.

Khouw et al. (Khouw et al., *2000*) reported that the foreign body response to degradable materials differs between rats and mice. van Minnen et al., (van Minnen et al., 2008) also suggested that it is possible that the response between rats and rabbits differs as well, due to the faster degradation in the rabbit. This may be related to differences at the enzymatic or cellular level, but also to the highly mobile and well vascularized skin of the rabbit, as compared to the rat.

3.1. Calcification

Mineralization or calcification (formation of various types of calcium phosphates such as apatite) is a well documented event in various medical devices, especially in those used in the cardiovascular field. Calcification is in fact, the most common macroscopic cause of failure in heart valves including those made of polyurethanes (Santerre et al., 2005). Even when calcification has been identified in heart valves, *in vitro* experiments on SPU showed little mineralization and associated exclusively to failure regions, indicating that the SPU's have a lower intrinsic capacity for calcification compared to bovine bioprosthesis (Bernacca et al., 1997).

3.2. Thrombosis

Plasma protein adsorption is well accepted as one of the first events to occur when blood is in contacts with a biomaterial. These adsorbed proteins mediate the subsequent interactions of cells and platelets with the surface and may induce thrombus formation, which remains one of the major problems associated with the long-term use of blood-contacting medical devices. The surface properties of the implanted materials are determinant in protein adsorption and biological interactions with the material. The effects of various physicochemical properties such as surface hydrophilicity/hydrophobicity balance, surface charge density, ability to form hydrogen bonds, and chemical composition of biomaterials on protein adsorption as well as subsequent blood platelet adhesion have been investigated (Xu et al., 2010).

Antithrombogenicity is one of the essential requirements for a vascular graft, but it is very difficult to achieve. There are two common approaches employed to attain this goal. One is to develop biomaterials with inherent antithrombogenicity or to use surface modified biomaterials with an anticoagulant. The other approach is to quickly and completely endothelialize the inner surface of the tubular scaffolds, thereby, reducing thrombogenicity (Yan et al., 2007).

Thrombosis is a leading cause of vascular graft failure in small-diameter prostheses, where it leads to decreased flow or occlusion. In addition to inducing acute or subacute failure of grafts, it may be a cause of late failure owing to thrombosis superimposed on stenosis due to other causes of vessel narrowing, such as intimal hyperplasia. Methods to improve vascular grafts (e.g., antithrombotic therapy) have been shown to be beneficial in decreasing graft occlusion after surgery. Agents known to inhibit thrombogenesis or promote anticoagulation (e.g., heparin, prostaglandin E1, hirudin, dipyridamole, tissue factor pathway inhibitor and aspirin) have also been bound to the lumen of the synthetic vessels (Wang et al., 2007; Lu et al., 2012).

3.3. Environmental stress cracking and metal ion oxidation

Traditionally, SPUs have been used as permanent devices such as pacemaker leads insulation and ventricular assisting devices. When used as pacemaker lead insulators, they substitute silicone rubbers and have been used as biostable polymer for outer or inner insulated coating of coaxial bipolar pacemakers. Unfortunately, decades of experience showed that they were degraded by environmental stress cracking (ESC) or metal ion oxidation (MIO) or even autooxidation (AO) within a period of 28 and 34 months.

Environmental stress cracking includes crack formation and propagation on the surface of the polyurethane (Santerre et al., 2005). However, this type of degradation is a combination of the *in vivo* chemical degradation with the presence of mechanical stresses. In other words, polymer chain scission caused by the chemical degradation, create microscopic defects that are augmented by the presence of mechanical loads, leading to the formation of cracks on the surface (Wiggins et al., 2003). ESC it is also enhanced by the presence of residual stresses in the polymeric surface introduced during manufacturing and not eliminated during polymer annealing (Santerre et al., 2005).

The generally accepted *in vivo* degradation MIO mechanism involves the presence of hydrogen peroxide (H_2O_2) produced by a variety of inflammatory cells (McBane et al., 2007; Chandy et al., 2009) and divalent metal ion such as Co^{2+} released from the lead. This reaction is known as the Haber-Weiss reaction and yields hydroxyl radicals that can attack α-methylene groups in the polyether (PTMO based polyurethanes) to render hydroperoxydes with decompose in the presence of divalent cations rendering carbonyl groups that can accelerate (catalyse) further this decomposition (Kehrer, 2000; Wiggins et al., 2001). Polyether diol based polyurethanes are prone to oxidation and environmental stress cracking (ESC) (Król, 2007). However, polycarbonate based-polyurethanes (PCNUs) have been proven superior to polyether and polyester PUs, especially in terms of reduced ESC and metal ion oxidation (MIO), although they are still susceptible to hydrolysis (McBane et al., 2007).

Environmental stress cracking, calcification and thrombosis only became evident after a sufficiently long-term implantation of several years. For biodegradable PUs, which are designed to degrade in a relative short period (several months), the effective degradation mechanisms are hydrolysis with or without the assistance of enzymatic catalysis (Chen et al., 2012).

4. *In vitro* degradation to simulate *in vivo* degradation

In vivo experimentation using an animal model is not always available for elucidating the degradation mechanism of polyurethanes. Instead, various *in vitro* experiments have been designed to simulate their *in vivo* degradation. Among these tests, hydrolytic degradation has been conducted using distilled water (at elevated temperature), strong acids and alkalies, and sometimes physiological conditions using Phosphate Buffer Saline (PBS) as degrading media.

4.1. Hydrolytic degradation

Degradation of segmented polyurethanes through hydrolysis depends strongly not only on the chemical composition of the soft segment, when is the major component, but also on the rigid segment chemistry. It is generally accepted that water absorption is a necessary condition for hydrolytic degradation of materials. Therefore, in a typical hydrolytic degradation test the rate of water absorption (or sample weight gain) can be correlated with sample weight loss (Mondal et al., 2012).

The presence of labile ester linkages in PCL containing polyurethanes makes them susceptible to degradation in the presence of water (Gunatillake et al., 1992; Nakajima-Kambe et al., 1999; Kannan et al., 2006). This type of reactions is catalysed by the presence of acids or alkaline compounds. In some cases, the acid is produced by the degradation of the soft segment; caproic acid in the case of PCL or lactic acid in the case of PLA. Polyester urethanes are more prone to hydrolytic degradation although they are more resistant to oxidative environments as can be observed in Table 2, where PCL based polyurethanes (BSPU1 and BSPU2) and a commercial polyether polyurethane (Tecoflex) are compared (Chan-Chan et al., 2010).

	H_2O	NaOH 5M	HCl 2N	H_2O_2 30 wt.%
BSPU1*	1.46 ± 0.08	63.42 ± 7.63	82.70 ± 2.60	13.08 ± 3.35
BSPU2*	6.15 ± 1.35	87.23 ± 4.76	52.65 ± 13.26	19.07 ± 7.01
Tecoflex	0.63 ± 0.3	1.66 ± 0.66	1.48 ± 1.62	2.16 ± 1.47

* BSPU's were prepared with PCL, HMDI and either butanediol (BSPU1) or dithioerythritol (BSPU2) as chain extenders.

Table 2. Polyurethane mass loss (%) after degradation under hydrolytic and oxidative accelerated conditions

Because of the susceptibility of the ester groups to hydrolysis, biodegradable poly(ester urethanes) degrade *in vitro* through bulk erosion via chain scission. During hydrolysis, new carboxylic acid groups are formed that auto-catalyze the degradation, leading to faster degradation in the bulk than at the surface. Thus, a decrease in molecular weight preceding the loss of mechanical properties and weight loss is typical for such degradation. In addition, an increase in crystallinity is observed, if the soft segment contains a crystalline fraction. Polyesters in the soft segment will, therefore, increase the effect of hydrolysis compared with polyether or polycarbonates (Ma et al., 2012).

Tanzi et al. (Tanzi et al., 1991) degraded various commercial polyurethanes used in the cardiovascular field among them Cardiothane 51, Pellethane 2363 80A, Estane 5714 Fl, Estane 58810 and Biomer. The degradation was conducted in water or alkaline borate buffer (pH 10) at 37°C, 60°C and 85°C from 96 h to 168 h. They found that after hydrolytic degradation in distilled water at 85°C for 96 h, borate buffer during 96 h at 60°C and borate buffer during 168 h at 37°C there were no changes in tensile properties although a reduction in molecular weight was reported.

Polyester urethanes based on methylene-bis(4-phenylisocyanate) (MDI), BD and polyadipate diol were prepared by Pretsch (Pretsch et al., 2009) and accelerated degradation studied in distilled water at 80°C where the degradation process was followed by DSC. It was found that the intensities of the melting peaks and therefore the crystallinity of the soft segments increase after one day. Then, two main degradation scenarios were proposed: first, a hydrolytic scission of polymer chains in the molten soft segments take place, which is accelerated by the "high" immersion temperature; second, and on top of it, there is an annealing effect. For example, the domains of segmented polyurethane elastomers may become unstable at high temperatures and mixing of hard and soft segments is enforced.

Wang et al. (Wang et al., 2011a) prepared segmented polyurethanes based on poly(D,L-lactic acid)diol, hexamethylene diisocyanate (HDI) and with either peperazine (SPU-P), 1,4-butanediol (SPU-O) or 1,4-butanediamine (SPU-A) as chain extenders. The degradation process was conducted in double distilled water at 37°C and 50 rpm. For these SPUs, acidic groups from the degradation of PDLLA and BD could reduce the pH value of medium, while the dissolution of the hard segment (amide group and carbamide) could alkalize the medium; after 12 weeks, the pH values of SPU-O, SPU-A and SPU-P were 2.57, 3.87 and 3.71, respectively. These results suggest that the chain extender can play a main role in the degradation mechanism as using an alkaline chain extender can neutralize the acidity, the hydrophilicity and hydrolysis sensitivity of these bonds.

4.2. Oxidative degradation

Polyether urethanes (PEU) are readily degraded by oxidative conditions (Stachelek et al., 2006). Furthermore, the presence of metallic ions such as cobalt accelerates this process (Gunatillake et al., 1992; Dumitriu, 2002; Santerre et al., 2005). The MIO mechanism was reproduced *in vitro* by immersing a lead into a hydrogen peroxide solution. In a different *in vitro* test, a sealed PEU tube containing cobalt metal in the center was immersed into a 3% hydrogen peroxide solution and MIO was observed on the inner surface of the tube. The cobalt ion and hydrogen peroxide react to form hydroxyl radicals, simulating the oxidative radicals present at the material-macrophage interface.

Takahara et al. (Takahara et al., 1991) degraded SPU's based on MDI, BD (50% rigid segment content) and various polyols using 0.1 M $AgNO_3$ oxidative solution. They found a reduction in mechanical strength of those SPU's based on PTMO due to surface cracking related to ether scission upon oxidation.

Suntherland et al. (Sutherland et al., 1993) degraded Pellethane 2363 80A using 10 mM HClO in phosphate buffer (PB) at 25°C. In addition, peroxynitrite (ONOO-) degradation was achieved via the oxidation of hydroxylamine in an oxygen atmosphere at elevated pH. They observed a significant reduction in molecular weight, increase in polydispersity index and an increasing content of oxygenated species on the polymer surface.

Tanzi et al. (Tanzi et al., 2000) studied the oxidative degradation of polyether (Pellethane 2363 80A) and polycarbonate (Corethane 80A, Bionate 80A and Chronoflex AL 80A) urethanes in 0.5 N nitric acid (acidic) and sodium hypochlorite (4% Cl_2, alkaline) up to 14 days at 50°C and under constant strain (100%). It was found that PEU were more degraded under alkaline oxidation (HClO) mainly in the absence of applied strain while PCU was more affected by HNO_3.

Our own work using Tecoflex as model PEU degraded in H_2O_2 did not show significant changes in FTIR absorptions and only small differences in the bands located at 3330 cm^{-1} and 1660 cm^{-1} were observed (see Figure 3), although this was clear when the polyol was tested alone. However, TGA revealed that their degradation temperature were lowered and the amorphous content determined by XRD only exhibited a little changes (Chan-Chan et al., 2010).

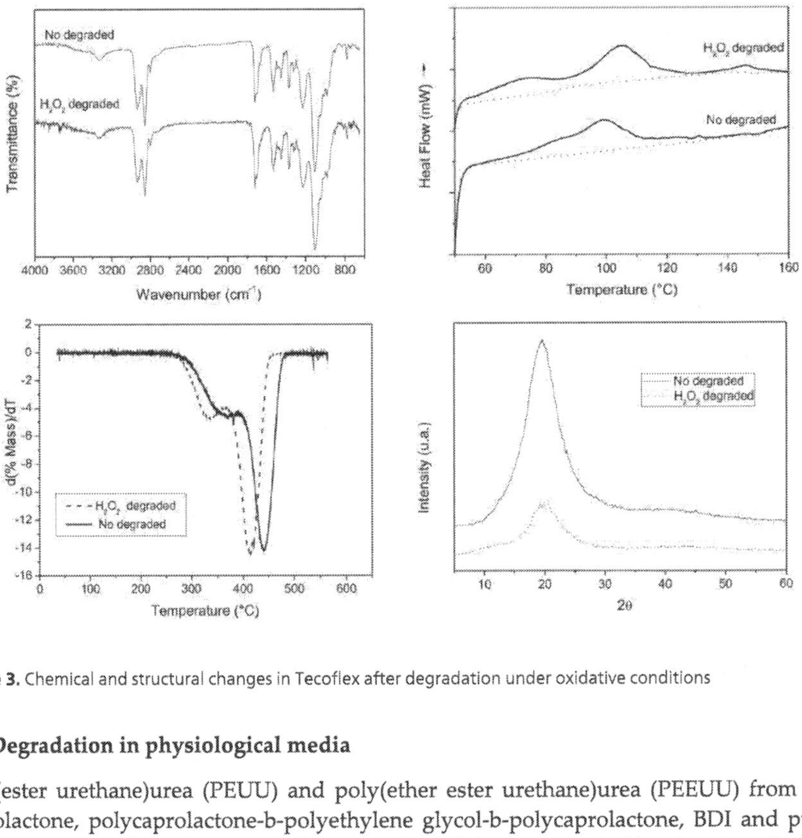

Figure 3. Chemical and structural changes in Tecoflex after degradation under oxidative conditions

4.3. Degradation in physiological media

Poly(ester urethane)urea (PEUU) and poly(ether ester urethane)urea (PEEUU) from polycaprolactone, polycaprolactone-b-polyethylene glycol-b-polycaprolactone, BDI and putrescine were prepared by Guan et al. (Guan et al., 2005a) and degraded in phosphate buffered saline (PBS, pH=7.4) at 37°C; scaffold degradation was related to the porosity and polymer hydrophilicity. The scaffolds exhibited progressive mass loss over the 8-week period ranging from 13.3% to 20.7% for PEUU scaffolds and from 25.4% to 47.3% for PEEUU scaffolds. In this study, the polymer films and scaffolds did not show evidence of an autocatalytic effect during the monitored degradation process. Furthermore, the presence of BDI and 1,4-butanediamine in the hard segment of PU yielded putrescine as degradation product, which is already present in the body and has been implicated as an important mediator of cellular growth and differentiation in response to growth factors.

Two gelatin based poly(ester urethane) were prepared by Sarkar et al. (Sarkar et al., 2006) using polyethylene lactate ester diol as a soft segment, and degraded in phosphate buffer saline solution (pH 7.4) at 37°C in a Biochemical Oxygen Demand (BOD) incubator shaker. It was found that the weight loss (up to 45.7% in 30 days) occurred due to the hydrolytic degradation of the gelatin based polyester urethane scaffold by PBS solution and it was proportional to the gelatin content.

Sarkar et al. (Sarkar et al., 2008) prepared segmented polyurethanes using polyethylene glycol (PEG) or poly caprolactone diol (PCL) as the soft segment while hexamethylene diisocyanate (HDI) or dicyclohexylmethane 4,4-diisocyanate (HMDI) were used with desaminotyrosyl tyrosine hexyl ester (DTH) as the chain extender in the rigid component. For degradation in PBS (0.1M, pH 7.4 containing 200 mg of sodium azide) samples were incubated at 37°C. It was found that PEG-based polyurethanes degrade at a faster rate compared with PCL-based polyurethanes due to their hidrophillicity and that this effect was marked when using high molecular weight PEG. It was also found that more amorphous SPU (i.e. exhibiting more phase mixing and therefore more urethane linkages H-bonded with the soft segment), such as those prepared with HMDI, degrade faster as they absorb more water.

Knight et al. (Knight et al., 2008) studied new hybrid thermoplastic polyurethane (TPU) system that incorporates an organic, biodegradable poly(D,L-lactide) soft block with a hard block bearing the inorganic polyhedral oligosilsesquioxane (POSS) moiety and degraded them in PBS buffer at 37 °C over a 2 months period. They found that less than 4% of the original mass elutes from the sample after a month in the buffer, most likely from chain ends on the surface of the sample undergoing hydrolysis. Although only a small mass loss was observed, the molecular weight of the samples dropped dramatically after only one week to 40% of the initial molecular weight.

Biodegradable ionic polyurethanes (PUs) were synthesized from methylene di-p-phenyl-diisocyanate (MDI), polycaprolactone diol (PCL-diol) and N,N-bis (2-hydroxyethyl)-2-aminoethane-sulfonic acid (BES) by Zhang et al. (Zhang et al., 2008). *In vitro* degradation of the PUs was evaluated by recording the samples' weight loss, molecular weight changes, and mechanical properties changes over time in PBS buffer solution at 67°C to accelerate degradation. Although there was a 20% molecular weight reduction, degradation rate was lower in those PUs containing sulfonic acid compared to PU's without this chain extender. This was explained in terms of their higher phase separation.

Segmented polyurethane based on poly(ε-caprolactone), ethyl lysine diisocyanate or hexamethylene diisocyanate in combination with ethylene glycol or ester from ethylene glycol and lactic acid (2-hydroxyethyl 2-hydroxypropanoate) were degraded *in vitro* (0.1 M phosphate buffered saline at 37°C in a shaken incubator set at 50 rpm (ASTM F 1635)) over a 1 year period (Zhang et al., 2008). It was found that all polyurethanes exhibited considerable molecular weight decrease over the test period and ester chain extender polyurethanes showed the highest mass loss and that it was directly proportional to hard segment not to the PCL used as soft segment.

Guelcher et al. (Guelcher et al., 2008) prepared injectable polyurethanes by two-component reactive liquid molding of low-viscosity quasi-prepolymers derived from lysine polyisocyanates and poly(3-caprolactone-co-DL-lactide-co-glycolide) triols and degraded porous discs by incubation in PBS at 37°C and 5% CO_2 for 2, 4, 6, and 8 months. They found that these polymers degrade by hydrolysis of ester linkages to yield α-hydroxy acids and soluble urethane fragments. Furthermore, the materials prepared from PCL triol exhibit minimal (e.g., <5%) degradation after 8 months. However, materials prepared from P6C3G1L (triol synthesized from a glycerol starter and a mixture of monomers comprising 60% caprolactone, 30% glycolide, and 10% DL-lactide) exhibit 15-27% mass loss after 8 months.

Multi-block poly(ether ester urethane)s consisting of poly[(R)-3-hydroxybutyrate] (PHB), poly(propylene glycol) (PPG), and poly(ethylene glycol) (PEG) were prepared by Loh et al. (Loh et al., 2007). The poly(PEG/PPG/PHB urethane) copolymer hydrogels were hydrolytically degraded in phosphate buffer at pH 7.4 and 37°C for a period of up to 6 months. The degradation products in the buffer were characterized by GPC, ^1H NMR, MALDI-TOF, and TGA. The results showed that the ester backbone bonds of the PHB segments were broken by random chain scission, resulting in a decrease in the molecular weight. In addition, the constituents of degradation products were found to be 3-hydroxybutyric acid monomer and oligomers of various lengths (n= 1–5).

Multiblock poly(ether ester urethane)s comprising of poly(lactic acid) (PLA), poly(ethylene glycol) (PEG), and poly(propylene glycol) (PPG) segments and hexamethylene diisocyanate were synthesized by Loh et al. (Loh et al., 2008). Their degradation process in pH 7.4 buffer solution (8.0 g of NaCl, 0.2 g of KCl, 1.44 g of Na_2HPO_4, and 0.24 g of $K_2H_2PO_4$ in 1 L of solution) was studied over a period of 3 months. Multi-modal GPC profiles of these polymers suggested that the polymer degrades in fragments with molecular weight of about 2000, 4000, 6000 and 8000 g/mol. These gels degraded at a much faster rate than the previously reported PEG-PPG-PHB poly(ester urethane) thermogels, which were reported to degrade over a period of 6 months.

Degradation of segmented poly(urethane urea)s (SPUUs) with hard segments derived only from methyl 2,6-diisocyanatehexanoate (LDI) and PCL, PTMC, P(TMC-co-CL), P(CL-co-DLLA) or P(TMC-co-DLLA) as soft segment was conducted by Asplund et al. (Asplund et al., 2008). For the hydrolysis study, sterile and nonsterile samples were placed in 40 mL PBS buffer solution (pH 7.4) and put in an oven at 37°C. Degradation was studied after 5, 10, 15, and 20 weeks and analyses performed in triplicate for each sample. The effect of sterilization was studied after 10 weeks of hydrolysis. Physical ageing was studied after 5 and 15 weeks at 50°C. They found that the degradation rate was dependant on the soft segment structure, with a higher rate of degradation for the polyester-dominating PUUs exhibiting a substantial reduction in intrinsic viscosity. A tendency of reduction of tensile strength and strain hardening was seen for all samples. Also, loss in elongation at break was detected, for PUU-P(CL-DLLA) it went from 1600% to 830% in 10 weeks. Gamma radiation caused an initial loss in inherent viscosity and induced more rapid hydrolysis compared with nonsterilized samples, except for PUU-PTMC.

Yeganeh et al. (Yeganeh et al., 2007) prepared epoxy terminated polyurethanes from glycidol and isocyanate-terminated polyurethanes made from poly(ε-caprolactone) (PCL) or poly(ethylene glycol) (PEG) and 1,6-hexamethylene diisocyanate. Degradation studies were performed using tris buffered saline solutions (TBS; 0.05, 0.1molL^{-1} NaCl, pH 7.4) and incubated at 37°C up to 6 months. They observed that degradation rates correspond to their water-absorbing ability, with faster degradation in the more absorbent polymers while the weight loss, due to hydrolytic degradation, increased as the amount of PEG content increased. A possible explanation is that following dissolution of some PEG segments, there will be an increase in the porosity of the blends, leading to a greater surface area for water to access the ester bonds of hydrophobic PCL, which dominates the degradation rate. Other possible explanations include an increase in the hydrophilicity of the surface, which accelerates degradation, or an increase in the mobility of the PCL molecules, which could also facilitate hydrolytic degradation. Also the rate of hydrolysis was raised with increasing time, which might result from the augmentation content of hydrophilic hydroxyl, amine, and carboxylic groups generated at the surface during degradation.

Wang et al. (Wang et al., 2008) prepared novel biodegradable and biocompatible poly(ester-urethane)s by *in situ* homogeneous solution polymerization of poly(3-caprolactone) diol, dimethylolpropionic acid (DMPA), and methylene diphenyl diisocyanate in acetone followed by solvent exchange with water. The hydrolytic degradation test was conducted on buffer solution (pH=7.4) at 37°C up to 12 weeks and showed that the degradation rate was little affected by the DMPA content in the range investigated, but was observed to be influenced by the hard segment content.

Hong et al. (Hong et al., 2010) synthesized poly(ester carbonate)urethane ureas (PECUUs) using a blended soft segment of poly(caprolactone) (PCL) and poly(1,6-hexamethylene carbonate) (PHC), 1,4- butane diisocyanate and putrescine as chain extender. They found that degradation of PECUUs in aqueous buffer (PBS at 37°C) and subcutaneous implantation in rats (Adult female Lewis rats) was slower than poly(ester urethane)urea but faster than poly(carbonate urethane)urea (PCUU). Over a period of 56 days, poly(ether urethane)ureas (PEUU) exhibited a 9% mass loss in addition to a reduction in inherent viscosity, while all of the PECUUs and PCUU did not show detectable loss of mass. *In vivo* it was observed that the majority of the PEUU scaffold was degraded, and loose connective tissue occupied the implant area with few observed putative macrophages. For the PECUU 50/50 scaffolds, more remnant material was seen with darker violet staining of the putative infiltrating macrophages and fibroblasts.

Chan-Chan et al. (Chan-Chan, et al. 2012) synthesized new polyester poly(urethane-urea)s and their molecular weight changes during PBS degradation were monitored by gel permeation cromatography (GPC) (see Figure 4). Significant weight loss was not observed at six months but bulk degradation was corroborated by this analytical technique.

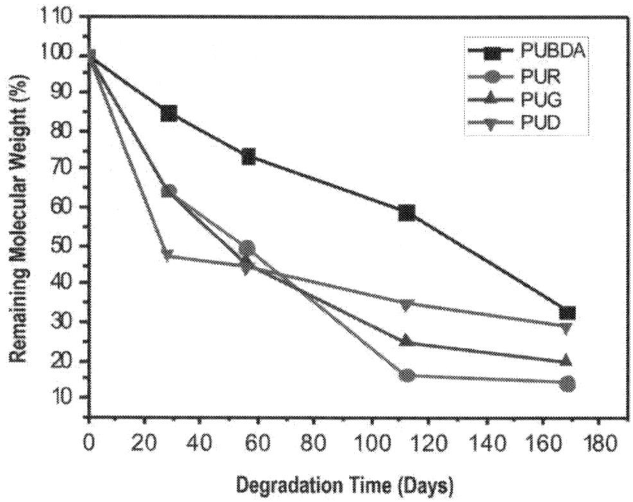

Figure 4. Molecular weight reduction in polyurethanes based on butanediamine (PUBDA), arginine (PUR), glycine (PUG) and aspartic acid (PUD).

4.4. Enzymatic degradation

Huang et al. (Huang et al., 1979) reported that a low molecular weight poly(ester-urea), poly(L-phenyl alanine/ethylene glycol/1,6-hexane diisocyanate), and a model diesterdiurea, dimethyl diphenyl alanine hexamethylene urea, were hydrolyzed by chymotrypsin at pH 8. They also observed degradation with papain latex (pH 6.5, PBS) of the model diesterdiurea.

Takahara et al. (Takahara et al., 1992) degraded SPU's based on MDI, BD and various polyols using papain (80 U/mL) and papain activating solution (0.05 M cysteine, 0.02 EDTA, pH=6.5) in sodium acetate buffer solution. In this study it was found that PEO based polyurethanes exhibited the larger mass loss from all the SPU's studied in addition to a reduction in Young's modulus and tensile strength due to a reduction in molecular weight.

Labow et al. (Labow et al., 1996) degraded in elastase (from human neutrophils or pancreatic porcine) a poly(ester-urea-urethane) containing [^{14}C]toluene diisocyanate (TDI), poly(caprolactone) and ethylenediamine as well as a poly(ether-urea-urethane) containing [^{14}C]TDI, poly(tetramethylene oxide) and ethylenediamine (ED). They used neutrophils, which contain elastolytic activity, as they are present during the inflammatory response. Ten-fold more radioactive carbon was released when porcine pancreatic elastase was incubated with [^{14}C]TDI/PCL/ED than when human neutrophil elastase was used. Ten-fold less radioactive carbon was released when [^{14}C]TDI/PTMO/ED was incubated with porcine pancreatic elastase (PPE) as compared to [^{14}C]TDI/PCL/ED. Radioactive carbon release data for [^{14}C]TDI/PCL/ED polymer incubated with trypsin, a possible contaminant in pancreatic porcine elastase showed no significant release of radioactive carbon by the same number of units of trypsin which would be present in the commercial PPE preparation used in the biodegradation experiments.

Skarja and Woodhouse (Skarja et al., 2001) studied degradable segmented polyurethanes containing a phenylalanine diester chain extender and degraded them in buffer chymotrypsin and trypsin solutions for up to 28 days. In this study it was found that the presence of phenylalanine resulted in an increased susceptibility to enzyme-mediated while the magnitude of degradation and erosion was highly variable and was dependent on soft segment type (PCL or PEO) and molecular weight (500-2000 g/mol).

It is well-known that the segmented poly(urethane ureas) prepared from 4,4-diphenylmethane diisocyanate, oligotetramethylene glycol, and diamines are not easily hydrolyzed by enzymes. This was further extended by Thomas and Jayabalan (Thomas et al., 2001) who reported that completely aliphatic poly(urethane urea) based on 4,4-methylene bis-cyclohexyl isocyanate/hydroxy terminated polybutadiene/1,6-hexamethylene diamine did not degrade in papain after 30 days at 37°C.

Labow et al. reported that cholesterol esterase cleaved polyetherurethanes at the most probable site susceptible to hydrolytic cleavage, which is the urethane bonds, resulting in the release of free amine (Labow et al., 2002). Santerre's group has also reported the degradation of polycarbonate polyurethanes with cholesterol esterase (Tang et al., 2002). Both the carbonate and urethane bonds were cleaved, resulting in many products ranging in molecular weight from 150 to 850 g/mol, as identified by GC–MS.

Yamamoto et al. (Yamamoto et al., 2007) degraded with different thiol proteases (papain, bromelain, and ficin) and Protease K and chymotrypsin, lysine diisocyanate (LDI) based poly(urethanes) and segmented poly(urethane ureas). For this, 1 mg of enzyme was added into the test tube coated with the polymer at 37°C and the total organic carbon (TOC) measured. From ^{1}H NMR results, it was evident that the pendant methyl ester group in LDI was rapidly hydrolyzed, followed by slow hydrolysis of urethane bonds in the backbone chain while the susceptibility of urea bonds to papain was very low. Before 50 h almost 30% of the PU has been degraded, with ethylene glycol exhibiting the highest rate of degradation; thiol proteases were most effective for all SPUUs. LDI/PTMO (Mw=2000 g/mol)/1,3-propylendiamine (PDA) (2/1/1), which does not contain degradable soft segments (caprolactone block), showed degradation by various proteases. This fact strongly suggests that the cleavage of the hard segment (urethane and/or urea) by these proteases occurred. For the SPUU the expected water-soluble degradation products are diamine, α-hydroxy caproic acid, and its low molecular oligomers, in addition to lysine derivatives.

Hafeman et al. (Hafeman et al., 2011) investigate the effects of esterolytic and oxidative conditions on scaffold degradation by incubating in 1 U/mL cholesterol esterase (CE), 1 U/mL carboxyl esterase (CXE), and 10 U/mL lipase (L) hydrogen peroxide (20 wt% hydrogen peroxide (H_2O_2) in 0.1 M cobalt chloride ($CoCl_2$), and buffer alone (0.5 M monobasic sodium phosphate buffer with 0.2% w/w sodium azide) and measured the mass loss for 10 weeks at 37°C. Polyurethane scaffolds were prepared by one-shot reactive liquid molding of hexamethylene diisocyanate trimer (HDIt) or lysine triisocyanate (LTI) and a polyol as hardener. Trifunctional polyester polyols of 900-Da molecular weight were prepared from a glycerol starter and 60% ε-caprolactone, 30% glycolide, and 10% D,L-lactide monomers (6C), ($t_{1/2}$ = 20 days) and 70% caprolactone, 20% glycolide, and 10% lactide (7C) ($t_{1/2}$ = 225 days) and stannous octoate catalyst. Incubation with esterases slightly accelerated degradation relative to PBS. Differences in degradation between the three candidate enzymes at any given time point were not significant. In contrast, incubation with medium that created an oxidative microenvironment had a more significant effect on the polyurethane degradation rate, especially for the LTI-based materials, except the 6C/HDIt (hexamethylene diisocyanate trimer) + PEG, which interestingly degraded faster in the presence of cholesterol and carboxyl esterase than in oxidative medium.

A new family of water borne polyurethanes (WBPU) were synthesized by Jiang et al. (Jiang et al., 2007) using isophorone diisocyanate (IPDI), polycaprolactone (PCL), polyethylene glycol (PEG) and BD:Lysine (1:1) as the chain extender. The polyurethane was then enzymatically degraded in PBS (pH = 7.4) with a solution mixture including PBS 60.0 ml, 0.1% $MgC1_2$ 15.0 ml and Lipase AK (10 mg/ml) 15.0 ml and then incubated with shaking for certain time at 55°C, which was the optimum temperature for enzyme activities of Lipase AK. An increased degradation was observed as decreasing of the amount of PEG in soft segments of WBPU, as judged from the change of tensile properties with time, owing to Lipase AK only interacting with PCL soft segments in these polymers structures. This result reveals that the degradation rate is proportional to the PCL content, and inverse proportion to the PEG content in the WBPUs. Depending on the PCL content, degradation started even at 6 h in the presence of Lipase AK.

A polyurethane was synthesized with LDI, PCL, and BD in the presence of dilaurate as catalyst by Han et al. (Han et al., 2009) and then degraded in PBS with a solution mixture including 4.0 mL PBS, 1.0 mL 0.1 wt.% $MgCl_2$ and 1.0 mL Lipase AK (10 mg/mL) in water at 50°C. It was found that loss mass decreased with increasing the PCL soft segment content in hydrolytic degradation in PBS. Because PCL is hydrophobic in comparison with the polar hard segment, increasing its content would decrease water uptake of PU films, and then decrease mass loss. In contrast, in the presence of Lipase AK the mass loss was observed to be increased with increasing the PCL soft segment content.

Biodegradable polyurethanes were prepared by Wang et al., using PLA-PEG-PLA as soft segment, and L-lysine ethyl ester diisocyanate (LDI) and 1,4-butanediol (BD) as rigid segment (Wang et al., 2011b). These polymers were degraded in PBS (0.1 M PBS with 0.9% NaCl and 0.02% NaN_3, pH 7.4, 6 and 5) and enzymatic (0.1mg/ml lipase from porcine pancreas in 0.1 M PBS with 0.9% NaCl and 0.02% NaN_3, pH 7.4) solutions at 37 °C to simulate *in vivo* dynamic tissue environment. PU samples demonstrated rapid degradation in 96 h (more than 90%) which might be attributed to hydrophilicity of PEG segments, low number-average molecular weight and microphase separation degree of these polyurethanes and enzyme functions. The enzymatic degradation rate was higher than hydrolytic degradation rate, verifying that Lipase from porcine pancreas can accelerate hydrolysis on these polyurethanes.

A series of pH-sensitive biodegradable polyurethanes (pHPUs) were designed and synthesized using pH-sensitive macrodiol (poly(ε-caprolactone)-hydrazone-poly-(ethylene glycol)-hydrazone-poly(ε-caprolactone) diol (PCL-Hyd-PEG-Hyd-PCL)), L-lysine ethyl ester diisocyanate (LDI) and L-lysine derivative tripeptide as chain extender by Zhou et al. (Zhou et al., 2011). The polyurethanes could be cleaved in acidic media (pH ~ 4-6) as well as degraded in PBS (100 mM, and pH 7.4) overnight at room temperature and enzymatic solution (Lipase AK (10 mg/mL, 2 mL) in PBS buffer solution with 0.1 wt % $MgCl_2$ (2 mL) and then incubated with cyclic shaking at 52.5°C). It was found that the hydrolysis rates of the two samples observed in Lipase AK PBS are higher than that in PBS i.e. 31.1% and 35.9% of weight loss are detected after hydrolytic and enzymatic degradation for 144 h of pHPU4 (pHPU prepared with LDI/macrodiol/tripeptide 3.15/2/1), respectively. The results indicate that the pHPUs are also facile to degrade in enzymatic solution, which is in agreement with reported literatures that Lipase AK is able to accelerate the PCL-based polymers biodegradation. Polymers with more pH sensitive macrodiol and lower crystallinity degraded even faster. The importance of studying these materials (pH-sensitive biodegradable polyurethanes) lies in the fact that they been used for intracellular multifunctional antitumor drug delivery (Zhou et al. 2012).

Elliott et al. (Elliott et al., 2002) determined mechanism of enzymatic degradation by HPLC/MS. Prior to product separation and identification, residual enzyme (chymotrypsin) was removed from the incubation solution samples. This process was necessary since the chymotrypsin could interfere with the accurate detection of the degradation products in the high performance liquid chromatography (HPLC) columns, and because proteins have a tendency to aggregate and then later precipitate during the gradient run, thereby causing additional difficulties in data acquisition. The results of the tandem mass spectrometry (MS/MS) analysis indicated that chymotrypsin may act to cleave urea bonds adjacent to L-phenylalanine residues. This is a significant finding since it confirms that the polyurethanes are susceptible to selective enzymatic degradation in the hard segment. Traditionally, this domain of the polyurethane has been considered a relatively stable group. The materials used in this study, however, were especially developed to encourage degradation of the hard segment rather than relying solely on degradation of the soft segment. Hence, the results of this study confirm that this goal was achieved. The cleavage of urea bonds by chymotrypsin is an important finding as it contradicts results of a previous study with similar chemistry that found that urea bonds adjacent to L-phenylalanine residues were not cleaved. However, since the level of chymotrypsin activity was not stated in the other study, it may be possible that the right conditions were not presented in order to degrade the urea bond (Elliott et al., 2002).

4.5. Lipid degradation

Lipid absorption has been reported to occur in many medical devices such as heart valves made of silicon, leading to their calcification. In addition, fatigue properties of SPU have been reduced by lipid absorption.

Takahara et al. (Takahara et al., 1992) degraded SPU's based on MDI, BD and various polyols using 0.25 % phophatidyl choline and 0.1% M cholesterol liposome solution during 28 days at 37°C. They found that SPU based on PDMS disintegrated under these conditions while PTMO based SPUs exhibited a severe reduction in tensile strength and elongation. These results were not related to the presence of a specific chemical group in the soft segment as PEO based SPU's were not affected.

4.6. Compost biodegradation

Synthetic poly(ester urethanes) are known to be degraded by microbes mainly due to the presence of ester linkages, being more susceptible those containing long chains rather than short polyester chains. Lactic acid based polyester urethanes have been degraded with compost inoculum (thermophilic-stage household waste compost was added to 100 ml of ASTM solution and the CO_2 evolved was followed by Hiltunen et al. (Hiltunen et al., 1997). The data showed that poly(ester-urethanes) did not biodegrade at 25°C but when the temperature was raised, biodegradation was accelerated. At 37°C the stereo structure of polymer chains had a strong effect on biodegradation. This temperature was below the glass transition temperature of poly(ester-urethanes) but about the same as the glass transition temperature of prepolymer chains. The lower the glass transition temperature of prepolymer, the faster the biodegradation. Urethane bonds probably break first, and after that the properties of lactic acid prepolymer chains determine the biodegradation behavior. All poly(ester-urethane) samples biodegraded well at 55°C, and the percentage of biodegradation varied between 45 and 77% in 55 days. At 60°C the poly(ester-urethanes) biodegraded well and they reached even higher levels of biodegradation than Biopac™ and lactic acid. The biodegradation varied from 45 to 77% in 55 days.

Polyurethanes based on MDI and PCL with different molecular weights were prepared by Watanabe et al. (Watanabe et al., 2009) and degraded by soil burial test at 28°C. It was found that biodegradation rate of the polyurethanes increased as the number of average molecular weight (Mn) of poly(caprolactone) diol used increased from 500 to 1000 (urethane content 11.9 to 7.6 wt % respectively), whereas it decreased as the Mn of poly(caprolactone) diol increased from 1200 to 2000 (4.2 wt % of urethane content). Furthermore, when 2000 PCL triol was used led to a high degradation ratio.

4.7. Thermal degradation of polyurethanes

Although the thermal degradation of both polyurethanes (PU) and segmented polyurethanes (SPU) has been extensively investigated due to their wide range of applications, studies on thermal decomposition of polyurethanes used specifically in biomedical field such as catheters, heart valves, vascular prostheses, etc., are less common as generally these materials are not subjected to high temperatures during their *in vivo* performance [Cervantes-Uc et al. 2009]. In some cases, these studies are used to investigate the composition and stability of the remaining material after the chemical hydrolysis and oxidation of SPU [Chan-Chan et al. 2010] as well as to determine the soft and hard segment ratio of polyurethanes.

The thermal degradation of polyurethanes allows determination of the proper conditions for manipulating and processing them and for obtaining high-performance products that are stable and free of undesirable by-products; if not processed properly, commonly by extrusion or by injection moulding, the PU's would generate toxic products to the human body, which is very critical in biomedical applications [Gomes Lage et al. 2001].

It is well known that polyurethanes are not thermal stable polymers and that the onset degradation temperature of the urethane bond depends on the type of isocyanate and alcohol used. It is a general rule that the more easily formed polyurethanes are less stable, i.e. more easily dissociated when compared with more difficulty formed ones [Petrovic et al. 1994]. Petrovic reported that the degradation temperature for these materials ranged from 120°C to 250°C depending on their structure [Petrovic et al. 1994]; however, literature reports processing temperatures closer to 180°C [Guignot 2002].

Polyurethanes are thermally degraded through three basic mechanisms. First, by urethane bond dissociation into its starting components (isocyanate and alcohol); secondly, by breaking the urethane bond with formation of primary amines, carbon dioxide and olefins; and finally, splitting the urethane bond into secondary amine and carbon dioxide [Petrovic et al. 1994; Cervantes-Uc et al. 2009].

5. Degradation mechanism

The nature of PU chemistry is central to understand why some PUs undergo faster degradation than others (Santerre et al., 2005). However, the degradation mechanism of polyurethanes depends on not only the PU chemistry structure but also the degradation environment, i.e. in the presence of water, acidic, alkaline or oxidative conditions, or in the presence of enzymes. Generally, the characterization of the by-products during the degradation of the polyurethane is the key to understand the mechanisms of degradation. Identification of degradation products is an important issue but of equal interest is the eventual toxicity of the degradation products. If the biomaterial degrades, either spontaneously or due to biological activity, components can leach into surrounding tissues and cause an inflammatory response if not easily metabolized by natural pathways. Therefore, it is compulsory to identify the major species produced at different stages of degradation and the kinetics of their formation (Azevedo et al., 2005).

Accelerated degradation has been used to determinate stability of non degradable polyurethanes (Gunatillake, 1992) but it can be used to provide valuable information about degradation mechanism of resorbable polyurethanes. In this context, both soluble products and solid residues can be studied with different analytical techniques and tests to determine their composition.

The main techniques used to evaluate the degradation of biomaterials can be divided into surface analysis (infrared spectroscopy, X-ray photoelectron spectroscopy, contact angle measurements), which are more appropriated to monitor the changes occurring in the first stages of degradation, and bulk analysis (determination of changes in molecular weight, weight loss, temperature transitions, mechanical properties) for characterizing the later stage of degradation (Azevedo et al., 2005).

In general, polyesterurethanes are susceptible to hydrolytic degradation because of ester groups in the soft segments while polyetherurethanes are susceptible to oxidative degradation. Furthermore, it has been observed that ester linkages hydrolyze about a magnitude faster than urethane linkages, and it has been shown that urea linkages hydrolyze faster than urethane, although at slightly acidic conditions. Figure 5 shows the possible mechanism of hydrolytic degradation of various functional groups present in polyurethanes.

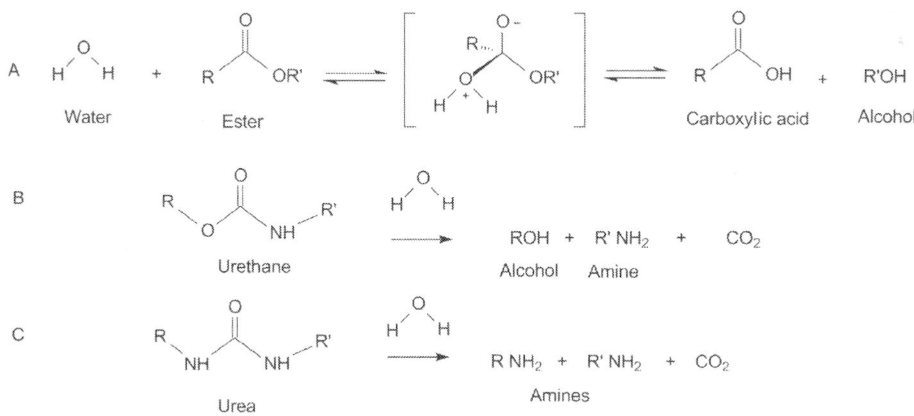

Figure 5. Hydrolytic degradation mechanism of polyesters (A), poly(urethane) (B) and poly(ureas) (C).

In spite of this, the degradation rate of the poly(ester urethane) based on PCL was found to be slow (i.e., 15% weight loss in 11 weeks (Wang et al., 2008). Furthermore, IR spectra for the degradation products of the LDI/PCL and LDI (lysine methyl ester diisocyanate)/P6C3G1L (triol synthesized from a glycerol starter and a mixture of monomers comprising 60% caprolactone, 30% glycolide, and 10% DL-lactide) materials after 2 and 8 months in PBS (Guelcher et al., 2008) shows an absorption band at approximately 1070-1050 cm^{-1}, which is assigned to C-O stretching vibrations in alcohols and carboxylic acids. This observation implies that the polyurethanes degrade by hydrolysis of ester linkages to yield α-hydroxy acids and is further supported by the appearance of the strong peaks at 1675-1650 cm^{-1}, which correspond to the COO asymmetric stretching vibration associated with carboxylic acid salts. Therefore, it is possible under these conditions that phosphate salts of carboxylic acids will form in the PBS solution due to the reaction of carboxylic acids with the basic phosphate salts present in PBS. Hydrolysis of LDI/PCL containing poyurethanes networks in sodium hydroxide solutions has been reported to yield L-lysine as a degradation product; however, the presence of L-lysine in the degradation products under physiological conditions was not confirmed.

Other studies reported the presence of lysine in the degradation products from lysine-derived polyurethanes networks.

Segmented polyurethane based on poly(ε-caprolactone), ethyl lysine diisocyanate or hexamethylene diisocyanate in combination with ethylene glycol or ester from ethylene glycol and lactic acid (2-hydroxyethyl 2-hydroxypropanoate) with greater hard segment content (HS) liberate higher amine concentrations during their degradation (Tatai et al., 2007). Amine concentration was determined using a spectrophotometer by acquiring the A_{570} (absorbance at 570 nm) of the test sample and by quantifying the detected concentration with use of the standard curve. This being expected, on the assumptions that PUs with higher HS contained more urethane bonds. To detect amine groups using this technique, a degradation product must undergo hydrolysis at its respective urethane linkage. Since this process is somewhat slower than that of ester bond hydrolysis, it seems that part of the degradation product may still contain urethane segments.

Hafeman et al. (Hafeman et al., 2011) investigate the effects of esterolytic and oxidative conditions on scaffold degradation by incubating in 1 U/mL cholesterol esterase (CE), 1 U/mL carboxyl esterase (CXE), and 10 U/mL lipase (L) hydrogen peroxide (20 wt% hydrogen peroxide (H_2O_2) in 0.1 M cobalt chloride ($CoCl_2$), and buffer alone (0.5 M monobasic sodium phosphate buffer with 0.2% w/w sodium azide) and analysed the degradation products by HPCL. Hydrolysis of ester bonds was anticipated to yield α-hydroxy acids (e.g., hydroxycaproic, lactic, and glycolic acids), which was confirmed by HPLC. The lysine triisocyanate (LTI) scaffolds produced more α-hydroxy acids than trimer of hexamethylene diisocyanate (HDIt) scaffolds. The 7C/LTI (triol synthesized from a glycerol starter and a mixture of monomers comprising 70% caprolactone, 30% glycolide and 10% lactide) formulation, which degraded more slowly due to the longer polyester half-life, yielded lower concentrations of α-hydroxy acids than the 6C/LTI (60% caprolactone, 30% glycolide and 10% lactide) formulation. Inclusion of polyethylene glycol (PEG) in the 6C/HDIt scaffold reduced the amount of α-hydroxy acids in the degradation medium due to the replacement of 50% of the polyester with PEG. Several unidentified peaks appeared in the HPLC spectra, which are conjectured to be adducts of α-hydroxy acids and either lysine or ethanolamine connected by urethane or urea bonds. Oxidation of urethane and urea bonds was predicted to yield lysine and ethanolamine from LTI scaffolds, and cyanuric acid from HDIt scaffolds. Both lysine and ethanolamine were detected in the degradation products from LTI scaffolds when incubated in PBS; however, cyanuric acid was not detected in the degradation products from HDIt scaffolds. The amount of lysine recovered from 6C/LTI scaffolds was significantly greater than that from 7C/LTI scaffolds after 14 weeks, which is consistent with the faster *in vitro* degradation of the 6C/LTI materials. At 36 weeks, 18% of the lysine incorporated in the 6C/LTI scaffolds was recovered, while 100% of the original mass had degraded to soluble degradation products. This suggests that the majority of the lysine was incorporated in soluble urethane and urea adducts with α-hydroxy adducts. The recovery of ethanolamine arises from the hydrolysis of the ester group in LTI and a urethane bond. Ethanolamine was not detected (<0.001 μg/mg polyurethane) until 14 weeks, and at later time points the ethanolamine concentration increased with time. The recovery of ethanolamine upon complete dissolution of the 6C/LTI scaffold at 36 weeks was 9%.

Suntherland et al. (Sutherland et al., 1993) degraded Pellethane 2363 80A with either HClO or ONOO. An oxidative reaction involving the ether or ester moieties of PEU would be reflected by a decrement in the urethane-aliphatic ester and/or aliphatic ether stretching peaks on ATR/FTIR analysis. Indeed, a substantial decrement in the aliphatic ether stretching at 1105-1110 cm^{-1} relative to the urethane-aliphatic ester peak at 1075 cm^{-1} has been observed in implanted material. In fact, the intensity of both aliphatic ether and urethane-aliphatic ester peaks decreases after long-term implantation, suggesting that both groups are oxidized *in vivo*. PEU previously exposed to HClO exhibited a decrement in the signal from the urethane-aliphatic ester.

FTIR has been used to determine the composition of residues after degradation. In this sense, hydrolytic degradation of polyester urethanes affects carbonyl bands at 1730 cm^{-1}. Soluble products of the ester scission are carboxyl acids and alcohols that can be observed between 2500 and 3500 cm^{-1}. Pérez et al. (Pérez et al., 2006) studies showed that urea bonds derived from amino acids can be hydrolyzed in basic conditions but after more prolonged period than ester groups, this degradation was monitored by capillary electrophoresis-ion trap-mass.

Oxidative degradation has been generally associated with poly(ether-urethane)s, since many studies have determined that these polymers degrade by mean alpha-hydrogen abstraction adjacent to oxygen in polyethers and polycarbonates (Christenson et al., 2004; Xie et al., 2009). In contrast, few works related to oxidative degradations on polyester polyurethanes has been done, and even less has studied the mechanism of degradation of polyurethane ureas. However recent studies about oxidative degradations of PCL and polyester poly(urethane urea)s PCL based have showed ester, urethane and urea groups are susceptible to oxidative degradation (Sabino, 2007; Sarkar et al., 2007; Hafeman et al., 2011). This mechanism is illustrated in Figure 6.

A $$-CH_2\text{-}O-CH_2- \;+\; \bullet OH \longrightarrow -CH_2\text{-}O-\dot{C}H- \;+\; H_2O$$

B $$-CH_2-OCOO-CH_2- \;+\; \bullet OH \longrightarrow -CH_2-OCOO-\dot{C}H- \;+\; H_2O$$

C $$-Ph-NHCOO-CH_2- \;+\; \bullet OH \longrightarrow -Ph-NHCOO-\dot{C}H- \;+\; H_2O$$

Figure 6. Mechanism of oxidative degradation by H_2O_2 in poly(ether urethanes) (A), poly(carbonate urethanes) (B) and aromatic polyurethanes (C).

Oxidative degradation using HClO is less commonly pursued but it may be clinically more relevant as hipochlorous anions can be produced by neutrophils. These conditions can be simulated *in vitro* and the suggested mechanisms of the polyurethane degradation can be depicted in Figure 7.

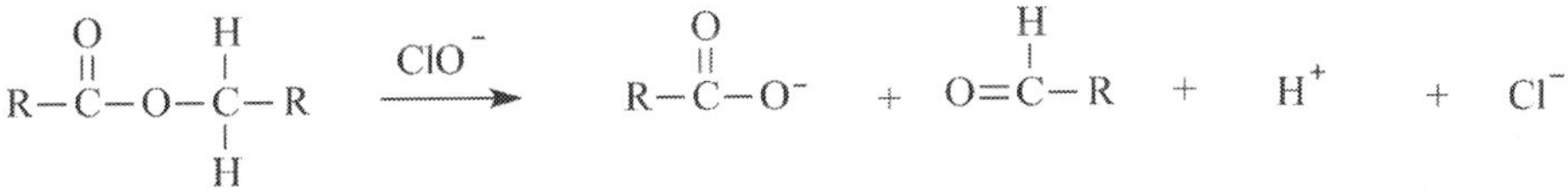

Figure 7. Mechanism of oxidative degradation in polyurethanes by means of HClO

Degradation of polyurethanes with H_2O_2 (30% v/v) does not seem to affect ester bonds but affect urea bonds as observed by FTIR. The wide band of 3650-3400 cm^{-1} and a small peak in 930 cm^{-1} corresponding to carboxylic acid confirm scission of urea groups as shown in Figure 8. Other bands such as those at 1298 cm^{-1} show some crosslinking by C-N bonds and an increase in PCL crystallinity, as the 1143 y 1189 cm^{-1} bands, corresponding to amorphous and crystalline PCL, changed (Chan-Chan, 2012).

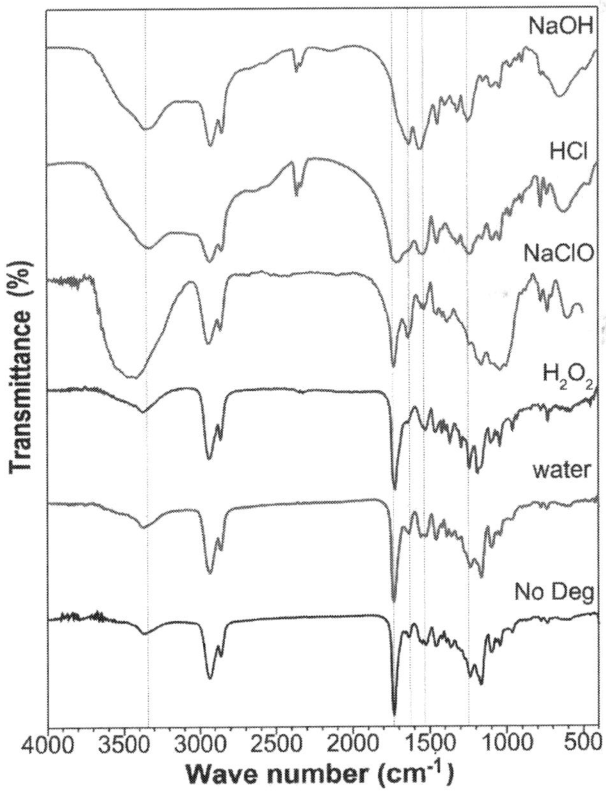

Figure 8. FTIR spectra of poly(urethane ureas) degraded in various media.

6. Conclusions

Polyurethanes are very versatile polymers that found application in the biomedical field, especially in cardiovascular applications. In spite of their good physicochemical and mechanical properties and acceptable biocompatibility they are prone to degradation under different conditions. These conditions range from hydrolysis, oxidation, metal induced oxidation, environmental stress cracking, enzyme-assisted degradation, etc. which can be found *in vivo* during the useful life of the device. In order to simulate these, *in vitro* approaches has been followed. Thanks to this information today it is well accepted that polyurethanes are no longer inert materials placed within the body. However, this disadvantage can be used to modulate their degradation to a rate that can be controlled mainly by their composition, and be used in the design of tissue engineering scaffolds.

Acknowledgements

This work was supported by CONACYT (México) Grant 79371, and Fondo Mixto CONACYT-Gobierno del Estado de Yucatán, Grants 108160 and 170132.

Author details

Juan V. Cauich-Rodríguez, Lerma H. Chan-Chan, Fernando Hernandez-Sánchez and José M. Cervantes-Uc

Centro de Investigación Científica de Yucatán A.C., Grupo de Biomateriales e Ingeniería de Tejidos, Mexico

References

[1] Abraham, G. A., A. Marcos-Fernández and J. San Román (2006). Journal of Biomedical Materials Research Part A 76(4): 729-736.

[2] Adhikari, R., P. A. Gunatillake, I. Griffiths, L. Tatai, M. Wickramaratna, S. Houshyar, T. Moore, R. T. M. Mayadunne, J. Field, M. McGee and T. Carbone (2008). Biomaterials 29(28): 3762-3770.

[3] Alferiev, I., N. Vyavahare, C. Song, J. Connolly, J. Travis Hinson, Z. Lu, S. Tallapragada, R. Bianco and R. Levy (2001). Biomaterials 22(19): 2683-2693.

[4] Aranguren M.I., J.F. González, M.A. Mosiewicki. (2012) Polymer Testing 31 (1): 7-15.

[5] Arjun, G. N. and P. Ramesh (2012). Journal of Biomed Materials Research Part A: Polymer Chemistry. DOI: 10.1002/jbm.a.34255

[6] Asplund, B., C. Aulin, T. Bowden, N. Eriksson, T. Mathisen, L. M. Bjursten and J. Hilborn (2008). Journal of Biomedical Materials Research Part B: Applied Biomaterials 86(1): 45-55.

[7] Azevedo, H. S. and R. L. Reis, Eds. (2005). Boca Raton, FL, CRC Press. pp.177-201.

[8] Basak, P. and B. Adhikari (2012). Materials Science and Engineering: C 32(8): 2316-2322.

[9] Baudis, S., S. C. Ligon, K. Seidler, G. Weigel, C. Grasl, H. Bergmeister, H. Schima and R. Liska (2012). Journal of Polymer Science Part A: Polymer Chemistry 50(7): 1272-1280.

[10] Bernacca, G. M., I. Straub and D. J. Wheatley (2002). Journal of Biomedical Materials Research 61(1): 138-145.

[11] Bernacca, G. M., T. G. Mackay, R. Wilkinson and D. J. Wheatley (1997). Journal of Biomedical Materials Research 34(3): 371-379.

[12] Cervantes-Uc J.M., J.I. Moo-Espinosa, J.V. Cauich-Rodríguez, A. Ávila-Ortega, H. Vázquez-Torres, A. Marcos-Fernández, J. San Román. (2009). Polymer Degradation and Stability 94 (10) 1666-1677.

[13] Chan-Chan, L. H., R. Solis-Correa, R. F. Vargas-Coronado, J. M. Cervantes-Uc, J. V. Cauich-Rodríguez, P. Quintana and P. Bartolo-Pérez (2010). Acta Biomaterialia 6(6): 2035-2044.

[14] Chan-Chan, L. H. (2012). PhD Thesis, Centro de Investigación Científica de Yucatán. Mérida, México.

[15] Chen, Q., S. Liang and G. A. Thouas (2012). Progress in Polymer Science. DOI: 10.1016/j.progpolymsci.2012.05.003.

[16] Christenson, E. M., J. M. Anderson and A. Hiltner (2006). Journal of Biomedical Materials Research Part A 76A(3): 480-490.

[17] Christenson, E. M., M. Dadsetan, M. Wiggins, J. M. Anderson and A. Hiltner (2004). Journal of Biomedical Materials Research Part A 69(3): 407-416.

[18] Da Silva, G. R. (2010). Polymer Degradation and Stability 95(4): 491-499.

[19] Ding, M., J. Li, H. Tan and Q. Fu (2012). Soft Matter 8(20): 5414-5428.

[20] Dumitriu, S. S. (2002). USA, Marcel Dekker.

[21] Elliott, S., J. Fromstein, J. PSanterre and K. Woodhouse (2002). Journal of Biomaterials Science, Polymer Edition 13(6): 691-711.

[22] Francolini, I., F. Crisante, A. Martinelli, L. D'Ilario and A. Piozzi (2012). Acta Biomaterialia 8(2): 549-558.

[23] Gogolewski, S. (1989). Colloid & Polymer Science 267(9): 757-785.

[24] Gomes Lage L and Y. Kawano (2001). Journal of Applied Polymer Science 79(5): 910-919.

[25] Guan, J., K. L. Fujimoto, M. S. Sacks and W. R. Wagner (2005a). Biomaterials 26(18): 3961-3971.

[26] Guan, J., M. S. Sacks, E. J. Beckman and W. R. Wagner (2002). Journal of biomedical materials research 61(3): 493-503.

[27] Guan, J., M. S. Sacks, E. J. Beckman and W. R. Wagner (2004). Biomaterials 25(1): 85-96.

[28] Guan, J. and W. R. Wagner (2005b). Biomacromolecules 6(5): 2833-2842.

[29] Guelcher, S. A., A. Srinivasan, J. E. Dumas, J. E. Didier, S. McBride and J. O. Hollinger (2008). Biomaterials 29(12): 1762-1775.

[30] Guignot C., N. Betz, B. Legendre, A. Le Moel and N. Yagoubi (2002). Journal of Applied Polymer Science 85(9): 1970-1979.

[31] Gunatillake, P. A., D. J. Martin, G. F. Meijs, S. J. McCarthy and R. Adhikari (2003). Australian journal of chemistry 56(6): 545-557.

[32] Gunatillake, P. A., G. F. Meijs, E. Rizzardo, R. C. Chatelier, S. J. McCarthy, A. Brandwood and K. Schindhelm (1992). Journal of Applied Polymer Science 46(2): 319-328.

[33] Hafeman, A. E., B. Li, T. Yoshii, K. Zienkiewicz, J. M. Davidson and S. A. Guelcher (2008). Pharmaceutical research 25(10): 2387-2399.

[34] Hafeman, A. E., K. J. Zienkiewicz, A. L. Zachman, H. J. Sung, L. B. Nanney, J. M. Davidson and S. A. Guelcher (2011). Biomaterials 32(2): 419-429.

[35] Han, J., B. Chen, L. Ye, A. Zhang, J. Zhang and Z. Feng (2009). Frontiers of Materials Science in China 3(1): 25-32.

[36] He, X., Z. Zhai, Y. Wang, G. Wu, Z. Zheng, Q. Wang and Y. Liu (2012). Journal of Applied Polymer Science 126(S1): E354-E361.

[37] Heijkants, R. G. J. C., R. V. Calck, T. G. Van Tienen, J. H. De Groot, P. Buma, A. J. Pennings, R. P. H. Veth and A. J. Schouten (2005). Biomaterials 26(20): 4219-4228.

[38] Hiltunen, K., J. V. Seppälä, M. Itävaara and M. Härkönen (1997). Journal of Polymers and the Environment 5(3): 167-173.

[39] Hong, Y., J. Guan, K. L. Fujimoto, R. Hashizume, A. L. Pelinescu and W. R. Wagner (2010). Biomaterials 31(15): 4249-4258.

[40] Huang, S. J., D. A. Bansleben and J. R. Knox (1979). Journal of Applied Polymer Science 23(2): 429-437.

[41] Jiang, G., X. Tuo, D. Wang and Q. Li (2012). Journal of Materials Science: Materials in Medicine DOI: 10.1007/s10856-012-4670-y.

[42] Jiang, X., J. Li, M. Ding, H. Tan, Q. Ling, Y. Zhong and Q. Fu (2007). European Polymer Journal 43(5): 1838-1846.

[43] Kannan, R. Y., H. J. Salacinski, M. Odlyha, P. E. Butler and A. M. Seifalian (2006). Biomaterials 27(9): 1971-1979.

[44] Kartvelishvili, T., G. Tsitlanadze, L. Edilashvili, N. Japaridze and R. Katsarava (1997). Macromolecular Chemistry and Physics 198(6): 1921-1932.

[45] Kehrer, J. P. (2000). Toxicology 149(1): 43-50.

[46] Khouw I. M. S. L., P. B. van Wachem, G. Molema, J. A. Plantinga, L. F. M. H. de Leij, M. J. A. van Luyn. (2000). Journal of Biomedical Materials Research 52: 439–446.

[47] Knight, P. T., K. M. Lee, H. Qin and P. T. Mather (2008). Biomacromolecules 9(9): 2458-2467.

[48] Król, P. (2007). Progress in materials science 52(6): 915-1015.

[49] Kuan, Y. H., L. P. Dasi, A. Yoganathan and H. L. Leo (2011). International Journal of Biomaterials Research and Engineering 1(1):1-17.

[50] Labow, R. S., D. J. Erfle and J. P. Santerre (1996). Biomaterials 17(24): 2381-2388.

[51] Labow, R. S., Y. Tang, C. B. Mccloskey and J. P. Santerre (2002). Journal of Biomaterials Science, Polymer Edition 13(6): 651-665.

[52] Li, Z., X. Yang, L. Wu, Z. Chen, Y. Lin, K. Xu and G. Q. Chen (2009). Journal of Biomaterials Science, Polymer Edition 20(9): 1179-1202.

[53] Liu, Q., S. Cheng, Z. Li, K. Xu and G. Q. Chen (2009). Journal of Biomedical Materials Research Part A 90(4): 1162-1176.

[54] Loh, X. J., S. H. Goh and J. Li (2007). Biomaterials 28(28): 4113-4123.

[55] Loh, X. J., Y. X. Tan, Z. Li, L. S. Teo, S. H. Goh and J. Li (2008). Biomaterials 29(14): 2164-2172.

[56] Lu, H., P. Sun, Z. Zheng, X. Yao, X. Wang and F.-C. Chang (2012). Polymer Degradation and Stability 97(4): 661-669.

[57] Lu, Y., L. Shen, F. Gong, J. Cui, J. Rao, J. Chen and W. Yang (2012). Polymer International 61(9): 1433-1438.

[58] Ma, Y., D. Yang, W. Shi, S. Li, Z. Fan, J. Tu and W. Wang (2012). Polymer Engineering & Science DOI:10.1002/pen.23269.

[59] Macocinschi, D., D. Filip, S. Vlad, M. Cristea and M. Butnaru (2009). Journal of Materials Science: Materials in Medicine 20(8): 1659-1668.

[60] Madhavan, K. and B. S. R. Reddy (2006). Journal of Polymer Science Part A: Polymer Chemistry 44(9): 2980-2989.

[61] Marcos-Fernández, A., G. A. Abraham, J. L. Valentín and J. S. Román (2006). Polymer 47(3): 785-798.

[62] May-Hernández, L., F. Hernández-Sánchez, J. L. Gomez-Ribelles and R. Sabateri - Serra (2011). Journal of Applied Polymer Science 119(4): 2093-2104.

[63] McBane, J. E., J. P. Santerre and R. S. Labow (2007). Journal of Biomedical Materials Research Part A 82A(4): 984-994.

[64] Mondal, S. and D. Martin (2012). Polymer Degradation and Stability 97(8): 1553-1561.

[65] Nakajima-Kambe, T., Y. Shigeno-Akutsu, N. Nomura, F. Onuma and T. Nakahara (1999). Applied microbiology and biotechnology 51(2): 134-140.

[66] Park, J. H., K. D. Park and Y. H. Bae (1999). Biomaterials 20(10): 943-954.

[67] Pérez, P., C. Simó, C. Neusüss, M. Pelzing, J. San Román, A. Cifuentes and A. Gallardo (2006). Biomacromolecules 7(3): 720-727.

[68] Petrovic Z., Z. Zavargo, J. H. Flynn, W. J. Macknight (1994). Journal of Applied Polymer Science 51(6):1087–1095.

[69] Pretsch, T., I. Jakob and W. Müller (2009). Polymer Degradation and Stability 94(1): 61-73.

[70] Sabino, M. A. (2007). Polymer Degradation and Stability 92(6): 986-996.

[71] Santerre, J. P., K. Woodhouse, G. Laroche and R. S. Labow (2005). Biomaterials 26(35): 7457-7470.

[72] Sarkar, D., J.-C. Yang, A. S. Gupta and S. T. Lopina (2009). Journal of Biomedical Materials Research Part A 90A(1): 263-271.

[73] Sarkar, D., J. C. Yang and S. T. Lopina (2008). Journal of Applied Polymer Science 108(4): 2345-2355.

[74] Sarkar, D. and S. T. Lopina (2007). Polymer Degradation and Stability 92(11): 1994-2004.

[75] Sarkar, S., A. Chourasia, S. Maji, S. Sadhukhan, S. Kumar and B. Adhikari (2006). Bulletin of Materials Science 29(5): 475-484.

[76] Schubert, M. A., M. J. Wiggins, J. M. Anderson and A. Hiltner (1997). Journal of Biomedical Materials Research 34(4): 493-505.

[77] Silvestri, A., P. Serafini, S. Sartori, P. Ferrando, F. Boccafoschi, S. Milione, L. Conzatti and G. Ciardelli (2011). Journal of Applied Polymer Science 122(6): 3661-3671.

[78] Skarja, G. and K. Woodhouse (2001). Journal of Biomaterials Science, Polymer Edition 12(8): 851-873.

[79] Skarja, G. and K. Woodhouse (1998). Journal of Biomaterials Science, Polymer Edition 9: 271-295.

[80] Spirkova, M., J. Pavlicevic, A. Strachota, R. Poreba, O. Bera, L. Kaprálková, J. Baldrian, M. Slouf, N. Lazic and J. Budinski-Simendic (2011). European Polymer Journal 47(5): 959-972.

[81] Stachelek, S. J., I. Alferiev, H. Choi, C. W. Chan, B. Zubiate, M. Sacks, R. Composto, I. W. Chen and R. J. Levy (2006). Journal of Biomedical Materials Research Part A 78(4): 653-661.

[82] Stachelek, S. J., I. Alferiev, J. Fulmer, H. Ischiropoulos and R. J. Levy (2007). Journal of Biomedical Materials Research Part A 82A(4): 1004-1011.

[83] Styan, K. E., D. J. Martin, A. Simmons and L. A. Poole-Warren (2012). Acta Biomaterialia 8(6): 2243-2253.

[84] Sutherland, K. and J. Mahoney (1993). Journal of Clinical Investigation 92(5): 2360.

[85] Takahara, A., A. J. Coury, R. W. Hergenrother and S. L. Cooper (1991). Journal of Biomedical Materials Research 25(3): 341-356.

[86] Takahara, A., R. W. Hergenrother, A. J. Coury and S. L. Cooper (1992). Journal of Biomedical Materials Research 26(6): 801-818.

[87] Tang, Y., R. Labow, I. Revenko and J. Santerre (2002). Journal of Biomaterials Science, Polymer Edition 13(4): 463-483.

[88] Tanzi, M., L. Ambrosio, L. Nicolais, S. Iannace, L. Ghislanzoni and B. Mambrito (1991). Clinical materials 8(1-2): 57-64.

[89] Tanzi, M. C., S. Farè and P. Petrini (2000). Journal of Biomaterials Applications 14(4): 325-348.

[90] Tatai, L., T. G. Moore, R. Adhikari, F. Malherbe, R. Jayasekara, I. Griffiths and P. A. Gunatillake (2007). Biomaterials 28(36): 5407-5417.

[91] Thomas, V. and M. Jayabalan (2001). Journal of Biomedical Materials Research 56(1): 144-157.

[92] Thomas, V. and M. Jayabalan (2009). Journal of Biomedical Materials Research Part A 89(1): 192-205.

[93] Thomson, T. (2005). Boca Raton, Fl, CRC. pp.35-53.

[94] van Minnen, B., M. B. M. V. Leeuwen, G. Kors, J. Zuidema, T. G. v. Kooten and R. R. M. Bos (2008). Journal of Biomedical Materials Research Part A 85A(4): 972-982.

[95] Wang, F., Z. Li, J. L. Lannutti, W. R. Wagner and J. Guan (2009). Acta Biomaterialia 5(8): 2901-2912.

[96] Wang, W., Y. Guo and J. U. Otaigbe (2008). Polymer 49(20): 4393-4398.

[97] Wang, X., P. Lin, Q. Yao and C. Chen (2007). World Journal of Surgery 31(4): 682-689.

[98] Wang, Y., C. Ruan, J. Sun, M. Zhang, Y. Wu and K. Peng (2011a). Polymer Degradation and Stability 96: 1687-1694.

[99] Wang, Z., L. Yu, M. Ding, H. Tan, J. Li and Q. Fu (2011b). Polymer Chemistry 2(3): 601-607.

[100] Ward, R., J. Anderson, R. McVenes and K. Stokes (2007). Journal of Biomedical Materials Research Part A 80(1): 34-44.

[101] Watanabe, A., Y. Takebayashi, T. Ohtsubo and M. Furukawa (2009). Journal of Applied Polymer Science 114(1): 246-253.

[102] Wiggins, M. J., J. M. Anderson and A. Hiltner (2003). Journal of Biomedical Materials Research Part A 66A(3): 463-475.

[103] Wiggins, M. J., B. Wilkoff, J. M. Anderson and A. Hiltner (2001). Journal of Biomedical Materials Research 58(3): 302-307.

[104] Xie, X., R. Wang, J. Li, L. Luo, D. Wen, Y. Zhong and C. Zhao (2009). Journal of Biomedical Materials Research Part B: Applied Biomaterials 89B(1): 223-241.

[105] Xu, L. C. and C. A. Siedlecki (2010). Journal of Biomedical Materials Research Part A 92(1): 126-136.

[106] Yamamoto, N., A. Nakayama, M. Oshima, N. Kawasaki and S.-i. Aiba (2007). Reactive and Functional Polymers 67(11): 1338-1345.

[107] Yan, Y., X. H. Wang, D. Yin and R. Zhang (2007). Journal of bioactive and compatible polymers 22(3): 323-341.

[108] Yeganeh, H., H. Jamshidi and S. Jamshidi (2007). Polymer International 56(1): 41-49.

[109] Zhang, C., X. Wen, N. R. Vyavahare and T. Boland (2008). Biomaterials 29(28): 3781-3791.

[110] Zhao, Q., A. McNally, K. Rubin, M. Renier, Y. Wu, V. Rose-Caprara, J. Anderson, A. Hiltner, P. Urbanski and K. Stokes (1993). Journal of Biomedical Materials Research 27(3): 379-388.

[111] Zhou, L., D. Liang, X. He, J. Li, H. Tan, J. Li, Q. Fu, Q. Gu. (2012) Biomaterias 33 (9): 2734-2745.

[112] Zhou, L., L. Yu, M. Ding, J. Li, H. Tan, Z. Wang and Q. Fu (2011). Macromolecules 44(4): 857-864.

Chapter 4

Substrates with Changing Properties for Extracellular Matrix Mimicry

Frank Xue Jiang

Additional information is available at the end of the chapter

http://dx.doi.org/10.5772/53547

1. Introduction

Cell-ECM interactions Fundamental to the success of using biomaterials in medical and health care applications, is the understanding of their interactions with biological tissues and systems. First step towards this end is the elucidation of cell-ECM interactions, which has attracted considerable interest in recent decade. Cellular decision-making process is driven by the internal genetic program and external factors comprising primarily other cells and extracellular matrix (ECM) via soluble factors and direct physical connections such as focal adhesion [1, 2]. Three key features of ECM have been identified of great significance in affecting cells, namely, chemical and biological composition, dimensionality (two- vs. three-dimensional), and physical properties [3-6]. These features can be sensed by cells via cell-ECM linkages, and the resulting signals subsequently follow intracellular pathways and trigger a cascade of events leading to alterations in gene expression and manifestation in phenotype. In contrast to the long recognized chemical composition and adhesive characteristics of the ECM, physical cues including topography, pore size, geometric patterns, and mechanical stiffness and their significance has just started being appreciated [7-10]. Whilst characteristics of ECM have profound effect on cells, cells may also actively exert impact on ECM by secretion of soluble factors or modify properties of ECM, or contribute to maintaining integrity or properties of ECM. At a larger scale, biological systems may actively interact with biomaterials to maintain or re-establish homeostasis.

Dynamic aspect of ECM To date, the majority of the substrates employed in cellular studies and other biological investigations have been of fixed mechanical stiffness and/or adhesive properties throughout cell culture. There is an increasing realization that a cell's microenvironment is dynamic and changing with time [11-13]. It is the case in both pathologic and normal tissues, at the tissue-implant interfaces, and during development and aging [14], especially for load-bearing and mechanically active tissues (e.g., heart, cartilage, lung) [15]. Not only do these changes naturally occur, but there are also benefits associated with them from a tissue engineering viewpoint, as highlighted in the series of discussions in the March 2005 issue of MRS (Materials Research Society) Bulletin [16, 17] and later studies. Whitesides [18] and Mrksich [19] and their coworkers among a number of investigators pioneered the work on engineering cell growth by using dynamic substrates. Their work and later reports on differential cell responses to materials with different properties suggest that it is beneficial for biomaterials to have controlled changing properties [20]. These facts make it very desirable for the bio-mimetic materials to have the capability of undergoing controlled remodeling with respect to time. They also raised caution in interpretation of the observations made from the majority of the biological studies, where properties of the substrates (e.g., culture flask, Petri dish, and hydrogels) remain unchanged throughout the process.

The scope of this work A significant number of reviews are available on the changes in soluble factors of ECM that may affect cellular behavior (e.g., [21]) and particularly on the changing environment in bioreactors [22] (e.g., nutrients concentration, oxygen level, temperature). Thus, they are not covered in this review. Moreover, flow conditions and the resulting traction forces, and their effect on certain cell types including blood cells (i.e., endothelial cells and red blood cells) have also been intensively reviewed and hence are not discussed here. This is also the case for mechanical forces, strain and stress applied directly to the cells (e.g., [23, 24]) in load-bearing tissues such as bone, cartilage and lung (for reference, see, e.g., [25-27]).

Therefore, this review is focused on the latest studies and current knowledge of two- or three- dimensional substrates with changing or dynamic mechanical and adhesive properties, design and conditions to trigger and achieve designed dynamics, and the impact of in-situ change of these properties on cell behavior, which provides guideline for design of biomaterials for their applications in medical and healthcare applications. Note that mechanical stiffness and elastic modulus were used interchangeably in this work.

2. Dynamic nature of the cell microenvironment

Normal tissues The micro-environment within which cells reside in natural tissues undergoes constant synthesis and degradation [16, 17], and has long been recognized as dynamic and changing [25, 28]. Although the composition of tissues generally remain tightly controlled in maturation, ECM remodeling constantly takes place [25, 29, 30], particularly when under hormonal stimulation or stress responses (e.g., [31]). Cells actively participate in tissue remodeling by secreting and mobilizing ECM molecules [32]. Alterations in ECM composition may result in changes in cell adhesion and /or tissue stiffness [33, 34] which further stimulates cellular responses. For instance, laminin component involved in cell adhesion to ECM is variant due to dynamics in exogenous factors [35]; normal cartilage shows elevated stiffness [36]; and vocal fold tissue exhibits dynamic viscoelastic properties [37]. Some specialized cell types can experience fast adhesion and detachment from ECM [38]. Changing ECM can also modify cell-cell interactions, further affecting cell behavior [39].

Pathological tissues Diseased tissue may possess properties such as mechanical stiffness different from those of the normal tissue [40, 41]. As a typical example, it has been found that tumor cells display enhanced movement towards ECM with lower mechanical rigidity, which is interesting considering the general stiffening phenomenon of tumor tissues [42], and biomechanical characteristics of tissues play a crucial role in tumor development [41]. It has also been shown that during the surgical procedures such as radio-frequency (RF) ablation, tissue properties can be modified [43]. Moreover, changes in ECM composition and relative quantity of ECM molecules can be correlated to pathology. For instance, ECM composition change that occurs during sub-epithelial tissue remodeling proved associated with asthma [33]. ECM remodeling in diseased heart valves is correlated to myofibroblast contractility [44], and certain cell types such as valvular interstitial cells can be activated and contribute to further tissue remodeling [45]. Additionally, ECM remodeling affects tissue mechanical properties in addition to inflammatory responses [46]. Moreover, mechanical forces, as experienced in traumatic brain injury or even under normal conditions, could potentially cause protein aggregation, giving rise to various diseases including neurodegenerative diseases [47]. Furthermore, early investigation of properties of central nervous (CNS) tissue under impact yielded modulus values with considerable variation. As an example, Fallenstein and coworker reported storage modulus of human brain tissue of 0.6 ~ 1.1 kPa under sinusoidal shear stress input mimicking head impact [48].

Development and aging During development, synthesis and degradation of ECM is a controlled process (e.g., [8, 49]), and mis-regulation contributes to many forms of diseases [30] Particularly, the microenvironment for embryonic and adult stem cells is regulated both temporally and spatially [2, 34], and is involved in various developmental processes including responses to soluble factors, cell differentiation, and morphogenesis [12]. ECM in musculoskeletal and other tissues adapts to increasing mechanical requirements by altering the size of tissue components [50] during development. Structural dynamics of ECM components such as collagen, laminin, and fibronectin coincides with estrous cycle and developmental progression [51]. Besides development, aging is also accompanied by changing ECM composition and structures. For example, in connective tissues, aging has been reported to be associated with increase in type I collagen content and decrease in both type III collagen and proteoglycans content, and with collagen fiber disruption and unraveling [50].

Tissue-implant interfaces With the growing interest in developing biomimetic materials for tissue engineering applications, tissue-implant interfaces have been subject to considerable research effort. Previous reports showed that cells can actively modify ECM at the interfaces, and cause drastic changes in tissue or construct mechanics using fibroblast-populated construct and other biomaterials [52, 53]. The study by Lee and co-workers suggested that dynamic moduli of an alginate material may be due to the bioactivities of the chondrocytes encapsulated in the scaffold [54]. In a similar study, different substrate composition and architecture gave rise to distinct levels of modulus increase owing to chondrocytes responses [55]. To take another example, smooth muscle cells (SMCs) in contact with engineered arterial construct displayed distinctive responses in protein synthesis and consequently the mechanical properties of ECM were significantly different [56]. Additionally, biodegradable materials used in various tissue engineering applications possess changing properties associated with specific degradation profiles.

Engineering advantages It has been suggested that temporal control over substrate or scaffold properties may entail great benefits in engineering cell growth. Among the notable examples is the stem cell differentiation and proliferation. A recent work showed enhanced hepatic functions from differentiated stem cells on softer substrates and improved expansion of undifferentiated cells on stiffer ones [57]. Therefore, it is promising to use stiffer substrates for optimal proliferations and subsequently soften them to gain better hepatic functions once differentiation completes. Langrana group also found that different neurite properties (e.g., axonal length and primary dendrite number) show differential preference towards substrate stiffness [58], suggesting the strategy of promoting nerve regeneration with scaffold of varying properties. Similar approach can be adopted to take advantage of differential cell responses (e.g., migration and functions) to adhesive properties.

The recurring indication from the above discussions is that *in vivo* ECM interacts with cells in many ways, and that the alteration in ECM composition or structures leads to changes in adhesive properties (hence cell adhesion) and/or mechanical properties. This potentially affects a variety of cell types and their properties and functioning, at different developmental stages, under normal or pathological conditions, or upon impact or injury. It also holds promises in offering novel approaches to tissue engineering applications. As a result, it is imperative to understand cellular responses to *changing substrate properties* for basic biology and biomimetic material (including biodegradable materials) design.

3. Types of dynamic substrates and stimuli

Here we consider two major classes of dynamic substrates that are based on self-assembled monolayer, or SAM, and hydrogels, as well as other types of substrates with surface or structural modifications. Since the focus of this work is on mimicking dynamic nature of ECM to examine cellular responses, those dynamic materials that are developed for other specific applications such as drug delivery [59] and do not involve changes that mimic dynamic ECM are beyond the scope of the review.

3.1. Self-assembled monolayer (SAM)

SAMs are formed by adsorption of molecules in solution or gas phase onto substrates in a spontaneous and organized fashion [60], and have emerged as an important candidate of materials in studying cellular responses to dynamic substrates [60, 61] where modifications could be made *in situ*. One of the major research focuses in this direction is to examine the effect of dynamic adhesive property of the substrate on cells, particularly by leveraging the ability to selectively capture or release cells upon application of a variety of stimuli (Table 1).

Type	Substrates	Properties changed and stimuli	Cell model	Observations and notes	Ref.
SAM	SAM incorporating O-silyl hydroquinone moiety	Adhesion on/off *Stimulus: electric potential*	3T3 fibroblasts	Modulation of cell adhesion and migration	[17, 61, 65]
	Electro-active quinine monolayer on Au	Adhesion on/off *Stimulus: electric potential*	3T3 fibroblasts	Selective release of adherent cells	[68]
	Azobenzene containing SAM on Au	Adhesion on/off *Stimulus: UV/visible light*	3T3 fibroblasts	Attachment and release of adherent cells Potential to control part of a single cell or groups of cells	[69]
Polymeric Hydrogel	MMP responsive polymer hydrogel network	Degradation of hydrogel *Stimulus: cell secreted MMP*	Human foreskin fibroblasts (HFFs)	Cell infiltration into the gel network with time	[74]
	Thermo-responsive polymer with photosensitive surface	Adhesion on/off *Stimulus: UV radiation and temperature*	CHO-K1 cells	Reversible control over cell adhesion Ability to control a population of cells	[72]
	poly(NIPAM-co-sodium acrylate) hydrogel films on rigid substrates.	Topographic change (swelling/de-swelling of gels) *Stimulus: temperature*	Porcine epithelial cells	Dynamic patterned substrates Reversible encapsulation of adherent cells	[73]
	DNA crosslinked PAM gels	Crosslinking density↑è Mechanical stiffness ↑, vice versa *Stimulus: ssDNA*	L929 & GFP fibroblasts	On dynamic substrate, L929 cells spread more than those on static stiff substrates (~23 kPa), while GFP fibroblasts respond differently to stiffening and softening of substrates Cell spreading and polarity (aspect ratio) respond differently to stiffness dynamics The range, starting point, and end point of change matter	[81, 83]
	DNA crosslinked PAM gels	Crosslinking density↓èstiffness ↓ *Stimulus: ssDNA*	Primary spinal cord cells	Neurite outgrowth respond to dynamic stiffness The trend in the response match that to the static stiffness except for primary dendrite length	[20]
	HA hydrogel	Crosslinking density change and ECM deposition è Mechanical stiffness change *Stimulus: hydrolysis or enzyme*	human mesenchymal stem cells (hMSCs)	Mechanical properties can be engineered with degradation Stiffness ↑ when degradation equals ECM deposition, and Stiffness ↓ at rapid degradtion Cellular responses to dynamic stiffness are different from static gels with the same initial or ending conditions	[78]

Type	Substrates	Properties changed and stimuli	Cell model	Observations and notes	Ref.
	Methacrylated HA hydrogel	UV exposure è stiffness ↑ *Stimulus: UV radiation and addition of reactive groups for*	hMSCs	Fate of hMSCs differentiate depends on the dynamics of stiffness change of substrates Adipogenic differentiation favored when cells is on the softer substrate long (stiffening at later times) Osteogenic differentiation when cells are on the stiffer substrate (stiffening at early times).	[79]
	Hydrogel based on PAM crosslinked by photosensitive reagent	Mechanical Stiffness (global or local)↓ *Stimulus: UV radiation*	3T3 fibroblasts	Stiffness decrease of 20-30% upon propose UV radiation Global stiffness decrease results in less spreading Localized softening to anterior and posterior area gives to differential responses	[76]
	PEG based hydrogel with photosensitive crosslinker	Mechanical Stiffness↓ Adhesive property change *Stimulus: UV radiation*	hMSCs and Valvular inter-stitial cells (VICs)	Valvular cell differentiation into myo-fibroblasts is inhibited by softening Good viability of hMSCs	[77]
Other types of substrates	Piezo-controlled substrate and AFM cantilever	Mechanical stiffness with cycling changes *Stimulus: stiffness clamp on AFM*	NIH 3T3	Apparent stiffness↑ leads to cells contraction rate↑ and contraction velocity↓ Changes took place instantaneously, and so did responses Responses were reversible, and consistent for same cell.	[84]
	Photo-active glass substrate with modifications	Adhesion on/off *Stimulus: UV radiation & pro- adhesive molecules*	HEK293, COS, NIH 3T3	Spatio-temporal control over cell adhesion Single cell control	[62]
	Substrates with photo-responsive caged peptides	Adhesion on/off *Stimulus: UV*	3T3 fibroblasts	Modifications of non-adhesive surfaces to adhesive ones	[63]
	PEG-modified ITO microelectrodes on glass substrates	Adhesion on/off *Stimulus: electric potential*	HepG2 (hepatic) and 3T3 cells, co-culture	Micro-patterned co-culture made possible	[85]
Promising materials*	Photo-crosslinked alginate hydrogel	Stiffness change; *Stimulus: light or hydrolysis*	Primary bovine chondrocytes	High survival rate for primary bovine chondrocytes Cellular responses to dynamic changes to be studied	[92]
	Gellan Gum hydrogel with both ionic crosslinking and	Stiffness, swelling, and degradation change *Stimulus: light or ion exchange*	NIH 3T3	Swelling and hydrolytic degradation vary with respect to crosslinking mechanism Stiffness may be changed quickly during photo-crosslinking process	[88]

http://dx.doi.org/10.5772/53547

Type	Substrates	Properties changed and stimuli	Cell model	Observations and notes	Ref.
	Methacrylated HA hydrogel with photo-crosslinker	Stiffening *Stimulus: UV radiation*	NIH3T3L HeLa Primary osteoblast	Good cell viability Cellular responses to dynamic changes to be studied	[80]
	PEG-based hydrogel incorporating CMP*	Softening *Stimulus: temperature or free CMP*	N/A	Cellular responses to dynamic changes to be studied	[90]
	PEG hydrogel (PEG vinyl sulfone crosslinked with PEG-diester-dithiol)	Softening *Stimulus: hydrolysis*	3T3 balb fibroblasts	Good cell viability in 3D gels Cellular responses to dynamic changes to be studied	[91]
	Resilin-like polypeptide (RLP) network crosslinked by THPP	Dynamic stiffness *Stimulus: oscillation*	N/A	Cellular responses to dynamic changes to be studied	[86]
	Thermo-reversible hydro-ferrogels (FGs)	Mechanical stiffness change *Stimulus: temperature change*	N/A	Cellular responses to dynamic changes to be studied	[89]

Note: ssDNA: single-stranded DNA; ↑ increase; ↓ decrease. *For 'Promising materials', most provides in vitro cyto-toxicity study, and cellular responses to dynamic properties remain to be investigated. N/A: not available

Table 1. A partial list dynamic substrates currently used in studying cell responses.

These stimuli, applied to initiate substrate dynamic, include light [62, 63], electricity [16, 17], pH, temperature, and others [16, 64] (Fig. 1). These approaches generally involve photochemical or electrochemical conversion, redox reactions, or stimulated configuration change of surface proteins, which leads to the attachment, detachment, shielding, or exposing of cell adhesion molecules, among which a popular choice is RGD peptide.

Mrksich group has been actively engaged in the development of SAM-based dynamic substrates by integrating surface chemistry, micro-patterning, and cell microenvironment engineering [17, 19, 61, 65, 66]. Based on an elegant design of SAM with electrochemically responsive group on a micro-patterned substrate, they first applied electrical stimulation to release 3T3 fibroblasts from designed areas on the substrate, and subsequently encouraged migration of neighboring cells to those areas with newly added adhesion molecules [65]. Refining this design by adding responsiveness to both negative and positive electric potentials, they demonstrated selective control over cell release [67] (Fig. 1C). Other groups have also engaged in the effort along this direction. By employing a hydroquinone terminated SAM based on redox reactions, Chan and colleagues proposed a SAM on gold surface that enables attachment and release of cell adhesion molecules such as those with RGD motif [68], and selectively released adherent 3T3 fibroblasts bound through RGD motifs but not those adherent based on hydro-phobic interactions (Fig. 1A). Reversibility of cell adhesion is attractive in studying cellular responses and cell-ECM interactions [60]. As an example, a surface chemistry involving azobenzene capable of switching between two configurations was utilized to expose or hide adhesion sites (e.g., RGD motif) upon photochemical stimulation[69] (Fig. 1B). While the finding is interesting, the long exposure of cells to UV may be problematic despite the reported negligible impact of light with wavelength over 320 nm on cells [63].

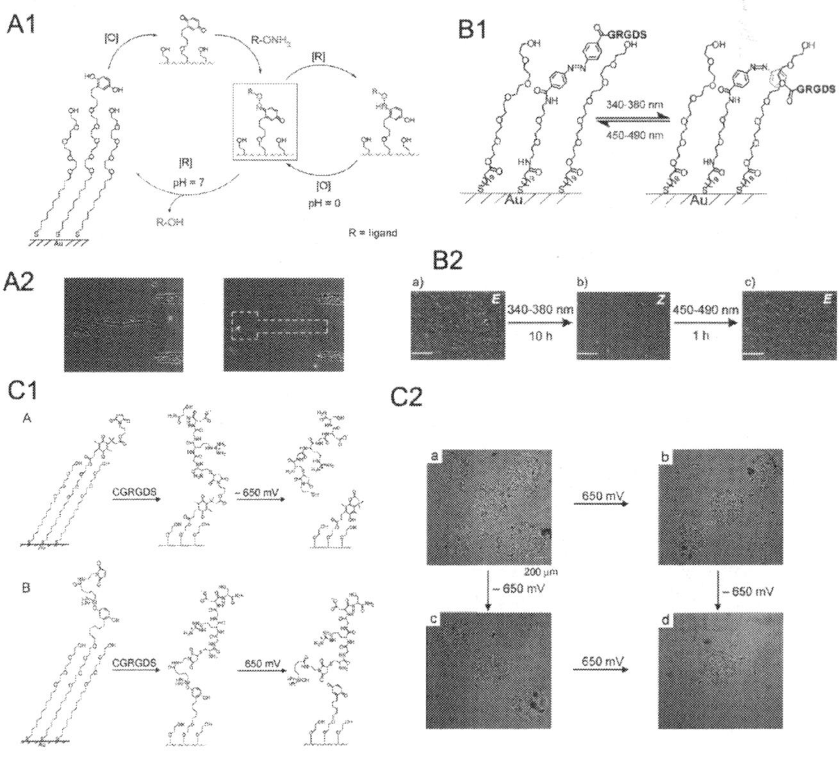

Figure 1. SAM-based dynamic substrates. (A) Schematic of the approach based on redox reaction (A1) by adjusting electrochemical potential, and cell detachment upon application of electric potential (A2). Extracted from [68]. (B) Schematic of altering configuration of azobenzene group under light of different ranges of wave length (B1) [69] and application to cell culture (B2) where NIH 3T3 fibroblasts initially adhere to adhesive surface (a) which was inhibited upon surface modification (b) followed by recovery of adhesion due to azobenzene configuration change (c). Extracted from [69]. (C) Illustration of a SAM that allows different modifications with positive and negative electric potential (C1) and its application in selective release of Swiss 3T3 cells (C2). Extracted from [67]. All with permission from publishers.

The above surveys part of the key advancement using SAM in modifying adhesion properties of the substrates mimicking those of natural cellular microenvironment. For a complete analysis of SAMs and their various applications, readers are referred to other reviews (e.g., [60, 70]). It suffices to point out that SAMs possess advantages in the precision (down to molecular level) of the control that can be applied in mechanistic studies [60, 66] of cell-ECM interactions, and are potentially useful for cell-based diagnostics among many applications. However, this approach has certain limitations. First, it mostly relies on coupling between electrical, chemical (including pH), mechanical, thermal, optical and biochemical (e.g., protein conformation) cues whose applicability under *in vivo* conditions is problematic. Next, the resulting changes in these studies are mostly of surface biochemical properties or of the presentation and biological activities of the surface ligands. Nevertheless, SAMs have greatly facilitated the probe and understanding of cell-ECM interactions and particular interplay between cells and ECM with dynamics in adhesive properties.

3.2. Polymeric hydrogels

Hydrogel materials are gaining popularity in the development of biomimetic materials, primarily due to the hydrated nature of natural ECM [14, 71]. Implantable hydrogel materials are increasingly being used in cardiovascular disease, nerve regeneration, and other conditions [59]. With careful design, hydrogel materials can have tunable materials properties, which have been demonstrated in a myriad of examples (Table 1). For instance, different than SAM-based approach, a polymer with both thermo- and photo-sensitivity was used to reversibly control adhesion of a group of cells [72]. Kim and colleagues took advantages of the thermo-responsive swelling behavior of copolymer between NIPAM and sodium acrylate, and created a hydrogel film that can be used to control cell encapsulation with surface topography [73]. Moreover, biomaterials responsive to the natural stimuli such as those experienced by biodegradable materials were found useful in mimicking biological events under physiological conditions, as illustrated in cell invasion to a MMP-responsive hydrogel scaffold [74]. This finding, among others, exemplifies the strategy of triggering material dynamic from bio-responsiveness to potential site- or disease-specific cues. The information from these studies is instrumental to the design of biodegradable materials in optimizing degradation profile for target cellular responses [75]. Naturally, in order to achieve desired outcome in adopting these strategies, it is important to gain thorough understanding of the natural environment, and minimize risks associate with biodegradable materials such as premature degradation, and potential toxicity of intermediate products from degradation.

Using a popular polyacrylamide hydrogel culture system with modifications that impart it with photo-sensitivity, Wang and colleagues [76] showed that upon UV induced substrates softening, spreading of 3T3 fibroblasts was hindered in contrast to that under static conditions (Fig. 2A). More interestingly, localized softening at anterior and posterior of cells yielded differential cellular morphology and migration responses [76]. Meanwhile, a PEG based polymer (PEGA) crosslinked by photosensitive crosslinker (PEGdiPDA) has been developed by Kloxin et al. [77], and used to lower gel stiffness upon UV exposure, which resulted in de-activation of myofibroblasts (Fig. 2B). Although UV radiation is preferentially avoided, these methods made possible high precision in applying changes of cellular mechanical microenvironment, and potentially allow creation of dynamic stiffness gradient.

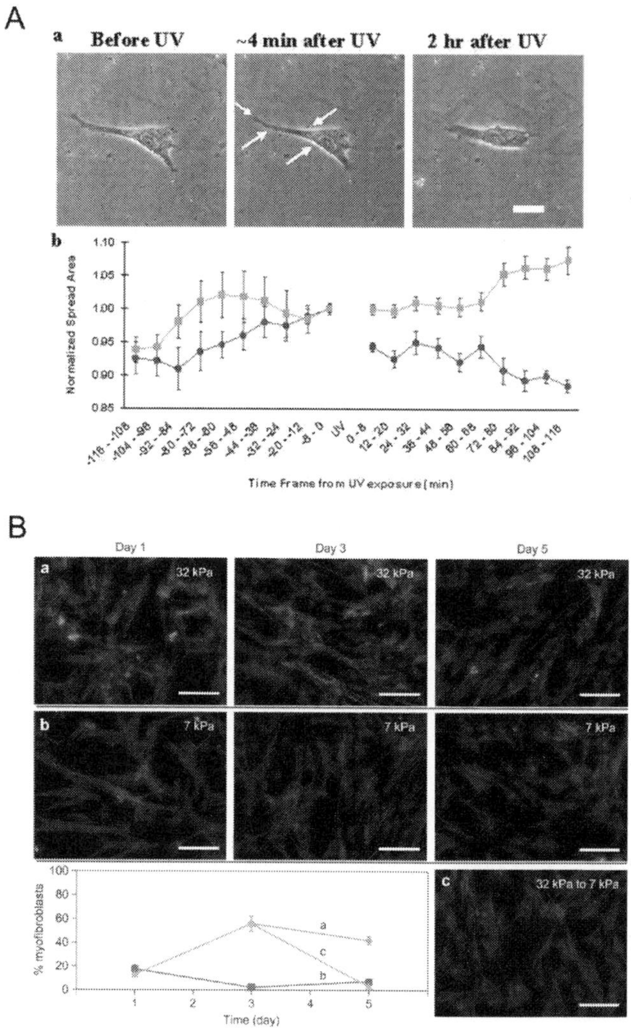

Figure 2. Photosensitive hydrogels and the study of cellular responses. (A) On a polyacrylamide hydrogel with photosensitive crosslinker, NIH 3T3 cells contract as indicated by projection area in response to UV-induced substrate softening. Extracted from [76]. (B) Valvular interstitial cells (VICs) on a PEG based hydrogel with photosensitive crosslinker displayed de-activation when UV radiation triggered substrate stiffness decrease. Extracted from [77]. Both with permission from publishers.

http://dx.doi.org/10.5772/53547

Similar observations were made by Burdick group for human mesenchymal stem cells (hMSCs) by using hyaluronic acid (HA) hydrogel degradable from hydrolytic and enzymatic reactions [78] (Fig. 3A). Very recently, a new material platform has been constructed by this group [79] and others [80] where the stiffness of a methacrylated hyaluronic acid hydrogel is increased via addition of photo-initiator and UV light exposure. In response to elevated stiffness, human mesenchymal stem cells (hMSCs) spread more and exert greater traction forces in hours (almost one magnitude of difference), and the rate of stiffness elevation dictates fate of cell differentiation towards adipogenic (slower) or osteogenic (faster) lineage. Their work highlighted that cellular behavior on dynamic gels is not the same as that on static gels with same initial or final properties, underlining the significance of dynamics in gel properties. This has been echoed in the concurrent work [81], where, for instance, the fibroblasts on 100% crosslinked hydrogels demonstrated different morphology from that on 100% crosslinked gels modified from 50% gels. Therefore, it is conceivable that the previous state of the cells and their ECM is also among the determinants of their current state, and that time dimension of ECM is of great importance.

Factors other than environmental conditions (e.g., light, pH, temperature) can also be delivered to stimulate dynamics in substrate properties. Incorporating DNA as crosslinker, Jiang and colleagues have developed a hydrogel system for cell attachment where mechanical properties of the substrates can be altered *in situ* in a controlled fashion when the cell culture is present [20, 81]. These DNA crosslinked hydrogels may also be designed to be potentially responsive to bio-stimuli, such as temperature or enzymes. Two representative cell types were chosen for the study of cell responses to dynamic substrate: fibroblasts whose sensitivity to mechanical cues is well documented, and neurons whose mechanosensing capability has recently just started being appreciated. The reports [20, 81] offered evidence that both cell types do respond to dynamic alternations in the mechanical characteristics of ECM, and suggest that the *alternations* in the mechanical stiffness may be involved in disease progression (Fig. 3B). It has been shown that the stiffness change resulting from pathological processes, may also aid in further progression of diseases [82].

The same material system was employed by Previera and co-workers, and they firstly proved the dual mechanical stimuli, namely strain and stiffness drop, during the dynamic processes, and secondly contrasted cell behavior to stiffness decrease with that for hardening of the substrates [83]. On hardening gels (from 12 kPa to 22 kPa), cells spread more than those on static substrates of higher stiffness (22 kPa), whereas on softening ones, they have greater spreading area than that on either starting or ending stiffnesses. In these studies, cell responses are determined by the range of rigidity change (due to crosslinking density), starting and ending rigidity, and specific cell properties (e.g., projection area vs. aspect ratio and protrusion for fibroblasts). The stress generation may also be involved in affecting cell behavior [83].

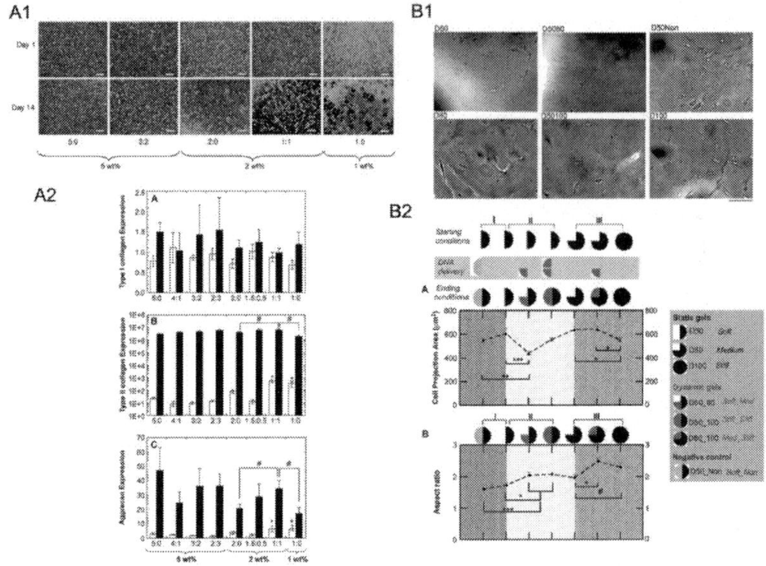

Figure 3. Substrates with dynamic mechanical stiffness and their application in cell culture studies. (A) Live/deal cell staining of human mesenchymal stem cells (hMSCs) in a hyaluronic acid hydrogel (A1) and the analysis of gene expression of type I/II collagen and aggrecan revealed that aside from type I collagen, both type II collagen and aggrecan exhibited an elevated level of expression on dynamic gels from the static ones. Extracted from [78]. With permission from publisher. (B) L929 fibroblast growth in a DNA crosslinked hydrogel with dynamic stiffness from crosslinking density change (B1) and the quantification of spreading area and aspect ratio (B2) showed that dynamic gels are significantly different from their static counterparts. Extracted from [81].

3.3. Other types of materials

The approach of employing polymeric hydrogel to study dynamic changes has certain limitations, one of which is the coupling of mechanical stiffness and forces (e.g., [83]). To address this concern and others, different from the approach by using SAM or polymeric hydrogel, AFM based method put forth by Webster and co-workers [84] probed cellular response to instant step change in stiffness excluding influence from stress or strain in the substrates (Table 1). It has been confirmed that indeed individual 3T3 fibroblasts are able to sense and respond to the stiffness in a scale of seconds as demonstrated in traction rate and contraction velocity [84]. However, this approach is most likely with inherent limitation in mimicking natural cell environment while remains an interesting tool in probing cellular responses to instantaneously change in stiffness. Additionally, this approach is applicable mostly to cells with dynamic morphology.

Common cell culture substrates (e.g., glass coverslip) modified with common photo-cleavable agents (NPE-TCSP) were shown to be useful for controlling cell adhesion selectively and temporally [62]. In this method, target areas were first irradiated to remove BSA known to prevent cell adhesion, and then pro-adhesive molecules (e.g., fibronectin) were added, followed by cell seeding. It is useful to study dynamics in interactions between single cells. Petersen et al. [63] used light to stimulate photosensitive surface modification resulting in uncovering of the RGD motif upon release of a caging group (Fig. 4A). In doing so, adhesion of 3T3 fibroblasts was first inhibited and then encouraged, although this process is not reversible. With a sequential activation of adhesive sites upon application of electric potential, a recent study [85] demonstrated the utility of substrates with ITO (indium tin oxide) microelectrodes modified with poly(ethylene glycol), or PEG, in co-culture of two cell types (hepatic cells and fibroblasts) in a controlled manner (Fig. 4B).

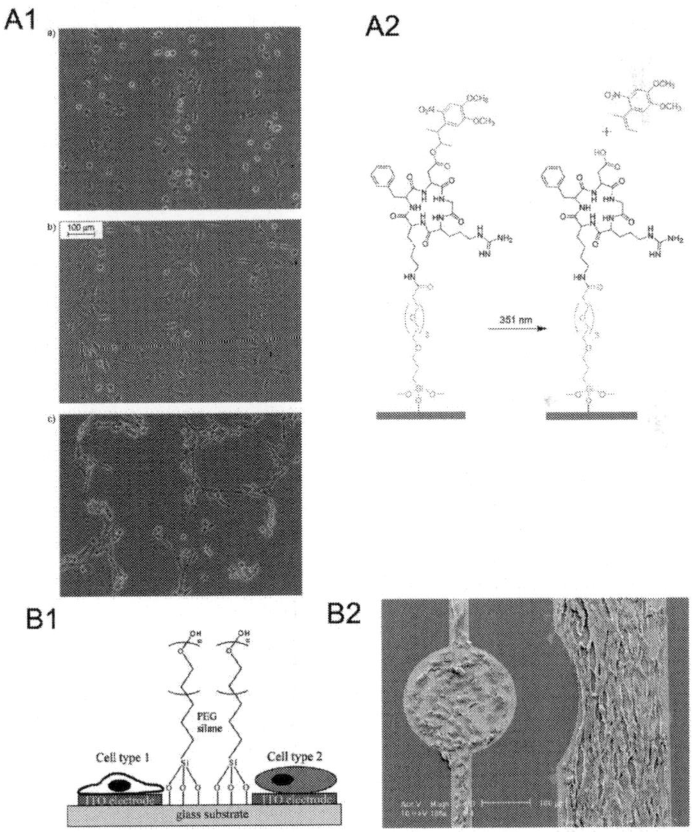

Figure 4. Cellular behavior in responses to other substrates with dynamic properties. (A) 3T3 fibroblasts grown in areas of patterned stripes (A1) generated from UV radiation based on the surface chemistry involving a caging group (A2). Extracted from [63]. (B) With PEG-modified ITO microelectrodes on glass substrates (B1), co-culture of two different cell types, HepG2 (hepatic) and 3T3 cells, was made possible. Extracted from [85]. Both with permission from publishers.

3.4. Promising substrates

By applying an oscillation to a resilin-like polypeptide network crosslinked by THPP (β-[Tris(hydroxymethyl) phosphino] propionic acid (betaine), Li and colleagues were able to observe dynamic mechanical stiffness of the gels varying with regard to oscillatory frequency mimicking the load from human vocal tissues [86] Oscillatory shear induced stiffening and softening of the collagen network might also serve as good substrates for mimicking cellular microenvironment particularly that in mechanically active tissues [87]. Ion concentration may be another stimulus to allow for temporal modification of hydrogel properties as exchange of ions between monovalant and divalant cations [88], and further work is needed to confirm it. Temperature-dependent substrate softening has been demonstrated by Krekhova et al. [89] and 3D complex with temperature-mediated crosslinking hence mechanical properties has been proposed by Stahl et al. [90], while applicability of these systems in mimicking cellular microenvironment remains to be seen. Zustiak et al. [91] reported mechanical stiffness drop, from approximately 1 kPa and at different rate, along with degradation of a poly(ethylene glycol), or PEG, hydrogel which might serve as not only drug delivery vehicle but also biomaterial construct, and they have offered preliminary evidence of good viability of 3T3 balb fibroblasts on the hydrogel substrate. Similarly, rigidity change from ~180 kPa to tens of kPa in 3-week period of degradation from a photo-crosslinked alginate hydrogels based on alginate methacrylation were presented by Jeon et al [92], and cyto-toxicity has been found to be low. In summary, these substrates holds promises as substrates with modifiable properties *in situ*, and need to be carefully tuned and evaluated for use as substrates with dynamic properties (Table 1). Other materials responsive to various stimuli including pH, temperature, and biochemical factors for a variety of applications, including can be found in the earlier reviews [64, 70, 93], and thus are not discussed in detail here due to the focus of the current analysis.

4. Design considerations and outlook

4.1. Dynamic properties of the substrates

As indicated in the discussions in Background and Motivation, the progression in changes of ECM properties is also critical in addition to changes *per se* in light of the observations in normal and pathological tissues, development and aging, and potential engineering benefits. Towards this end, rate of change (e.g., gradual vs. abrupt), range of change (e.g., small perturbation vs. drastic modifications), and change profile (e.g., monotonic increase vs. fluctuation) characterizing the nature of changes and their impact on cellular processes are subject to research effort, apparently adding to the complexity of the problem (Table 1). Take biodegradable material (e.g., [94]) as an example. It would be relevant to understand how mechanical and adhesive characteristics evolve with degradation and how the degradation profile affects the changes in the cellular micro-environment. Experimental design along this line may include, for example, different rates of release of RGD motif decreasing adhesiveness while keeping the same range of change (e.g., half of the total RGD presenting sites), or altering the range of change while maintaining the same rate of change. Furthermore, it is not clear at this point whether cellular responses to opposite changes (e.g., increase vs. decrease in adhesiveness or rigidity) of substrate properties are symmetric, thus their behavior to one direction of dynamic alterations may not be a reliable predictor of that to the opposite changes.

4.2. Potential effect on cell-cell interactions

Changes in adhesive or mechanical properties of ECM can stimulate cells, which, in response, secrete soluble factors and ECM molecules, and this further impacts neighboring cell types. Additionally, some cell types such as neurons may use other cells (e.g., astroglia) as substrate [95], and stiffness change of 'underlying' cells per se due to ECM stimulation may give rise to further alternations thanks to cell-cell interactions. For instance, during asthma, ECM stiffening contributes to stiffness increase of airway smooth muscle (ASM) cells, which potentially affects other cell types in the close proximity [33].

4.3. Design parameters for biomaterials and outlook

The design parameters of dynamic substrates from current studies are summarized in Table 2, which includes, but are not limited to, material system to consider (e.g., SAM or polymeric hydrogel materials), nature of change (mechanical stiffness or adhesion), rate of change (e.g., transient or gradual change, controllability of the rate of change), range of change (e.g., at different stiffness range) as well as potential issues in further investigations and applications to medical and healthcare applications. If the interest is in understanding the cellular behavior to mechanical stiffness alone, then an AFM based approach might be more attractive [84] as others will involve stress or strain as part of the stiffening or softening process. If precise control over stiffness range is desired, the DNA crosslinked PAM hydrogel system will serve the purpose better [20, 58, 81]. Polymeric hydrogel materials with controllable degradation profile and hence mechanical stiffness dynamics during degradation (e.g., [88]) will serve the purpose best when biodegradable materials are applied. Some of the material systems do offer unique benefits such as reversible property change or without using environmental factors (by applying oscillation, crosslinker, or ssDNA).

Meanwhile, there are inherent limitations to each of the material system under discussion (Table 2). UV exposure generally causes concern to its impact on cellular activities despite the findings of little impact from a number of studies based on a range of biological assays. Under physiological conditions, application of certain cues (e.g., ssDNA, light, or ion) might be too difficult or it might be greatly limited (e.g., temperature triggered changes). However, it is still possible to find ways to apply these cues with careful design. For instance, ssDNA design based on pre-screening using BLAST search against targeted specie or tissue type may minimize the chance of interfering with normal biological activities. Local heating/cooling may be carefully applied to induce dynamic changes to achieve cellular responses. Three dimensional system may better mimic natural cellular micro-environment than their 2D counter parts.

Stimuli	Material system	Nature of change	Range of change	Rate of change	Invasiveness of stimulus and potential issues	Ref.
Ion	Ion-crosslinked GC hydrogel	Stiffness	~22 to ~17 kPa (with chemical crosslinking)	N/A	Under physiological conditions, divalent ions exchanged by mono-valent ones	[88]
Light	Hydrogels based on PAM crosslinked by photosensitive agents	Softening	Stiffness: 5.5~7.2 kPa	Approximately 0.5~0.6 kPa/ min	UV exposure for 3 min UV radiation with low energy density Depth of penetration and limit on dose	[76]
	Methacrylated HA hydrogel	Stiffening Irreversible change	Stiffness: ~3 to ~30 kPa	Approximately 9 kPa/hr (short term); 2 kPa/day (long-term)	UV exposure for a few min Potential toxicity of photoinitiator Depth of penetration and limit on dose	[79]
	Photo-crosslinked methacrylated Gellan Gum hydrogel	Stiffness; Swelling Hydrolytic degradation	Stiffness: a few kPa to 22 kPa (by physical crosslinking)	Approximately 20 kPa/ min	UV exposure for one min Depth of penetration and limit on dose	[88]
	Methacrylated HA hydrogel with photo-crosslinker	Stiffness; Irreversible change	Stiffness: 1.6 to 3.8 kPa; 3-12 kPa	Approximately 0.1 or 0.3 kPa/min (during gelation)	UV exposure for a few min Potential toxicity of photo-initiator Depth of penetration and limit on dose	[80]
	PEG based hydrogel with photosensitive crosslinker	Stiffness↓ Adhesive property Irreversible change	N/A	N/A	Depth of penetration and limit on dose	[77]
DNA	DNA crosslinked PAM system	Stiffening & softening, potentially coupled with strain/stress Reversible change	Stiffness: ~5.9 to 22.9 kPa Stress > 0.5 Pa	Up to 8.5 kPa/ day	No differentiation in cellular responses between forces, stress, and stiffness Potentially interference from DNA with bio-activity (e.g., as anti-sense DNAs), and potential issue with DNase BLAST search against target specie & tissue type	[20, 81, 83]
AFM/ stiffness clamp	Piezo-controlled substrates and AFM stiffness clamp	Instantaneous change in stiffness Unidirectional	Stiffness: 3.6 to 90 nN/μm	Step change (instantane-ously)	Applicable only to cells with dynamic morphology	[84]
Hydro-lysis	Photo-crosslinked alginate hydrogel	Softening due to degradation	Stiffness: ~25 to ~180 kPa	7-8 kPa/ day	In sample preparation (with cells), UV exposure for 10 mins	[92]
	HA hydrogel	Stiffening & structure change	Stiffness: e.g., ~5 to 30 kPa for one case	0.7 kPa/ day	Dense crosslinking may impede cellular growth limited by diffusion& concentration of radicals	[78]
	PEG hydrogel (PEG vinyl sulfone	Softening due to degradation	Stiffness: from ~1 kPa-3 kPa to very low	From ~900 Pa/day to 500 Pa/day	Good cell viability Hydrogel degraded in 16 hours	[91]

http://dx.doi.org/10.5772/53547

Stimuli	Material system	Nature of change	Range of change	Rate of change	Invasiveness of stimulus and potential issues	Ref.
	crosslinked with PEG-diester-dithiol)					
Temperature	Thermo-reversible hydro-ferrogels (FGs)	Stiffening due to structural transition	Stiffness: ~28-24 kPa for 2ºC change at 37ºC	A few kPa for 1ºC of temperature change	Temperature change needs to be defined to be relevant to cell culture	[89]
	PEG-based hydrogel incorporating CMP*	Stiffness change Due to temperature and free CMP	Stiffness (indirect measurement)	N/A	Temperature change needs to be defined to be relevant to cell culture Bio-compatibility of free CMP	[90]
Oscillation	Resilin-like polypeptide based elastomer	Stiffness change due to oscillation	Storage modulus between 0.5 and 10 kPa	Highly dynamic	Strain/stress that is associated with oscillation	[86]

Note: N/A: not available; This is a partial list of the current work under examination.

Table 2. Design considerations in constructing dynamic substrates mimicking extra-cellular matrix (ECM).

A few new material system have been identified with the potential as dynamic cell culture platform as well as choice of biomaterials (Table 1). Many of them have demonstrated good cyto-compatibility, and investigation of impact of *in situ* changes to cells will be desired.

5. Concluding remarks

There is an increasing recognition of the discrepancy between static nature of the current cell culture substrates or scaffolds and the dynamics in ECM in natural or diseased tissue, during development and aging, or at tissue-scaffold interfaces. This has motivated the development of materials with controlled changing properties that mimic those of ECM. An array of stimuli, including environmental factors (temperature, pH, light, electrical potential) and non-environmental cues including enzyme and DNA, have been implemented to trigger dynamics in a number of material platform such as SAMs, polymeric hydrogels and other substrates with surface chemistry and modifications.

To date, most of the effort along this line has been devoted to *in vitro* models, and *in vivo* studies of the effect of dynamic tissue properties on cellular behavior are still rather limited, which awaits further development in cell biology and proper tools such as imaging techniques [12, 14, 29].

Understanding the interplay between cells and the extracellular matrix (ECM) including its dynamic aspect is fundamental to biology, development, aging and pathology, and can aid in the design of biomaterials. Ultimately, the system enabling both spatial and temporal control [96] of cells would be most relevant in terms of bio-mimicry and tissue engineering applications. Some of the potential directions include creating dynamic adhesive gradient to guide cell migration or neurite outgrowth at desired time point, constructing scaffolds with suitable mechanical rigidity to inhibit glia cell growth (thus hinder scar formation) while promoting nerve regeneration with compliance gradient, and developing dynamic platform for stem cell harvesting and differentiation for cell-based therapies.

Acknowledgments

The helpful discussions and advice from Langrana group at Rutgers University, New Jersey, USA as well as previous collaborators are greatly appreciated.

Author details

Frank Xue Jiang

Address all correspondence to: Frank.Jiang@unilever.com

Unilever Research & Development, Shanghai, P.R. China

Any opinions, findings, and conclusions or recommendations expressed in this material are those of the authors and do not necessarily reflect the views of the Unilever, its management or employees. The authors have no financial interest in this publication and receive nothing in exchange for providing this review.

References

[1] Keatch RP, Schor AM, Vorstius JB, Schor SL. Biomaterials in regenerative medicine: engineering to recapitulate the natural. Curr Opin Biotechnol 2012;23:579-582.

[2] Brizzi MF, Tarone G, Defilippi P. Extracellular matrix, integrins, and growth factors as tailors of the stem cell niche. Curr Opin Cell Biol 2012;(Epub).

[3] Cukierman E, Pankov R, Stevens DR, Yamada KM. Taking cell-matrix adhesions to the third dimension. Science 2001;294:1708-1712.

[4] Geiger B. Cell biology. Encounters in space. Science 2001;294:1661-1663.

[5] Discher DE, Mooney DJ, Zandstra PW. Growth factors, matrices, and forces combine and control stem cells. Science 2009;324:1673-1677.

[6] Page H, Flood P, Reynaud EG. Three-dimensional tissue cultures: current trends and beyond. Cell Tissue Res 2012;(Epub).

[7] Benjamin M, Hillen B. Mechanical influences on cells, tissues and organs - 'Mechanical Morphogenesis'. Eur J Morphol 2003;41:3-7.

[8] Mammoto T, Ingber DE. Mechanical control of tissue and organ development. Development 2010;137:1407-1420.

[9] Tee SY, Bausch AR, Janmey PA. The mechanical cell. Curr Biol 2009;19:R745-748.

[10] Zhu C, Bao G, Wang N. Cell mechanics: mechanical response, cell adhesion, and molecular deformation. Annu Rev Biomed Eng 2000;2:189-226.

[11] Lu P, Takai K, Weaver VM, Werb Z. Extracellular matrix degradation and remodeling in development and disease. Cold Spring Harb Perspect Biol 2011;3.

[12] Rozario T, Desimone DW. The extracellular matrix in development and morphogenesis: A dynamic view. Dev Biol 2010;341:126-140.

[13] Xu R, Boudreau A, Bissell MJ. Tissue architecture and function: dynamic reciprocity via extra- and intra-cellular matrices. Cancer Metastasis Rev 2009;28:167-176.

[14] Goody MF, Henry CA. Dynamic interactions between cells and their extracellular matrix mediate embryonic development. Mol Reprod Dev 2010;77:475–488.

[15] Godier AF, Marolt D, Gerecht S, Tajnsek U, Martens TP, Vunjak-Novakovic G. Engineered microenvironments for human stem cells. Birth Defects Res C Embryo Today 2008;84:335-347.

[16] Lahann J, Langer R. Smart Materials with Dynamically Controllable Surfaces. MRS Bulletin 2005;30:185-188.

[17] Mrksich M. Dynamic Substrates for Cell Biology. MRS BULLETIN 2005;30:180-184.

[18] Abbott NL, Gorman CB, Whitesides GM. Active Control of Wetting Using Applied Electrical Potentials and Self-Assembled Monolayers. Langmuir 1995;11:16-18.

[19] Yousaf MN, Houseman BT, Mrksich M. Using electroactive substrates to pattern the attachment of two different cell populations. Proc Natl Acad Sci U S A 2001;98:5992-5996.

[20] Jiang FX, Yurke B, Schloss RS, Firestein BL, Langrana NA. Effect of dynamic stiffness of the substrates on neurite outgrowth by using a DNA-crosslinked hydrogel. Tissue Eng Part A 2010;16:1873-1889.

[21] Peerani R, Zandstra PW. Enabling stem cell therapies through synthetic stem cell-niche engineering. J Clin Invest 2010;120:60-70.

[22] Dawson E, Mapili G, Erickson K, Taqvi S, Roy K. Biomaterials for stem cell differentiation. Adv Drug Deliv Rev 2008;60:215-228.

[23] Haq F, Keith C, Zhang G. Neurite development in PC12 cells on flexible micro-textured substrates under cyclic stretch. Biotechnol Prog 2006;22:133-140.

[24] Nicodemus GD, Bryant SJ. Mechanical loading regimes affect the anabolic and catabolic activities by chondrocytes encapsulated in PEG hydrogels. Osteoarthritis Cartilage 2010;18:126-137.

[25] Brown L. Cardiac extracellular matrix: a dynamic entity. Am J Physiol Heart Circ Physiol 2005;289:H973-974.

[26] Jung Y, Kim SH, Kim YH. The effects of dynamic and three-dimensional environments on chondrogenic differentiation of bone marrow stromal cells. Biomed Mater 2009;4:055009.

[27] Tschumperlin DJ, Boudreault F, Liu F. Recent advances and new opportunities in lung mechanobiology. J Biomech 2010;43:99-107.

[28] Gorstein F. The dynamic extracellular matrix. Hum Pathol 1988;19:751-752.

[29] Dallas SL, Chen Q, Sivakumar P. Dynamics of assembly and reorganization of extracellular matrix proteins. Curr Top Dev Biol 2006;75:1-24.

[30] Daley WP, Peters SB, Larsen M. Extracellular matrix dynamics in development and regenerative medicine. J Cell Sci 2008;121:255-264.

[31] Shi YB, Fu L, Hsia SC, Tomita A, Buchholz D. Thyroid hormone regulation of apoptotic tissue remodeling during anuran metamorphosis. Cell Res 2001;11:245-252.

[32] Sivakumar P, Czirok A, Rongish BJ, Divakara VP, Wang YP, Dallas SL. New insights into extracellular matrix assembly and reorganization from dynamic imaging of extracellular matrix proteins in living osteoblasts. J Cell Sci 2006;119:1350-1360.

[33] An SS, Kim J, Ahn K, Trepat X, Drake KJ, Kumar S, et al. Cell stiffness, contractile stress and the role of extracellular matrix. Biochem Biophys Res Commun 2009;382:697-703.

[34] Reilly GC, Engler AJ. Intrinsic extracellular matrix properties regulate stem cell differentiation. J Biomech 2010;43:55-62.

[35] Chen ZL, Indyk JA, Strickland S. The hippocampal laminin matrix is dynamic and critical for neuronal survival. Mol Biol Cell 2003;14:2665-2676.

[36] Coles JM, Zhang L, Blum JJ, Warman ML, Jay GD, Guilak F, et al. Loss of cartilage structure, stiffness, and frictional properties in mice lacking Prg4. Arthritis Rheum 2010;62:1666-1674.

[37] Kutty JK, Webb K. Tissue engineering therapies for the vocal fold lamina propria. Tissue Eng Part B Rev 2009;15:249-262.

[38] Siu MK, Cheng CY. Dynamic cross-talk between cells and the extracellular matrix in the testis. Bioessays 2004;26:978-992.

[39] Hui EE, Bhatia SN. Micromechanical control of cell-cell interactions. Proc Natl Acad Sci U S A 2007;104:5722-5726.

[40] Discher DE, Janmey P, Wang YL. Tissue cells feel and respond to the stiffness of their substrate. Science 2005;310:1139-1143.

[41] Yu H, Mouw JK, Weaver VM. Forcing form and function: biomechanical regulation of tumor evolution. Trends Cell Biol 2011;21:47-56.

[42] Zaman MH, Trapani LM, Sieminski AL, Mackellar D, Gong H, Kamm RD, et al. Migration of tumor cells in 3D matrices is governed by matrix stiffness along with cell-matrix adhesion and proteolysis. Proc Natl Acad Sci U S A 2006;103:10889-10894.

[43] Bharat S, Techavipoo U, Kiss MZ, Liu W, Varghese T. Monitoring stiffness changes in lesions after radiofrequency ablation at different temperatures and durations of ablation. Ultrasound Med Biol 2005;31:415-422.

[44] Walker GA, Masters KS, Shah DN, Anseth KS, Leinwand LA. Valvular myofibroblast activation by transforming growth factor-beta: implications for pathological extracellular matrix remodeling in heart valve disease. Circ Res 2004;95:253-260.

[45] Rabkin-Aikawa E, Farber M, Aikawa M, Schoen FJ. Dynamic and reversible changes of interstitial cell phenotype during remodeling of cardiac valves. J Heart Valve Dis 2004;13:841-847.

[46] Dobaczewski M, Gonzalez-Quesada C, Frangogiannis NG. The extracellular matrix as a modulator of the inflammatory and reparative response following myocardial infarction. J Mol Cell Cardiol 2010;48:504-511.

[47] Hachiya NS, Kozuka Y, Kaneko K. Mechanical stress and formation of protein aggregates in neurodegenerative disorders. Med Hypotheses 2008;70:1034-1037.

[48] Fallenstein GT, Hulce VD, Melvin JW. Dynamic mechanical properties of human brain tissue. J Biomech 1969;2:217-226.

[49] Latimer A, Jessen JR. Extracellular matrix assembly and organization during zebrafish gastrulation. Matrix Biol 2010;29:89-96.

[50] Silver FH, DeVore D, Siperko LM. Invited Review: Role of mechanophysiology in aging of ECM: effects of changes in mechanochemical transduction. J Appl Physiol 2003;95:2134-2141.

[51] Yamada O, Todoroki J, Takahashi T, Hashizume K. The dynamic expression of extracellular matrix in the bovine endometrium at implantation. J Vet Med Sci 2002;64:207-214.

[52] Marquez JP, Genin GM, Pryse KM, Elson EL. Cellular and matrix contributions to tissue construct stiffness increase with cellular concentration. Ann Biomed Eng 2006;34:1475-1482.

[53] Bellows CG, Melcher AH, Aubin JE. Contraction and organization of collagen gels by cells cultured from periodontal ligament, gingiva and bone suggest functional differences between cell types. J Cell Sci 1981;50:299-314.

[54] Lee B, Han L, Frank EH, Chubinskaya S, Ortiz C, Grodzinsky AJ. Dynamic mechanical properties of the tissue-engineered matrix associated with individual chondrocytes. J Biomech 2010;43:469-476.

[55] Appelman TP, Mizrahi J, Elisseeff JH, Seliktar D. The differential effect of scaffold composition and architecture on chondrocyte response to mechanical stimulation. Biomaterials 2009;30:518-525.

[56] Crapo PM, Wang Y. Physiologic compliance in engineered small-diameter arterial constructs based on an elastomeric substrate. Biomaterials 2010;31:1626-1635.

[57] Li L, Sharma N, Chippada U, Jiang X, Schloss R, Yarmush ML, et al. Functional modulation of ES-derived hepatocyte lineage cells via substrate compliance alteration. Ann Biomed Eng 2008;36:865-876.

[58] Jiang FX, Yurke B, Firestein BL, Langrana NA. Neurite outgrowth on a DNA crosslinked hydrogel with tunable stiffnesses. Ann Biomed Eng 2008;36:1565-1579.

[59] Ulijn RV, Bibi N, Jayawarna V, Thornton PD, Todd SJ, Mart RJ, et al. Bioresponsive hydrogels. Materials Today 2007;10:40-48.

[60] Robertus J, Browne WR, Feringa BL. Dynamic control over cell adhesive properties using molecular-based surface engineering strategies. Chem Soc Rev 2010;39:354-378.

[61] Mrksich M. Using self-assembled monolayers to model the extracellular matrix. Acta Biomater 2009;5:832-841.

[62] Nakanishi J, Kikuchi Y, Takarada T, Nakayama H, Yamaguchi K, Maeda M. Photoactivation of a substrate for cell adhesion under standard fluorescence microscopes. J Am Chem Soc 2004;126:16314-16315.

[63] Petersen S, Alonso JM, Specht A, Duodu P, Goeldner M, del Campo A. Phototriggering of cell adhesion by caged cyclic RGD peptides. Angew Chem Int Ed Engl 2008;47:3192-3195.

[64] Mendes PM. Stimuli-responsive surfaces for bio-applications. Chem Soc Rev 2008;37:2512-2529.

[65] Yeo WS, Yousaf MN, Mrksich M. Dynamic interfaces between cells and surfaces: electroactive substrates that sequentially release and attach cells. J Am Chem Soc 2003;125:14994-14995.

[66] Mrksich M. A surface chemistry approach to studying cell adhesion. Chem Soc Rev 2000;29:267 - 273.

[67] Yeo WS, Mrksich M. Electroactive self-assembled monolayers that permit orthogonal control over the adhesion of cells to patterned substrates. Langmuir 2006;22:10816-10820.

[68] Chan EW, Park S, Yousaf MN. An electroactive catalytic dynamic substrate that immobilizes and releases patterned ligands, proteins, and cells. Angew Chem Int Ed Engl 2008;47:6267-6271.

[69] Liu D, Xie Y, Shao H, Jiang X. Using azobenzene-embedded self-assembled monolayers to photochemically control cell adhesion reversibly. Angew Chem Int Ed Engl 2009;48:4406-4408.

[70] Nandivada H, Ross AM, Lahann J. Stimuli-responsive monolayers for biotechnology. Progress in Polymer Science 2010;35:141-154.

[71] Gillette BM, Jensen JA, Wang M, Tchao J, Sia SK. Dynamic hydrogels: switching of 3D microenvironments using two-component naturally derived extracellular matrices. Adv Mater 2010;22:686-691.

[72] Edahiro J, Sumaru K, Tada Y, Ohi K, Takagi T, Kameda M, et al. In situ control of cell adhesion using photoresponsive culture surface. Biomacromolecules 2005;6:970-974.

[73] Kim J, Yoon J, Hayward RC. Dynamic display of biomolecular patterns through an elastic creasing instability of stimuli-responsive hydrogels. Nat Mater 2010;9:159-164.

[74] Lutolf MP, Lauer-Fields JL, Schmoekel HG, Metters AT, Weber FE, Fields GB, et al. Synthetic matrix metalloproteinase-sensitive hydrogels for the conduction of tissue regeneration: engineering cell-invasion characteristics. Proc Natl Acad Sci U S A 2003;100:5413-5418.

[75] Baker BM, Nerurkar NL, Burdick JA, Elliott DM, Mauck RL. Fabrication and modeling of dynamic multipolymer nanofibrous scaffolds. J Biomech Eng 2009;131:101012.

[76] Frey MT, Wang YL. A photo-modulatable material for probing cellular responses to substrate rigidity. Soft Matter 2009;5:1918-1924.

[77] Kloxin AM, Benton JA, Anseth KS. In situ elasticity modulation with dynamic substrates to direct cell phenotype. Biomaterials 2010;31:1-8.

[78] Chung C, Beecham M, Mauck RL, Burdick JA. The influence of degradation characteristics of hyaluronic acid hydrogels on in vitro neocartilage formation by mesenchymal stem cells. Biomaterials 2009;30:4287-4296.

[79] Guvendiren M, Burdick JA. Stiffening hydrogels to probe short- and long-term cellular responses to dynamic mechanics. Nat Commun 2012;3:792.

[80] Hachet E, Van Den Berghe H, Bayma E, Block MR, Auzely-Velty R. Design of Biomimetic Cell-Interactive Substrates Using Hyaluronic Acid Hydrogels with Tunable Mechanical Properties. Biomacromolecules 2012;(Epub).

[81] Jiang FX, Yurke B, Schloss RS, Firestein BL, Langrana NA. The relationship between fibroblast growth and the dynamic stiffnesses of a DNA crosslinked hydrogel. Biomaterials 2010;31:1199-1212.

[82] Huang S, Ingber DE. Cell tension, matrix mechanics, and cancer development. Cancer Cell 2005;8:175-176.

[83] Previtera ML, Trout KL, Verma D, Chippada U, Schloss RS, Langrana NA. Fibroblast morphology on dynamic softening of hydrogels. Ann Biomed Eng 2012;40:1061-1072.

[84] Webster KD, Crow A, Fletcher DA. An AFM-based stiffness clamp for dynamic control of rigidity. PLoS One 2011;6:e17807.

[85] Shah S, Lee JY, Verkhoturov S, Tuleuova N, Schweikert EA, Ramanculov E, et al. Exercising spatiotemporal control of cell attachment with optically transparent microelectrodes. Langmuir 2008;24:6837-6844.

[86] Li L, Teller S, Clifton RJ, Jia X, Kiick KL. Tunable mechanical stability and deformation response of a resilin-based elastomer. Biomacromolecules 2011;12:2302-2310.

[87] Kurniawan NA, Wong LH, Rajagopalan R. Early stiffening and softening of collagen: interplay of deformation mechanisms in biopolymer networks. Biomacromolecules 2012;13:691-698.

[88] Coutinho DF, Sant SV, Shin H, Oliveira JT, Gomes ME, Neves NM, et al. Modified Gellan Gum hydrogels with tunable physical and mechanical properties. Biomaterials 2010;31:7494-7502.

[89] Krekhova M, Lang T, Richter R, Schmalz H. Thermoreversible hydroferrogels with tunable mechanical properties utilizing block copolymer mesophases as template. Langmuir 2010;26:19181-19190.

[90] Stahl PJ, Romano NH, Wirtz D, Yu SM. PEG-based hydrogels with collagen mimetic peptide-mediated and tunable physical cross-links. Biomacromolecules 2011;11:2336-2344.

[91] Zustiak SP, Leach JB. Hydrolytically degradable poly(ethylene glycol) hydrogel scaffolds with tunable degradation and mechanical properties. Biomacromolecules 2010;11:1348-1357.

[92] Jeon O, Bouhadir KH, Mansour JM, Alsberg E. Photocrosslinked alginate hydrogels with tunable biodegradation rates and mechanical properties. Biomaterials 2009;30:2724-2734.

[93] Mano JF. Stimuli-Responsive Polymeric Systems for Biomedical Applications. Advanced Engineering Materials 2008;10:515-527.

[94] Kim S, Chung EH, Gilbert M, Healy KE. Synthetic MMP-13 degradable ECMs based on poly(N-isopropylacrylamide-co-acrylic acid) semi-interpenetrating polymer networks. I. Degradation and cell migration. J Biomed Mater Res A 2005;75:73-88.

[95] Lu YB, Franze K, Seifert G, Steinhauser C, Kirchhoff F, Wolburg H, et al. Viscoelastic properties of individual glial cells and neurons in the CNS. Proc Natl Acad Sci U S A 2006;103:17759-17764.

[96] Yousaf MN. Model substrates for studies of cell mobility. Curr Opin Chem Biol 2009;13:697-704.

Section 2

Biocompatibility Studies

Chapter 5

Overview on Biocompatibilities of Implantable Biomaterials

Xiaohong Wang

Additional information is available at the end of the chapter

http://dx.doi.org/10.5772/53461

1. Introduction

A biomaterial is any material that comprises whole or part of a living structure or biomedical device which performs, augments, or replaces a natural function to improve the quality of life of the patients [1]. Over the past fifty years biomaterials has been developed as a science with various forms of implants/medical devices, and have been widely used to replace and/or restore the function of traumatized or degenerated tissues or organs. As a life-saving and life-improving option for countless patients, biomaterials have been paid more and more attention during the last decade. Only in the United States, more than 13 million implant/medical devices implanted annually. As a result, the impact factor of the journal of "Biomaterials" has boomed from 2.489 to 7.404 from the year 2001 to 2012.

The implant/medical device scope of biomaterials ranges from simple implants like intraocular lenses (which restore sight to millions of cataract patients every year), sutures, wound dressings, decellular matrices, bone plates, joint replacements to more complex materials like biosensors, catheters, pacemakers, blood vessels, artificial heart (that provide both mechanical and biological functions in a body), left ventricular assist devices and prosthetic arterial grafts. According to the resources and properties biomaterials can be assorted into autografts, allografts, organic polymers, such as natural collagen, fibrin, chitosan, hyaluronan, heparin, cellulose, and synthetic polyurethane (PU), polyester, metal, such as aluminium, steel, titanium, inorganic salts, such as calcium phosphate, hydroxyapatite, and their compounds or derivatives. There are more than one hundred different biomaterials which have been applied *in vivo*. All biomaterials when implanted into a body initiate a host response that reflects the first steps of tissue repair. The host/biomaterial interactions which follow implantation of any prosthesis or device are a series of complex events that have not been well defined. Generally, host reactions following implantation of biomaterials include injury, blood–material interactions, provisional matrix formation, acute inflammation, chronic inflammation, granulation tissue development, foreign body reaction, and fibrosis/fibrous capsule development [2]. There are numerous types of host responses to a broad spectrum of biomaterials.

When considering a biomaterial for implantation or medical use, the first and most important requirement is nontoxic, nonimmunogenic, chemically inert/active, and acceptable by the human body. Biocompatible in most cases means that the biomaterials must not form thrombi in the blood system, result in tumors in the surround tissues, or be immediately attacked, encapsulated, or rejected by the body [3]. According to the host responses to implantable biomaterials, there are many different kinds of biocompatibilities, including local tissue responses, such as necrosis, repulsion, infection, inflammation, calcification, scar, cyst, amalgamation, thrombus, tumor, cancer, and whole body responses, such as fever, toxicity, circulation impediment, nerve anesthesia, malformation, etc. The overall biocompatibilities including cyto-compatibility, hemo-compatibility, and tissue-compatibility, are often evaluated using histological sections, cell markers, and metabolite measurements. Sometimes, polymers with similar chemical characteristics behave differently in certain situations. For example, polyethylene and ultrahigh molecular weight polyethylene behave differently as orthopedic biomaterials for knee and hip replacement [4]. Until present, most of the implantable biomaterials trigger acute or/and chronic inflammatory responses in the body. These reactions can totally black a biomaterial and even lead to huge disasters or personal misfortunes. Among the numerous types of host responses, early interactions between implants and inflammatory cells are probably mediated by a layer of host proteins on the biomaterial surface. Franz and coworkers have described several typical host responses of implantable biomaterials (Figure 1) [5]. This model can be used as a reference for evaluation of an implantable biomaterial when it is implanted shortly *in vivo*.

In this chapter, I will focus on the *in vivo* host responses about twenty common used biomaterials which cover nearly every tissue and organ in human body. Advanced biologic techniques have been employed in determining the mechanisms behind observed macroscopic or microscopic responses. An understanding of the molecular and cellular events which occur in response to implantable biomaterials may allow us to manipulate responses and design more biocompatible, bioactive and functional biomaterials for clinical applications, such as regenerative medicine and controlled releasing drugs.

2. Allografts

Allograft (also called homograft) is a tissue/organ graft from one individual to another of the same species with a different genotype [6]. It has been successfully used in various medical procedures for more than 150 years. Approximately 1500000 allografts are transplanted each year for a variety of life-saving and life-enhancing surgeries. For example, skeletal grafts for patients with bone defects from cancer or traffic accidents; cornea transplants to help restore sight; heart valves to replace damaged heart tissues; skin grafts to save the lives of burn victims, and tendon replacements to help people with more active lives [7].

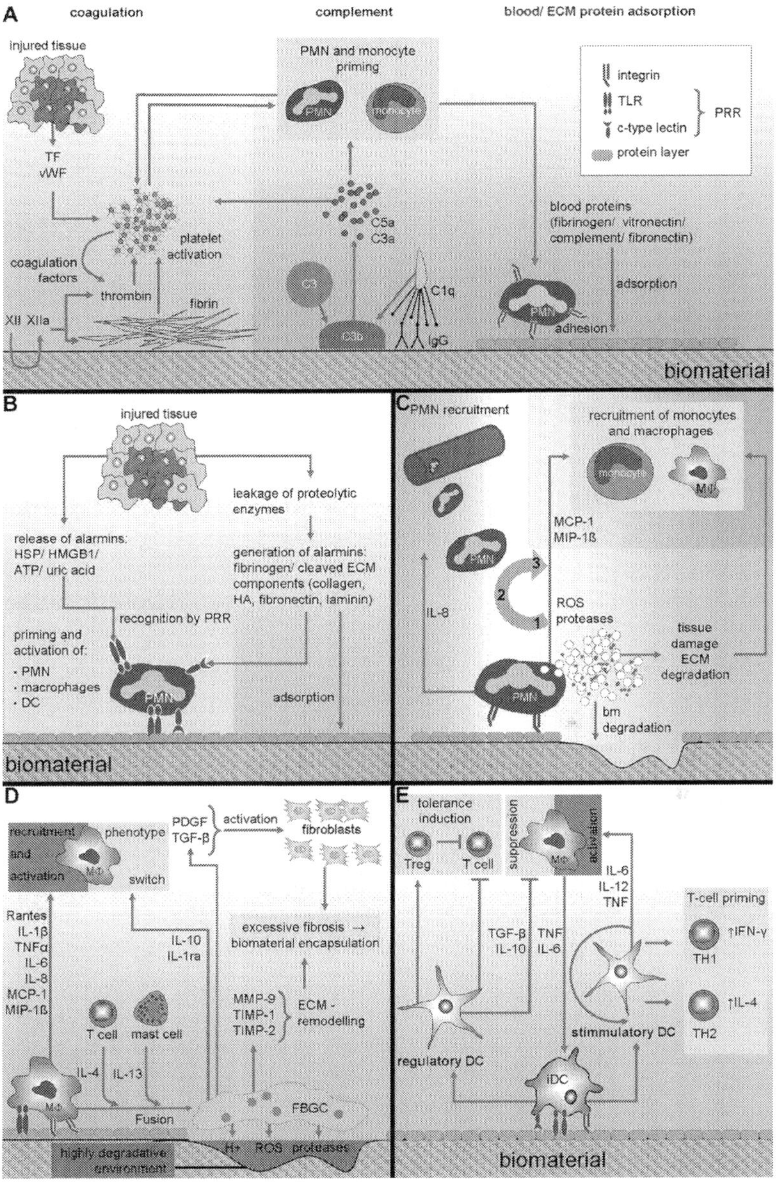

Figure 1. Immune response toward biomaterials. (A) Adsorption of blood proteins and activation of the coagulation cascade, complement and platelets result in the priming and activation of polymorphonuclear leukocytes (PMNs), monocytes and resident macrophages. (B) Danger signals (alarmins) released from damaged tissue additionally prime the immune

cells for enhanced function via pattern recognition receptor (PRR) engagement. (C) The acute inflammatory response is dominated by the action of PMNs. PMNs secrete proteolytic enzymes and reactive oxygen species (ROS), corroding the biomaterial surface. Interleukin (IL)-8 released from PMNs enhances PMN influx and priming. In the transition from acute to chronic inflammation, PMNs stop secreting IL-8 in favor of cytokines promoting immigration and activation of monocytes and macrophages. (D) Macrophages are the driving force of chronic inflammation. Constant release of inflammatory mediators like tumor-nekrose-faktor alpha (TNFα), IL-6, and monocyte chemotactic protein (MCP)-1 results in permanent activation of macrophages. Fusion-inducing stimuli like IL-4 and IL-13 promote the fusion of macrophages to foreign body giant cells (FBGCs,) which form a highly degradative environment on the biomaterial surface. Furthermore, FBGC promote extracellular matrix (ECM) remodeling and fibroblast activation resulting in excessive fibrosis and biomaterial encapsulation. (E) Macrophage-derived cytokines and pattern recognition receptor engagement activate dendritic cells (DCs) on the biomaterial surface. Depending on the nature of the stimulus, DCs mature to either immunogenic or tolerogenic subtypes, amplifying or suppressing the inflammatory response [5].

Compared with autografts which come from the same bodies and are only available in limited amounts, allografts are more readily available and accompany with less risk and postoperative morbidity. The healing times is therefore shorter and less painful for a patient with no second surgical site is required (as there is when an autograft is utilised). Currently, the use of allograft tissues is increasingly popular all over the world, with widespread orthopaedic surgeons and debilitating musculoskeletal conditions. Nearly one tissue/organ donor can save or improve the lives of up to 60 people. Especially, Musculoskeletal Transplant Foundation, the world's largest tissue bank, provides allograft tissue and biologic solutions for ligament reconstruction [8]. Meanwhile bone and soft tissue allografts from the SteriGraft™ line has been in existence for over 13 years and has helped doctors and their patients with over one hundred thousand successful transplantations. Before transplantation, a blood sample from the donor is normally tested in case any infected diseases, such as human immunodeficiency virus (HIV), Hepatitis, and Syphilis [9].

Specially, decellularized tissue/organ matrices derived from allografts have been used since the 1940s to support tissue repair and replacement. Their popularity has grown sharply during the last decade with the advent of tissue engineering [10]. At present, decellularized tissues/organs have been successfully used in a variety of tissue/organ regenerative medicines. The efficiency of cell removal from a tissue/organ is dependent on the origin of the tissue/organ and the specific physical, chemical, and enzymatic methods that are used. Each of these treatments affects the biochemical composition, tissue ultrastructure, and mechanical behavior of the remaining extracellular matrix (ECM) scaffold, which in turn, affect the host response to the material [11].

3. Collagen and gelatin

Collagen is one of the most prevalent proteins in the connective tissue of animals and constitutes approximately 25% of total body protein in vertebrates. It therefore is an important biomaterial in medical, dental, and pharmacological fields. After the immunogens in the collagen molecules are dislodged, collagen has excellent biocompatibilities either *in vitro* or *in vivo*. Collagen is capable of being cross-linked into solid or lattice-like gels. Resorbable forms of collagen have been used to dress oral, skin or some of the other soft tissue wounds, for closure of graft and extraction sites, and to promote healing [12]. During *in vivo* implantation, collagen irritates slight inflammation accompanying with some scar tissues.

A collagen sponge obtained from Beijing Yierkang Biengineering Development Center China was implanted subcutaneously in rats for time periods up to 8 weeks (Figure 2) [13]. One week after implantation, slight inflammation with some lymphocytes, myofibrils and fibroblasts were observed. The appearance of myofibrils and fibroblasts indicated that scar tissue was developed (Figure 2A). Two weeks after implantation, fibrous tissue was formed with scattered macrophage and lymphocyte cells in the fibrous layer. Newly formed blood vessels appeared in the implant site while the collagen sponges were completely resorbed (Figure 2B). Four weeks after implantation, the thin fiber layer had changed into wavelike scar tissue and tightly connected with the surrounding muscles. Capillaries were evident in the new fibrous scars (Figure 2C). Six weeks after implantation, scar tissue in the collagen samples was mature (Figure 2D). Eight weeks after implantation, the wave-like scar tissue in the collagen samples became thinner with some lipocytes and vacuoles (Figure 2E) [13].

Collagen compounds, such as collagen/chitosan, collagen/hyaluronan, have been investigated extensively during the past several decades. The biocompatibilities of these compounds depend largely on the incorporated constituents. For example, a corneal collagen cross-linked with riboflavin and ultraviolet radiation-A has been used for keratoconus repair of a 29-year-old woman with some good results [14]. In some instances, it is more competing to use a compound to improve the mechanical properties of the collagen based biomaterials. For example, a porous implantable dexamethasone-loaded polylactide-co-glycolide (PLGA) microspheres/collagen glucose sensors [15] and a mitomycin C (MMC) delivery system (MMC-film), incorporating polylactide (PLA)–MMC nanoparticles in a composite film from blends of collagen–chitosan–soybean phosphatidylcholine (SPC) with a mass ratio of 4:1:1 have been explored with no sign of internal infection and fibrous encapsulation in any animals after 20 days of implantation [16].

Gelatin is a mixture of peptides/proteins produced by partial hydrolysis of collagen extracted from the skin, boiled crushed bones, connective tissues, organs and some intestines of animals such as domesticated cattle, chicken, horses hooves, and pigs [17]. Gelatin possesses a better biocompatibility than its ancestry collagen. Alloimplants of bone matrix gelatin are effective in the treatment of bone defects with a low risk of complication such as rejection or infection [18]. Aqueous gelatin solution is an amorphous natural hydrogel in which cells can be encapsulated, extruded and deposited at desired positions. Unlike collagen hydrogel, gelatin hydrogel holds a special gelation property around 20℃. In Tsinghua University the author's own group, this property has been explored extensively for rapid prototyping (RP) (or additive manufacturing) of three-dimensional (3D) complex geometrical structures with computer-aided design channel models [19-24]. Until now, a hybrid hierarchical 3D construct consisting both synthetic polyurethane PU and natural cell/ gelatin-based hydrogel with interconnected macro-channels has been produced via a double nozzle RP technique at a low temperature (-28℃). These constructs have demonstrated excellent in vivo biocompatibilities [23,25]. This technique holds the potential to be widely used in the future complex tissue/organ manufacturing areas.

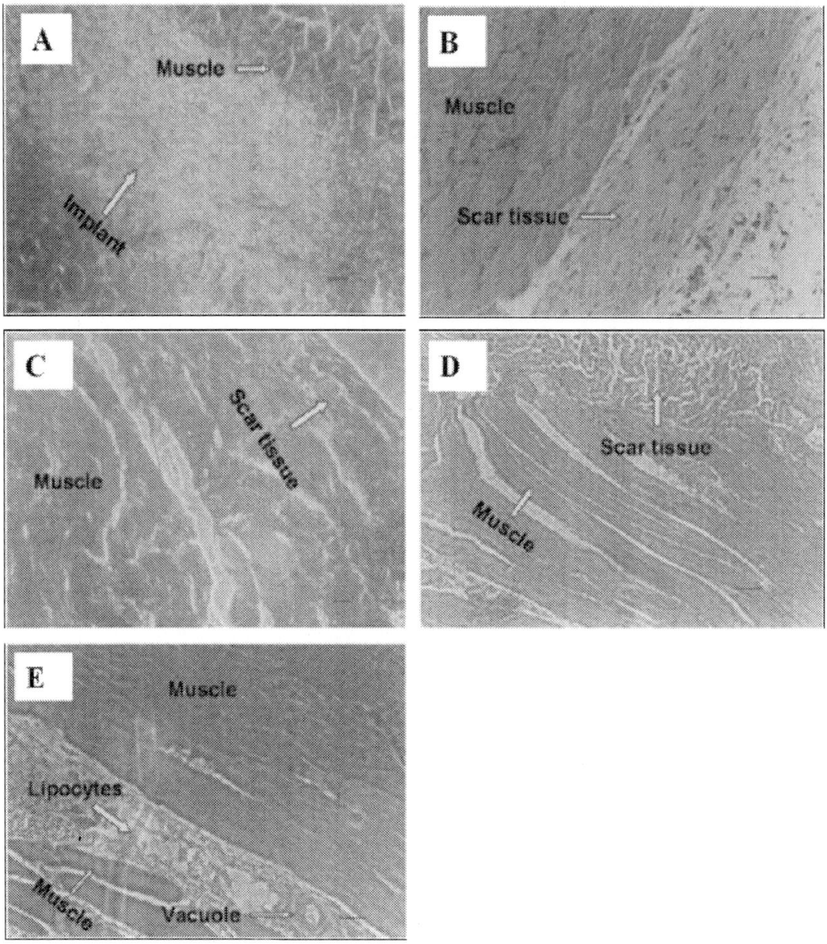

Figure 2. Light-microscope evaluation of the tissue response to collagen sponges with hamatoxylin-eosin (HE) staining: (A) 1 week after implantation; (B) 2 weeks after implantation; (C) 4 weeks after implantation; (D) 6 weeks after implantation; (E) 8 weeks after implantation. The scale bar indicates a distance of 50μm in (A), (C), and (D), and a 25μm in (B) and (E) [13].

Combination of gelatin microspheres/scaffolds with other biomaterials, such as collagen, alginate, chitosan, hyaluronan, and fibrin has also been explored extensively. For example, a gelatin microsphere containing basic fibroblast growth factor and preadipocytes, is essential to achieve a engineered fat tissue [26]. A PLGA microparticles containing an anticancer agent paclitaxel was formulated for the treatment of lung cancers [27]. Gelatin hydrogel incorporating hepatocyte growth factor induced angiogenic change around the implanted hydrogel [28]. A silk fibroin/gelatin composite scaffold was implanted into subcutaneous pockets on male Sprague-Dawley rats with a slight inflammation reaction. By day 30, the scaffold had been completely infiltrated and organized by fibroblasts and inflamed cells. The greater the gelatin concentration in the scaffold, the faster the degradation rate [29].

4. Fibrin

Fibrin (also called Factor Ia) is a fibrous, non-globular protein involved in the clotting of blood. It is formed from fibrinogen by the protease thrombin, and is then polymerised to form a hemostatic plug or clot (in conjunction with platelets) over a wound site [30]. The clot fibrin can be naturally degraded by proteolytic enzymes from the fibrinolytic system, such as plasmin [31,32]. *In vivo*, fibrin(ogen) plays an important role in hemostasis, inflammation, signal transduction, platelet activation, wound healing, osteoinductive and angiogenesis [33-36]. The food and drug administration (FDA) in American has approved commercially made fibrin sealants in 1998 [37].

During the last decade, autologous fibrin-based matrices have demonstrated great potential as being used as tissue engineered replacements, such as heart valves [38-40], cartilages [41], and blood vessels [42]. Immunohistochemistry and ECM assay demonstrated that the fibrin scaffolds can be completely absorbed *in vivo* in about 3 months with low granulomatous inflammation (Figure 3) [43-46]. Farhat and coworkers have evaluated whether a fibrin glue spray technique enhances cell seeded acellular matrix (ACM) repopulation in a porcine bladder model. The *in vivo* central fibrosis results indicated that while fibrin glue enhanced cellular organization on ACM *in vitro*, factors supporting seeded cell survival are lacking [47].

On the other hand, spatio-temporal controlled delivery of bioactive molecules within fibrin has been expanded rapidly. Various states of fibrin, such as scaffold, sheets, microparticles and fibrin-coated drug particles have been used as drug delivery systems [48,49]. Growth factors, such as vascular endothelial growth factor (VEGF) and transforming growth factor-β (TGF-β) can easily bind to the fibrin molecules and be controlled released subsequently by diffusion [50-56]. In the future, autologous fibrin may play an important role in customized clinical applications, such as anti-immune drug delivery systems and human tissue/organ constructions to avoid any negative host reactions [57].

5. Dextran and its derivatives

Dextran, a high-molecular-weight polymer of d-glucose, formed by sucrose enzymes on the cell surface of certain lactic acid bacteria in the mouth adhere to the tooth surfaces and produce dental plaque. Uniform molecular weight dextrans (named for their average molecular weight) from Leuconostoc mesenteroides with specific preparations has been used for over 50 years in plasma volume expansion, thrombosis prophylaxis, peripheral blood flow enhancement and for the rheological improvement of, for instance, artificial tears [30,58]. Dextrans with an average molecular weight of 1000 to 2 million g/mol are commercially available for research purposes [59]. Two preparations of dextran with lower fractions (40000 and 70000 g/mol) are suitable for nontoxic clinical use [60]. However, high fractions of dextrans can produce erythrocyte aggregation, impaired microcirculation, and a clinical picture akin to shock and certain other diseases.

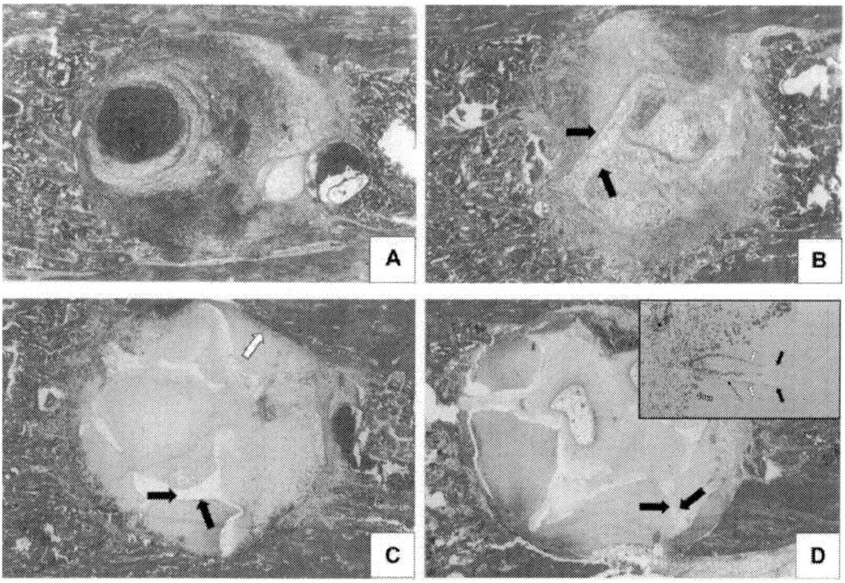

Figure 3. Eleven-day Masson's trichrome (MT) staining sections of a fibrin scaffold. (A) Untreated defects and (B) defects containing empty scaffolds were filled with new bone tissue. However, no reparative bone was observed in the center of defects containing (C) scaffolds filled with fibrin (low T) and (D) scaffolds filled with fibrin (high T). (Inset) Patches of multinucleated giant cells (striped arrow) were observed at the scaffold interface in all scaffold-containing groups. Black arrows point to areas occupied by the scaffold, whereas white arrows point to the advancing bone front. Field width 5.2 mm, inset field width 0.2 mm [46].

During 1990-1994, extensive toxicologic evaluations indicate that small-volume infusions of 7.5% NaCl/6% dextran 70 (HSD) at the proposed therapeutic dose of 4 mL/kg, present little risk as implantable biomaterials [61,62]. Dextran hydrogels have offered good opportunities as protein delivery systems or tissue engineering scaffolds because of an inherent biocompatibility [63]. The hydrophilic, soft and rubbery properties of the dextran hydrogels ensure minimal tissue irritation and a low tendency of cells and proteins to adhere to the hydrogel surface [59]. Althogh dextran itself is not toxic, some of the methods used for crosslinking the polymer may result in toxic byproducts. For example, the toxicity of dialdehyde cross-linked dextran/gelatin hydrogel can be detected in fibroblast and endothelial cell cultures. Subcutaneous implantation studies in mice showed that the foreign body reaction seen around the implanted hydrogel samples was moderate and became minimal upon increasing implantation time [64]. A methacrylate-derivatized dextran hydrogel also shows good *in vitro* biocompatibilities [65].

More recently another effect of dextran, namely that of antithrombogenesis, has been recognized [66]. Dextran sulfate, a dextran derivative, its effects on coagulation has already been proven [67]. It has been reported that dextran sulfate has been found to activate the polymerization of fibrin monomer, ATIII, conversion of prekallikrein to kallikreinand fibrinolysis. Kallikrein, the conversion of fibrinogen to fibrin appears to be inhibited by dextran sulfate. These effects are, *inter alia,* concentration dependent [67,68]. Meanwhile, a dextran sulphate sodium model of colitis has demonstrated several correlations of this biomaterial with human inflammatory bowel disease [69]. Furthermore, a lauric acid modified dextran-agmatine bioconjugate (Dex-L-Agm) was prepared by 1,1'-carbonyldiimidazole activation and the nucleophilic reaction between tosyl of tosylated dextran and primary amine of agmatine was found to be highly cytocompatible without causing hemolysis and red blood cell aggregation [70].

6. Hyaluronan

Hyaluronan (also called hyaluronic acid or hyaluronate, HA) is a natural anionic, viscoelastic and hygroscopic glycosaminoglycan, discovered in 1934, by Karl Meyer and his assistant, John Palmer in the vitreous of bovine eyes [71]. As one of the chief components of the ECM, hyaluronan distributes widely throughout connective, epithelial, and neural tissues. It is unique among glycosaminoglycans in that it is nonsulfated, forms in the plasma membrane instead of the Golgi, and can be very large in molecular weight (often reaching the millions) [72]. HA plays several important organizational roles in the ECM by binding with cells and other protein components through specific and nonspecific interactions [73] and is responsible for various functions within the ECM such as cell growth, proliferation, differentiation, migration [74], and even some malignant tumors [76].

Basically hyaluronan is a highly non-toxic, non-antigenic and non-immunogenic polysaccharide, owing to its high structural homology across species, and poor interaction with blood components [77,78]. The FDA in American has approved the use of hyaluronic acid for certain eye surgeries, such as cataract removal, corneal transplantation, and detached retina [79]. People take hyaluronic acid for various joint disorders (lubricant agents), lip fillers, "youth fountains", and even wound healing catalysts [80]. Nowadays various hyaluronan hydrogels have been used to delivery drugs and cell growth factors [81,82]. There are some evidence show that fragmented hyaluronan stimulates the expression of inflammatory genes by a variety of immune cells at the injury site. With the protein-bonding abilities, hyaluronan fragments signal through both Toll-like receptor (TLR) 4 and TLR2 as well as CD44 to stimulate inflammatory genes in inflammatory cells. Hyaluronan presents on the epithelial cell surface can provide protection against tissue damage from the environment by interacting with TLR2 and TLR4 [83-85]. It is well known that accumulation and turnover of ECM components are the hallmarks of tissue injury. Current model of hyaluronic acid appear in the early stages of wound healing is to physically make room for white blood cells, which mediate the immune response and at least in part, reduce collagen deposition and therefore lead to reduced scarring [86]. This hypothesis is in agreement with the research of West and coworkers, who have showed that in adult and late gestation fetal wound healing process, removal of HA results in fibrotic scarring [87].

HA can be modified through several different ways, such as chemically esterify its carboxylic groups with some types of alcohol. The physico-chemical properties of the new biopolymers allow the preparation of many biomaterials with different biocompatibilities for various medical applications [88]. Shen and coworkers implanted hyaluronan hydrogel and periodate oxidated hyaluronan hydrogel in ischemic myocardium and found rapid degradation rates, low quantity of inflammation-mediating cells, thin fibrous capsules with dense blood vessels around the hydrogels at week 2 [89]. Praveen and coworkers used HA/polyvinyl alcohol (PVA) coating membrane to minimize the problems related to protein deposition and fibrous tissue formation on an implanted glucose sensor [90]. HA hydrogels modified with laminin could support cell infiltration, angiogenesis, and simultaneously inhibit the formation of glial scar after being implanted into the lesion of the cortex [91]. Compared with pure gelatin hydrogen, HA/gelatin composite has a better compatibility and contiguity with the surrounding brain tissue with no inflammatory reaction and fibrous encapsulation [92]. Intravitreal implants of hyaluronic acid esters represent useful biocompatible and biodegradable properties for a potential drug delivery system in the treatment of posterior segment ocular diseases [93]. A cross-linked HA hydrogel that contained a covalently bound derivative of the anti-proliferative drug MMC was synthesized and evaluated *in vitro* and *in vivo*. This hydrogel has strong potential as anti-fibrotic barriers for the prevention of post-surgical adhesions [94]. Two injectable thiolated HA derivatives were coupled to four alpha, beta-unsaturated ester and amide derivatives of poly(ethylene glycol) (PEG) 3400 and were found that the encapsulated cells can retained their original fibroblast phenotype and secreted ECM in vivo [95]. A fibrin/HA composite gel with autologous chondrocytes has been synthesized for tracheal reconstruction. Histologically, the grafts showed no signs of inflammatory reaction and were covered with ciliated epithelium [96].

7. Heparin

Heparin (from Ancient Greek ηπαϱ (hepar), liver), a highly sulfated glycosaminoglycan, is widely used as an injectable anticoagulant, and has the highest negative charge density of any known biological molecule [97]. Heparins are involved in different pathways of the coagulation cascade with anticoagulant, antithrombotic, profibrinolytic, anti-aggregative, as well as anti-inflammatory effects [98]. As stated in the fibrin section, the primary anticoagulant effect of heparin is through the suppression of thrombin-dependent amplification of the coagulation cascade, and inhibition of thrombin-mediated conversion of fibrinogen to fibrin [99].

Heparin holds the ability to relieve pain, inhibit clotting and inflammation, restore blood flow, enhance healing, and can be a useful addition to a range of available treatments for burn wounds [100]. Unfractioned heparin exhibits a broad spectrum of immunomodulating and anti-inflammatory properties, by inhibiting the recruitment of neutrophils and reducing pro-inflammatory cytokines in the treatment of inflammatory bowel disease [101]. Low-molecular-weight heparin can reduce or prevent development of signs/symptoms associated with post-thrombotic syndrome [102]. Heparin has been widely used to form an inner anti-coagulant surface on various experimental and medical devices such as membranes [103,104], tubes and renal dialysis machines [105,106].

Although heparin is used principally in medicine for anticoagulation, its true physiological role in the body remains unclear. Blood anti-coagulation is usually achieved by heparan sulfate proteoglycans which derive from endothelial cells stored within the secretory granules of mast cells and only released into the vasculature at sites of tissue injury [107]. Rather than anticoagulation, the main role of heparin may be defense at such sites against invading bacteria and other foreign materials [108]. A thiol-modified heparin in the Extracel-HP® mimics heparan sulfate proteoglycans also normally presents in the ECM and regulates the *in vivo* growth factor release for a functional microvessel network development [109]. A well-known adverse effect of heparin therapy is thrombocytopenia, a serious, immune system–mediated complication with significant mortality (Figure 4) [110-112].

8. Alginate

Alginate, is a salt of alginic acid (medical-dictionary.thefreedictionary.com), and an anionic polysaccharide distributed widely in the cell walls of brown algae, where it, through binding water, forms a viscous gum (Wikipedia, the free encyclopedia). Sodium alginate (composed of mannuronic and guluronic (G) dimmers) is a biocompatible and biodegradable polymer, and has been widely used in cell encapsulation technology, though the biocompatibility of the alginates in relation to their composition is a matter of debate [113]. In the molecules of sodium alginate the primary block guluronic acid contains available carboxylic acid groups that allow the alginate to be reversibly crosslinked by divalent cations, such as Ca^{+2} and Mg^{+2}, to form a relatively stable hydrogel [114,115]. Clinically, water-soluble alginates are useful as materials for dental impressions. Calcium alginates have been widely used as a base material to encapsulate glucose-sensing pancreatic islets that secrete insulin into the lymphatic system to reverse the effects of insulin-dependent diabetics [116]. Some investigators have utilized alginates to promote the viability of encapsulated cells [117]. Alginate-poly-L-lysine-alginate (APA) microcapsules continue to be the most widely studied device for the immuno-protection of transplanted therapeutic cells [118]. Alginate-chitosan-alginate (ACA) microcapsules have been developed as a device for the transplantation of living cells with protein adsorption onto the surface of microcapsules immediately upon implantation [119].

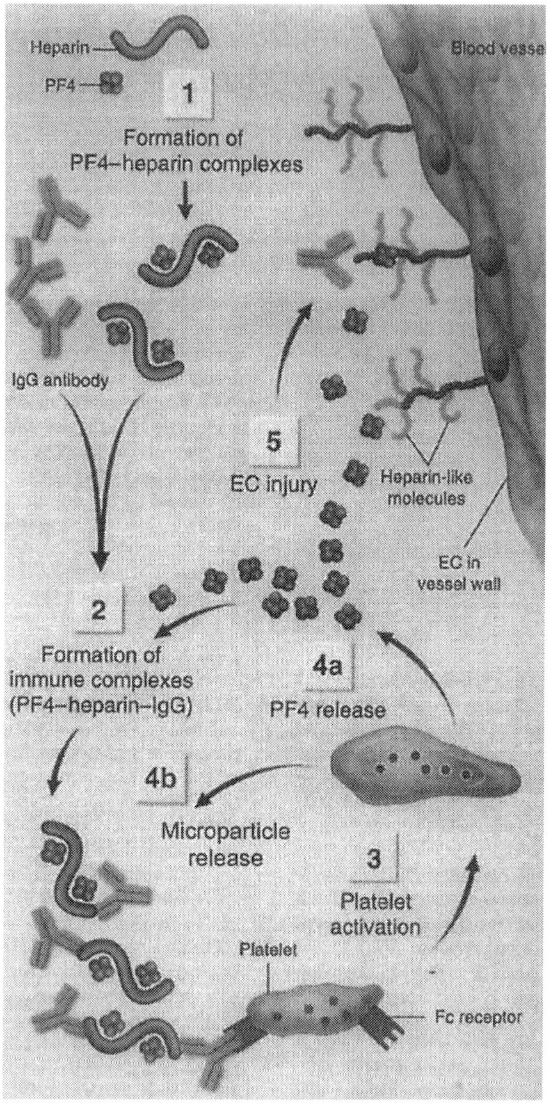

Figure 4. Model of pathogenesis of heparin-induced thrombocytopenia (HIT). Heparin binds with Platelet factor 4 (PF4), which exposes neoepitopes on PF4 and leads to antibody production (1). Heparin-PF4-IgG immune complexes form (2), and IgG in multimolecular complex triggers platelet activation via binding to Fc receptors (3). Activated platelet releases additional PF4 (4a) and prothrombotic platelet microparticles (4b). Thrombotic risk is further promoted by binding of PF4 to heparin-like molecules on endothelial cells (EC), contributing to immune system–mediated endothelial damage (5) [112].

http://dx.doi.org/10.5772/53461

9. Chitin, chitosan and their derivatives

Chitosan is a naturally occurring linear polysaccharide, consisting of glucosamine and N-acetyl-glucosamine, normally made of deacetylated chitin which is the structural polymer found in the shells of crabs and shrimp (lobster, squid, some yeast and mould), by N-deacetylation using strong alkali [120]. More than 40 years have lapsed since this biomaterial had aroused the interest of the scientific community around the world for its potential biomedical applications [121]. Until now chitosan possess a number of commercial and biomedical applications in wound dressing, drug delivery and tissue engineering. For example, chitosan based scaffold biomaterials have demonstrated versatile properties to promote the epithelial and soft tissue regeneration in the body [122,123]. Chitosan patches in various sizes that have been cleared by the FDA are a topical hemostat for moderating severe bleeding. Nevertheless, an obvious disadvantage of this implantable, absorbable biomaterial is that chitosan initiates serious host inflammation reactions (Figure 5) [13,124]. Additionally, chitosan is bioadhesive and has the ability to transiently open tight junctions in the nasal epithelia, thereby permitting drugs to diffuse through this barrier. Advantages of this nasal route of administration include: a higher permeability of the nasal mucosa than in the gastrointestinal tract; a low degree of pre-systemic metabolism; and a high level of patient compliance, compared to injectable systems [125].

It is very interesting that when the number of N-acetyl-glucosamine units in a chitin/chitosan mixture is higher than 50%, the biopolymer is termed chitin. 50% deacetylated chitosan has a less inflammation reaction than the others when they are implanted *in vivo* [126]. Cross-linking of chitosan membrane using genipin and some other chemical agents can increase the membrane's ultimate tensile strength but significantly reduced its strain-at-fracture and swelling ratio [127]. In the author's own group, an ammonia treated chitosan sponge was implanted subcutaneously in rats for 8 weeks (Figure 5). One week after implantation, the chitosan sponges were entirely retained and wrapped with a layer of purulent cells. The purulent cells had infiltrated the outside chitosan sponges (Figure 5A). Two weeks after implantation, the encapsulated purulent layer was enlarged at the periphery of chitosan sponges. More acute inflammatory cells had infiltrated the chitosan sponges and there was no sign of biodegradation of the chitosan sponges (Figure 5B). Four weeks after implantation, the chitosan sponges still maintained their porous structure. A much thicker purulent layer and more acute inflammatory cells were found around or in the chitosan sponges (Figure 5C). Six weeks after implantation, most of the chitosan still maintained their scaffold integrity with numerous interspersed purulent cells. Some purulent cells even formed large channels throughout the chitosan sponges (Figure 5D). Eight weeks after implantation, purulent cell infiltrations had further increased in the chitosan sponges. Some collapsed matrix structures were detected at the outer margins of the implants and more channel structures were found between the remnants of chitosan lamellae (Figure 5E).

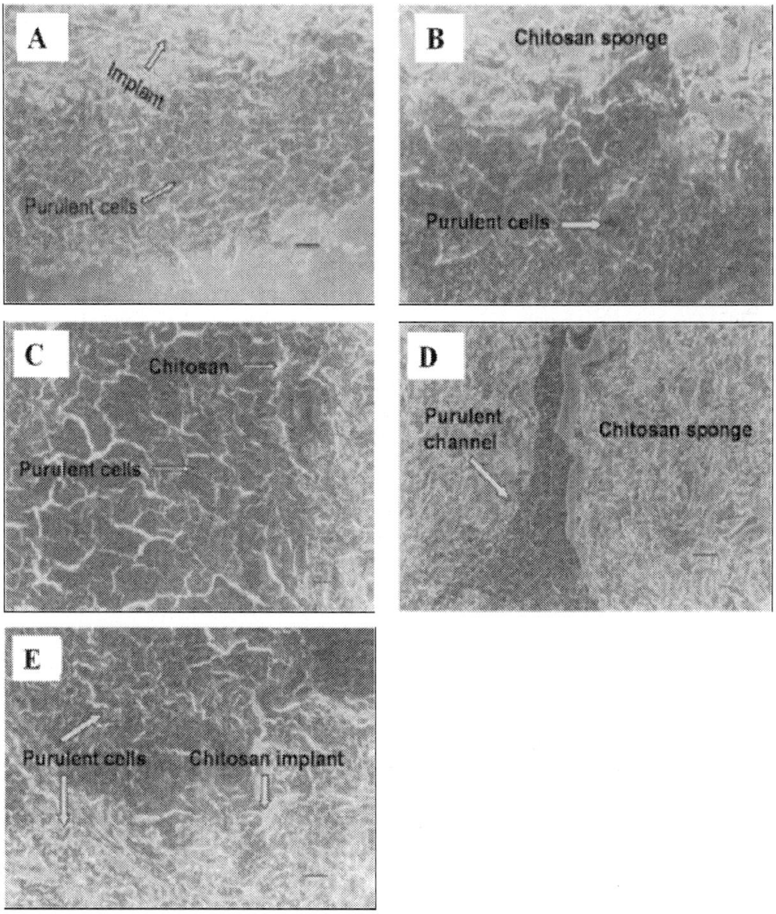

Figure 5. Light-microscope evaluation of the tissue response to chitosan sponges with HE staining: (A) 1 week after implantation; (B) 2 weeks after implantation; (C) 4 weeks after implantation; (D) 6 weeks after implantation; (E) 8 weeks after implantation. The scale bar indicates a distance of 50 μm in (A), (C), and (D), and a 25 μm in (B) and (E) [13].

Also in this author's own group, a series of bone repair materials were fabricated by adding three chitosan derivatives, such as phosphorylated chitin (P-chitin), phosphorylated chitosan (P-chitosan), and disodium (1→4)-2-deoxy-2-sulfoamino-β-D-glucopyranuronan (S-chitosan) into two kinds of biodegradable calcium phosphate cements (CPCs). All the chitosan derivatives can greatly improve the mechanical properties and reduce the biodegradation rates of the CPCs. At least six totally different tissue responses were detected when the implants were examined in tibial and radial defects of rabbits. Large bone defect (9 mm in length for radii and 3 mm in depth and diameter for tibias) repair in rabbits with the P-chitosan incorporated CPCs exhibits excellent tissue compatibilities with no any adverse or negative effects, such as fibrous encapsulation, osteolysis, hyperplasia, and inflammation, no matter the concentrations of P-chitosan is high or low (Figure 6) [128,129]. Tissue responses to P-chitin are highly sensitive (Figure 7) [130,131]. Three different bone formation types in the resorption lacuna of the P-chitin incorporated CPCs due to the P-chitin concentrations were found during the 22 weeks implantation. The first is that with low P-chitin content trabeculae formed directly from the implant (Figure 7A). The second is that with middle P-chitin content cartilages formed from the outside of fibers before they turned into trabeculae (Figure 7B, 7C). The third is that with high P-chitin content callous formed from the outside of fibers before they turned into trabeculae (Figure 7D, 7E). P-chitin content has a negative relationship with the biodegradation rate of the cements. However, the degradation rates are compatible with the ingrowth of trabeculae. A mild foreign-body reaction in the high P-chitin content sample during the first three time spans did not impair its placement by a newly formed bone. The generally properties of these biomaterials have met the main requirements for bone repair (Figure 7) [130,131]. Different from the above mentioned bone repair types, tissue responses to water-soluble S-chitosan, prepared from chitin by successive N-deacetylation, specific carboxylation at C-6 and sulfonation, was rather obtuse. No inflammation or other negative response was found in the S-chitosan containing samples (S-CPCs). After 4 weeks implantation, newly formed trabeculae contacted with the implant directly in the lower S-chitosan sample, while a thin layer of fibers formed between the newly formed bone and the implant in the higher S-chitosan samples [132,133]. These results indicate that the concentrations and functional groups in a linear polysaccharide play a key role in determining the ultimate biocompatibilities of an implantable biomaterial. In addition, as a derivative of chitin, chitosan initiates blood coagulation while S-chitosan inhibits blood coagulation when they are used as hemo-contact biomaterials.

Recently, chitosan and its derivatives have been widely used in skin wound, burn and disease treatments. For instance, a chitosan-gelatin-hyaluronic acid scaffold was found flexible with good mechanical properties when it was used as artificial skin substitutes [134]. A bacterial cellulose synthesized by Acetobacter xylinum and modified by chitosan was found to be optimal in providing wound dresses with a moist environment for wound healing [135]. When an artificial chitosan skin regenerating template was implanted subcutaneously it showed a similar inflammatory pattern as Integra, a two-layer skin regeneration system, constructed of a matrix of crosslinked fibers [136,137].

With the combination with other natural polymers, such as collagen, gelatin, hyaluronan, fibrin, the strong host inflammation reactions of chitosan can be reduced to a certain degree. It was found that a bioactive glass-chitosan composite containing 17% (wt%) chitosan produced by a freeze-drying process and implanted in the femoral condyl of an ovariectomised rat can promote a highly significant bioactive and osteoinductive property [138-140]. The ultimate biocompatibility of a chitosan compound depends largely on the ratio of the different components. Host tissues, such as smooth muscle and hepatic tissue have a similar response to the chitosan containing collagen/chitosan mixtures [141]. A collagen/chitosan matrix crosslinked by agent 1-ethyl-3-(3-dimethylaminopropyl)-carbodiimide in a N-hydroxysuccinimide and 2-morpholinoethane sulfonic acid buffer system has exhibited improved blood and cell compatibilities than the pure chitosan samples [142,143].

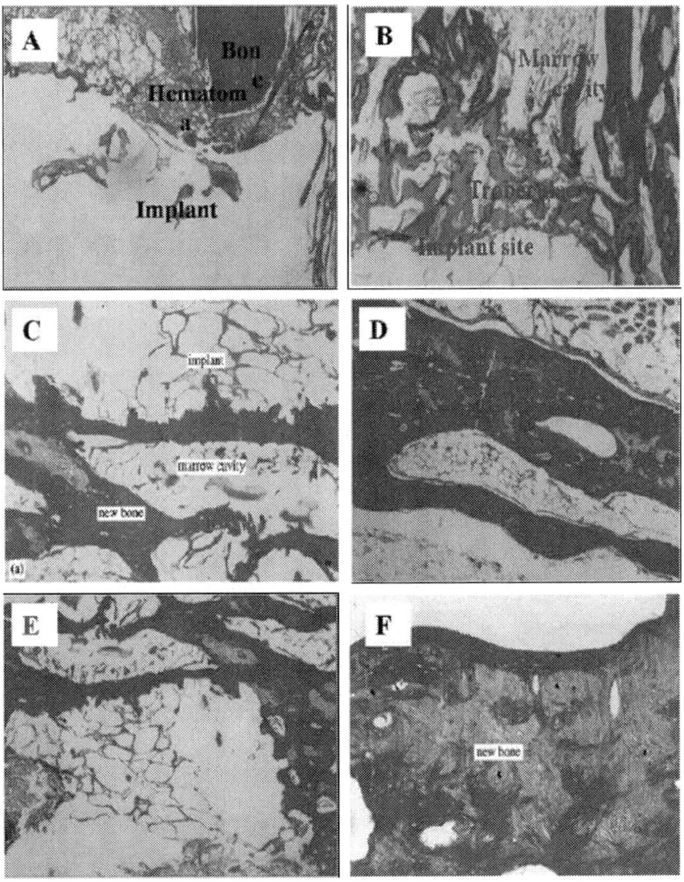

Figure 6. Tissue responses to the P-chitosan incorporated CPC specimen at different time points with MT staining. (A) 1 week after implantation in the high P-chitosan content (0.12 g/mL) sample with very little hematoma. (B) 4 weeks after implantation in the high P-chitosan content (0.12 g/mL) sample newly formed woven bone clearly appeared with tightly bonding between the implant and host bone. No macrophage was found around the implant. The implant was directly changed into new trabeculae after degradation. (C) 12 weeks after implantation newly formed long bone in the low P-chitosan content (0.02 g/mL) sample. (D) 12 weeks after implantation newly formed long bone in the middle P-chitosan content (0.07 g/mL) sample. (E) 12 weeks after implantation newly formed long bone in the high P-chitosan content (0.12 g/mL) sample. Trabeculae formed after the implant was gobbled up (infiltrated) by body fluid. Clear evidence of remodeling around the implant surface was displayed. (F) 22 weeks after implantation the newly formed dense trabeculae in the high P-chitosan content (0.12 g/mL) sample [129].

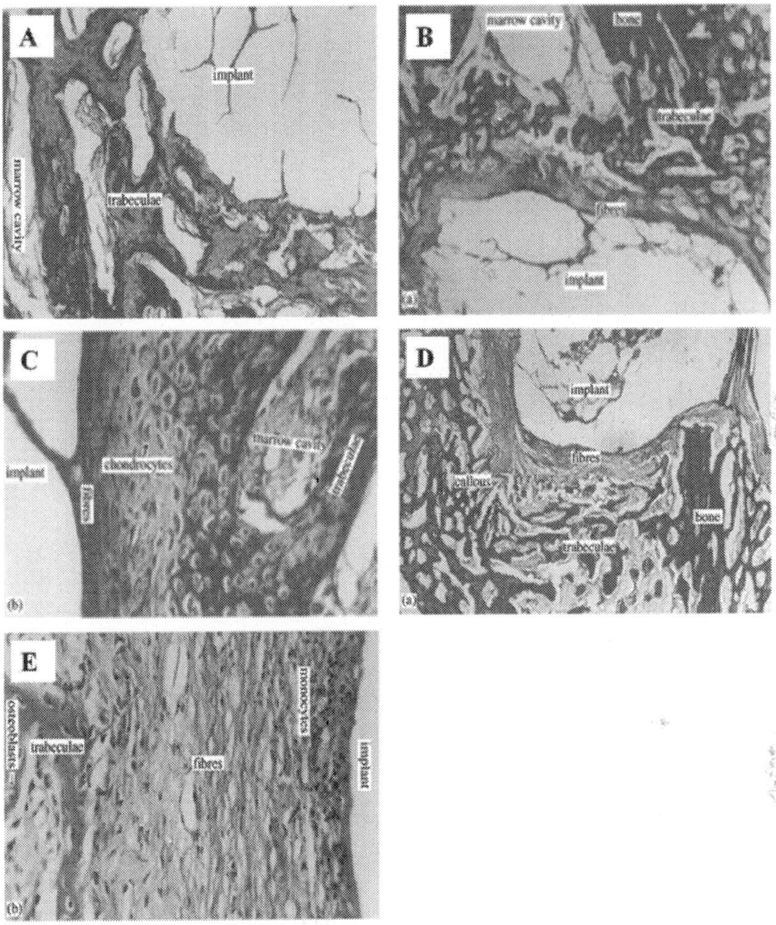

Figure 7. Tissue responses to the P-chitin incorporated CPC specimen 4 weeks after implantation. (A) P-chitin: 0.02 g/mL with MT staining. Magnification ×100. (B) P-chitin: 0.08 g/mL with MT staining. Magnification ×40. (C) A magnification of (B) with MT staining. Magnification ×400. (D) P-chitin: 0.14 g/mL with MT staining. Magnification ×40. (E) A magnification of (D) with HE staining. Magnification ×400 [131].

Current advances in some drug delivery systems make it possible to improve the therapeutic efficacy and minimized the side effects associated with toxicity of the drug. Chitosan has shown promise in the development of non-parenteral delivery systems for challenging drugs. For example, a 5-Fluorouracil (5-FU) loaded scaffold composed of chitosan fibers were prepared by a modified wet spinning technique [144]. Thermosensitive hydrogel composed of chitosan and glycerophosphate is proposed to be the potential candidate of *in situ* gel-forming implant for long-term drug delivery [145]. However, unpredictable body responses to the chitosan systems as stated above can complicate their applications to some degree. The composite chitosan-collagen-soybean phosphatidylcholine film impregnated with MMC-PLA-nanoparticles for treatment of hepatocellular carcinoma in mice has exhibited some special characteristics compared with pure chitosan delivery systems. In vivo, the growth of the tumors were inhibited considerably and dose-dependently by the MMC-film ($P<0.05$) with no any signs of vice reactions, such as inflammation, infection, and fibrous encapsulation after 20d of implantation [16,146,147]. Thus a careful balance between the immune reaction and drug effectiveness is needed when a chitosan pertaining template is used for biomedical applications.

10. Polyglycolide (PGA), Polylactide (PLA) and poly(Lactic-co-Glycolic Acid) (PLGA)

Polyglycolide also named polyglycolic acid (PGA) is a biodegradable, thermoplastic polymer and the simplest linear, aliphatic polyester which contains the ester functional group in it's main chain [148]. It can be prepared starting from glycolic acid by means of polycondensation or ring-opening polymerization. PGA has been known since 1954 as a tough fiber-forming polymer. Owing to its hydrolytic instability, its use has initially been limited [149]. *In vivo*, PGA initiates a marked host reaction around the implantations. This leads to the development of a foreign body response that comprises an initial acute inflammatory phase and a subsequent chronic inflammatory phase. For example, when a synthetic PGA scaffold seeding with adult-derived or somatic lung progenitor cells from mammalian lung tissue was implanted in an immunocompetent host, a serious foreign body response totally altered the integrity of the developing lung tissue [150].

Polylactic acid or polylactide (PLA) is another thermoplastic aliphatic polyester derived from renewable resources, such as corn starch, tapioca products, and sugarcanes [30]. A poly(L-lactide) (PLLA) coil stent has ever been implanted in pigs with no stent thrombosis and late restenosis [151]. However, PLA, as well as PLLA, and poly(D,L-lactide) (PDLA), induces a strong inflammatory response when they are implanted in the body due to their acidic products [152]. Aframian and coworkers implanted tubular PLLA, PGA coated with PLLA (PGA/PLLA), or nothing (sham-operated controls) in Balb/c mice either beneath the skin on the back, and found that inflammatory reactions were shorter and without epithelioid and giant cells in the sham-operated controls. Tissue responses to PLLA and PGA/PLLA scaffolds are generally similar in areas subjacent to skin in the back and oral cavity. Biodegradation proceeded more slowly with the PLLA tubules than with the PGA/PLLA tubules. No significant changes in clinical chemistry and hematology were seen due to the implantation of tubular scaffolds. [153]. It was reported that, after the PLLA segments were swallowed *in vivo* by phagocytes, cell damage and cell death were obvious. The highest numbers of necrotic cells were observed on day 2 [154]. These reactions can result in an unexpected risk for patients and have strongly limited in clinical applications of this kind of biomaterials.

To date, numerous strategies have been investigated to overcome body reactions induced by this kind of biomedical devices [155]. As a result, most of the PLA, PLLA, and PDLA have been used as a composite or compound with some other biomaterials. For example, a PLLA and poly(ethylene oxide) (PEO) blend has been prepared by mechanical mixture and fusion of homopolymers [156]. A biodegradable star-shaped 8 arms PEG-b-PLLA block copolymer was synthesized by Nagahama and coworkers to create a novel implantable soft material with drastically lowered crystallinity, increased swelling ability, and desirable mechanical properties [157,158].

Currently PGA, PLA and their copolymers, such as poly(lactide-co-caprolactone) (PLCA), poly(glycolide-co-caprolactone) (PGC), and poly (glycolide-co-trimethylene carbonate) are widely used as biomaterials for the synthesis of absorbable sutures and tissue engineering scaffolds in the biomedical field [159,160]. For example, a resorbable PLGA bone fixation implanted in craniofacial patients in 1996 resulted in 0.2 percent significant infectious complications, 0.3 percent device instability, and 0.7 percent self-limiting local foreign-body reactions [161]. As long-term implants, the toxicity of the accumulated acidate products made the situations even worse [162]. Until the present, most of the implanted PGA, PLA and PLGA related biomaterials still encounter an immune tissue response due to tissue trauma during implantation and the presence of foreign body reactions [163]. Surface coating has become one of the research hot points for the implantable devices with poor biocompatibilities. For instance, the biocompatibilities of some artificial polymer devices, such as heart valves, stents and vascular prosthesis that come into contact with bodily tissues or fluids particularly blood, have been improved by Venkatraman and coworkers with endothelialization surface layers [164,165].

Similarly, when a polyvinyl acetate (PVA)/PLGA microsphere was implanted into the subcutaneous tissue of rats, acute inflammation with neutrophils was found at day 3. Chronic inflammation with multinucleate giant cells, fibrosis, and mixed inflammatory cells was found at day 30. Mineralization around the implant was found at day 60 [166]. On the contrary, a dexamethasone/PLGA microsphere system can suppress the inflammation reaction by a fast releasing of dexamethasone [167]. A highly monodisperse and smooth PLGA-paclitaxel microspheres against malignant brain tumors were fabricated using an electrohydrodynamic atomization (EHDA) process [168]. In addition, PLA, PGA and PLGA can be tailored to meet mechanical performance and resorption rates required for applications ranging from non-structural drug delivery applications, nanoparticles (nanofibers), to resorbable screws and anchors [169,170].

11. Polycaprolactone (PCL)

Polycaprolactone (PCL) is a biodegradable polyester with a low melting point of around 60°C and a glass transition temperature of about −60°C. It is commonly used as an additive for resins or starch to improve their processing characteristics, lower their costs, and change their properties (e.g. impact resistance), or as a plasticizer in the manufacture of special polymers (e.g. Pus) [30]. PCL has been approved by the FDA for specific applications, such as a drug delivery devices, sutures, or adhesion barriers. It has been widely used as a scaffold material for tissue engineering with mismatched mechanical properties and slow degradation rate [171,172]. In rats the *in vivo* degradation of PCL is about 3 years [173].

Various categories of drugs have been encapsulated in PCL, in microsphere, nanosphere or bulk states, for targeted drug delivery and for controlled drug release [174-176]. For example, a PCL scaffold modified by grafting nerve growth factor (NGF) and Tirofiban (TF) has been used as nerve conduits to promote the regeneration of sciatic nerves [177]. Low molecular weight PCL pieces can be ingested and digested ultimately by phagocyte and giant cell without any cumulate vice-products (Figure 8) [178-180].

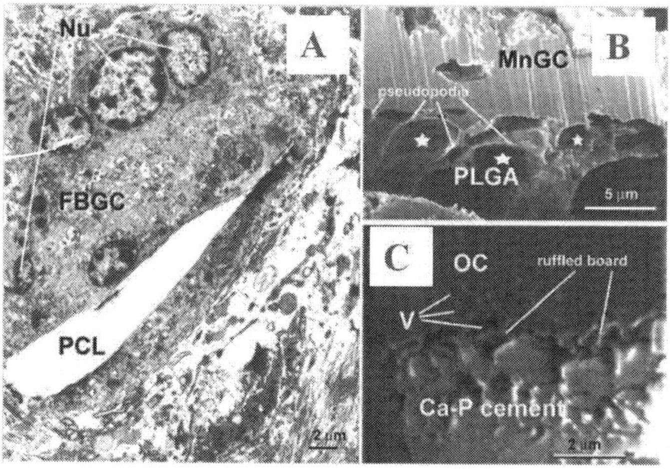

Figure 8. Micrographs illustrating extracellular degradation of biomaterials by macrophage fused multinuclear giant cells. (A) A foreign body giant cell (FBGC) engulfed a fragment of poly(epsilon-caprolactone), PCL polymer *in vivo*. Nu, nuclei of FBGC. The PCL polymer was dissolved during sample preparation. Transmission electron microscopy (TEM), bar = 2 µm. (B) *In situ* cross-section of the interface between a multinuclear giant cell (MnGC) and PLGA film. Note the pseudopodia of the MnGC penetrated deep inside the surface of PLGA film and formed sealed compartments. PLGA polymers are eroded within the compartments. Focused ion beam (FIB) microscopy, bar = 5 µm. (C) *In situ* cross-section of the interface between an osteoclasts-like cell (OC) and calcium phosphate cement. Note the typical ruffled board of OC and vesicles (V) secreting from OC to the sealed extracellular space. FIB microscopy, bar = 2 µm [162].

12. Polyurethane (PU)

PU is a series of biomaterials that contains urethane radical and offers the greatest versatility in compositions and properties of any family of polymers. Especially, a few specific elastomeric PU compositions have demonstrated a combination of toughness, durability, biocompatibility and biostability for being used as implantable medical devices, which is not achieved by any other available materials [181]. Because urethane is available in a very broad hardness range (e.g. eraser-soft to bowling-ball-hard), it allows the engineer to replace rubber, plastic and metal with the ultimate goals in abrasion resistance and physical properties. During the last half century, PUs have become and remained the most valuable implantable elastomers for uses requiring toughness, durability, biocompatibility and biostability [182]. With their inherently stable in the body environment, some of the PUs have been widely used in medical applications such as synthetic heart valves, vascular grafts, and pacemaker electrodes. However, these usages of PUs have been limited by three major complications: calcification, thrombosis, and chemical degradation [183].

In the 1970s and 1980s as the PUs became recognized as the blood contacting material and were used in a wide range of cardiovascular devices in long-term implants, they fell under scrutiny with the failure of pacemaker leads and breast implant coatings in the late 1980s. According to the manufacturer's report, high voltage coil fracture and PU defects were the predominant causes of lead failure [184,185]. During the next decade PUs had been extensively researched for their relative sensitivity to biodegradation and the desire to further understand the biological mechanisms for *in vivo* implantation [186,187]. Some investors have seeded autologous sheep blood outgrowth endothelial cells (BOECs) on a cholesterol (Chol)-modified PU (PU-Chol) heart valve leaflet to result in an intact, shear-resistant endothelium that would promote resistance to thrombosis [188]. Because of the complex behavior of implantable PUs in the body environment, special attention to the choice of the constituted components must be paid for designing and manufacturing the PU-containing devices. Subsequent treatment during sterilization, storage, implantation, *in vivo* operation and explantation also determine the performance and provide the means for assessing the efficacy of the PUs implants [189].

The most prominent disadvantages of PUs being used as artificial heart valves include mineralization, environmental stress-cracking and oxidation. While the mechanisms of these forms of degradation are not fully understood, an awareness of their causes and effects that leads to all of the long-term functionality is required for the sophisticated PU-based devices of today and tomorrow [190-191]. Over the last half century, extremely efforts have been paid in the biomedical research field to improve the biocompatibilities and biodurability of the PU implants, but only resulted in very little clinical effects [192-194].

In the later 1990s a number of new bioresorbable materials with all the versatility of PUs in terms of physical properties and biocompatibility have been yielded. AorTech Biomaterials was set up in 1997 to commercialise a range of medical grade PUs developed by the Australian research group (Commonwealth Scientific and Industrial Research Organization, CSIRO). The company estimates that the worldwide market for surgical heart valve products is worth more than $1bn (€705 m) and to be growing at a rate of 8% a year. Meanwhile, the market for catheter-delivered heart valves is worth around $200 m (€141 mm) [195]. In the authors' own group in Tsinghua University, China, a novel PU made of PCL, PEG, and 1,6-hexmethyldiisocyanate has been synthesized. The hydrolytic degradation property of the PU can be highly tuned by changing the composition and structure of copolymers, such as PEG and PCL. When this kind of PU was used as a small-caliber (1.2 mm inner diameter) vein and nerve repair grafts it demonstrated excellent antithrombogenicity and superior biocompatibility (Figure 9) [196,197].

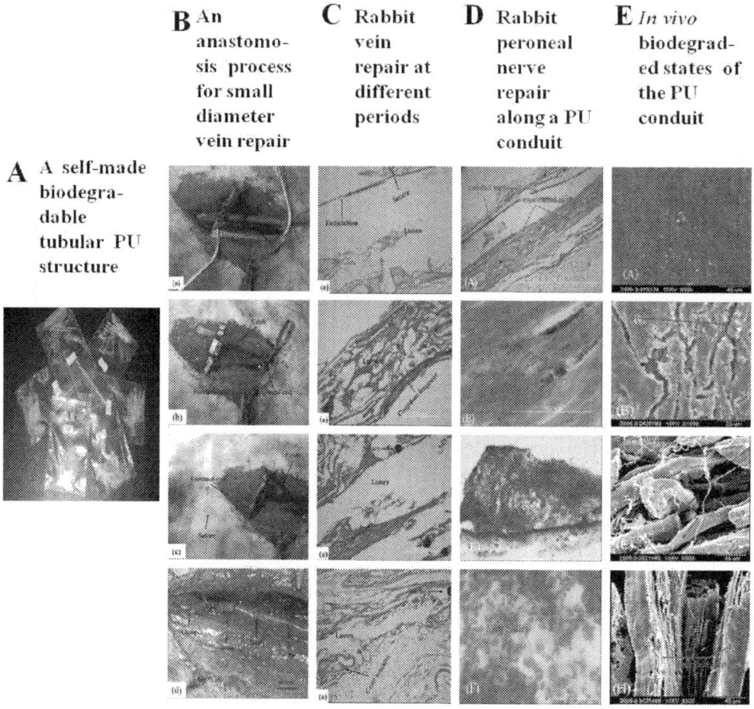

Figure 9. An implantable small-diameter nerve and blood vessel repair PU conduit. (A) PU conduits with different inner diameters. (B) The PU conduit was connected to the vein of a rabbit. (C) The vein defect repair processes with a very thin layer of fibrin-platelet deposition. (D) The nerve repair processes in rabbits with growing myelinated axons. (E) The PU conduits degraded gradually *in vivo* in 12 weeks [196, 197].

13. Polytetrafluoroethylene (PTFE)

Polytetrafluoroethylene (PTFE), Discovered in 1938 by Roy J. Plunkett, is a synthetic high-molecular-weight compound consisting wholly of carbon and fluorine with numerous applications [198]. The best known brand name of PTFE is Teflon made by DuPont Co. It is insoluble in all normally used organic solvents, not biodegradable *in vivo* and can suffer high temperatures as 260 ℃ permanently. Clinically, PTFE has been widely used as a large blood vessels repair materials.

A 5 year research using PTFE-Gore-Tex grafts mainly for superficial femoral occlusion has been conducted. The majority of the grafts were inserted in an elderly poor risk group of patients with critical ischaemia of the lower limb. The overall cumulative patency at 2 years was 29% falling to 18% at 5 years. Perioperative angiographic indicated that inflammatory reaction is the only risk factor significantly affecting the cumulative graft patency. The presence of diabetes was found to have a significant detrimental effect on limb salvage [199]. A permanently implantable left ventricular assist device, made of Dacron velour, Teflon felt, and Teflon-coated polyester fiber sutures, has been tested in chronic animal experiments. *In vivo* experiments demonstrated that all components elicited mild to moderate inflammatory reactions. Tissue responses to PTFE are rather passivated. Hematocele occurred only when the components were implanted in the aorta with direct blood contact and exposed to arterial blood pressures [200]. An 8 cm long PTFE prosthesis was implanted into defects of the abdominal aorta of dogs, and the following changes were found: the blood flow through the vascular prosthesis induced a shortening of the blood clotting time and a slight increase in the prothrombin consumption. It has a favourable effect of the sealing of pores in the prosthesis and covering its internal surface with a fibrin membrane [201].

14. Silicone

Silicon is a metal in the same column as carbon in the periodic table with the symbol Si and atomic number 14 [30]. It is the most abundant element on earth and does not occur naturally in its pure metallic state. Dimethylsiloxane is the building block for most medical-grade silicone products, including breast implants. This FDA Grade Silicone sheeting is commonly used in applications where food or consumables are present. For more than 20 years silicone miami breast implants have gone through a lot of changes since their first uses. After the mid-1980s many reports concerns the rupture rate of the thinner-shell products, the risk of subsequent breast cancer, and the connective-tissue diseases or symptoms in women with silicone gel-filled breast implants appeared. In the United States a moratorium (in place since 1992) on the use of these prostheses has been maintained by the pressure of overwhelming litigation. At the same time, Australian authorities also restricted the availability of silicone breast implants. Huge damages awarded by United States courts forced Dow Corning, manufacturer of a large percentage of breast prostheses, to file for Chapter 11 bankruptcy in May 1995 [202].

As with any implantable medical devices or drugs, the risk of possible adverse effects must always be weighed against the ability to provide benefits. A great deal of safety research combined with more than 40 years of clinical experience has proven the efficacy and relative safety of the silicone gel breast implants. A rough estimate of implant shell rupture rate is ~10% at 10 years with both biocompatibility and biodurability problems [203]. A fibroconnective tissue capsule was found around all the samples [204]. The capsule formed around implanted mammary prosthesis is highly differentiated and organized, consisting of three layers: interface layer in three variations, intermediate fibrous layer of dense rough collagen fibers and light elongated cells with oval nucleus between them and adventitious layer. Between the fibers of the interface and the middle strata intra- and extracellular silicone droplets and bulks were observed, representing the location where further pathological processes can take place [205]. It is said by Dr. Sidney Wolfe, director of Public Citizen's Health Research Group, in a statement that: "Public Citizen continues to oppose the FDA's 2006 decision to return silicone breast implants to the market for cosmetic use in women for augmentation. The agency's newer information about the risk of implant-associated lymphoma and the previously known risks are serious enough to warrant advising women against having these implanted."

On March 9, 2012 a new silicone breast implant, which joins the two other silicone breast implants on the market - one made by Allergan and the other by Mentor, was approved by the FDA of the United States of America. Recommended monitoring after initially silicone breast implantation is 3 years and then every two years thereafter. In a review Roach and coworkers concern the importance of length and time on physicochemical interactions between living tissue and biomaterials that occur on implantation. The review provides detailed information on material host interactions, dynamic material/cell surface states, surface chemistry and topological roles during the first stage of implant integration, namely protein adsorption. Generally, after the first contact of material with host tissue a state of flux due to protein adsorption, cell adhesion and physical and chemical alteration of the implanted material is followed (Figure 10) [206]. This model can answer many questions concerning the conformational form and bound proteins and therefore has instruction meanings in new implantable biomaterial design field.

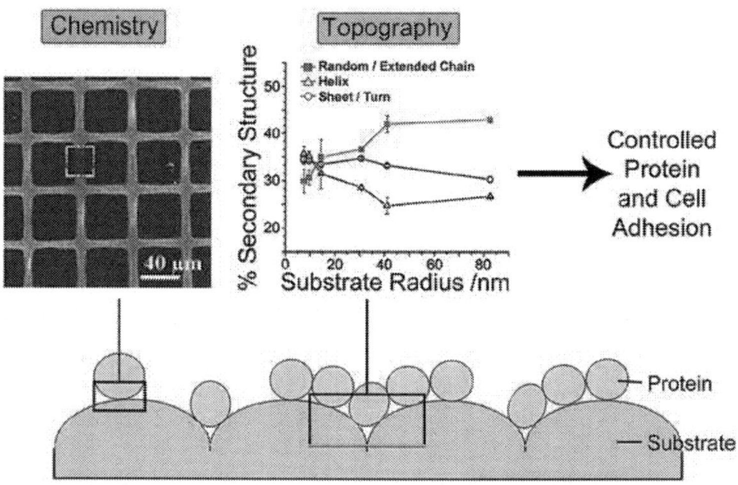

Figure 10. Schematic of protein–surface interactions: Chemistry—adsorption onto biotinylated stripes which appear white, whilst adsorption is hindered on square oligoethylene-glycol regions, the white box shows an intentionally bleached area Topography—albumin adsorption onto hydrophilic silica spheres of varying dimensions as a model of surface curvature [206].

Beside the breast implants a silicon-silk transistor about one millimeter long and 250 nanometers was created. So far the technique has been tested on mice with no adverse effects. Electrical, bending, water dissolution, and animal toxicity studies suggest that this approach might provide many opportunities for future biomedical devices and clinical applications [207]. A silicone catheter attached to a 2-5 x 1-3 cm stainless steel chamber with a self sealing injection port had been intravenously for antimicrobial chemotherapy. Peripheral venous access had become unsatisfactory in all of patients, and six had required central venous catheterisation [208]. More recently, a silicon-based neural probe with microfluidic channels was developed [209].

Origins of controlled release of implantable drug delivery dates back to 1964 when silicone implants were used to prolong a drug effect. Over 40 years, the progress to a safe, effective and acceptable implant system has been slow. The critical factors in implant research which need to be addressed include: erodibility, reproducibility, lack of irritation and carcinogenicity, lack of dose dumping, duration and pulses. While it is possible to surgically implant and remove drug-containing devices or polymeric matrices, the requirement for such intervention could have a significant negative impact on the acceptability of a product candidate. In recent years, two implant systems have been approved for human use; (a) a silicone-based device (Norplant^R), and (b) a system based on lactide/glycolide copolymers to release a luteinizing hormone - releasing hormone (LHRH) agonist for treatment of male reproductive tract tumours. This drug delivery approach is very appealing for a number of classes of drugs, particularly those that cannot be given via the oral route, and drug candidates whose therapeutic index is relatively large [210].

15. Aluminium (or aluminum) and ceramics

Aluminium (or aluminum) is the third most abundant chemical element (after oxygen and silicon in the boron group) with symbol Al and atomic number 13. It is one of the typical metals which has been widely used as hard tissue repair materials with unique properties, such as strong mechanical strength, not soluble and degradable in body fluid under normal circumstances, combined in over 270 different minerals, low density (weight) and corrosion resistances [211]. It is generally accepted that metallic implant materials with higher strength/modulus ratios are more favorable for hard tissue repair due to a combined effect of high strength and reduced stress-shielding risk [212].

Al alloys, such as Al-silicon (Si), Al-platinum, and Al-titanium (Ti) are widely used in implantable engineering structures and components where light weight or corrosion resistance is required except for blood-contacting surfaces [213,214]. For example, an implantable double-sided electrode microdevices, called flexible nerve plates, with a prototype of Al layer could reduce the number of insertion sites and thereby the insertion trauma during implantation of neural prostheses [215]. A Ti-6Al-4vanadium (V) alloy was selected as the ceramic-to-metal seal because its excellent mechanical properties and favorable biocompatibility [216]. The first-generation of implantable left ventricular assist devices (LVADs) were Ti-Alvanadium alloy pulsatile, volume-displacement pumps. The modern LVAD era began with the introduction of the HeartMate X (vented electric) VE in 1998 [217]. These devices can provide excellent circulatory support and improve survival until heart transplantation. However, they have many application limitations, such as a large volume, an excessive surgical dissection, a large diameter driveline, a noisy pump operation, and particularly a limited mechanical durability. Other complications include bleeding, infections and thromboembolic events. During the succeeding decade, vast improvements in pump design resulted in a new crop of LVADs, whose attributes are transforming LVAD therapy into a kind of standard of care for end-stage heart failure [218]. LVAD therapy has now evolved into a solution which is strikingly superior to optimal medical therapy [219, 220].

It was reported that changes in the porous hydroxyapatite and Al oxide orbital implant densities may correspond to healing and maturation of soft tissues surrounding and penetrating the implants [221]. The thermal oxidation behavior of Al ion implanted Ti nitride films has been studied in dry oxygen atmosphere and found that Al implantation caused the oxidation rate of TiN films to slow down at the initial stage of oxidation [222]. Until recently, there is limited evidence regarding comparative effectiveness of various hip implant bearings, especially metal on metal or ceramic on ceramic implants compared with traditional metal on polyethylene or ceramic on polyethylene bearings [223].

For clinical applications, it is an important character that the metal devices do not cause mental or body uncomfortable reactions, such as delaminate or infiltrate ions to the surround tissues. For example, a defibrillator is a medical device that generates and delivers a shock to the heart of someone in cardiac arrest. Although this device can save lives, there are risks involved, for both the patient and the first responders. One risk associated with defibrillator use is that of burns. Certain transdermal medication patches contain aluminum backings, and when they come in contact with the defibrillator paddle, can cause minor burns to the patient. Accidental shocks to others can occur when first responders accidentally contact with the patient who is being defibrillated. The only objects that should touch the patient during treatment are the defibrillator paddles held by the administrator of the procedure. Sometimes internally implanted defibrillators discharge shocks when they are unnecessary. When this occurs, it can cause pain and promote a dangerous heart rhythm. In addition, the event can be emotionally disturbing and frightening. Doctor can recalibrate the device to minimize the risk of additional unnecessary shocks, and offer suggestions on how to manage these rare events [224].

A ceramic is an inorganic, nonmetallic solid material possessing strong mechanical properties prepared by the action of heat and subsequent cooling [30]. The uses of ceramics have been revolutionizing the biomedical field in deployment as implants for humans during the past three decades. In the search to improve the biocompatibility and mechanical strength of skeletal implant materials, attention has been directed towards the potential use of ceramic composites [225]. Since 1975 alumina ceramic has proven its bioinertness and have been accepted in biomedical applications, some alumina ceramic, such as Al2O3 has been characterized with high hardness and high abrasion resistance. Noiri and coworkers evaluated the biocompatibility of alumina-ceramic material histopathalogically for eight weeks by implanting in the eye sockets of albino rabbits with no signs of implant rejection or prolapse of the implanted pieces. After a period of four weeks of implantation, fibroblast proliferation and vascular invasion were noted. By the eighth week, tissue growth was observed in the pores of the implants [226].

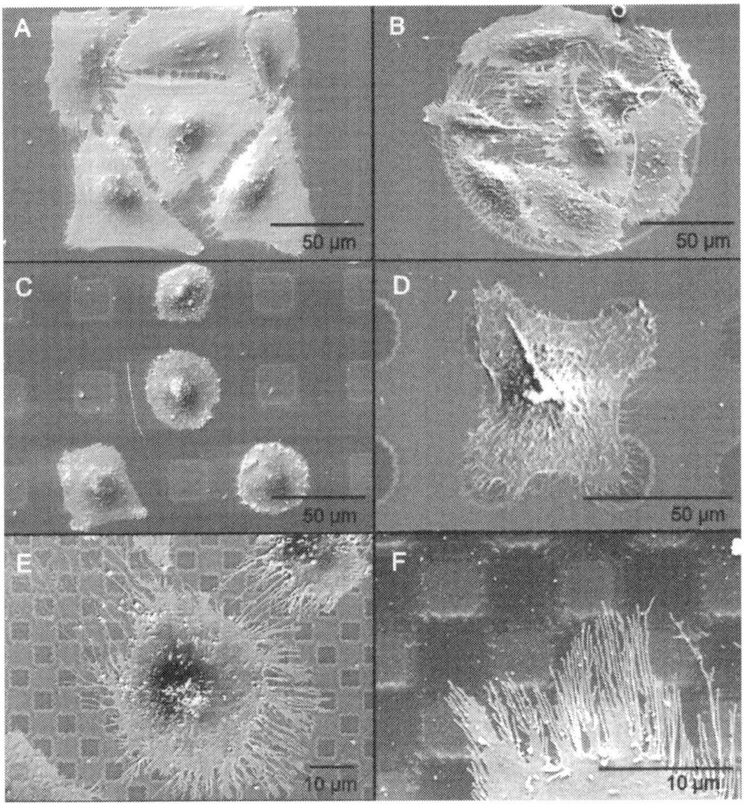

Figure 11. Scanning electron microscopy (SEM) images of SaOS-2 cells cultured for 48 h on micropatterned Ti (A-D) and diamond-like carbon (DLC) (E, F) surfaces. Large-sized (125 µm) squared (A) and circular (B) features facilitated the adhesion of several cells on one Ti island with the cells aligning themselves along the edges of the cell-friendly material. The cells adhering to mediumsized Ti islands no longer conformed strictly to the geometrical shape of the patterns (C) but particularly on circularly patterned surfaces, star-like cellular morphologies appeared (D). On small-sized inverse DLC samples, the cell bodies non-selectively covered large micro-patterned areas (E), but their filopodia clearly showed a preference for DLC trying to avoid bare Si circles (F) [222].

16. Conclusion remarks

Biocompatible is a vital important aspect for an implantable biomaterial. Among the numerous types of host responses to a broad spectrum of biomaterials, those with no adverse or negative effects, such as, fibrous encapsulation, osteolysis, hyperplasia, and inflammation are among the most expectant ones. As advances are made in biomaterial science and technology, new implants/medical devices will be continually explored, alternatives to conventional implants will become more and more effective, and hence more and more attractive. In an effort to provide the best clinical outcomes for the patients, we need to develop the best candidates with minimum invasive surgery times and unnecessary health risks. In the future, design and manufacture immuno or low-immuno implantable biomaterials according to or mimicking the patients' own ingredients, such as blood components, ECMs, tissues and organs, will be possible. For an implantable biomaterial biocompatibility should be always put into the primary importance position no matter it is used as a temporary scaffold, a permanent template, or a drug delivery vihicle.

Acknowledgment

Work in the authors' laboratory is supported by the State Key Laboratory of Materials Processing and Die & Mould Technology, Huazhong University of Science and Technology (No. 2012 - P03), the National Natural Science Foundation of China / the Research Grants Council of Hong Kong (NSFC/RGC, No. 50731160625), the National Natural Science Foundation of China (No. 81271665 & 30970748), the National High Tech 863 Grant (No. 2009AA043801), the Finland Distinguished Professor program (FiDiPro) of Tekes (No. 40041/10), and the Cross-Strait Tsinghua Cooperation Basic Research (No.2012THZ02-3).

Author details

Xiaohong Wang[1,2,3]

Address all correspondence to: wangxiaohong@tsinghua.edu.cn

1 Key Laboratory for Advanced Materials Processing Technology, Ministry of Education & Center of Organ Manufacturing, Department of Mechanical Engineering, Tsinghua University, Beijing, P.R. China

2 Business Innovation Technology (BIT) Research Centre, School of Science and Technology, Aalto University, Aalto, Finland

3 State Key Laboratory of Materials Processing and Die & Mould Technology, Huazhong University of Science and Technology, Wuhan, P.R. China

References

[1] Tathe A, Ghodke M, Nikalje AP. International Journal of Pharmacy and Pharmaceutical Sciences (Int J Pharm Pharm Sci) 2010;2(4):19-23.

[2] Anderson JM, Rodriguez A, Chang DT. Foreign body reaction to biomaterials. Seminars in Immunology 2008;20 (2): 86–100.

[3] Patel NR, Gohil PP. A review on biomaterials: scope, applications & human anatomy significance. International Journal of Emerging Technology and Advanced Engineering 2012; 2(4): 91-101.

[4] Seal BL, Otero TC, Panitch A. Polymeric biomaterials for tissue and organ regeneration. Materials Science and Engineering R 2001;34:147-230.

[5] Franz S, Rammelt S, Scharnweber D, Simon JC. Immune responses to implants - A review of the implications for the design of immunomodulatory biomaterials. Biomaterials 2011;32(28): 6692-6709

[6] Medical-dictionary.thefreedictionary.com.

[7] The American Heritage® Medical Dictionary Copyright © 2007, 2004 by Houghton Mifflin Company.

[8] www.indeed.com.

[9] www.bonebank.com/cancellous-bone.html

[10] Hodde J. Naturally occurring scaffolds for soft tissue repair and regeneration. Tissue Eng 2002;8:295-308.

[11] Thomas W. Gilbert, Tiffany L. Sellaro, Stephen F. Badylak. Decellularization of tissues and organs. Biomaterials 2006;27(19): 3675–3683.

[12] Patino MG, Neiders ME, Andreana S, Noble B, Cohen RE. Collagen as an implantable material in medicine and dentistry. J Oral Implantol. 2002;28(5):220-225.

[13] Wang XH, Yan YN, Zhang RJ. A comparison of chitosan and collagen sponges as heostatic sressings. J Bioact Compat Polym 2006;21(1)::39-54.

[14] George D. Kymionis, Michael A. Grentzelos, Alexandra E. Karavitaki, Zotta Paraskevi, Sonia H. Yoo, Ioannis G. Pallikaris. Combined corneal collagen cross-linking and posterior chamber toric implantable collamer lens implantation for keratoconus. Ophthalmic Surgery, Lasers and Imaging. DOI: 10.3928/15428877-20110210-0.

[15] Ju YM, Yu B, West L, Moussy Y, Moussy F. A dexamethasone-loaded PLGA microspheres/collagen scaffold composite for implantable glucose sensors. J Biomed Mater Res A. 2010;93(1):200-210.

[16] Hou ZQ, Sun Q, Wang Q, Han J, Wang Y, Zhang QQ. *In vitro* and *in vivo* evaluation of novel implantable collagen–chitosan–soybean phosphatidylcholine composite film for the sustained delivery of mitomycin C. Drug Dev Res 2009;70(3):169-219.

[17] Ward, A.G.; Courts, A. The Science and Technology of Gelatin. New York: Academic Press. 1977;ISBN0-12-735050-0.

[18] Kakiuchi M, Hosoya T, Takaoka K, Amitani K, Ono K. Human bone matrix gelatin as a clinical alloimplant. A retrospective review of 160 cases. Int Orthop 1985;9(3): 181-188.

[19] Wang XH, Yan YN, Zhang RJ. Gelatin-based hydrogels for controlled cell assembly. In: Ottenbrite RM, ed. Biomedical Applications of Hydrogels Handbook. New York: Springer, 2010;269-284.

[20] Wang XH, Yan YN, Zhang RJ Rapid prototyping as a tool for manufacturing bioartificial livers. Trends Biotechnol 2007;25:505-513.

[21] Wang XH, Yan YN, Zhang RJ. Recent trends and challenges in complex organ manufacturing. Tissue Eng Part B 2010;16:189-197.

[22] Wang XH, Zhang QQ. Overview on "Chinese–Finnish workshop on biomanufacturing and evaluation techniques. Artificial Organs 2011;35(10):E191- E193.

[23] Wang XH. Intelligent freeform manufacturing of complex organs. Artificial Organs. 2012; doi:10.1111/j.1525-1594.2012.01499.x.

[24] Xu W, Wang XH, Yan YN, Zhang RJ. A polyurethane-gelatin hybrid construct for manufacturing implantable bioartificial livers. J Bioact Compat Polym 2008;23(5): 409-422.

[25] He K, Wang XH. Rapid prototyping of tubular polyurethane and cell/hydrogel construct. J Bioact Compat Polym 2011;26(4):363-374.

[26] Kimura Y, Ozeki M, Inamoto T, Tabata Y. Adipose tissue engineering based on human preadipocytes combined with gelatin microspheres containing basic fibroblast growth factor. Biomaterials 24 (2003) 2513–2521.

[27] Liu J. Controlled trans-lymphatic delivery of chemotherapy for the treatment of lymphatic metastasis in lung cancer. A thesis submitted in conformity with the requirements for the degree of Doctor of Philosophy Institute of Medical Science, University of Toronto, 2008.

[28] Ozeki M, Ishii T, Hirano Y, Tabata Y. Controlled release of hepatocyte growth factor from gelatin hydrogels based on hydrogel degradation. J Drug Target. 2001;9(6): 461-471.

[29] Yang Z, Xu LS, Yin F, Shi YQ, Han Y, Zhang L, Jin HF, Nie YZ, Wang JB, Hao X, Fan DM, Zhou XM. *In vitro* and *in vivo* characterization of silk fibroin/gelatin composite scaffolds for liver tissue engineering. J Dig Dis. 2012 Mar;13(3):168-178.

[30] en.wikipedia.org/wiki, the free encyclopedia.

[31] Radosevich M, Goubran H, Burnouf T. Fibrin sealant: scientific rationale, production methods, properties and current clinical use. Vox Sang 1997;72:133-143.

[32] Dunn C, Goa K. Fibrin sealant: a review of its use in surgery and endoscopy. Drugs 1999;58:863-886.

[33] Weisel JW. Fibrinogen and fibrin. Adv Protein Chem 2005;70:247-299.

[34] Clark RA. Fibrin is a many splendored thing. J Invest Dermatol 2003;121(5):xxi-xxii.

[35] Herrick S, Blanc-Brude O, Gray A, Laurent G. Fibrinogen. Int J Biochem Cell Biol 1999;31(7):741-746.

[36] Abiraman S, Varma HK, Umashankar PR, John A. Fibrin glue as an osteoinductive protein in a mouse model. Biomaterials 23 (2002) 3023–3031.

[37] Hwang TL, Chen MF. Randomized trial of fibrin tissue glue for low output enterocutaneous fistula. Br J Surg 1996;83:112.

[38] Flanagan TC, Sachweh JS, Frese J, Schnöring H, Gronloh N, Koch S, Tolba RH, Schmitz-Rode T, Jockenhoevel S. *In vivo* Remodeling and Structural Characterization of Fibrin-Based Tissue-Engineered Heart Valves in the Adult Sheep Model. Tissue Engineering Part A. 2009;15(10):2965-2976.

[39] Mol A, Driessen NJ, Rutten MC, Hoerstrup SP, Bouten CV, and Baaijens FP. Tissue engineering of human heart valve leaflets: a novel bioreactor for a strain-based conditioning approach. Ann Biomed Eng 2005;33(12):1778-1788

[40] Barsotti MC, Felice F, Balbarini A, Di Stefano R. Fibrin as a scaffold for cardiac tissue engineering. Biotechnol Appl Biochem 2011;58(5):301-310.

[41] Peretti GM, Xu JW, Bonassar LJ, Kirchhoff CH, Yaremchuk MJ, Randolph MA. Review of injectable cartilage engineering using fibrin gel in mice and swine models. Tissue Eng. 2006;12(5):1151-1168.

[42] Daniel D. Swartz, James A. Russell, Stelios T. Andreadis. Engineering of fibrin-based functional and implantable small-diameter blood vessels. Am J Physiol Heart Circ Physiol 2005;288: H1451–H1460.

[43] Sameem M, Wood TJ, Bain JR. A systematic review on the use of fibrin glue for peripheral nerve repair. Plast Reconstr Surg. 2011;127(6):2381-2390.

[44] Koch S, Flanagan TC, Sachweh JS, Tanios F, Schnoering H, Deichmann T, Ellä V, Kellomäki M, Gronloh N, Gries T, Tolba R, Schmitz-Rode T, Jockenhoevel S. Fibrin-polylactide-based tissue-engineered vascular graft in the arterial circulation. Biomaterials 2010;31(17):4731-4739

[45] Guéhennec LL, Daculsi G. A review of bioceramics and fibrin sealant. LE.u Lroe pGeuanéh Ceenlnles ca netd aMl.aterials 2004;8: 1-11.

[46] Jeffrey M. Karp JM, Sarraf F, Shoichet MS, Davies JE. Fibrin-filled scaffolds for bone-tissue engineering: An *in vivo* study. J Biomed Mater Res 2004;71A:162–171.

[47] Farhat WA, Chen J, Sherman C, Cartwright L, Bahoric A, Yeger H. Impact of fibrin glue and urinary bladder cell spraying on the in-vivo acellular matrix cellularization: a porcine pilot study. Can J Urol 2006;13(2):3000-3008.

[48] Breen A, O'Brien T, and Pandit A. Fibrin as a Delivery System for Therapeutic Drugs and Biomolecules. Tissue Eng Part B Rev. 2009;15(2):201-214.

[49] Briganti E, Spiller D, Mirtelli C, Kull S, Counoupas C, Losi P, Senesi S, Di Stefano R, and Soldani G. A. Composite fibrin-based scaffold for controlled delivery of bioactive pro-angiogenetic growth factors. J Control Release 2010;142(1):14-21.

[50] Jockenhoevel S, Zund G, Hoerstrup SP, Chalabi K, Sachweh JS, Demircan L, Messmer BJ, Turina M. Fibrin gel -- advantages of a new scaffold in cardiovascular tissue engineering. Eur J Cardiothorac Surg 2001;19(4):424-430.

[51] Catelas I, Dwyer JF, and Helgerson S. Controlled release of bioactive transforming growth factor Beta-1 from fibrin gels *in vitro*. Tissue Eng Part C Methods 2008;14(2): 119-128.

[52] Arkudas A, Pryymachuk G, Hoereth T, Beier JP, olykandriotis E, Bleiziffer O, Horch RE, and Kneser U. Dose- Finding Study of Fibrin Gel-Immobilized Vascular Endothelial Growth Factor 165 and Basic Fibroblast Growth Factor in the Arteriovenous Loop Rat Model. Tissue Eng Part A. 2009;15(9):2501-2511.

[53] Breen A, O'Brien T, and Pandit A. Fibrin as a Delivery System for Therapeutic Drugs and Biomolecules. Tissue Eng Part B Rev 2009;15(2):201-214.

[54] Sakiyama SE, Schense JC, and Hubbell JA. Incorporation of heparin-binding peptides into fibrin gels enhances neurite extension: an example of designer matrices in tissue engineering. FASEB J. 1999;13(15):2214-2224.

[55] Zisch AH, Schenk U, Schense JC, Sakiyama-Elbert SE, and Hubbell JA. Covalently conjugated VEGF—fibrin matrices for endothelialization. J Control Release 2001;72(1-3):101-113.

[56] Hara T. Bhayana B, Thompson B, Kessinger CW, McCarthy AKR, Weissleder R, Lin CP, Tearney GJ, Jaffer FA. Molecular Imaging of Fibrin Deposition in Deep Vein Thrombosis Using Fibrin-Targeted Near-Infrared Fluorescence. J Am Coll Cardiol Img 2012;5:607–1551.

[57] Ruszymah BH. Autologous human fibrin as the biomaterial for tissue engineering. Med J Malaysia 2004;59(Suppl B):30-31.

[58] Roderick P, Ferris G, Wilson K, Halls H, Jackson D, Collins R, Baigent C. Towards evidence-based guidelines for the prevention of venous thromboembolism: systematic reviews of mechanical methods, oral anticoagulation, dextran and regional anaesthesia as thromboprophylaxis. Health Technol Assess 2005;9(49):iii-iv, ix-x, 1-78.

[59] Sophie R. Tomme V, Hennink WE. Biodegradable dextran hydrogels for protein delivery applications. Expert Review of Medical Devices 2007;4:147-164.

[60] Atik M. Dextran 40 and Dextran 70. A Review. Arch Surg 1967;94(5):664-672.

[61] Dubick MA, Wade CE. A review of the efficacy and safety of 7.5% NaCl/6% dextran 70 in experimental animals and in humans. J Trauma 1994;36(3):323-330.

[62] Richardson D, Deakin CD. Hypertonic saline dextran. A review of current literature. Scand J Trauma Resusc Emerg Med. 2004;12:142-146.

[63] Sun GM, Zhang XJ, She Y-I, Sebastian R, Dickinson LE, Fox-Talbot K, Reinblatt M, Steenbergen C, Harmon JW, Gerecht S. Dextran hydrogel scaffolds enhance angiogenic responses and promote complete skin regeneration during burn wound healing. PNAS 14, 2011;108:20975-20980.

[64] Draye J-P, Delaey B, Van de Voorde A. Van Den Bulcke A, Reu BD, Schacht E. *In vitro* and *in vivo* biocompatibility of dextran dialdehyde cross-linked gelatin hydrogel films. Biomaterials 1998;19(18):1677-1687.

[65] Groot CJD, Van Luyn MJA, Van Dijk-Wolthuis WND, Cadée JA, Plantinga JA, Otter WD, Hennink WE. *In vitro* biocompatibility of biodegradable dextran-based hydrogels tested with human fibroblasts. Biomaterials 2001;22(11):1197-1203.

[66] Dhaneshwar SS, Kandpal M, Gairola N, Kadam SS. Dextran: a promising macromolecular drug carrier. Indian Journal of Pharmaceutical Sciences 2006;68(6):705-714.

[67] de Belder AN. Dextran. Hndbooks from Amersham bosciences. 18-1166-12, Edition Amersham AA. Biosciences AB 2003. (ww.amershambiosciences.com.)

[68] Roderick P, Ferris G, Wilson K, Halls H, Jackson D, Collins R, Baigent C. Towards evidence-based guidelines for the prevention of venous thromboembolism: systematic reviews of mechanical methods, oral anticoagulation, dextran and regional anaesthesia as thromboprophylaxis. Health Technology Assessment 2005;9:(49): Editor-in-Chief: Professor Tom Walley. 2005 Crown Copyright.

[69] Solomon L, Mansor S, Mallon P, Donnelly E, Hoper M, Loughrey M, Kirk S, Gardiner K. The dextran sulphate sodium (DSS) model of colitis: an overview. Comp Clin Pathol 2010;19:235–239.

[70] Yang J, Liu Y, Wang H, Liu L, Wang W, Wang C, Wang Q, Liu W. The biocompatibility of fatty acid modified dextran-agmatine bioconjugate gene delivery vector. Biomaterials 2012;33(2):604-613.

[71] Necas J, Bartosikova L, Brauner P, Kolar J. Hyaluronic acid (hyaluronan): a review. Veterinarni Medicina, 2008;53(8):397–411.

[72] Frasher JRE; Laurent TC; Laurent UBG. Hyaluronan: its nature, distribution, functions and turnover. Journal of Internal Medicine 1997;242(1):27–33.

[73] Kahmann JD, O'Brien R, Werner JM, Heinegard D, Ladbury JE, Campbell ID, Day AJ. Localization and characterization of the hyaluronan-binding site on the Link module from human TSG-6. Structure 2000;8:763–774.

[74] Jaracz S, Chen J, Kuznetsova LV, Ojima I. Recent advances in tumor-targeting anticancer drug conjugates. Bioorganic and Medicinal Chemistry 2005;13:5043–5054.

[75] Paiva P, Van Damme MP, Tellbach M, Jones RL, Jobling T, Salamonsen LA. Expression patterns of hyaluronan, hyaluronan synthases and hyaluronidases indicate a role for hyaluronan in the progression of endometrial cancer. Gynecologic Oncology 2005;3:202.

[76] Lokeshwar, VB, Lopez LE, Munoz D, Chi A, Shirodkar SP, Lokeshwar SD, Escudero DO, Dhir N, Altman N. Antitumor activity of hyaluronic acid synthesis inhibitor 4-methylumbelliferone in prostate cancer cells. Cancer Res 2010;70(7):2613–2623.

[77] Amarnath LP, Srinivas A, Ramamurthi A. *In vitro* hemocompatibility testing of UV-modified hyaluronan hydrogels. Biomaterials 2006; 27(8):1416–1424.

[78] Jansen K., van der Werff JFA, van Wachem PB, Nicolai JPA, de Leij L.F.M.H, van Luyn MJA. A hyaluronan-based nerve guide: *in vitro* cytotoxicity, subcutaneous tissue reactions, and degradation in the rat. Biomaterials 2004;25(1):483–489.

[79] Rah MJ. A review of hyaluronan and its ophthalmic applications. Optometry 2011;82(1):38-43.

[80] De Andrés Santos MI, Velasco-Martín A, Hernández-Velasco E, Martín-Gil J, Martín-Gil FJ. Thermal behaviour of aqueous solutions of sodium hyaluronate from different commercial sources". Thermochim Acta 1994;42:153-160.

[81] Peattie RA, Rieke E, Hewett E, Fisher RJ, Shu XZ, Prestwich GD. Dual growth factor-induced angiogenesis *in vivo* using hyaluronan hydrogel implants.Biomaterials 2006; 27 (9):1868-1875,

[82] Pike DB, Cai S, Pomraning KR, Firpo MA, Fisher RJ, Shu XZ, Prestwich GD, Peattie RA. Heparin-regulated release of growth factors *in vitro* and angiogenic response *in vivo* to implanted hyaluronan hydrogels containing VEGF and Bfgf. Biomaterials 2006;27(30):5242–5251.

[83] Dianhua Jiang, Jiurong Liang, Paul W. Noble. Hyaluronan as an Immune Regulator in Human Diseases. Physiol Rev January 1, 2011 vol. 91 no. 1 221-264.

[84] Jiang DH, Liang JR, Noble PW. Hyaluronan in Tissue Injury and Repair. Annual Review of Cell and Developmental Biology 2007;23:435-461.

[85] Luke R. Bucci, Amy A. Turpin. Will the real hyaluronan please stand up? Journal of applied nutrition 2004;54(1):10-33.

[86] Longaker MT, Chiu ES, Adzick NS, Stem M, Harrison MR, Stem R. Studies in fetal wound healing: V. A prolonged presence of hyaluronic acid characterizes fetal wound fluid. Annals of Surgery. Annals of Surgery 1991;213(4):292-296.

[87] Chen WYJ, Giovanni Abatangelo G. Functions of hyaluronan in wound repair. Wound Repair and Regeneration 1999;7(2):79-89.

[88] Benedetti L, Cortivo R, Berti T, Berti A, Pea F, Mazzo M, Moras M, Abatangelo G. Biocompatibility and biodegradation of different hyaluronan derivatives (Hyaff) implanted in rats. Biomaterials 1993;14(15):1154-1160.

[89] Shen X, Tanaka K, Takamori A. Coronary arteries angiogenesis in ischemic myocardium: biocompatibility and biodegradability of various hydrogels. Artificial Organs 2009;33(10):781-787.

[90] Praveen SS, Hanumantha R, Belovich JM, Davis BL. Novel hyaluronic acid coating for potential use in glucose sensor design. Diabetes Technol Ther. 2003;5(3):393-399.

[91] Hou SP, Xu QY, Tian WM, Cui FZ, Cai Q, Ma J, Lee I-S. The repair of brain lesion by implantation of hyaluronic acid hydrogels modified with laminin. Journal of Neuroscience Methods 2005;148:60–70.

[92] Zhang T, Yan YN, Wang XH, Xiong Z, Lin F, Wu RD, Zhang RJ. Three-dimensional gelatin and gelatin/hyaluronan hydrogel structures for traumatic brain injury. J Bioact Compat Polym 2007;22(1):19-29.

[93] Avitabile T, Marano F, Castiglione F, Bucolo C, Cro M, Ambrosio L, Ferrauto C, Reibaldi A. Biocompatibility and biodegradation of intravitreal hyaluronan implants in rabbits. Biomaterials 2001;22(3):195-200.

[94] Li H, Liu Y, Shu XZ, Gray SD, Prestwich GD. Synthesis and biological evaluation of a cross-linked hyaluronan-mitomycin C hydrogel. Biomacromolecules 2004;5(3): 895-902.

[95] Zheng SX, Liu Y, Palumbo FS, Luo Y, Prestwich GD. In situ crosslinkable hyaluronan hydrogels for tissue engineering. Biomaterials 2004;25(7-8):1339-1348.

[96] Hong HJ, Lee JS, Choi JW, Min BH, Lee HB, Kim CH. Transplantation of Autologous Chondrocytes Seeded on a Fibrin/Hyaluronan Composite Gel Into Tracheal Cartilage Defects in Rabbits: Preliminary Results. Artif Organs 2012;30:1525-1594.

[97] Cox M; Nelson D. Lehninger, Principles of Biochemistry. Freeman 2004:1100.

[98] Jang I-K, Hursting MJ. When Heparins Promote Thrombosis : Review of Heparin-Induced Thrombocytopenia. Circulation 2005;111:2671-2683.

[99] Moore BR, Hinchcliff KW. Heparin: a review of its pharmacology and therapeutic use in horses. J Vet Intern Med. 1994;8(1):26-35.

[100] Saliba Jr MJ . Heparin in the treatment of burns: a review. Burns 2001;27(4):349-358.

[101] Papa A, Danese S, Gasbarrini A, Gasbarrini G. Review article: potential therapeutic applications and mechanisms of action of heparin in inflammatory bowel disease. Aliment Pharmacol Ther. 2000 Nov;14(11):1403-1409.

[102] Hull RD, Liang J, Townshend G. Long-term low-molecular-weight heparin and the post-thrombotic syndrome: a systematic review. Am J Med. 2011;124(8):756-765.

[103] Wang XH, Yan YN, Lin F, Xiong Z, Wu RD, Zhang RJ, Lu QP. Preparation and characterization of a collagen/chitosan/heparin matrix for an implantable bioartificial liver. J Biomater Sci.Polym Ed 2005;16(9):1063-1080.

[104] Yu X, Bichtelen A, Wang XH, Yan YN, Lin F, Xiong Z, Wu RD, Zhang RJ, Lu QP. Collagen/chitosan/heparin complex with improved biocompatibility for hepatic tissue engineering. J Bioact Compat Polym 2005;20(1):15-28.

[105] Matthew D. Mitchell, Barbara Jo Anderson, Kendal Williams & Craig A. Umscheid. Heparin flushing and other interventions to maintain patency of central venous catheters: a systematic review. Journal of Advanced Nursing 2009;65(10), 2007–2021.

[106] Ziakas PD, Pavlou M, Voulgarelis M. Heparin treatment in antiphospholipid syndrome with recurrent pregnancy loss: a systematic review and meta-analysis. Obstet Gynecol. 2010;115(6):1256-1262.

[107] Marcum JA, McKenney JB, Galli SJ, Jackman RW, Rosenberg RD. Anticoagulantly active heparin-like molecules from mast cell-deficient mice. Am J Physiol 1986;250 (5 Pt 2): H879–888.

[108] Nader HB, Chavante SF, dos-Santos EA, Oliveira FW, de-Paiva JF, Jerônimo SMB, Medeiros GF, de-Abreu LRD, Leite EL, de-Sousa-Filho JF, Castro RAB, Toma L, Tersariol ILS, Porcionatto MA, Dietrich CP. Heparan sulfates and heparins: similar compounds performing the same functions in vertebrates and invertebrates? Braz J Med Biol Res 1999;32(5):529–538.

[109] Peattie RA, Rieke E, Hewett E, Fisher RJ, Shu X Z, Prestwich G D. Dual growth factor-induced angiogenesis *in vivo* using hyaluronan hydrogel implants. Biomaterials 2006;27(9):1868-1875.

[110] Swanson JM. Heparin-induced thrombocytopenia: a general review. J Infus Nurs. 2007;30(4):232-240.

[111] Hong MS, Amanullah AM. Heparin-induced thrombocytopenia: a practical review. Rev Cardiovasc Med 2010;11(1):13-25.

[112] James L. Januzzi, Jr. and Ik-Kyung Jang. Heparin Induced Thrombocytopenia: Diagnosis and Contemporary Antithrombin Management. J Thromb Thrombolysis 1999;7:259–264.

[113] Strand BL, Ryan L, Veld PI, Kulseng B, Rokstad AM, Skjåk-Bræk G., Especik T. Cell Transplant. 2001;10:263-275.

[114] Yao R, Zhang RJ, Yan YN, Wang XH. *In vitro* angiogenesis of 3D tissue engineered adipose tissue. J Bioact Compat Polym 2009;24(1):5-24.

[115] Yao R, Zhang RJ, Wang XH. Design and evaluation of a cell microencapsulating device for cell assembly technoloty. J Bioact Compat Polym 2009;24(1):48-62.

[116] Soon-Shiong P, Heintx RE, Merideth N, Yao QX, Yao Z, Zheng T, Murphy M, Moloney MK, Schmedhl M, Harris M, Mendez R, Mendez R, Sanford PA. The Lancet 1994;16:950-951.

[117] Li RH. Materials for Immunoisolated Cell Transplantation. Advanced Drug Delivery Reviews 1998;33:87–109.

[118] Tam SK, de Haan BJ, Faas MM, Hallé JP, Yahia L, de Vos P. Adsorption of human immunoglobulin to implantable alginate-poly-L-lysine microcapsules: effect of microcapsule composition. J Biomed Mater Res A. 2009;89(3):609-615.

[119] Xie H-G, Li X-X, Lv G-J, Xie W-Y, Zhu J, Luxbacher T, Ma R, Ma X-J. Laboratory of Biomedical MaterialDevelopment of an implantable alginate scaffold for the treatment of spinal cord trauma. J Biomed Mater Res 2010;92A:1357–1365.

[120] Shahidi F, Synowiecki J. Isolation and characterization of nutrients and value-added products from snow crab (Chionoecetes opilio) and shrimp (Pandalus borealis) processing discards. Journal of Agricultural and Food Chemistry (American Chemical Society) 1991;39 (8): 1527–1532.

[121] Khor E, Lim LY. Implantable applications of chitin and chitosan. Biomaterials. 2003; 24(13):2339-2349.

[122] Yang T-L. Chitin-based Materials in tissue engineering: applications in soft tissue and epithelial organ. Int J Mol Sci 2011;12:1936-1963.

[123] Costa-Pinto AR, Reis RL, Neves NM. Scaffolds based bone tissue engineering: the role of chitosan. Tissue Engineering Part B 2011;17(5):1-17.

[124] Wang XH, Yu X, Yan YN, Zhang RJ. Liver tissue responses to gelatin and gelatin/chitosan gels. J Biomed Mater Res 2008;87A(1):62-68.

[125] Michael Dornish, VP research and development, and Dr Are Kristiansen, commercial development manager, at FMC BioPolymer.

[126] Khor E, Lim LY. Implantable applications of chitin and chitosan. Biomaterials 2003;24 (13):2339–2349.

[127] Mi FL, Tan YC, Liang HC, Huang RN, Sung HW. *In vitro* evaluation of a chitosan membrane cross-linked with genipin. J Biomater Sci Polym Ed 2001;12(8):835-850.

[128] Wang XH, Ma JB, Wang YN, He BL. Structural characterization of phosphorylated chitosan and their applications as effective additives of calcium phosphate cements. Biomaterials 2001;22(16):2247-2255.

[129] Wang XH, Ma JB, Wang YN, He BL. Skeletal repair in radii and tibias of rabbits with phosphorylated chitosan reinforced calcium phosphate cements. Biomaterials. 2002;23(21):4167-4176.

[130] Wang XH, Ma JB, Wang YN, He BL. Reinforcement of calcium phosphate cements with phosphorylated chitin. Chin J Polym Sci 2002;4:325-332.

[131] Wang XH, Ma JB, Feng QL, Cui FZ. Skeletal repair in rabbits with calcium phosphate cements incorporated phosphorylated chitin. Biomaterials 2002;23(23):4591-4600.

[132] Wang XH, Ma JB, Feng QL, Cui FZ. *in vivo* evaluation of S-chitosan enhanced calcium phosphate cements. J Bioact Compat Polym 2003;18(4):259-271.

[133] Wang XH, Feng QL, Cui FZ, Ma JB. The effects of S-chitosan on the physical properties of calcium phosphate cements. J Bioact Compat Polym 2003;18(1):45-57.

[134] Liu H, Mao J, Yao K, Yang G, Cui L, Cao Y. A study on a chitosan-gelatin-hyaluronic acid scaffold as artificial skin *in vitro* and its tissue engineering applications. J Biomater Sci Polym Ed 2004;15(1):25-40.

[135] Fu LN, Wang W, Yu LJ, Zhang SM, Yang G. Fabrication of novel cellulose/chitosan artificial skin composite. Materials Science Forum 2009;610-613:1034-1038.

[136] Yussof SJM, Halim AS, Saad AZM, Jaafar H. Evaluation of the Biocompatibility of a Bilayer Chitosan Skin Regenerating Template, Human Skin Allograft, and Integra Implants in Rats. International Scholarly Research Network (ISRN) Materials Science Volume 2011:1 -7.

[137] Schulz III JT, Tompkins RG, Burke JF. Artificial Skin. Annual Review of Medicine 2000;51:231-244.

[138] Jebahi S, Oudadesse H, Bui XV, Keskes H, Rebai T, Feki A, Feki H. Repair of bone defect using bioglass-chitosan as a pharmaceutical drug: An experimental study in an ovariectomised rat model. African Journal of Pharmacy and Pharmacology 2012; 6(16):1276-1287.

[139] Zhang Z, Cui HF. Biodegradability and Biocompatibility Study of Poly(Chitosan-g-lactic Acid) Scaffolds. Molecules 2012;17:3243-3258.

[140] Li XG, Yang ZY, Zhang AF, Wang TL, Chen WC. Repair of thoracic spinal cord injury by chitosan tube implantation in adult rats. Biomaterials 2009;30(6):1121–1132.

[141] Wang XH, Yan YN, Xiong Z, Lin F, Wu RD, Zhang RJ, Lu, QP. Preparation and Evaluation of Ammonia Treated Collagen/Chitosan matrices for Liver Tissue Engineering. J Biomed Mater Res Part B Appl. Biomater 2005;75B: 91-98.

[142] Wang XH, Li DP, Wang WJ, Feng QL,Cui FZ, Xu YX, Song XH, van der Werf M. Crosslinked collagen/chitosan matrix for artificial livers, Biomaterials 2003;24(19): 3213-3220.

[143] Wang XH, Cui FZ, Feng QL, Li JC, Zhang YH. Preparation and characterization of collagen/chitosan matrices as potential biomaterials. J Bioact Compat Polym 2003;18(6):453-467.

[144] Denkbas EB, Seyyal M, Piskin E. Implantable 5-fluorouracil loaded chitosan scaffolds prepared by wet spinning. Journal of Membrane Science 2000;172(1):33-38.

[145] Sun J, Jiang G, Qiu T, Wang Y, Zhang K, Ding F. Injectable chitosan-based hydrogel for implantable drug delivery: body response and induced variations of structure and composition. J Biomed Mater Res A. 2010;95(4):1019-1027.

[146] Feng M, Han J, Sun Q, Hou ZQ, Zhang QQ. Preparation and Evaluation of Implantable Chitosan-Collagen-Soybean Phosphatidylcholine Film Impregnated with Mitomycin C-PLA- Nanoparticles. 2009. ICBBE 2009. 3rd International Conference on Bioinformatics and Biomedical Engineering

[147] Sun JL, Jiang GQ, Wang YJ, Ding FX. Thermosensitive chitosan hydrogel for implantable drug delivery: Blending PVA to mitigate body response and promote bioavailability. Journal of Applied Polymer Science 2012;125(3):2092-2101.

[148] Rosato DV, Rosato DV; Rosato MV. Plastic product material and process selection handbook, Elsevier, 2004 p.85, ISBN 978-1-85617-431-2.

[149] Gilding DK; Reed AM. Biodegradable polymers for use in surgery - polyglycolic/poly (lactic acid) homo- and copolymers: 1. Polymer 1979;20(12):1459–1464.

[150] Cortiella J, Nichols JE, Kojima K, Bonassar LJ, Dargon P, Roy AK, Vacant MP, Niles JA, Vacanti CA. Tissue-Engineered Lung: An *In vivo* and *In Vitro* Comparison of Polyglycolic Acid and Pluronic F-127 Hydrogel/Somatic Lung Progenitor Cell Constructs to Support Tissue Growth. Tissue Engineering 2006;12(5):1213-1225.

[151] Tamai H, Igaki K, Kyo E, Kosuga K, Kawashima A, Matsui S, Komori H, Tsuji T, Motohara S, Uehata H. Initial and 6-month results of biodegradable poly-l-lactic acid coronary stents in humans. Circulation. 2000;102:399-404.

[152] Bergsma JE, Bos RRM, Rozema F R, Jong W, Boering G. Biocompatibility of intraosseously implanted predegraded poly(lactide): an animal study. Journal of Materials Science: Materials in Medicine 1996;7(1):1-7.

[153] Aframian DJ, Redman RS, Yamano S, Nikolovski J, Cukierman E, Yamada KM, Kriete MF, Swaim WD, Mooney DJ, Baum BJ. Tissue compatibility of two biodegradable tubular scaffolds implanted adjacent to skin or buccal mucosa in mice. Tissue Eng 2002;8(4):649-659.

[154] Lam KH, Schakenraad JM, Esselbrgge H, Esselbrugge H, Feijen J, Nieuwenhuis P. The effect of phagocytosis of poly(L-lactic acid) fragments on cellular morphology and viability. Journal of Biomedical Materials Research 1993;27:1569-1577.

[155] Onuki Y, Bhardwaj U, Papadimitrakopoulos F, Burgess DJ. A review of the biocompatibility of implantable devices: current challenges to overcome foreign body response. J Diabetes Sci Technol. 2008;2(6):1003–1015.

[156] Zoppi RA, Duek EAR, Coraça DC. Barros PP. Preparation and characterization of poly (L-lactic acid) and poly(ethylene oxide) blends. Materials Research 2001;4(2): 117-125.

[157] Nagahama K, Ohya Y, Ouchi T. Synthesis of Star-shaped 8 arms Poly(ethylene glycol)-Poly(L-lactide) Block Copolymer and Physicochemical Properties of Its Solution Cast Film as Soft Biomaterial. Polymer Journal 2006;38:852–860.

[158] van der Lei B, Bartels HL, Nieuwenhuis P, Wildevuur CR. Microporous, complaint, biodegradable vascular grafts for the regeneration of the arterial wall in rat abdominal aorta. Surgery. 1985 Nov;98(5):955-963.

[159] Middleton J; Tipton A. Synthetic biodegradable polymers as medical devices". Medical Plastics and Biomaterials Magazine. 1998 Retrieved 2006-07-04.

[160] Muhonen J, Suuronen R, Oikarinen VJ, Sarkiala E, Happonen R-P. Effect of polyglycolic acid (PGA) membrane on bone regeneration around titanium implants inserted in bone sockets. Journal of Materials Science: Materials in Medicine 1994; 5(1):40-42.

[161] Taylor MS, Daniels AU, Andriano KP, Heller J. Six bioabsorbable polymers: *in vitro* acute toxicity of accumulated degradation products. J Appl Biomater 1994;5(2): 151-157.

[162] Anderson JM. Biological responses to materials. Annu Rev Mater Res. 2001;31:81–110.

[163] van der Giessen WJ, Lincoff AM, Schwartz RS, van Beusekom HM, Serruys PW, Holmes DR, Jr, Ellis SG, Topol EJ. Marked inflammatory sequelae to implantation of biodegradable and nonbiodegradable polymers in porcine coronary arteries. Circulation. 1996;94(7):1690–1697.

[164] Venkatraman S, Boey YC. Patent: Implantable article, method of forming same and method for reducing thrombogenicity. IPC8 Class: AA61F206FI, USPC Class: 623001440, Publication date: 2012-07-12, Patent application number: 20120179242.

[165] Bhardwaj U, Papadimitrakopoulos F, Burgess DJ. A review of the development of a vehicle for localized and controlled drug delivery for implantable biosensors. J Diabetes Sci Technol. 2008;2(6):1016-1029.

[166] Onuki Y, Bhardwaj U, Burgess PDJ. A Review of the Biocompatibility of Implantable Devices: Current Challenges to Overcome Foreign Body Response. J Diabetes Sci Technol 2008;2(6):1003-1015.

[167] Bhardwaj U, Sura R, Papadimitrakopoulos F, Burgess DJ. Controlling acute inflammation with fast releasing dexamethasone-PLGA microsphere/PVA hydrogel composites for implantable devices. Journal of Diabetes Science and Technology 2007;1(1):8-17.

[168] Ranganath SH, Yang A, Chan YY, Huang J, Krantz WB, Wang CH. Implantable hydrogel beads entrapping PLGA-paclitaxel microspheres: exploring the effects of near-zero order drug release for intracranial chemotherapy. AIChE Annual meeting, Philadelphia, USA, November 2008.

[169] Corey JM, Gertz CC, Wang BS, Birrell LK, Johnson SL, Martin DC, Feldman EL. The design of electrospun PLLA nanofiber scaffolds compatible with serum-free growth of primary motor and sensory neurons. Acta Biomater 2008;4(4):863-875)

[170] Leach KJ, Takahashi S, Mathiowitz E. Degradation of double-walled polymer microspheres of PLLA and P(CPP:SA)20 : 80. II. *In vivo* degradation. Biomaterials 1998;19(21):1981-1988.

[171] Liu JY, LiRen L, Wei Q, Wu JL, Liu S, Wang YJ, Li GY. Toward better bone repair. 2012 Society of Plastics Engineers (SPE).

[172] Woodruff MA, Hutmacher DW. The return of a forgetten polymer – polycaprolactone in the 21st century. Progress in Polymer Science 2010;35:1217-1256.

[173] Sun HF, Mei L, Song CX, Cui XM, Wang PY. The *in vivo* degradation, absorption and excretion of PCL-based implant. Biomaterials 2006;27(9):1735–1740.

[174] Sinha VR, Bansal K, Kaushik R, Kumria R, Trehan A. Poly- -caprolactone microspheres and nanospheres: an overview. International Journal of Pharmaceutics 2004;278(1):1-23.

[175] LaVan AD, McGuire T, Langer R. Smallscale systems for *in vivo* drug delivery. Nature Biotechnology 2003; 21:1184-1191.

[176] Orosz KE, Gupta S, Hassink M, Abdel-Rahman M, Moldovan L, Davidorf FH, Moldovan NI. Delivery of antiangiogenic and antioxidant drugs of ophthalmic interest through a nanoporous inorganic filter. Molecular Vision 2004;10:555-565.

[177] Chung TW, Yang MC, Tseng CC, Sheu SH, Wang SS, Huang YY, Chen SD. Promoting regeneration of peripheral nerves in-vivo using new PCL-NGF/Tirofiban nerve conduits. Biomaterials. 2011;32(3):734-743.

[178] Pitt CG, Schinder A. Capronor-A biodegradable delivery system for levonorgestrel. In: Zatachini GL, editor. Long-acting contraceptive systems. Philadelphia: Harpen and Row; 1984. p.63–84.

[179] Woodward SC, Brewer PS, Moatamed F, Schindler AK Pitt CG. The intracellular degradation of poly (epsilon-caprolactone). J Biomed Mater Res 1985;19(4):437–44.

[180] Xia Z, Triffitt JT. A review on macrophage responses to biomaterials. Biomed. Mater 2006;1:R1–R9.

[181] Coury AJ, Slaikeu PC, Cahalan PT, Stokes KB, Hobot CM. Factors and interactions affecting the performance of polyurethane elastomers in medical devices. J Biomater Appl 1988 Oct;3(2):130-179.

[182] Coury AJ, Stokes KB, Cahalan PT, Slaikeu PC. Biostability considerations for implantable polyurethanes. Life Support Syst 1987;5(1):25-39.

[183] Maisel WH. Increased failure rate of a polyurethane implantable cardioverter defibrillator lead. Pacing Clin Electrophysiol. 2002;25(6):877-878.

[184] Hauser RG, Cannom D, Hayes DL, Parsonnet V, Hayes J, Ratliff N 3rd, Tyers GF, Epstein AE, Vlay SC, Furman S, Gross J. Long-term structural failure of coaxial polyurethane implantable cardioverter defibrillator leads. Pacing Clin Electrophysiol. 2002;25(6):879-882.

[185] Stokes KB, Church T. Ten-year experience with implanted polyurethane lead insulation. Pacing Clin Electrophysiol. 1986;;9(6 Pt 2):1160-1165.

[186] Santerre JP, Woodhouse K, Laroche G, Labow RS. Understanding the biodegradation of polyurethanes: From classical implants to tissue engineering materials. Biomaterials 2005;26(35):7457-7470.

[187] Wang Y. Patent: Implantable Medical Devices Fabricated From Polyurethanes With Grafted Radiopaque Groups. IPC8 Class: AA61F282FI, USPC Class: 623 134, Class name: Prosthesis (i.e., artificial body members), parts thereof, or aids and accessories therefor arterial prosthesis (i.e., blood vessel) having marker (e.g., color, radiopaque, etc.), Publication date: 2009-10-15, Patent application number: 20090259297

[188] Stachelek SJ, Alferiev I, Connolly JM, Sacks M, Hebbel RP, Bianco R, Levy RJ. Cholesterol-modified polyurethane valve cusps demonstrate blood outgrowth endothelial cell adhesion post-seeding *in vitro* and *in vivo*. Ann Thorac Surg. 2006 Jan;81(1):47-55.

[189] Ghanbari H, Viatge H, Kidane AG, Burriesci G, Tavakoli M, Seifalian AM. Polymerica heart valves: new materials, emerging hopes. Trends in Biotechnology 2009;27(6): 359-367.

[190] Gwendolyn M.R. Wetzels, Leo H. Koole. Photoimmobilisation of poly(N-vinylpyrrolidinone) as a means to improve haemocompatibility of polyurethane biomaterials. Biomaterials 1999;20(20):1879}-1887.

[191] Staniszewska-Kuś J, Paluch D, Krzemień-Dabrowska A, Zywicka B, Solski L.Tissue reaction to implanted polyurethane designed for parts of the artificial heart. Polym Med. 1995;25(3-4):3-18.

[192] Stachelek SJ, Alferiev I, Choi H, Chan CW, Zubiate B, Sacks M, Composto R, Chen I-W, Levy R. Prevention of oxidative degradation of polyurethane by covalent attachment of di-tert-butylphenol residues. J Biomed Mater Res A. 2006;78(4):653-661).

[193] Stachelek SJ, Alferiev I, Connolly JM, Sacks M, Hebbel RP, Bianco R, Levy RJ. Cholesterol-modified polyurethane valve cusps demonstrate blood outgrowth endothelial cell adhesion post-seeding in vitro and *in vivo* Ann Thorac Surg. 2006 Jan; 81(1):47-55.

[194] Alferiev I, Stachelek SJ, Lu ZB, Fu AL, Tiffany L. Sellaro TL, Jeanne M. Connolly JM, Richard W. Bianco RW, Michael S. Sacks MS, Robert J. Levy RJ. Prevention of polyurethane valve cusp calcification with covalently attached bisphosphonate diethylamino moieties. J Biomed Mater Res A. 2003; 66(2):385-395.

[195] Chris Smith. Implantable PU developer wins heart valve patent. Time:2009-08-03.

[196] Yan YN, Wang XH, Yin DZ, Zhang RJ. A new polyurethane/heparin vascular graft for small-caliber vein repair. J Bioact Compat Polym 2007;22(3):323-341.

[197] Yin DZ, Wang XH, Yan YN, Zhang RJ. Preliminary studies on peripheral nerve regeneration using a new polyurethane conduit. J Bioact Compat Polym 2007;22(2): 143-159.

[198] Fluorine and health:molecular imaging, biomedical materials and pharmaceuticals. Edited by Alain Tressaud and Günter Haufe. Elsevier, UK

[199] Williams MR, Mikulin T, Lemberger J, Hopkinson BR, Makin GS. Five year experience using PTFE vascular grafts for lower limb ischaemia. Ann R Coll Surg Engl. 1985; 67(3):152–155.

[200] von Recum AF, Imamura H, Freed PS, Kantrowitz A, Chen ST, Ekstrom ME, Baechler CA, Barnhart MI. Biocompatibility tests of components of an implantable cardiac assist device. J Biomed Mater Res. 1978 Sep;12(5):743-765.

[201] Głowiński S, Worowski K. Local activation of blood coagulation by polyester prostheses implanted into defects of the abdominal aorta of dogs. Polim Med 1977;7(4): 241-243.

[202] Renwick SB. Silicone breast implants: implications for society and surgeons. Med J Aust 1996 Sep 16;165(6):338-341.

[203] Daniels AU. Silicone breast implant materials. Swiss Med Wkly. 2012;142:doi: 10.4414/smw.2012.13614.

[204] Dewan PA, Condron SK, Morreau PN, Byard RW, Terlet J. Plastic migration from implanted central venous access devices. Arch Dis Child 1999;81:71–72.

[205] Pavlov St, Guidoin R, Marinov G. Histological organization of the capsulae formed around implanted silicone breast prostheses. Acta Morphologica et Anthropologica 01/2004; 9:50-57.

[206] Roach P, Eglin D, Rohde K, Perry CC. Modern biomaterials: a review—bulk properties and implications of surface modifications. Journal of Materials Science: Materials in Medicine 2007;Volume 18, Number 7 , 1263-1277.

[207] Kim D-H, Kim Y-S, Amsden KJ, Panilaitis B, Kaplan DL, Omenetto FG, Zakin MR, Rogers JA. Silicon electronics on silk as a path to bioresorbable, implantable devices. Appl Hys Lett 2009;95:133701.

[208] Stead RJ, Davidson TI, Duncan FR, Hodson ME, Batien JC. Use of a totally implantable system for venous access in cystic fibrosis. Thorax 1987;42:149-150.

[209] Guo K, Pei WH, Li XQ, Gui Q, Tang RY, Liu J, Chen HD. Fabrication and characterization of implantable silicon neural probe with microfluidic channels. Science China Technological Sciences January 2012;55 No(1): 1–5.

[210] Danckwerts M, Fassihi A. Implantable Controlled Release Drug Delivery Systems: A Review. 1991;17(11):1465-1502.

[211] Shakhashiri BZ. Chemical of the Week: Aluminum. SciFun.org. University of Wisconsin. http://scifun.chem.wisc.edu/chemweek/PDF/Aluminum.pdf. Retrieved 2012-03-04 (17 March 2008).

[212] Takami Y, Yamane S, Makinouchi K, Niimi Y, Sueoka A, Nosé Y. Evaluation of platelet adhesion and activation on materials for an implantable centrifugal blood pump 1. Artif Organs 1998;;22(9):753-758.

[213] Polmear IJ. Light Alloys, Arnold, 1995.

[214] Tseung ACC, King WJ, Wan BYC. An encapsulated, implantable metal-oxygen cell as a long-term power source for medical and biological applications. Medical and Biological Engineering and Computing 1971;9(3):175-184.

[215] Stieglitz T. Flexible biomedical microdevices with double-sided electrode arrangements for neural applications. Sensors and Actuators A 2001;90:202-211.

[216] Jiang GQ, Mishler D, Davis R, Mobley JP, Schulman JH. Zirconia to Ti-6Al-4V braze joint for implantable biomedical device. J Biomed Mater Res Part B: Appl Biomater 2005;72B: 316–321.

[217] Dowling RD, Park SJ, Pagani FD, Mohr F-W. HeartMate VE LVAS design enhancements and its impact on device reliability. European Journal of Cardio-Thoracic Surgery 2004; 25(6):958–963.

[218] Takami Y, Yamane S, Makinouchi K, Niimi Y, Sueoka A, Nosé Y. Evaluation of platelet adhesion and activation on materials for an implantable centrifugal blood pump. Artif Organs 1998;22(9):753-758.

[219] Spiliopoulos K, Giamouzis G, Karayannis G, Karangelis D, Koutsias S, Kalogeropoulos A, Georgiopoulou V, Skoularigis J, Butler J, Triposkiadis F. Current Status of Mechanical Circulatory Support: A Systematic Review. Cardiology Research and Practice 2012, Article ID 574198, 12 pages, doi:10.1155/2012/574198.

[220] Myllymaa S. Novel micro- and nano-technological approaches for improving the performance of implantable biomedical devices. Publications of the University of Eastern Finland. Dissertations in Forestry and Natural Sciences. University of Eastern Finland. P122.

[221] Olga L; Péter B; George KS; Jósef B. Porous hydroxyapatite and aluminium-oxide ceramic orbital implant evaluation using CBCT scanning: a method for *in vivo* porous structure evaluation and monitoring. International Journal of Biomaterials 2012; 1-9: 1687-8787.

[222] Mitsuo A, Aizawa T. Thermal oxidation and characterization for surface layers of Al implanted TiN films. Ion Implantation Technology Proceedings, 1988 International Conference on IEEE. Date of Conference: Dec 1999;2:865–868.

[223] Sedrakyan A, Normand ST, Dabic S, Jacobs S, Graves S, Marinac-Dabic D. Comparative assessment of implantable hip devices with different bearing surfaces: systematic appraisal of evidence. BMJ 2011;343:d7434-7446.

[224] Defibrillator Risks eHow.com http://www.ehow.com/facts_7438880_defibrillator-risks.html#ixzz23258fVn4.

[225] Thamaraiselvi TV, Rajeswari S. Biological evaluation of bioceramic materials - a review. Trends Biomater Artif Organs 2004;18(1);9-17.

[226] Patel NR, Gohil PP. A Review on biomaterials: scope, applications & human anatomy significance. International Journal of Emerging Technology and Advanced Engineering 2012; 2(4), 2250-2259.

Chapter 6

In Vitro Blood Compatibility of Novel Hydrophilic Chitosan Films for Vessel Regeneration and Repair

Antonello A. Romani, Luigi Ippolito,
Federica Riccardi, Silvia Pipitone, Marina Morganti,
Maria Cristina Baroni, Angelo F. Borghetti and
Ruggero Bettini

Additional information is available at the end of the chapter

http://dx.doi.org/10.5772/52706

1. Introduction

Tissue and organ failure treatments include drug therapy, surgical repair, medical devices but they do not always provide satisfactory restoration of organ function.

At present transplants represent an actual solution in treating organ failure once overcoming immunological rejection even though its application is largely affected by the paucity of available donors.

In the vascular field, currently, the best graft performance is given by saphenous vein autografts [1] whose main failures are related to thrombosis development, emboli production, and intimal hyperplasia. Synthetic non-bio-resorbable vascular prosthesis (such as Dacron® or extended-PTFE) exhibited very low incidences of thrombosis or hyperplasia and showed good clinical results in medium- and large-diameter graft sites.

Strategies based on polymeric materials (synthetic or natural) appear to be a valid alternative for the production of tissue graft materials. However, synthetic polymers are not able to induce any biological response leading to tissue regeneration, due to the lack of biomimetic activity. On the contrary, natural biodegradable polymeric supports, resembling extracellular matrix component, can provide a useful platform in tissue engineering and regenerative medicine applications [2-5].

Among them, chitosan (CS) a biodegradable [6], non-amphiphilic polymer of D-glucosamine obtained by partial de-acetylation of chitin [3], has shown interesting properties including biomimetism due to the similarity of is structure with that of glycosaminoglycanes [4].

Kind *et al.* [7] reported that it promotes plasma protein adsorption, platelet adhesion and activation, and thrombus development [8]. The positive charged CS surface induces a great degree of platelet adhesion. In fact, the Food and Drug Administration approved its use as haemostatic dressing for reducing haemorrhage [9-11]. Furthermore, it has been reported that the negatively charged-modified surface of CS prolongs clot formation after re-calcification of plasma [12].

Up to now, few data, often conflicting, on the haemocompatibility of negative charged-modified surfaces of CS films are available[11,12].

In a previous work [13] we developed novel CS hydrogel prepared in the presence of phosphate salts and relatively high amount of disaccharides such as D-(+)raffinose or D-(+)saccharose and investigated the physico-chemical characteristics as well as the cytocompatibility of films obtained with this hydrogel. These sugars were not retained in the final structure of the film but were able to act as viscosity modifiers during the solidification/gelation process. The interference of salts and disaccharides resulted in smooth, amorphous film with improved hydrophilicity and cytocompatibility compared to CS films produced with the same procedure but in low viscosity milieu. Differentiated human cells showed a great affinity for these sugar-modified chitosan (smCS) films, thus suggesting their candidature as promising biomaterial for tissue regeneration and repair.

The aim of the present study was to investigate qualities and aspects of the haemocompatibility (platelet activation, haemolysis and activation of coagulation cascade) of smCS films produced according to Bettini*et al.* [13]. Moreover, the cytotoxicity of fragmented smCS was investigated in view of its bio-resorbability.

These films were compared to materials able to activate platelets and induce thrombus formation such as plastic (standard polystyrene for cell culture) and glass (cover slips) as well as a material able to trigger cell death such as latex.

2. Methods

2.1. Production of sugar-modified chitosan films

Chitosan solution was prepared as described in [13]. Briefly, four grams of chitosan powder (Chitosan 95/50 HMC^{+}, Germany) were dissolved in a 1% (w/v) acetic acid aqueous solution until complete dissolution. Dibasic sodium phosphate (7.5 mM), sodium dihydrogen phosphate (22 mM), potassium dihydrogen phosphate (1.5 mM), sodium chloride (125 mM) and-potassium chloride (2mM) were then sequentially added. The solution was filtered under vacuum using a 0.8 μM filter. Finally, D-(+) raffinose pentahydrate (290 mM) or D-(+) sucrose (290 mM) were added to the solution and allowed to dissolve for 2 hours under gentle stirring. About one mL of this solution was poured into a circular mould (1 cm diameter) and dried at 45 °C for 45 minutes in a ventilated oven. The obtained dry film was placed in a 5% (w/v) KOH aqueous solution for 12 hours then, washed in distilled water until neutrality of the wash water.

2.2. Wettability

Contact angle measurements were performed at room temperature with a goniometer (AB Lorentzen & Wettre, Germany) on the surface of smC film in comparison to glass cover slip and plastic (standard polystyrene culture plates) to evaluate the wettability of the surface. Briefly, a drop (4 μL) of human serum was placed on the surface of the specimen. Images of the serum drop were recorded within 10 seconds of deposition by means of a digital camera (FinPix S602 Zoom, Fuji film, Japan). Digital pictures were analysed by ImageJ 1.43v software (NIH, USA) for angle determination. At least five measurements, taken at different positions on each specimen, were carried out on both left and right side of the drop and averaged.

2.3. Atomic force microscopy

Atomic force microscopy (AFM) images of the films were analysed by AFM Nanoscope IIIA (Digital Instruments Inc., USA). Point probe silicon cantilever tip was used in contacting mode by the accompanying software to determine the surface roughness of investigated surfaces. The roughness parameters of each sample was evaluated on three scanned areas of 10μm x 10μm each.

2.4. Procurement and processing of blood perfusates

This procedure was conducted in accordance with the tenets of the Declaration of Helsinki. Following the indication of Italian DLgs no.196/03 (Codex on Privacy) in order to guarantee the respect of the privacy of the patients and the confidentiality of the donors' information. Blood (3.5-4 mL/test) was drawn by venipuncture from four healthy volunteers and added with tri-sodium citrate (0.109 M, 3.2% final concentration) in a 9:1 volumetric ratio to prevent coagulation. Whole blood was used for the haemolysis and thrombus formation tests.

Platelet-rich plasma (PRP) was obtained by centrifugation (400xg for 10 minutes, at room temperature) while platelet-poor plasma (PPP, platelets less than 10.000/μL) by centrifugation at 2000xg for 20 minutes at room temperature.

Coagulation- and factor XII-assays were performed with platelet-poor plasma isolated from whole blood. For platelet function studies, PRP was volume adjusted with PPP to obtain a final physiologic stock platelet count of 3 10^5 platelets μL^{-1}.

2.5. Cell proliferation

Human endothelial cells derived from foetal umbilical vessels (HUVEC) were provided by the American Type Culture Collection (Rockville, MD, USA). Cell monolayer were cultured in complete medium (D-MEM containing antibiotics and 10% foetal calf serum) supplemented with 50 μg mL^{-1} of endothelial cell growth factor (Sigma-Aldrich, USA) and kept in a incubator at 37 °C in a water-saturated atmosphere with 5% CO_2. Endothelial cells were seeded onto smCS films as well as on tissue culture plates (TCPS, Corning, USA) or glasses (20x20 mm, ForLab, Carlo Erba, Italy,) at a density of 1–2.5 10^4 cells cm^{-2} in 24-well plates. After 1, 3 and 7 days, the monolayer was rinsed twice with phosphate buffer solution, PBS, and cells detached from the substrate by 0.02% trypsin in PBS. The number of adherent cells was then, counted with a Burkerhaemocytometer.

2.6. Cell morphology

For morphological characterization, endothelial cells cultured on smCS films were examined by contrast-phase microscopy. After 7 days, the cell monolayer adherent to the film was gently washed with PBS three times. Then, the film was fixed with 2.5% glutaraldehyde in PBS for 1 h at 4 °C. After thorough washing with PBS, the cells were dehydrated through graded alcohol series and positioned under the microscope (Zeiss AxioPhot, Germany) for observation and image recording (Zeiss AxioCam, Germany).

2.7. Cytotoxicity test

Endothelial cells were grown until confluence. The smCS films was cut in small pieces (0.5x0.5 mm) and placed in direct contact with the cell layer for 72 hours. Cells were detached and the resulting suspension was counted in a Burkerhaemocytometer after proper dilution.

Duplicate cell counts on each suspension from 3 culture wells were performed for each substrate investigated. Not less than 50 cells were scored for each counting. Counts from triplicate seeding differed by not more than 10% among replications throughout the experiments.

2.8. Haemolysis assay

Two positive controls, copper and deionised water, and a negative control, glass cover slip, were used in this study, SmCS films were dried and washed three times with PBS and then sterilized by soaking in 75% (v/v) ethanol for 15 minutes. Then, washed 5 times in sterile PBS and kept in the same buffer until use. Thereafter, the samples were put in vacutainers containing sodium citrate (0.109 M, 3.2% w/v final concentration) (Greiner Bio-One International AG, Austria) in which 3.5 mL of healthy volunteers blood was finally collected. The substrates were incubated with blood at 37 °C, with gentle shaking twice every 30 minutes. After 3 hours, 1.5 mL of each vacutainer was centrifuged at 740xg for 10 minutes at room temperature. The obtained pellet was re-centrifuged at 3000g for 15 minutes at room temperature. The haemolyses was quantified on a ADVIA 2120 system (Siemens-Bayer, Germany) using a colorimetric assay.

2.9. Coagulation assays

Human whole blood (3.5 mL) from a healthy volunteer was collected and mixed with an aqueous solution containing sodium citrate, then the human whole blood was centrifuged at 1500g for 15 min at room temperature to separate the blood corpuscles, and the resulting

PPP was used to study the coagulating ability of the CS film. All tests were performed on *IL Coagulation and ELECTRA™* system (Instrumentation Laboratories, USA). The level of Prothrombin Time (PT), activated Partial Thromboplastin Time (aPTT) and Thrombin Time (TT) were determined by using three different kits (Instrumention Laboratory USA): HemosIL ™ RecombiPlasTin 2G is a high sensitivity thromboplastin reagent based on recombinant human tissue factor (RTF) for the quantitative determination in human citrate plasma of Prothrombin Time (PT); HemosIL ™ SynthASil is a high synthetic phospholipids reagent for the *in vitro* determination of APTT (Activated Partial Thromboplastin Time). After incubation at 37 °C for an optimized period of time, calcium is added to trigger the coagulation process and the time required for clot formation is determined; HemosIL ™ Thrombin Time was used for the determination of TT in human citrated plasma.

2.10. Erythrocytes adhesion assay

SmCS films (10x10 mm) were equilibrated in PBS for 1 hour at 37°C. A washed-erythrocytes stock suspension containing 3 10^5 mL^{-1} was poured on plastic and smCS film surfaces and incubated for 30 minutes. The incubation volume was kept low (100 μL) to (a) minimize the floating population of erythrocytes and (b) maintain the total erythrocytes count at a level such as to prevent saturation-levels of adhesion and (c) to prevent other still suspended erythrocytes from contacting the surface. After that, the specimens were rinsed with PBS, fixed with glutaraldheyde and detached from surface with 1% sodium dodecyl sulphate, SDS. Ten microliters of recovered erythrocytes suspension were counted with a Burkerhaemocytometer.

2.11. Platelet adhesion assay

SmCS films, plastic and cover slips glass were sterilized with 75% (v/v) ethanol solution. Then air dried under a laminar-flow hood and rehydrated with 1 mL of sterile PBS for 1 hour. The surfaces were overlaid with 300 μL PRP at 37 °C for 2 hours. Then, the films were washed three times in PBS with mild shaking to remove non- or poorly-adherent platelets. After that, the specimens were rinsed with PBS, fixed with glutaraldheyde and cells detached with 1% SDS. Ten microliters of recovered platelets suspension were counted with a Burkerhaemocytometer.

3. Platelet immunofluorescence

The platelet count in the PRP was adjusted to 300.000 μL-1 by dilution with homologous PPP. After 1 hour contact of PRP with the different specimens at 37 °C, samples were washed with PBS, followed by fixation with 3% (w/v) paraformaldehyde and incubated with 1% (w/v) BSA in PBS. Labelling of the platelets was performed with a mouse monoclonal antibody CD62P (anti-P-Selectin, Santa Cruz Biotechnology, USA) at dilution 1:100, followed by 1:200 diluted monoclonal goat anti-mouse IgG antibody, FITC conjugated (Sigma-Aldrich, USA).

4. Platelet morphology

The platelets-coated testing surfaces were fixed with freshly prepared 2.5% glutaraldehyde for 20 minutes. After washing with PBS, the samples were dehydrated in a graded-ethanol series (50, 70, 90, and 100% v/v) for 15 minutes each and allowed to dry at room temperature. The platelet-attached surfaces were carbon sputter coated under vacuum to a thickness of 100–200 Å and examined at 10 kV using a Cambridge StereoScan 200 microscope (Cambridge Scientific Instruments, UK).

4.1. Platelet aggregation

The blood samples were collected in tubes containing PPACK (D-phenylalanyl-L-prolyl-L-arginine chloromethyl ketone) as anticoagulant. Platelet aggregation was measured by means of light transmission aggregometry using Born's turbidimetric procedure and the PPACK-4 Platelet Aggregation Chromogenic Kinetic System (Helena Laboratories, USA). Briefly, 250 μL of PRP were incubated with specimen surfaces for 10 (baseline) and 60 minutes. Thereafter,the PRP were placed in a cuvette containing a metal stir bar in the absence or in the presence (positive control) of the pro-aggregation agent, adenosine diphosphate (ADP) 20 μM. Upon the addition of ADP the platelets started to aggregate thus increasing light transmission through the sample. The degree of platelet aggregation was expressed as the maximum percentage change in light transmission from PPP used as baseline. The obtained values were expressed as mean of two measurements.

4.2. Complement activation assay

The test, based on Complement Reagents Kit (Siemens Healthcare Diagnostic, Germany) was performed on BCT Siemens coagulometer (Siemens, Germany). The test focused on the ability of the complement system to lyse a standard suspension of sheep erythrocytes, sensitized with a rabbit anti-serum against sheep erythrocytes. Briefly, 1 mL of fresh blood samples previously incubated for 1 hour with different substrates were incubated with sensitized erythrocytes to investigate the complement activation. Diminished levels of individual components (e.g. due to prior activation by a foreign surface) result in a prolongation of the time taken for lyses. The time necessary for the lyses of a defined amount of erythrocytes is used as basis for determining the complement activity [14,15]. The results were evaluated using a reference curve prepared by a serial dilution of standard plasma with isotonic saline to give 100% of complement activity, 75% (75% of plasma + 25% saline) down to 10% of complement activity (10% of plasma + 90% saline).

4.3. Statistical analysis

Data were expressed as means ± standard deviation (SD). Where not differently stated, measurements were conducted at least in triplicate. Chi-square test or Student's t-test on unpaired data was used to assess the statistical significance of the difference between the results obtained from the tested specimens (Kaleida-Graph, Synergy Software, USA). Statistical significance was assumed at a confidence level of 95% ($p < 0.05$).

http://dx.doi.org/10.5772/52706

5. Results

5.1. Physico-chemical characterisation

As already stated, the sugar added to the chitosan solution during the preparation of the smCS film was not retained in the final structure of the film.This assumption was mainly based on FT-IR spectra analysis for the identification of the absorption bands relevant to vibration of functional groups of chitosan [13]. The addition of phosphate salts and D-(+) raffinose to the chitosan solution used for film preparation led to non dramatic modifications in the IR spectrum of chitosan. The observation of the 1700-1500 cm^{-1} region evidenced that the amide I band (C=O in amide group) wavenumber was lower than the value for chitosan powder (1664 cm^{-1}) for all the prepared films and particularly for those prepared from a solution that did not contain the sugar [13]. This was interpreted as the result of a lower mobility of the C=O group in the film due to its involvement in the week bound formation in the solid structure. The incorporation of phosphate salts and significant amount of sugar in the chitosan solution used for film preparation reduced this effect. On the other hand, the amino group band of films prepared from a solution that did not contain the sugar was at a lower wavenumber (1588 cm^{-1}) than from chitosan powder (1592 cm^{-1}), while it was practically unchanged in film prepared from the sugar containing solution (1590 cm^{-1}).

D-(+)raffinose FT-IR spectrum evidenced characteristics bands at 2936 and 1649 cm^{-1}. Interestingly, no trace of this bands was found in the FT-IR spectra of the chitosan film prepared from solutions containing D-(+)raffinose. Similarly, no trace of the characteristic series of peaks between 2994 and 2914 cm^{-1} of the sucrose powder was found in the spectrum of the film prepared from a solution containing a high amount of sucrose [13].

These observations allowed to conclude that the excipients added to chitosan in the film forming solutions though not retained in the solid film, interact or interfere with chitosan chains during the film formation likely acting as viscosity modifiers during the solidification/gelation process.

5.2. Wettability

Contact angle measurements were performed by using serum droplets on plastic surface and on smCS film. As expected, plastic showed the least wettable surfaces with significantly higher contact angle (50° ± 6.3) compared to smCS film (15° ± 0.1) (Chi Square $P< 0.001$), thus confirming the high hydrophilicity of smCS [13].

The hydrophilicity of the sm CS film was also investigated by measuring the swelling index in water at the equilibrium according to the following equation:

$$S_w = \frac{W_s - W_d}{W_s} x100 \quad (1)$$

where W_s and W_d represent the weight of the fully hydrated and the dry film respectively. The smCS film afforded a degree of swelling at the equilibrium more than 3 order of magnitude (1285%) higher than that of the dry film. These data confirmed the very high hydrophilicity of the films obtained by adding a sugar to the solution used for the film preparation.

5.3. Roughness (AFM measurements)

The AFM analysis (Figure 1) revealed that the plastic specimen exhibited rather low surface roughness (average= 28 nm) in contrast to smCS film that showed a roughness approximately 1.7-fold higher, around 50 nm. It is interesting to note the almost regular appearance of groove and pits in smCS compared with plastic surface.

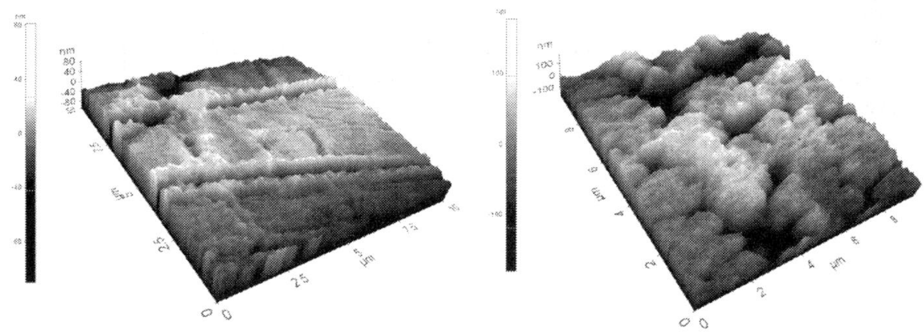

Figure 1. AFM 3D image of (A) standard colture plastic dish (plastic) and (B) smCS film surfaces.

5.4. Adhesion and proliferation assay of endothelial cells

As shown in Figures 2 endothelial cells attached (A), extended and proliferated (B) very well on all surfaces tested. Cell attachment (panel A, C) and proliferation (panel B) on smCS films were comparable to control cells grown on standard tissue culture surface (plastic). Contrast-phase microscopy showed that cells were well attached to the different surfaces and closely packed maintaining their original shapes. Moreover, endothelial cells did not evidence any morphological indication of cell death 72 hours after seeding (panel D). The counts of cells showed little variation for the three surfaces used. In the case of plastic surface (control) the growth of HUVEC reached the values of 21284 ± 650 cm^{-2}, while in the case of smCS reached a lower value of 19805 ± 305 cm^{-2} similarly to that obtained on glass surface (19543 ± 1050 cm^{-2}).

5.5. In vitro cytotoxicity

Data relevant to cell growth in the presence of small pieces of smCS film or latex (positive control) are reported in Table 1.

http://dx.doi.org/10.5772/52706

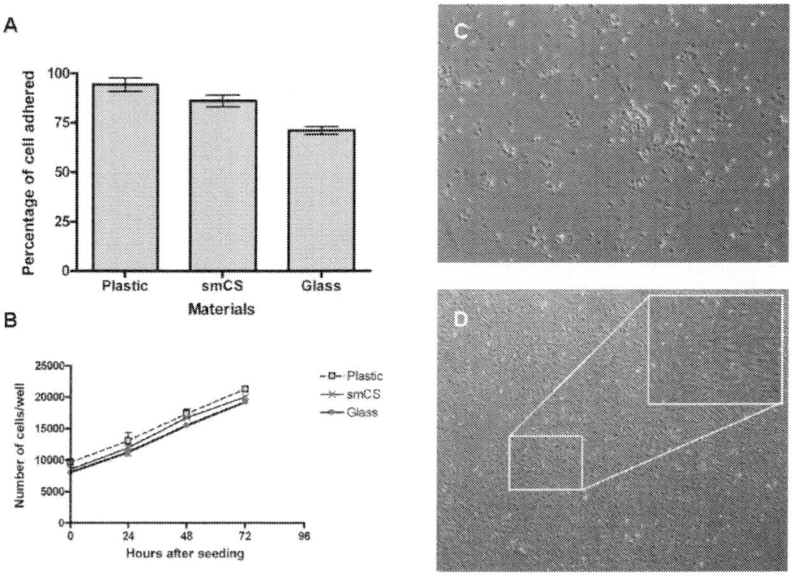

Figure 2. A) Percentage of cells adhered after 24 hoursand (B) proliferation assay of endothelial cells on the different surfaces tested. Pictures taken at the optical microscope, in phase contrast (40x), showing the morphology of endothelial cells 8 hours (C) and 72 hours (D) after seeding on smCS film.

	Hours after seeding			
	0	24	48	72
Latex (positive control)	10000 (± 239)	924 (± 25)	714 (±16)	50 (±30)
smCS	10000 (± 360)	11420 (±100)	16070 (±290)	18570 (± 200)

Table 1. Number of endothelial cells attached to the different substrates in the presence of latex or smCS film fragments.

The initial plating corresponds to the number of cells attached to the substrate 6 hours after their inoculation into the well. The measured plating efficiency was around 95%. When smCS fragments were present in the colture medium, a progressive increase of cell numbers was observed, while in the presence of latex a progressive detachment was noticed with almost all plated cells detached from the substratum after 72 hours.

These results indicate that smCS film were not cytotoxic while latex, as expected, was found to markedly affect endothelial cell survival.

5.6. Haemolysis assay

Haemolysis of red blood cells was used to evaluate the membrane damaging potential of the surface of the smCS film. Two positive controls, distilled water and Copper, and one negative control, glass, were used.

Substrate	Haemolysis (%)
Distilled Water	97 (± 5)
Copper	7 (± 2)
Plastic	5 (± 1)
Glass	2 (± 1)
smCS	1 (±3)

Table 2. Percentage of haemolysis measured on different substrates. The tested surfaces were incubated with whole blood for 1 hour. Distilled water was used as positive control.

As shown in Table 2, distilled water resulted in about 100% haemolysis, while Copper led to 7% haemolysis. Glass and smCS film caused negligible haemolysis (within the experimental error) indicating very low membrane damaging properties of smCS material.

5.7. Blood coagulation assay

The effects of the biomaterial on coagulation process were tested by means of the (aPTT), the (PT) and the (TT) selected as reliable measurements of the capacity of blood to coagulate through the intrinsic, extrinsic and common coagulation mechanisms, respectively. As shown in Figure 3 the values obtained for PT, TT and aPTT were similar to those observed for human plasma, thus indicating that all materials tested, including smCS, did not affect coagulation pathways.

Substrates	Erythrocyte lyses time (seconds)
Plasma (*control*)	35.4
Glass	38.2
Plastic	35.7
SmCS film	36.3

Table 3. Erythrocyte lyses time determined by plastic and glass surfaces in comparison with smCS film.

5.8. Complement activation assay

The erythrocyte lyses time observed and reported in Table 3 shows no significant difference among the material studied and the control. The data presented demonstrate that smCS is a nonreactive biomaterial that does not directly activate complement.

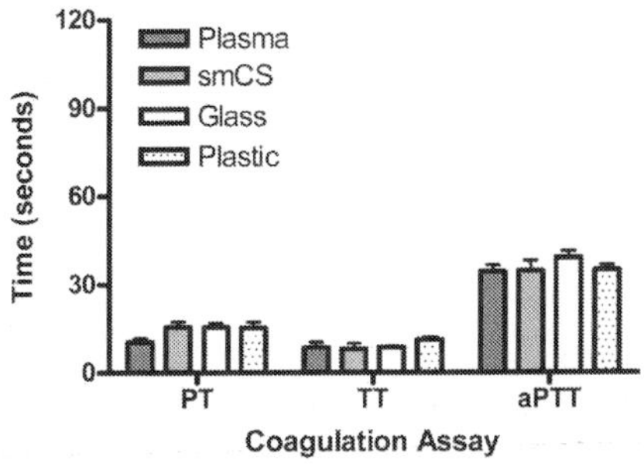

Figure 3. Effect of the different surfaces on coagulation time tested by means of the (aPTT), the (PT) and the (TT).

5.9. Erythrocytes and platelets adhesion assay

In Table 4 the number of cells detached with SDS from the different surfaces after adhesion test is reported. The smCS film presented a lower overall erythrocyte and platelet adhesion in comparison to plastic surface.

Materials	Surface Adhesion	
	Erythrocytes μL⁻¹	***Platelets μL⁻¹***
Plastic	40 (± 13)	11(± 8)
Glass	-	17 (± 6)
smCS	8 (±3)	5 (± 2)

Table 4. Numbers of erythrocytes and platelets adhered to the studied surfaces

The test showed a high significant difference in the number of adhered erythrocytes on materials studied ($p<0.0001$): the erythrocyte adhesion on smCS film was about 5 fold less than the adhesion on plastic surface. A similar behaviour was observed for platelets. In fact, the platelets recovered from plastic and glass surfaces ranged from 2 to 3 fold more than platelets recovered from smCS film surface.

5.10. Platelet aggregation

This test was performed in order to investigate the ability of plastic, smCS film and glass surfaces to induce platelet aggregation. The presence of ADP (adenosine-diphosphate, a pro-aggregation agent) determined a normal profile of platelets aggregation (range 90-95% after 5 minutes of incubation). Subsequently, the influence of materials on platelets aggregation in the absence of ADP was studied. There were no differences between materials observed at baseline (10 min) and after 1 hour of incubation with the substrates. smCS induced slightly higher aggregation of platelets (5-6%) compared to plastic (2.5%) or glass (less than 2%). However, these differences have to be considered with caution, as the coefficient of variation estimated with human plasma in the absence of ADP was around 10%.

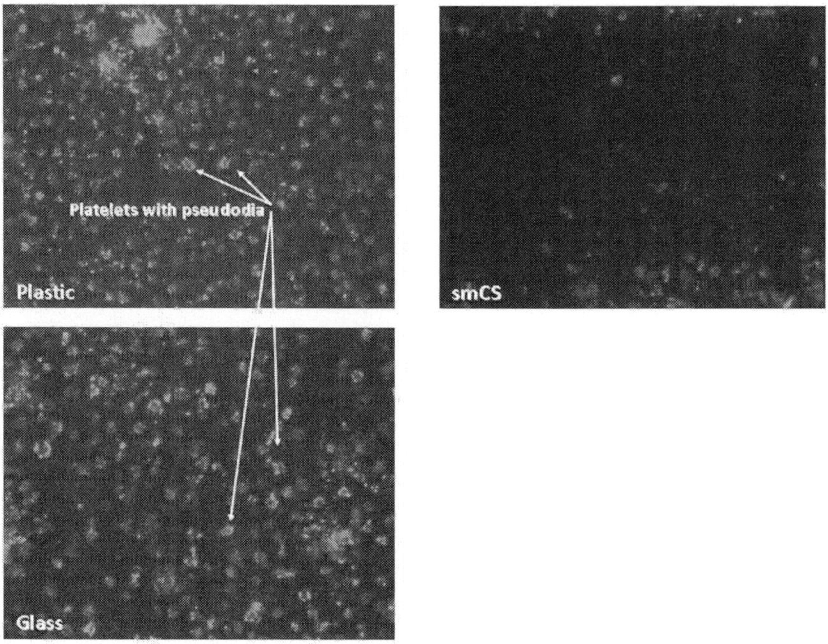

Figure 4. Fluorescence microscopy (100X)images of human platelets immunodecorated with CD62P (p-Selectin). Arrows indicate the presence of pseudopodia.

5.11. Platelet activation assay

Platelet activation was studied by the membrane expression of P-Selectin using the CD62P antibody. The expression of P-Selectin was evident on platelets adherent to plastic and glass surfaces and was negligible on platelets settled on smCS films (Figure 4).

On glass and plastic (see arrows) the analysis of morphology showed several fully spread platelets expressing pseudopodia with the occurrence of focal clumps. This was also evident when platelets were examined by Scanning Electron Microscopy, SEM (Figure 5B and 5C).

SmCS film (Figure 5A) induced very limited morphological changes over the 90 minutes of contact: platelets remained mostly discoid without the occurrence of pseudopodia.

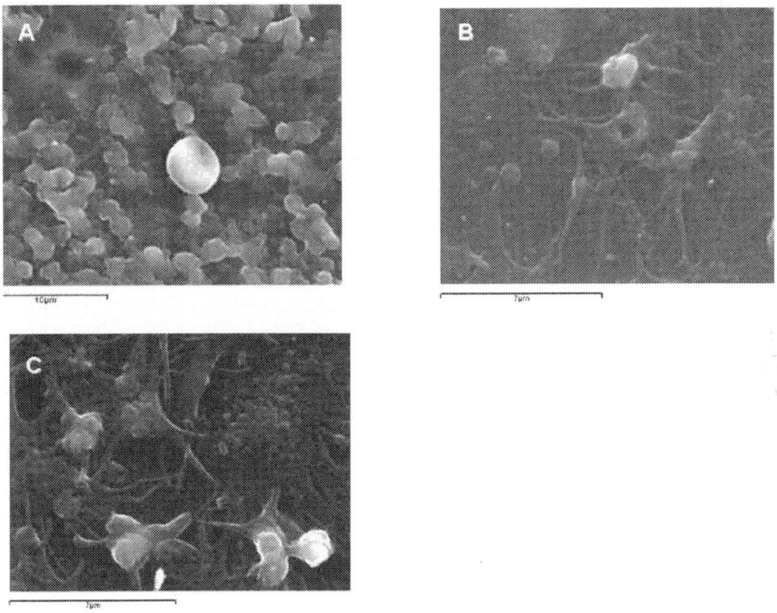

Figure 5. Scanning Electron Microscope images of platelets incubated onto different surfaces. (A) smCS film, (B) plastic, (C) glass.

6. Discussion

For blood-contact applications, haemocompatibility is largely determined by specific interactions with blood and its components [16]. Many, if not all, blood-contacting biomaterials are able to cause different undesired host responses like thrombosis, inflammatory reactions and infections.

The coagulation system and platelets are the main factors for thrombus formation on biomaterials and represent a major unmet problem in the design of vascular implants and blood-handling systems [17].

It is known that the endothelium is an active organ that maintains vessels integrity and prevent thrombosis and intimal hyperplasia [18,19]. Hence, biomaterials able to promote *in situ* endothelialisation of implants would be highly desirable.

Studies involving in vitro endothelialisation of grafts with cultured endothelial cells prior to implantation have shown that a confluent endothelium is able to prevent trombogenic complications and improves long-term patency [20,21]. Thus, taking into account that the endothelialisation of the blood-contacting polymeric materials is an important pre-requisite for the success of the synthetic vascular grafts [22] we firstly investigated the ability of endothelial cells to adhere and proliferate on smCS film. The results obtained, in agreement with those shown in our previous paper [13], showed only little variation among the surfaces tested (glass, plastic and smCS). However, from the cell proliferation and morphology it was very difficult to discriminate difference in cytophilicity among the surfaces tested. Furthermore, the presence of fragmented smCS did not induce any decrement in the total number of endothelial cells compared to latex that, on the contrary, strongly affected cell survival.

The high hydrophilicity of smCS, indicated by the low contact angle, could ease the interaction with the bipolar extra-cellular matrix proteins such as fibronectin and vitronectin. Furthermore, the reduced cationic nature, due to a water shell does not allow anionic proteins such as collagen and fibronectin to dissociate from CS surface in a physiological environment. This aspect is in agreement with the conclusion of [23] who reported that a hydrophilic surface is good for anti-non-specific protein adsorption. It was recently reported that the affinity for water of the cell-material interface seems to be a chief parameter in controlling cell adhesion, migration and differentiation [24].

Stevens and George [25] recognized that cells are sensitive to microscale patterns of chemistry and topography, and Dalby [26] noted that cell behaviour is directly influenced by the surface structures such as grooves, pits, or ridges.

In this paper AFM images of smCS films evidenced a topographically patterned surface. In the light of the above reported literature, this observation can be used to speculate about the enhanced adhesion and proliferation of vascular cells compared to conventional, CS films previously observed in [13].

Surface properties such as wettability, surface topography and charge are known to affect endothelial cells attachment and growth [8]likely by altering the rate of the amount of adsorbed proteins and their conformational changes [27,28]. The effect of surface materials on erythrocyte aggregation and platelet adhesion/activation becomes a chief parameter in haemocompatibility studies.

Several years ago Malette and co-workers [29] ascribed the pro-coagulation properties of chitosan to the negative charged surface of erythrocytes, while [30]showed that chitosan may induce the adhesion of erythrocytes.

In the present study, the surface of smCS films induced only a limited erythrocytes agglomeration, thus indicating that smCS surface neither captures erythrocytes nor forms a three-dimensional network structure with these cells.

The lack of erythrocyte aggregation may be likely due to a polymer chains rearrangement that masks the cationic nature of chitosan surface. Such a rearrangement can be ascribed to the larger amount of water in smCS films as described in [13].

One of the most important findings of this work is the observed difference in platelets morphology seeded on smCS in comparison with glass or plastic. On the latter surfaces platelets appeared flat with interconnecting pseudopodia coupled to strong P-Selectin membrane expression.On the contrary, the platelets on smCS films were discoidal, and neither pseudopodia formation nor a P-Selectin membrane translocation was observed.

This finding could be attributed to a new conformation of the adsorbed plasma proteins on glass or plastic that could have facilitated platelet aggregation. Indeed, it is well known that the surface topography can induce a spatial reorganization of adsorbed proteins as well as how this phenomenon occurs [31]. In contrast, when the adsorbed proteins maintain their native state, they do not support platelet adhesion and aggregation [32].

The absence of platelet activation on smCS surfaces suggests this outcome.

As far as the surface morphology is concerned, it has been reported that platelets adhere in similar manner on smooth and rough surfaces when tested under static conditions [33]. Similarly Ward et al. [34] concluded that it is not the roughness *per se* which affects the platelet adhesion.

One decade ago, Suzuki and Minami [35,36] showed that Chitosan depleted complement proteins from plasma, suggesting that chitosan activates complement. A greater depletion of complement activity was seen for a highly de-acetylated form of chitosan [36]. It is however, important to note that the results obtained about the complement activation were based on binding and depletion assays. This complement depletion can equally be explained by assuming a tight binding to the chitosan surface without activation [37].

The results presented here indicate that although large amounts of serum were deposited on smCS surface no activation of the complement system occurred, suggesting that the complement is not directly activated by the smCS surface in the process of blood coagulation.

Haemolysis testing of biomaterials has been advocated for, and used in, standard biological safety testing of materials for more than 30 years. The results of test for haemolysis should be considered with care even if they represent the only recommended test for some medical devices as stated in Part 4 of ISO 10993 guideline.

Different papers have reported that chitosan promotes surface-induced haemolysis likely through an electrostatic interactions [38]. In the present work, in the presence of whole blood smCS triggered less than 5% of haemolysis that, along with the low erythrocyte adhesion, indicates a wide safety margin in blood contacting applications and suitability for vascular implants.

In the process of haemostasis, the activation of platelet adhesion and aggregation could represent an initial and critical step. Here we showed that the surface of smCS films does not interfere with coagulation mechanism and supportswell endothelial cell adhesion and proliferation even if [39] reported that the haemostatic mechanism of chitosan may be independent of the classical coagulation cascade.

7. Conclusion

In this paper we demonstrated that the simple introduction of a viscosity modifier, such as a polysaccharide, during the process of production of chitosan films affords chitosan structures (smCS film) with improved capability to induce surface endothelialisation.

This structure is moreover, characterized by a high degree of haemocompatibility and does not induce clots formation.

These findings are of particular interest as they add new information with respect to the presently available literature and they put new light on the use of chitosan for producing surfaces that has to get in contact with blood.

As a matter of fact, the haemostatic properties of chitosan have to be considered more carefully as we have demonstrated that they could be dramatically reduced by an improvement of the hydrophilicity of the chitosan film surface.

Finally, from the results presented in this work, we can conclude that the sugar modified chitosan film could be envisaged as a new material for the design the luminal portion of vessel prosthesis based on a natural and bio-resorbable polymer.

Acknowledgements

This work was partially supported by a grant from Emilia Romagna Region, Italy, through its *"Programma di Ricerca Regione Università 2007-2009 Area Tematica 1B Medicina Rigenerativa"*.

Author details

Antonello A. Romani[1], Luigi Ippolito[2], Federica Riccardi[3], Silvia Pipitone[4], Marina Morganti[5], Maria Cristina Baroni[5], Angelo F. Borghetti[4] and Ruggero Bettini[6]

*Address all correspondence to: bettini@unipr.it

1 Department of Surgery, O.U. of General Surgery, University Hospital of Parma, Italy

2 Department of Pathology and Medicine of Laboratory University Hospital Parma, Italy

3 Department of Medicine poly-specialist 2, University Hospital Parma, Italy

4 Department of Experimental Medicine, University of Parma, Italy

5 Department of Internal Medicine and Biomedical Sciences, University of Parma, Italy

6 Department of Pharmacy, University of Parma, Italy

References

[1] Baig K, Fields RC, Gaca J, Hanish S, Milton LG, Koch WJ, Lawson JH. A porcine model of intimal-medial hyperplasia in polytetrafluoroethylene arteriovenous grafts. J Vasc Access 2003;4:111–7.

[2] Hardy JG, Scheibel TR. Silk-inspired polymers and proteins. Biochem. Soc. Trans. 2009;37:677–81.

[3] Chandy T, Sharma CP. Chitosan--as a biomaterial. Biomater Artif Cells Artif Organs 1990;18:1–24.

[4] Shi C, Zhu Y, Ran X, Wang M, Su Y, Cheng T. Therapeutic Potential of Chitosan and Its Derivatives in Regenerative Medicine. Journal of Surgical Research 2006;133:185–92.

[5] Ramshaw JAM, Peng YY, Glattauer V, Werkmeister JA. Collagens as biomaterials. J Mater Sci Mater Med 2009;20 Suppl 1:S3–8.

[6] Hirano S, Tsuchida H, Nagao N. N-acetylation in chitosan and the rate of its enzymic hydrolysis. Biomaterials 1989;10:574–6.

[7] Kind GM, Bines SD, Staren ED, Templeton AJ, Economou SG. Chitosan: evaluation of a new hemostatic agent. Curr Surg 1990;47:37–9.

[8] van Wachem PB, Hogt AH, Beugeling T, Feijen J, Bantjes A, Detmers JP, van Aken WG. Adhesion of cultured human endothelial cells onto methacrylate polymers with varying surface wettability and charge. Biomaterials 1987;8:323–8.

[9] Wedmore I, McManus JG, Pusateri AE, Holcomb JB. A special report on the chitosan-based hemostatic dressing: experience in current combat operations. J Trauma 2006;60:655–8.

[10] Ong S-Y, Wu J, Moochhala SM, Tan M-H, Lu J. Development of a chitosan-based wound dressing with improved hemostatic and antimicrobial properties. Biomaterials 2008;29:4323–32.

[11] Rao SB, Sharma CP. Use of chitosan as a biomaterial: studies on its safety and hemostatic potential. J. Biomed. Mater. Res. 1997;34:21–8.

[12] Sagnella S, Mai-Ngam K. Chitosan based surfactant polymers designed to improve blood compatibility on biomaterials. Colloids Surf B Biointerfaces 2005;42:147–55.

[13] Bettini R, Romani AA, Morganti MM, Borghetti AF. Physicochemical and cell adhesion properties of chitosan films prepared from sugar and phosphate-containing solutions. Eur J Pharm Biopharm 2008;68:74–81.

[14] Kolde HJ, Deubel R. Development of a rapid kinetic assay for the function of the classical pathway of the complement system and for C2 and C4. J Clin Lab Immunol 1986;21:201–7.

[15] Siedentopf HG, Lauenstein K, Fischer H. [On the automatic registration of hemolysis by serum complement and lysolecithin]. Z. Naturforsch. B 1965;20:569–74.

[16] Angelova N, Hunkeler D. Rationalizing the design of polymeric biomaterials. Trends Biotechnol. 1999;17:409–21.

[17] Brash JL. The fate of fibrinogen following adsorption at the blood-biomaterial interface. Ann. N. Y. Acad. Sci. 1987;516:206–22.

[18] Risler NR, Cruzado MC, Miatello RM. Vascular remodeling in experimental hypertension. ScientificWorldJournal 2005;5:959–71.

[19] Myhre HO, Halvorsen T. Intimal hyperplasia and secondary changes in vein grafts. Acta Chir Scand Suppl 1985;529:63–7.

[20] Pollara P, Alessandri G, Bonardelli S, Simonini A, Cabibbo E, Portolani N, Tiberio GA, Giulini SM, Turano A. Complete in vitro prosthesis endothelialization induced by artificial extracellular matrix. J Invest Surg 1999;12:81–8.

[21] Crombez M, Chevallier P, Gaudreault RC, Petitclerc E, Mantovani D, Laroche G. Improving arterial prosthesis neo-endothelialization: application of a proactive VEGF construct onto PTFE surfaces. Biomaterials 2005;26:7402–9.

[22] Bengtsson L, Rådegran K, Haegerstrand A. A new and simple technique to achieve a confluent and flow resistant endothelium on vascular ePTFE-grafts using human serum. Eur J Vasc Surg 1994;8:182–7.

[23] Lu D, Lee S, Park K. Calculation of Solvation Interaction Energies for Protein Adsorption on Polymer Surfaces. J Biomat Sci-Polym E 1991;3:127–47.

[24] Ayala R, Zhang C, Yang D, Hwang Y, Aung A, Shroff SS, Arce FT, Lal R, Arya G, Varghese S. Engineering the cell-material interface for controlling stem cell adhesion, migration, and differentiation. Biomaterials 2011;32:3700–11.

[25] Stevens MM, George JH. Exploring and engineering the cell surface interface. Science 2005;310:1135–8.

[26] Dalby M, McCloy D, Robertson M, Wilkinson C, Oreffo R. Osteoprogenitor response to defined topographies with nanoscale depths. Biomaterials 2006;27:1306–15.

[27] Steele JG, Dalton BA, Johnson G, Underwood PA. Adsorption of fibronectin and vitronectin onto Primaria and tissue culture polystyrene and relationship to the mechanism of initial attachment of human vein endothelial cells and BHK-21 fibroblasts. Biomaterials 1995;16:1057–67.

[28] Burmeister J, Vrany J, Reichert W, Truskey G. Effect of fibronectin amount and conformation on the strength of endothelial cell adhesion to HEMA/EMA copolymers. J. Biomed. Mater. Res. 1996;30:13–22.

[29] Malette WG, Quigley HJ, Gaines RD, Johnson ND, Rainer WG. Chitosan: a new hemostatic. Ann. Thorac. Surg. 1983;36:55–8.

[30] Klokkevold PR, Lew DS, Ellis DG, Bertolami CN. Effect of chitosan on lingual hemostasis in rabbits. J. Oral Maxillofac. Surg. 1991;49:858–63.

[31] Curtis A, Wilkinson C. New depths in cell behaviour: reactions of cells to nanotopography. Biochem Soc Symp, vol. 65, 1999, pp. 15–26.

[32] Tanaka M, Motomura T, Kawada M, Anzai T. Blood compatible aspects of poly (2-methoxyethylacrylate)(PMEA)--relationship between protein adsorption and platelet adhesion on PMEA surface. Biomaterials 2000; 21:1471-1481.

[33] Zingg W, Neumann AW, Strong AB, Hum OS, Absolom DR. Effect of surface roughness on platelet adhesion under static and under flow conditions. Can J Surg 1982;25:16–9.

[34] Ward CA, Forest TW. On the relation between platelet adhesion and the roughness of a synthetic biomaterial. Ann Biomed Eng 1976;4:184–207.

[35] Minami S, Suzuki H, Okamoto Y, Fujinaga T, Shigemasa Y. Chitin and chitosan activate complement via the alternative pathway. Carbohyd Polym 1998;36:151–5.

[36] Suzuki Y, Okamoto Y, Morimoto M, Sashiwa H, Saimoto H, Tanioka S, Shigemasa Y, Minami S. Influence of physico-chemical properties of chitin and chitosan on complement activation. Carbohyd Polym 2000;42:307–10.

[37] Marchand C, Bachand J, Perinet J, Baraghis E, Lamarre M, Rivard GE, De Crescenzo G, Hoemann CD. C3, C5, and factor B bind to chitosan without complement activation. J Biomed Mater Res A 2010;93A:1429–41.

[38] Yao K, Li J, Yao F. Chitosan-Based Hydrogels: Functions and Applications. 1st ed. CRC Press; 2011.

[39] Khan T, Peh K, Ch'ng H. Mechanical, bioadhesive strength and biological evaluations of Chitosan films for wound dressing. J Pharm Pharm Sci 2000;3:303–11.

Chapter 7

Amelioration of Blood Compatibility and Endothelialization of Polycaprolactone Substrates by Surface-Initiated Atom Transfer Radical Polymerization

Shaojun Yuan, Gordon Xiong, Ariel Roguin,
Swee Hin Teoh and Cleo Choong

Additional information is available at the end of the chapter

http://dx.doi.org/10.5772/52646

1. Introduction

Attempts to develop synthetic vascular grafts for the replacement of diseased vascular sections have been an area of active research over the past decades [1]. However, thrombosis formation as a result of platelet adhesion to the luminal surface of synthetic graft and restenosis caused by host inflammatory remain a challenge, especially for small-diameter (<6 mm) graft replacement [2,3]. Therefore, the haemocompatibility of the biomaterial used in the graft is a prerequisite for clinical success. As the result, various strategies have been developed to improve the blood compatibility of biomaterial surfaces, including the surface immobilization of anti-coagulants such as heparin [4] and sulfated silk fibroin [5], the incorporation of polyethylene oxide or negatively charged side chains [6,7], and surface passivation with protein layers, such as albumin [8]. Despite the efficacy of these approaches in preventing acute thrombogenesis, concerns remain on the drug elution lifespan, with possible consequence of late thrombosis [9]. To avoid undesirable blood-material interaction, the seeding of autologous endothelial cells (ECs) onto the luminal surface of the graft is considered to be an ideal approach to increase the patency of synthetic grafts [10]. Many studies have indicated that endothelial cells release factors that regulate thrombogenesis and platelet activation [11], while delayed or absent stent endothelialization has been implicated in late thrombosis and adverse clinical outcomes [13]. Thus, rapid endothelialization of vascular grafts is of great importance for blood-contacting vessels for long-term patency.

Due to its slow degradation rates in vivo (2-4 years) [14], good mechanical strength, and biocompatibility with vascular cell types [15,16], polycaprolactone (PCL) is currently being extensively investigated as scaffolds for vascular tissue engineering applications [17-21]. However, the intrinsic hydrophobicity and poor cytocompatibility of PCL substrates lead to poor affinity for cell adhesion, thereby restricting their applications as blood-contacting devices. Consequently, surface modification of PCL is necessary to improve cell adhesion and proliferation. Functional polymer brushes containing reactive hydroxyl (-OH), carboxyl (-COOH) or amine (-NH2) groups have been successfully grafted onto the PCL surfaces using γ-ray irradiated, ozone or photo-induced polymerization grafting to introduce hydrophilicity [9,16,22-24]. These flexible reactive groups on the polymer brushes are well-suited to conjugate bioactive macromolecules for improved cytocompatibility. However, γ-ray irradiated, ozone or photo-induced polymerization grafting of polymer brushes has several limitations, including low density of grafting due to steric hindrance, uncontrollable graft yield of polymer brushes, and undesired formation of a covalent bond between reactive groups on the polymer brushes and the surface [25]. Hence, an alternative grafting approach that allows control over brush density, polydispersity and composition is desired.

One such alternative is the use of surface-initiated atom transfer radical polymerization (ATRP) approach to covalently graft polymer brushes in a tunable and controllable manner [26]. This approach allows the preparation of well-defined dense polymer brushes containing reactive pendant groups (e.g. -OH, -COOH, or epoxide groups), and provides highly reactive binding sites for functional biomolecules.[27] As a result, surface-initiated ATRP provides a promising approach to fabricate PCL substrates with well-defined polymer brushes of controlled length and density, as well as tunable grafting density of biomacromolecules. However, to the best of our knowledge, only few studies have been devoted to modifying biodegradable polyester polymers using surface-initiated ATRP to improve their cytocompatibility or blood compatibility [27,28]. Also, the functionality of the attached cells was not thoroughly investigated in those studies.

As such, the aim of the current study is to utilize the surface-initiated ATRP method to tailor PCL substrates with dense functional P(GMA) brushes and high-density immobilized gelatin to improve their properties for cell attachment and proliferation. Each functionalization step was ascertained by XPS, AFM and water contact angle measurements. The cytocompatibility of the functionalized PCL substrates was evaluated using human umbilical vein endothelial cells (HUVECs) and the effect of different surface properties on the regulation of the thrombogenicity of the attached cells was also investigated.

2. Materials and methods

2.1. Materials

Polycaprolactone pellets (PCL, average M_n45000), 1,6-hexanediamine (98%), glycidyl methacrylate (GMA, >97%), 2-bromoisobutyrl bromide (BIBB, 98%), 2,2'-bipyridine (bpy, 98%), dichloromethane (anhydrous, >99.8%), triethylamine (TEA, 98%), isopropyl alcohol, hexane (anhydrous, >95%), copper (I) bromide (CuBr, 99%), copper (II) bromide ($CuBr_2$, 98%), and gelatin (Porcine skin, Type A) were obtained from Sigma-Aldrich Chemical Co. (St. Louis,

MO, USA), and were used without further purification. GMA was passed through a silica gel column to remove the inhibitor, and stored under a nitrogen atmosphere at -4°C. All the other chemical reagents and solvents were used as received. Human Umbilical Vein Endothelial cells (HUVECs, ATCC CRL-1730™) were purchased from American Type Culture Collection (Manassas, VA, USA). Cell culture medium (MCDB131), heparin and paraformaldehyde (4%, v/v) were obtained from Sigma-Aldrich Chemical Co. Medium supplements, such as Foetal Bovine Serum (FBS), penicillin, amphotericin, bovine brain extract, streptomycin, and Trypsin-EDTA (0.25%), were purchased from Life Technologies (Carlsbad, CA, USA). LIVE/DEAD® cell viability assay reagent, AlamarBlue™ reagent, tissue thromboplastin (human brain extract), ellagic acid, and DAF-FM Diacetatewere obtained from Life Technologies. The P-selectin assay reagents were obtained from Serotec Co. (Kidlington, Oxford, UK).

2.2. Aminolysis of PCL film substrates and immobilization of ATRP initiator

Polycaprolactone (PCL) films were prepared by solution casting method using previously established methods [29]. Briefly, 5 g of the PCL pellets was dissolved in 40 ml of dichloromethane to form the PCL solution. The polymer solution was then cast onto the glass substrate with predetermined thickness using the automatic film applicator (PA-2105, BYK). The solvent was removed at room temperature by slow evaporation over a 24 h period, and was further dried in a vacuum oven for another 24 h at 35 °C to obtain the translucent PCL films with a thickness of about 150 μm. The resultant PCL films were cut into round-shaped specimens with a diameter of 2 cm. The activation of PCL substrates was performed by aminolysis treatment using a procedure previously described [30,31]. Briefly, the PCL films were immersed in a 10% (w/w) 1,6-hexanediamine and isopropanol mixture at 40 °C for a predetermined time. After aminolysis treatment, the resultant PCL-NH_2 surfaces were thoroughly rinsed with copious amounts of deionized water and isopropanol, respectively, to remove free 1,6-hexanediamine, and dried in a vacuum oven at 30 °C for 24 h.

The introduction of an alkyl halide ATRP initiator on the PCL-NH_2 surface was accomplished through the reaction of the amino groups with 2-bromoisobutyrate bromide (BIBB) [32]. The PCL-NH_2 films were immersed in 30 ml of anhydrous hexane solution containing 1.0 ml (7.2 mmol) of triethylamine (TEA). After 30 min of degassing with nitrogen, the reaction mixture was cooled in an ice bath, and 0.89 ml (1.65g, 7.2 mmol) of BIBB was added dropwise via a syringe. The reaction was allowed to proceed with gentle stirring at 0 °C for 2 h and then at room temperature for 12 h. The resulting surface (referred to as the PCL-Br surface) was washed thoroughly with copious amounts of hexane, ethanol, and finally deionized water, in that order, and was subsequently dried in a vacuum oven under reduced pressure at ambient temperature overnight.

2.3. Surface-initiated ATRP of GMA and immobilization of gelatin

For the grafting of P(GMA) brushes from the PCL-Br surfaces, surface-initiated ATRP of GMA was performed using a [GMA (3 ml)]:[CuBr]:[$CuBr_2$]:[Bpy] molar feed ratio of 100:1.0:0.2:2.0 in 5 ml of methanol/water mixture (5/1, v/v) at room temperature in a Pyrex® tube. The reaction was allowed to proceed for 0.5 to 3 h to produce the PCL-*g*-P(GMA) surfaces. After the prescribed reaction time, the films were removed and washed sequentially with copious amount of methanol and deionized water, followed by immersing in methanol for about 48 h to ensure the complete removal of the physically-adsorbed reactants or polymers. For the immobilization of gelatin onto the pendant epoxide groups of the P(GMA) brushes, the PCL-*g*-P(GMA) films were incubated in 10 ml of the phosphate buffered saline (PBS, pH 7.4) containing 3 mg/ml gelatin. The reaction was allowed to proceed at room temperature for 24 h under continuous stirring to produce the corresponding PCL-*g*-P(GMA)-*c*-gelatin surfaces. After the reaction, the gelatin-immobilized PCL films were washed thoroughly with PBS solution and deionized water to remove the physically adsorbed (reversibly-bound) gelatin, prior to being dried in a vacuum oven under reduced pressure overnight.

2.4. Grafting density of the P(GMA) brushes and immobilized gelatin

The grafting density of the P(GMA) brushes and the amount of immobilized gelatin on the PCL substrates was determined by the grafting yield (GY) using the following equation [27,33]:

$$GY = \frac{W_a - W_b}{A} \tag{1}$$

Where Wa and Wb are the weights of the dry film after and before graft polymerization (or immobilization of gelatin) respectively, and A is the film area (about 3.2 cm^2). For each GY measurement, a minimum of three pieces of PCL films was used and the resulting values were averaged.

2.5. Surface characterization

The composition of the functionalized PCL films was determined by X-ray photoelectron spectroscopy (XPS). All the XPS spectra were recorded on a Krato AXIS Hsi spectrometer with a monochromatic Al Kα X-ray source (1486.6 eV photons), using procedures similar to those described previously [32]. The N 1s core-level signal can be used as an indicator of the immobilized gelatin. The [N]/[C] ratio, as determined from the sensitivity-factor-corrected N 1s and C 1s core-level XPS spectral area, indicated the relative abundance of the immobilized gelatin on the PCL substrates. Static water contact angles of the functionalized PCL film surfaces were measured at 25 °C and 60% relative humidity using a sessile drop method with 3 μl water droplets on a FTÅ 200 contact angle goniometer (First Ten Angstroms Inc., Portsmouth, VA, USA). The contact angles reported were the mean values from four substrates, with the value of each substrate obtained by averaging the contact angles for at least three surface locations. The surface topography of the functionalized PCL substrates was investigated by atomic force microscope (AFM). A multimode scanning probe microscope equipped with a NanoscopeIIIa controller (Digital Instrument, Santa Barbara, USA) was used to capture the AFM images in air. 10 μm scans were recorded in tapping mode with a silicon cantilever. The drive amplitude was about 300 mV, and the scan rate was between 0.5 and 1.0 Hz. The arithmetic mean of the surface roughness (R_a) was determined by Nanoscope software.

2.6. Cytocompatibility of the functionalized PCL substrates

Human umbilical vein endothelial cells (HUVECs, ATCC CRL-1730™) were cultured in gelatin-coated T25 flasks containing MCDB131 cell culture medium supplemented with Foetal Bovine Serum, 0.2% Bovine Brain Extract, 0.25 ug/ml amphotericin, 0.1 mg/ml heparin, 100 U/ml penicillin, and 100 ug/ml streptomycin, in a CO_2 environment at 37°C. The MCDB131 medium was changed every other day. Upon 90% culture confluency, cells were harvested by trypsinization using 0.25% Trypsin-EDTA. ECs between 4-6 passages were used for subsequent experiments.

2.6.1. Cell proliferation

The pristine and functionalized PCL films were sterilized by immersing into 75% (v/v) ethanol solution for 60 min, and then rinsed thrice with sterile PBS, followed by MCDB131 medium incubation overnight. Gelatin-coated coverslips (0.1%) were used as positive controls. Cell viability and proliferation was determined using the AlamarBlue™(AB) assay. 0.5 ml of EC cell suspension (2×10^4 cells/ml) was seeded into each well of 24-well plate containing the pristine and functionalized PCL films, and incubated in a 5% CO_2 environment at 37 °C for 1, 3, 5 and 7 days. The cell culture medium was changed every other day. After the predetermined incubation period, culture media was removed from the wells, and 0.5 ml of the AB solution (10% AB solution in culture media without FBS) was added to the wells. The plates were incubated in a 5% CO_2 atmosphere at 37ºC for 4 h and the fluorescence density was measured using a microplate reader (Model 680, Bio-Rad Laboratories, Inc. Hercules, CA, USA) at an excitation wavelength of 570 nm and an emission wavelength of 580 nm. Cell numbers were calculated using standards derived from seeding known quantities of cells and correlating with fluorescence emission.

2.6.2. Cell imaging

In vitro qualitative analysis of cell coverage and viability was performed using a LIVE/DEAD® viability/cytotoxicity assay to assess the extent of endothelialization on the functionalized PCL surfaces. For this procedure, calcein AM (4 mM in anhydrous dimethyl sulfoxide, DMSO) and EthD-1 (2 mM in DMSO/H2O, 1:4 v:v) were added to PBS (1:1000 ratio) to produce a LIVE/DEAD® staining solution. The cell-seeded PCL samples, obtained after 7 days cell culture, were first washed thrice with PBS to eliminate the nonadherent cells, followed by staining using 0.1 ml of LIVE/DEAD staining solution. After incubation in a 5% CO_2 atmosphere at 37°C for 30 min, the samples were visualized by Nikon Image Ti fluorescence microscope (emission at 515 nm and 635 nm (Nikon Instruments, Tokyo, Japan) to acquire fluorescent images using NIS-Elements Br software.

2.7. Blood compatibility of the bare and endothelialized PCL substrates

The hemolysis rate, coagulant activity, nitric oxide (NO) production, and platelet activation of the bare and endothelialized PCL films with various surface functionalizations were investigated to evaluate their blood compatibility. The endothelialized PCL substrates were obtained by culturing ECs (2×10^4 cells/ml) for 7 days on surface-functionalized PCL films using the procedures described above.

2.7.1. Hemolysis rate test

The pristine and functionalized PCL samples were immersed in diluted blood solution containing 2% fresh anticoagulated (ACD) human blood and 98% physiological salt solution and incubated at 37o for 1 h. After centrifugation at 3000 rpm for 5 min, the absorbance of solution was recorded as D_t. Under the same conditions, the solution containing 2% ACD blood and 98% physiological salt solution was used as a negative reference, and the solution containing 2% ACD blood and 98% distilled water was used as a positive reference. These absorbances were recorded as D_{nc} and D_{pc}, respectively. The hymolysis rate α of the samples was calculated using the following equation:

$$\alpha = \frac{D_t - D_{nc}}{D_{pc} - D_{nc}} \times 100\% \tag{2}$$

2.7.2. Coagulation assays

Whole blood of healthy human volunteers was mixed with 3.8% sodium citrate at a volume ratio of 1:9. The blood was centrifuged at 3000 rpm for 15 min at room temperature to obtain platelet-poor plasma (PPP). Aliquots of 500 μl of PPP were added to be in contact with the surfaces of each bare or endothelialized PCL substrate for 10 min at 37°C. The PPP was then collected and added with tissue thromboplastin (human brain extract) for prothrombin time (PT) tests, or added with a partial thromboplastin reagent (ellagic acid) for activated partial thromboplastin time (APTT) test. Subsequently, the fibrin clot formation time was determined by an automatic coagulation analyzer (Sysmex CA-7000). PPP that was not exposed to the PCL substrates was used as a blank sample.

2.7.3. Nitric Oxide (NO) secretion by HUVEC

ECs cultured for 7 days on the functionalized PCL substrates were washed twice with PBS and incubated at 37°C with trypsin-EDTA (0.25%) solution for cell detachment. The resultant ECs were serum-starved overnight in serum-free medium. After incubation with fresh serum-free medium for 6 h, the medium was removed and DAF-FM diacetate (Molecular Probes, D-23842) was added to the medium to effect a final concentration of 10 μM.The medium was subsequently incubated at 37 °C for 1 h, followed by detection of fluorescence using Glomax 20/20 luminometer equipped with a blue fluorescent module. The end product of DAF-FM diacetate and NO is a benzotriazole derivative with a fluorescence excitation and emission maxima of 495 and 515 nm, respectively. Fluorescence units were normalized to cell numbers.

2.7.4. Platelet activation determination by P-selectinassay

Platelet activation by the bare and endothelialized PCL substrates was investigated using the P-selectin (CD62P) assay. Briefly, 100 μl of fresh human platelet rich plasma (PRP) was incubated with the bare or 7-day endothelialized PCL substrates at 37°C for 2 h. At the end of incubation, the films were thoroughly with copious amounts of PBS solution thrice, followed by adding 40 μl of anti-CD62P (1:100, v:v) to each film, and then incubated at 37°C for 1 h. After being washed thrice with PBS solution, the films were each incubated with 40 μl of horseradish peroxidase-conjugated sheep anti-mouse polycolonal antibody at 1:100 (v:v) at 37°C for 1 h. Subsequently, the PCL films were reacted with 150 μl of 3,3',5, 5'-tetramethylbenzidine (TMB) chromogenic solution for 10 min. The color reaction was terminated by adding of 100 μL of 1 M H_2SO_4, and the optical densities (OD) were measured at 450 nm using a Varioskan Flash Microplate Reader (Thermo Fisher Scientific, Waltham, MA, USA).

2.7.5. Gene and protein expression of vWF and activity of MMP-2 in ECs cultured on functionalized PCL surfaces

For the real-time qPCR of von Willebrand factor (vWF) and matrix metalloproteinase-2 (MMP-2), the total RNA was extracted from the ECs after 7 days in culture, reverse-transcribed into cDNA and analyzed as described above. The expression of vWF and MMP-2 was normalized to the housekeeping ribosomal protein L27 (rpl27). Endothelial cells treated with 10 ng /ml TNF were used as positive controls.

For the immunoblot detection of vWF protein, cells were lyzed in protein lysis buffer (0.1% sodium dodecyl sulfate, 0.5% triton X-100, and 0.5% sodium deoxycholate dissolved in pH 7.4 PBS) and resolved using a denaturing 10% SDS-PAGE. The proteins were then blotted onto a nitrocellulose membrane and after blocking with 5% non-fat milk in tris-buffered saline with 0.1% Tween (TBS-T), the membrane stained using a rabbit anti-human vWF antibody at 1:5000 and subsequently with anti-rabbit HRP-conjugated antibody at 1:10,000 in TBS-T. The vWF was then visualized after developing using chemiluminescence on X-ray film. For determination of the MMP-2 activity, proteins were extracted from trypsinized cells using the protein lysis buffer before resolving by electrophoresis through a 10% SDS-PAGE copolymerized with 0.1% gelatin as substrate for enzymatic digestion. The molecular sizes of gelatinolytic activities were determined using protein standards (Fermentas, Prestained PAGE rulers). Upon completion of gel running, the gel was incubated with 100 mL renaturation buffer containing 2.5% triton X-100 for 1 h at room temperature with agitation. The gel was subsequently incubated in 100 ml of development buffer containing 50 mM Tris base, 200 mM NaCl, 5 mM CaCl2, and 0.02% Brij-35 overnight at 37 °C. Developed gel was then stained by the Coomassie Blue and gelatinolytic activities of MMP-2 were determined by the transparent bands appeared at the molecular weight of approximately 68 kDa and 98 kDa, respectively.

2.8. Statistical analysis

Each experiment was carried out with four replicates (n = 4), and the data are presented as mean ± standard deviation (SD) unless of otherwise stated. Statistical analysis was carried out by means of one-way analysis of variance (ANOVA) with Tukey's post hoc test. The confidence levels of 95% ($p<0.05$) and 99% ($p<0.01$) were used and no adjustments were made for multiple comparisons.

3. Results and discussion

Polycaprolactone (PCL) films with gelatin-coupled poly(glycidyl methacrylate) (P(GMA)) brushes were prepared via the following reaction sequence (Fig. 1): (a) active amine groups were introduced to the PCL film surfaces by the aminolysis reaction, (b) the immobilization of an alkyl bromide ATRP initiator was achieved via TEA-catalyzed condensation reaction between the amine groups on the aminolyzed PCL substrates and 2-bromoisobutyryl bromide (BIBB), (c) well-defined P(GMA) functional brushes were covalently grafted from the ATRP initiator-immobilized PCL surface via surface-initiated ATRP of GMA, (d) cell-adhesive gelatin was directly coupled to the pendant active epoxide groups of the grafted P(GMA). Details of each functionalization step are discussed below.

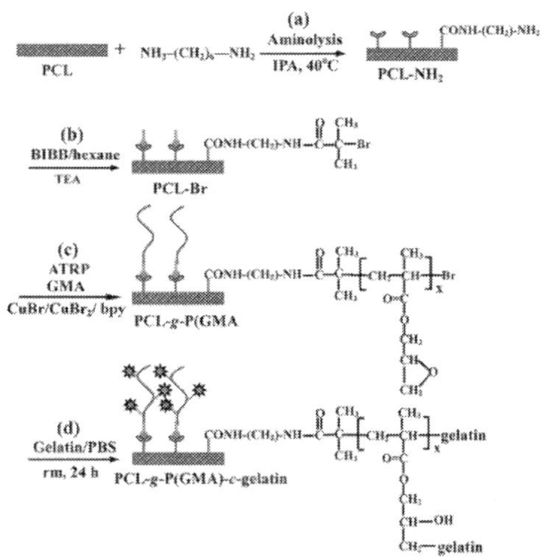

Figure 1. Schematic illustration of the process of (a) aminolysis of PCL substrates to introduce the free amino groups (the PCL-NH$_2$ surface), (b) immobilization of a alkyl bromine-containing initiator via condensation reaction to give the PCL-Br surface, (c) surface-initiated atom transfer radical polymerization (ATRP) of GMA from the PCL-Br surface to produce the PCL-*g*-P(GMA) surface, and (d) subsequently covalent conjugation of gelatin to obtain the PCL-*g*-P(GMA)-c-gelatin surface.

http://dx.doi.org/10.5772/52646

3.1. Aminolysis of PCL substrates and immobilization of ATRP initiator

Aminolysis represents an easy-to-perform chemical technique to engraft amino groups along the polyester chains, and hence has been widely used in the surface modification of scaffolds for tissue engineering applications [31,34]. In this study, PCL substrates were activated by the aminolysis reaction to introduce active amine groups. The relative amount of amine groups on the aminolyzed PCL (defined as the PCL-NH_2) surface was quantitatively determined by XPS measurements. The [N]/[C] ratio, as determined from sensitivity factor-corrected N 1s and C 1s core-level XPS spectral area, increases with the aminolysis time and reaches the maximal value after 1 h, which is estimated to be about 0.043 (Figs. 2). The result is consistent with the data reported previously [35,36]. As a degradation reaction, the aminolysis reaction is found to proceed preferentially at the amorphous regions of polymer in diamine solution during the initial period [37]. At longer aminolysis time, the decrease in bound amine groups may be caused by chain scission, formation of oligomers and other low mass fragments that are removed from the surface during reaction and the rinsing process [38]. Thus, the optimal aminolysis time for PCL film was found to be 1 h, and this reaction time was chosen for the subsequent surface modification and cell studies.

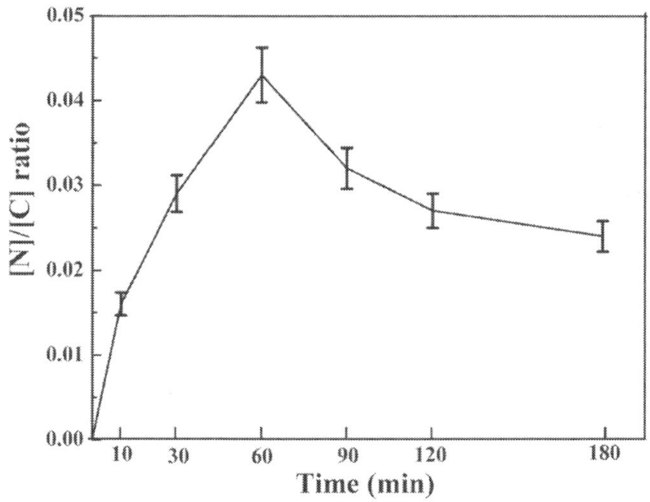

Figure 2. The [N]/[C] ratio of the aminolyzed PCL surface as a function of aminolysis time determined by XPS measurements. The analysis reaction of PCL films proceeded at 40°C in 10 wt% 1,6-hexanediamine/2-propanol solution. Error bars represent the standard deviation over separate measurement on three PCL films. The optimized aminolysis time was observed at 1 h with the [N]/[C] ratio of 0.043.

The chemical composition of the PCL film surfaces at various stages of surface modification was ascertained by XPS. Figs. 3a-3c show the respective wide scan, C 1s, and N 1s core-level spectra of the pristine PCL and PCL-NH_2 surfaces from 1 h of aminolysis. The C 1s core-level spectra of the pristine PCL can be curve-fitted into three peak components with binding energies (BEs) at about 284.6, 286.2 and 288.7 eV, attributable to the C-H, C-O, and O=C-O species, respectively (Fig. 3b) [39]. The area ratio of [C-H]:[C-O]:[O=C-O] is around 5.0:1.1:1.0 (Table 1), which is in good agreement with the theoretical value of 5:1:1 for the polycaprolactone structures. The appearance of N 1s signal in the wide scan spectrum (Fig. 3a) and an additional peak component at 285.5 eV, attributable to C-N species, in the curve-fitted C 1s core-level spectrum (Fig. 3c) indicate the successful introduction of amine groups onto the PCL substrates after 1 h of aminolysis. The only peak component found at the BE of 399.6 eV in the N 1s core-level spectrum is associated with the free amine group on the PCL-NH_2 film surface (Fig. 3c') [38]. The decrease in static water contact angles of the PCL substrates from 93 ± 2° to 66 ± 3° is consistent with the presence of amine groups on the PCL-NH_2 surface (Table 1). The amine groups on the aminolyzed PCL surface can not only improve surface hydrophilicity, but also offer the active sites for further functionalization.

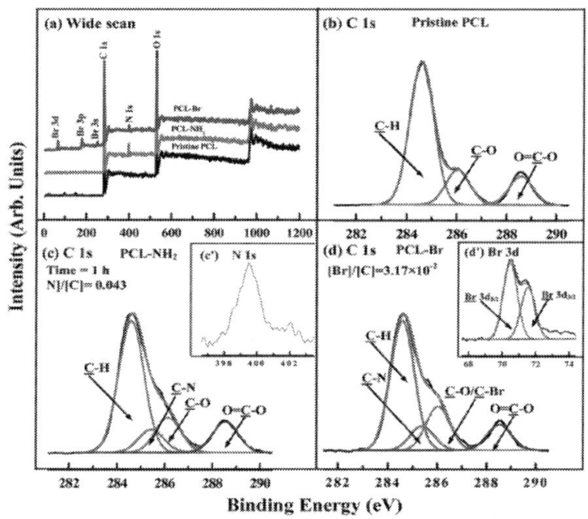

Figure 3. Wide scan and C 1s core-level curve-fitted XPS spectra of the (a,b) pristine PCL, (c,d) PCL-NH_2 from 1 h of aminolysis, and (e,f) PCL-Br surfaces. Insets of (d') and (f') correspond to the N 1s and Br 3d core-level XPS spectra of the PCL-NH_2 and PCL-Br surfaces, respectively.

The immobilization of a uniform monolayer of initiators on the solid surface is indispensible in the surface-initiated ATRP process [40]. An alkyl bromide ATRP initiator was introduced onto the PCL-NH_2 surface via TEA-catalyzed condensation reaction to produce the PCL-Br surface. Successful introduction of the alkyl bromide-containing ATRP initiator onto the PCL substrates can be deduced from the appearance of three additional signals with BEs at about 70, 189 and 256 eV, attributable to Br 3d, Br 3p, and Br 3s, respectively, in the wide scan spectrum of the PCL-Br surface (Fig. 3a) [41]. The [Br]/[C] ratio, as determined from the Br 3d and C 1s core-level spectral area ratio, was about 3.17×10^{-2} (Fig. 3d). The corresponding Br 3d core-level spectrum of the PCL-Br surface with a Br $3d_{5/2}$ BE of 70.4 eV is consistent with the presence of the alkyl bromide species [41] (Fig. 3d'). The alkyl bromide-immobilized PCL surface became more hydrophobic, as static water contact angle increased noticeably to 85 ± 3° (Table 1).

Sample	GY[g](μg/cm²) (mean± SD[h])	[Br]/[C][i]	[N]/[C][i]	Surface composition[j] (molar ratio)	WCAk (degree)
Pristine PCL[a]	–	–	–	[C-H]:[C-O]:[O=C-O] = 5:1.1:1.0 (5:1:1)	93±2
PCL-NH$_2$ [b]	–	–	0.043	[C-H]:[C-N]:[C-O]:[O=C-O] = 4.7:0.7:1.1:1.0	66±3
PCL-Br[c]	–	3.17×10^{-2}	–	[C-H]:[C-N]:[C-O/C-Br]:[O=C-O] = 5.0:0.9:1.7:1.0	85±3
PCL-*g*-P(GMA)1[d]	6.31±1.32	9.29×10^{-3}	–	[C-H]:[C-O]:[O=C-O] = 3.8:2.8:1.0 (3:3:1)	62±4
PCL-*g*-P(GMA)2[e]	14.76±2.63	4.72×10^{-3}	–	[C-H]:[C-O]:[O=C-O] = 3.1:3.0:1.0 (3:3:1)	61±5
PCL-*g*-P(GMA)1-*c*-gelatin[f]	2.63± 0.52	–	0.169	[C-H]:[C-N]:[C-O]:[O=CNH]: [O=C-O] = 3.5:1.5:1.7:1.0:0.7	37±3
PCL-*g*-P(GMA)2-*c*-gelatin[pala]	3.79±0.73	–	0.203	[C-H]:[C-N]:[C-O]:[O=CNH]: [O=C-O] = 3.1:1.5:1.4:1.0:0.3	35±4

[a] Pristine PCL refers to the cleaned PCL film after rigorous wasing with alcohol/water solution and deionized water, [b] PCL-NH$_2$ was obtained after 1 h of aminolysis in a 10% (w/w) 1,6-hexanediamine/isopropanol solution at 40 °C, [c] PCL-Br was obtained after the PCL-NH$_2$ surface reacted with 2-bromoisobutyryl bromide (BIBB) in dried hexane containing 1:1 (molar ratio) BIBB and triehtylamine (TEA), [d,e] Reaction conditions: [GMA]:[CuBr]:[CuBr$_2$]:[bpy]=100:1:0.2:2 in methanol-water solution (1:1, v:v) at room temperature for 1 and 3 h to produce the PCL-*g*-P(GMA)1 and PCL-*g*-P(GMA)2 surfaces, respectively, [f] Reaction conditions: the PCL-*g*-P(GMA)1 and PCL-*g*-P(GMA)2 surfaces incubated in PBS (pH 7.4) solution containing the gelatin at a concentration of 3 mg/mL at room temperature for 24 h, [g] GY denotes the grafting yield, and is defined as GY = (W_a - W_b)/A, where W_a and W_b corresponds to the weight of the dry films before and after grafting of polymer brushes, respectively, and A is the film area (about 3.2 cm2
), [h] SD denotes standard deviation, [i] Determined from the corresponding sensitivity factor-corrected element core-level spectral area ratios, [j] Determined from the curve-fitted C 1s core-level spectra. Theoretical values are shown in parentheses. [k] WCA denotes static water contact angles.

Table 1. Grafting yield, surface composition, and water contact angles of the pristine PCL and surface-functionalized PCL surfaces

3.2. Surface-initiated ATRP of GMA and immobilization of gelatin

P(GMA) is an effective surface linker to immobilize biomolecules, such as proteins, antibodies or enzymes, for tissue engineering applications [42]. Fig. 4 shows the respective wide scan, C 1s and Br 3d core-level spectra of the PCL-*g*-P(GMA) surface from 1 and 3 h of ATRP reaction. The C 1s core-level spectra of the PCL-*g*-P(GMA) surface can be curve-fitted into three peak components with BEs at 284.6, 286.2 and 288. 7 eV, attributable to C-H, C-O and O=C-O, respectively (Figs. 4b and 4d). For the PCL-*g*-P(GMA)1 surface from 1 h of ATRP, the area ratio of [C-H]:[C-O]:[O=C-O] is about 3.8:2.8:1.0 (Table 1), which is slightly different from the theoretical value of 3:3:1 for the GMA unit structure. The deviation in peak component area ratio of C 1s core-level spectrum of the PCL-g-P(GMA)1 surface suggests that the thickness of the P(GMA) brushes is less than the probing depth of XPS technique (about 8 nm in an organic matrix) [41]. Increasing the reaction time to 3 h leads to a [C-H]:[C-O]:[O=C-O] ratio of about 3.1:3.0:1.0, which is close to the theoretical value of the GMA repeat unit structure (Table 1). This is an indication that the P(GMA) brushes were thicker than the probing depth of XPS technique. It has been reported that the thickness of the P(GMA) brushes grafted on the silicon surface is around 30 nm after 3 h of ATRP of GMA under similar reaction conditions [41]. The presence of P(GMA) brushes leads to decrease in static water contact angles to 62 ± 4° and 61 ± 5°, respectively, for the PCL-*g*-P(GMA)1 and PCL-*g*-P(GMA)2 surfaces, owing to the presence of hydrophilic epoxide groups [43].

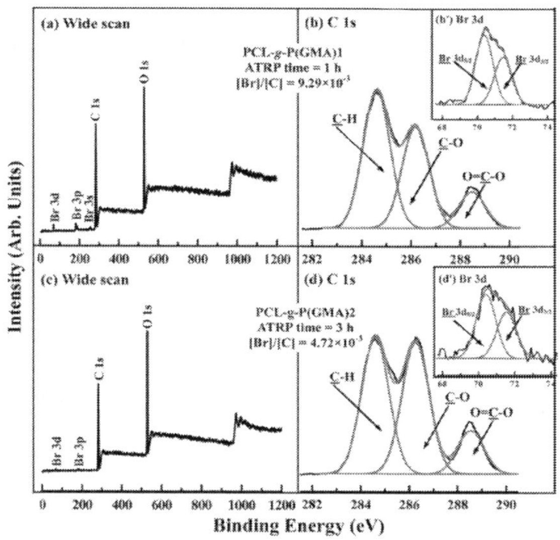

Figure 4. Wide scan, C 1s, and Br 3d core-level curve-fitted XPS spectra of the (a,b,b') PCL-*g*-P(GMA)1 from 1 h of ATRP reaction and (c,d,d') PCL-*g*-P(GMA)2 from 3 h of ATRP reaction. Successful grafting of P(GMA) polymer brushes can be deduced from the area ratios of [C-H]:[C-O]:[O=C-O] peak components comparable to the theoretical value of 3:3:1 of GMA molecular structure.

The grafting yield (GY) was measured to evaluate the kinetics of polymer chain growth in this study. As shown in Fig. 5, an approximate linear increase in GY of the grafted P(GMA) chains with polymerization time could be observed for the PCL-Br surface. The result suggests that the chain growth from the PCL-Br surface proceeds in a controlled and well-defined manner. The GY values of the PCL-*g*-P(GMA)1 and PCL-*g*-P(GMA)2 surfaces are about 6.31 ± 1.32 and 14.76 ± 2.63 μg/cm² (Table 1), respectively. The persistence of the Br 3d core-level signal (Figs. 4b' and 4d') is consistent with the fact that the living chain end from the ATRP process involves a dormant alkyl halide group, which can be readily reactivated to initiate the block copolymerization [25]. However, the molecular weight and molecular weight distribution of the surface-graft polymers cannot be determined with sufficient accuracy without precise cleavage of the grafted P(GMA) from the film surfaces [27].

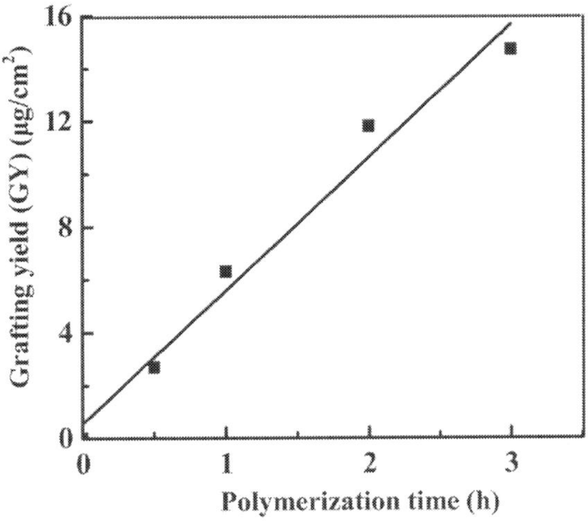

Figure 5. A linear relationship between the grafting yield (GY) of the P(GMA) brushes with the surface-initiated ATRP time. The polymer chain growth was tuneable by varying reaction time.

Nucleophilic reactions involving $-NH_2$ moieties of biomolecules and the pendant epoxide groups have been widely reported [44]. In this work, cell-adhesive gelatin was directly coupled to the pendant epoxide groups of the PCL-*g*-P(GMA)1 and PCL-*g*-P(GMA)2 surfaces to produce the corresponding PCL-*g*-P(GMA)1-*c*-gelatin and PCL-*g*-P(GMA)2-*c*-gelatin surfaces, respectively. Fig. 6 shows the respective wide scan, C 1s and N 1s core-level spectra of the gelatin-immobilized PCL surfaces. The corresponding curve-fitted C 1s core-level spectrum was composed of five peak components with BEs at about 284.6, 285.5, 286.2, 288.2 and 289.1 eV, attributable to the C-H, C-N, C-O, O=CNH, and O=C-O species [41], respectively (Figs. 6b and 6d). The C-N peak component is associated with the linkages in gelatin itself, as well as the linkage between P(GMA) and gelatin. The O=CNH peak component is ascribed to the peptide bonds in gelatin. The above results and the appearance of a strong N 1s signal with BE at 399.6 eV (Figs. 6b' and 6d'), characteristic of amine species, are consistent with the fact that gelatin has been covalently immobilized on the P(GMA) brushes. The surface wettability of the PCL substrates is significantly improved after the immobilization of gelatin, as water contact angles decrease to $37 \pm 2^{\circ}$ (for the PCL-*g*-P(GMA)1-*c*-gelatin) and $35 \pm 3^{\circ}$(for the PCL-*g*-P(GMA)2-*c*-gelatin) (Table 1). It is reported that gelatin contains large amount of glycine (Gly) and proline (Pro) which are hydrophilic amino acids [45]. The hydroxyl groups (-OH) generated in the ring-opening reaction of epoxide groups by coupling of gelatin could have also contributed to the lower water contact angle of the PCL-*g*-P(GMA)-*c*-gelatin surfaces.

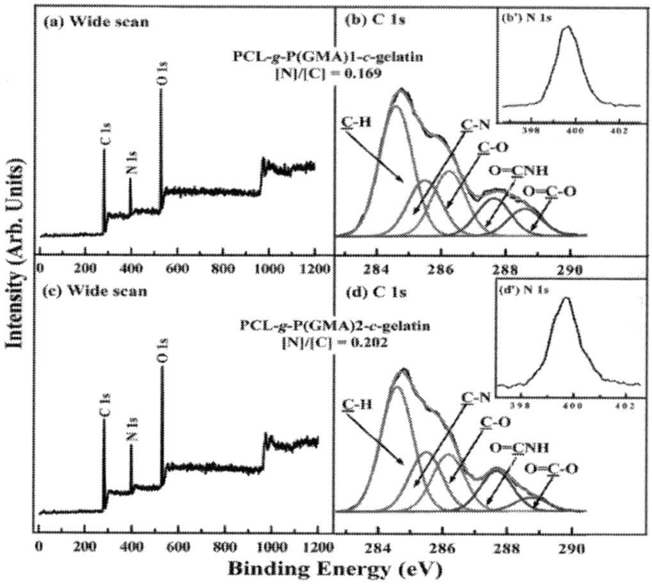

Figure 6. Wide scan, C 1s and N 1s core-level curve-fitted XPS spectra of the (a,b,b') PCL-*g*-P(GMA)1-*c*-gelatin and (c,d,d') PCL-*g*-P(GMA)2-c-gelatin surfaces. The appearance of two additional peak components of C-N and O=CNH as a result of immobilized gelatin.

3.3. Surface topography

The changes in topography of the PCL film surfaces after each functionalization step were investigated by AFM. Fig. 7 shows the representative AFM height images of the pristine PCL and functionalized PCL surfaces with scanned areas of 10 μm × 10 μm. The pristine PCL film surface is relatively uniform and smooth with a root-mean-square surface roughness values (R_a) of about 19 nm (Fig. 7a). After the aminolysis treatment, the R_a value increases to 31 nm (Fig. 7b). The observation that aminolysis caused a noticeable increase in surface roughness is in agreement with the findings by other groups [30,35]. The existence of shallow pits is probably the result of the penetration of hexanediamine molecules into the PCL films, since it has been previously reported that the aminolysis reaction can take place at a depth of around 50 μm [23,30]. After graft polymerization of GMA, obvious increases in R_a values are observed on the PCL-*g*-P(GMA)1 (56.9 nm, Fig. 7c) and PCL-*g*-P(GMA)2 surfaces (67.8 nm, Fig. 7e), and characteristic fiber-like features of polymer brushes are also visible on the P(GMA)-grafted film surfaces (Figs. 7c and 7e). The subsequent coupling of gelatin to P(GMA) brushes resulted in a further slight increase in surface roughness of PCL substrates, as R_a values increase to 59 nm and 71.5 nm for the PCL-*g*-P(GMA)1-*c*-gelatin and the PCL-*g*-P(GMA)2-*c*-gelatin surfaces, respectively.

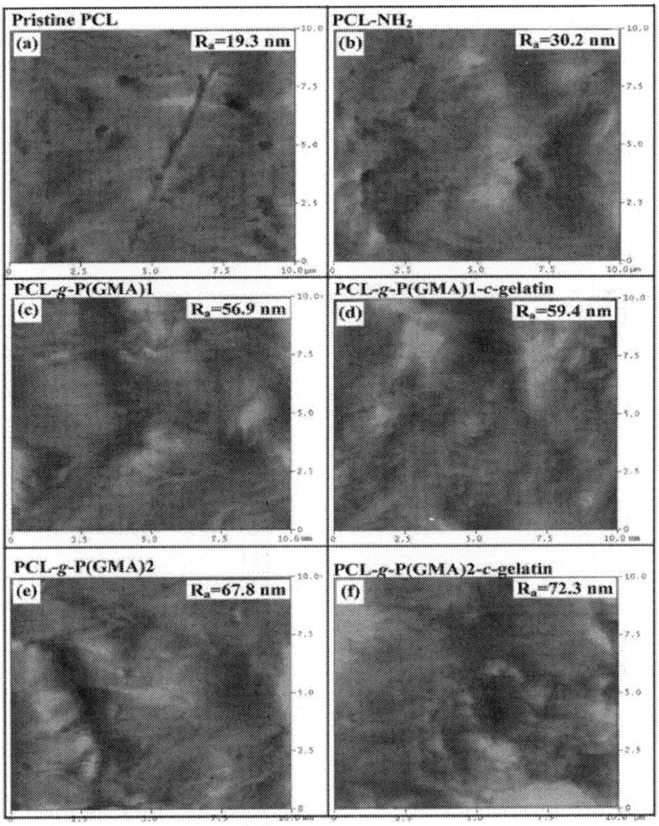

Figure 7. Representative two-dimensional (2D) AFM images of (a) pristine PCL, (b) PCL-NH_2, (c) PCL-*g*-P(GMA)1 from 1 h of ATRP, (d) PCL-*g*-P(GMA)2 from 3 h of ATRP, (e) PCL-*g*-P(GMA)1-*c*-gelatin, and (f) PCL-*g*-P(GMA)2-*c*-gelatin surfaces. The arithmetical mean roughness (R_a) of different PCL substrates was obtained from a scan size of 10 μm × 10 μm.

3.4. Endothelial cells proliferation and surface endothelialization

The adhesion and proliferation of endothelial cells (ECs) on the functionalized PCL surfaces was quantitatively determined by the AlamarBlueTM (AB) assay, and the results are shown in Fig. 8. The pristine PCL film surface is the least conducive for supporting cellular growth, since only a marginal increase in cells over 7 days of culture was observed. The cells attached to the PCL-NH_2 film surfaces showed a slight improvement in proliferation as compared to the pristine PCL surface. This result is consistent with previous findings that the presence of amine groups on the PCL surfaces leads to a positive effect on cell proliferation [23,30], albeit to a limited extent. Despite the improvement in surface hydrophilicity and roughness, the grafting of P(GMA) brushes onto the PCL film surfaces did not lead to an enhancement in EC proliferation behavior, which is probably associated with the cytotoxicity and mutagenicity of epoxides groups to ECs [46]. Besides the fact that polymer surfaces with moderate hydrophilicity of water contact angles in a range of 30-70° and rougher nano-topography are favorable for cell attachment and proliferation [47,48], other factors (e.g. biological cues) may also be required for positive cell interaction with material substrates. This hypothesis was confirmed by the observation that the gelatinized P(GMA)-grafted PCL substrates exhibited higher affinity and proliferation for ECs as compared to the surfaces that did not contain the bioactive gelatin motifs.

In fact, the proliferation rates of the PCL-*g*-P(GMA)1-*c*-gelatin and PCL-*g*-P(GMA)2-*c*-gelatin surfaces were comparable to that of the gelatin-coated coverslips (positive controls). Cell proliferation on the gelatin-immobilized PCL surfaces was not only significantly enhanced, but also found to be positively correlated to the amounts of immobilized gelatin. The PCL-*g*-P(GMA)2-*c*-gelatin surface exhibited more pronounced enhancement in cell adhesion and proliferation than that of the PCL-*g*-P(GMA)1-*c*-gelatin surface, as the longer ATRP reaction time allowed for more gelatin to be attached. This result suggests that an increase in surface density of the immobilized gelatin can lead to an increase in EC proliferation over time. This phenomenon is probably associated with the fact that immobilization of gelatin provides many epitopes or ligands for cell adhesion molecules, such as integrins, thus mimicking the natural extracellular environment that is favorable for EC adhesion, spreading and proliferation.

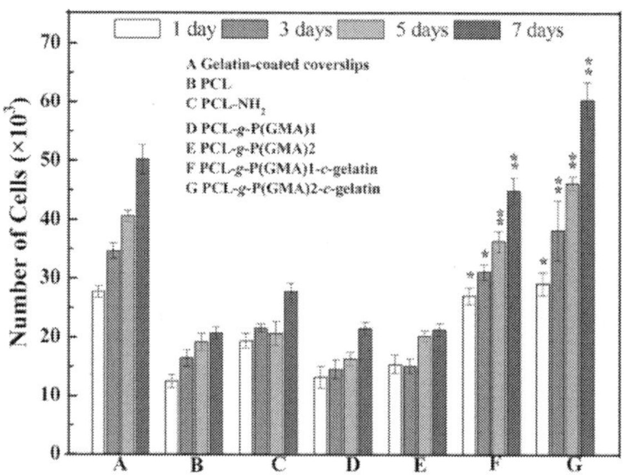

Figure 8. EC proliferation profile on the gelatin-coated coverslips, pristine PCL and functionalized PCL surfaces after 1, 3, 5 and 7 days of incubation at 37 °C in a 5% CO_2 atmosphere as determined by the AlamarBlue™(AB) assay. Data presented as means ± SD. *$p<0.05$ and **$p<0.01$ refers to statistically significant difference compared with the pristine PCL surface. The cell proliferation rate of ECs seeded on the gelatin-immobilized surfaces was significantly improved as compared to that the pristine PCL film.

The visualization of EC coverage on the functionalized PCL surfaces enabled a good assessment of the efficacy of endothelialization over the entire surface. Fig. 9 shows the representative fluorescence images of LIVE/DEAD-stained ECs on the pristine PCL and functionalized PCL film surfaces after 7 days of culture. The sparse coverage of ECs on the pristine PCL substrate further confirmed the unfavorable surface properties of pristine PCL for cell adhesion and proliferation (Fig. 9b). Poor endothelialization was also observed on the aminolyzed PCL (Fig. 9c) and P(GMA)-grafted PCL surfaces (Fig. 9d). The dense growth of ECs on the PCL-*g*-P(GMA)1-*c*-gelatin (Fig. 9e) and PCL-*g*-P(GMA)2-*c*-gelatin (Fig. 9f) surfaces indicated significant improvement in EC coverage for the gelatin-immobilized PCL surfaces. The observed denser coverage of ECs on the PCL-*g*-P(GMA)2-*c*-gelatin surfaces compared to the PCL-*g*-P(GMA)1-*c*-gelatin surfaces meant that the efficacy of endothelialization is positively correlated to the amount of immobilized gelatin. Taken together, the results suggest that the higher the surface concentration of immobilized gelatin, the better the endothelialization efficacy of the material within a given period of time.

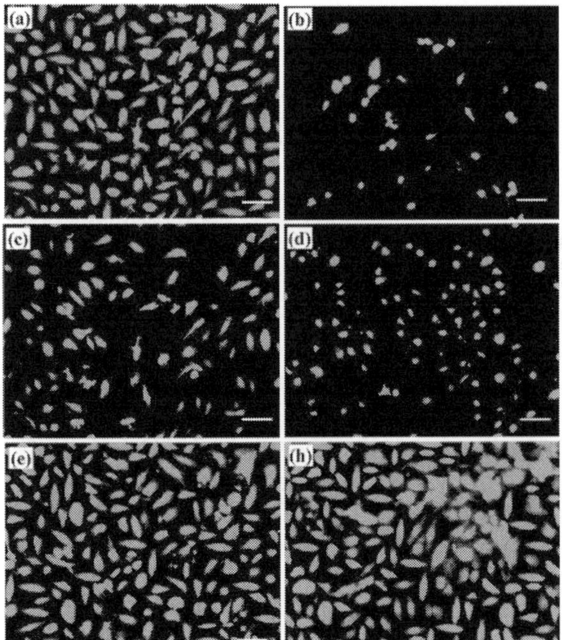

Figure 9. Fluorescence images of LIVE/DEAD-stained ECs on the (a) gelatin-coated coverslips (positive control), (b) pristine PCL, (c) PCL-NH_2, (d) PCL-*g*-P(GMA)2, (e) PCL-*g*-P(GMA)1-*c*-gelatin, and (f) PCL-*g*-P(GMA)2-*c*-gelatin after 7 days of incubation in cell suspension (2×10^4 cells/ml). Scale bar: 20 μm. Rapid endothelialization was observed for the gelatin-immobilized PCL surfaces.

3.5. Blood compatibility tests

3.5.1. *Hemolysis rate test*

Hemolysis rate is an important factor for characterization of the blood compatibility. The lower the hemolysis rate, the better the blood compatibility. Figure 10 shows the hemolysis rate of the pristine PCL and surface-functionalized PCL samples. It can be seen that the hemolysis rate of the functionalized samples has no substantial improvement. The gelatin-immobilized PCL substrates even show a relatively higher hemolysis rate than those of the pristine PCL and PCL-NH_2 surfaces. This result is consistent with the previous findings that gelatin exhibited somewhat hemostatic effect by nature. However, the hemolysis rate of the gelatin-immobilized PCL samples is approximately 3%, far below the accepted threshold value of 5% for biomaterial applications. Thus, the gelatin-immobilized samples can be used as a novel material with good hemocompatibility.

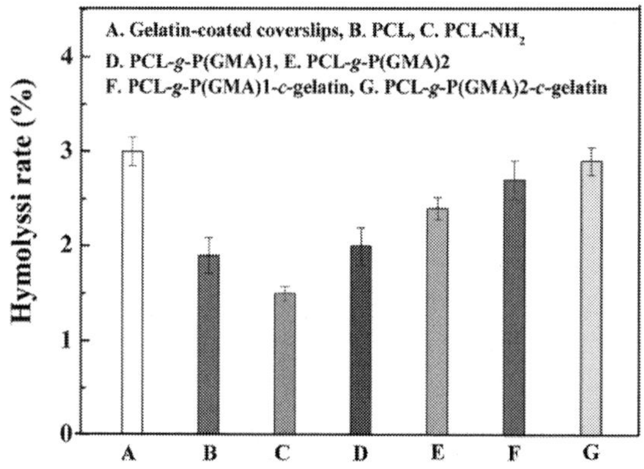

Figure 10. Hemolysis rate of the pristine PCL and surface-functionalized PCL samples. Data presented as means ± SD, n=3.

3.5.2. *Coagulation activity on the bare and endothelialized PCL substrates*

Blood coagulation, particularly under conditions of relatively low flow, has been recognized to be one of the main problems of vascular occlusion [49]. Activated coagulation factors influence the clotting time through extrinsic, intrinsic and common coagulation pathways. Prothrombin time (PT) is used to evaluate deficiencies in the extrinsic factor, and represents the time for blood plasma to clot after the addition of thromboplastin (activator of the extrinsic pathway) [50]. Activated partial thromboplastin time (APTT) is used to evaluate the intrinsic factors, such as VIII, IX, XI and XII, and common coagulation pathway factors V, X and II [50]. Both PT and APTT are commonly used to screen for adverse activation of the coagulation pathways on the vascular grafts and to evaluate their haemocompatibility in vitro. Thus, the PT and APTT values of the bare and endothelialized PCL substrates with various surface modifications were measured.

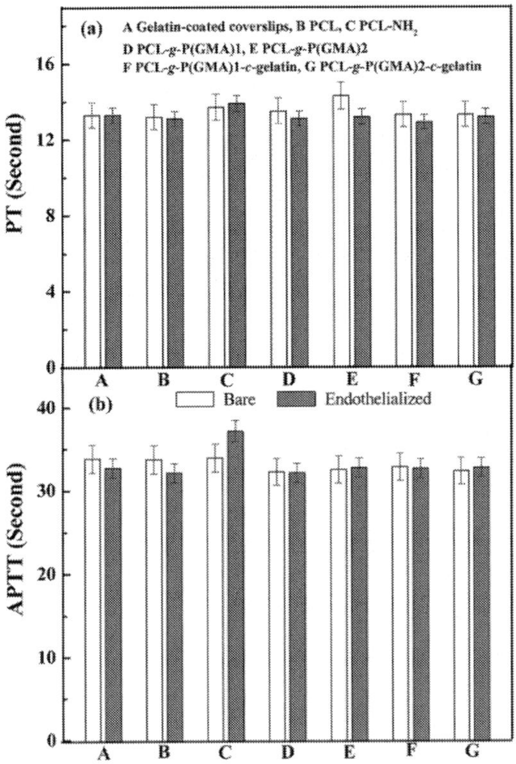

Figure 11. a) PT and (b) APTT results (means ± SD, n=3) for the bare and endothelialized surfaces of pristine PCL and surface-functionalized PCL, The coagulation activity of all surfaces was found to be in the normal range for healthy blood plasma.

The normal ranges of PT and APTT for healthy blood plasma are 12.0-14.5 s and 27.0-35.6 s, respectively [51]. Fig. 11 shows the PT and APTT results for the bare and endothelialized PCL substrates with various surface modifications. For the bare PCL and functionalized PCL surfaces, both PT and APTT are all within the normal ranges of coagulation time. None of the PCL substrates was found to affect the coagulation pathways significantly, and no discernible effect was observed in the presence of gelatin, indicating that surface functionalization activated neither the intrinsic nor the extrinsic coagulation pathways. This result is consistent with previous findings by other groups that cell-adhesive proteins and peptides do not affect or convert blood coagulation pathways [51,52]. Even after the PCL substrates were coated with a layer of ECs, the PT and APTT readings were still observed to be in the normal range of the clinical reference, and no significant differences in the coagulant activities were observed on the endothelialized PCL substrates as compared to the bare PCL substrates. The results also revealed that the ECs cultured on the pristine PCL and functionalized PCL surfaces remained unactivated and did not exhibit procoagulation phenotypes. Hence, it could be concluded that the presence of the monolayer of ECs had no effect on the intrinsic or the extrinsic coagulation pathways.

3.5.3. Nitric Oxide (NO) production

Nitric oxide (NO) is an important regulator of vascular tone and platelet adhesion and the continuous NO release by ECs prevents thrombogenesis [53]. In this study, the NO secretion of the ECs on the pristine PCL and functionalized PCL surfaces were measured. As shown in Fig. 12, the amount of NO secreted by ECs on the gelatin-immobilized PCL surfaces was significantly higher than those on the pristine PCL, aminolyzed and P(GMA)-grafted PCL surfaces. The amount of NO production of the ECs seeded on the PCL-*g*-P(GMA)2-*c*-gelatin surface was around 2-fold higher than that on the PCL-*g*-P(GMA)1-*c*-gelatin surfaces, indicating that the improved NO production observed for ECs grown on the gelatin-immobilized PCL surface may be positively correlated to the amount of covalently immobilized gelatin. The above results suggest that a high density of immobilized gelatin led to the enhancement in NO secretion.

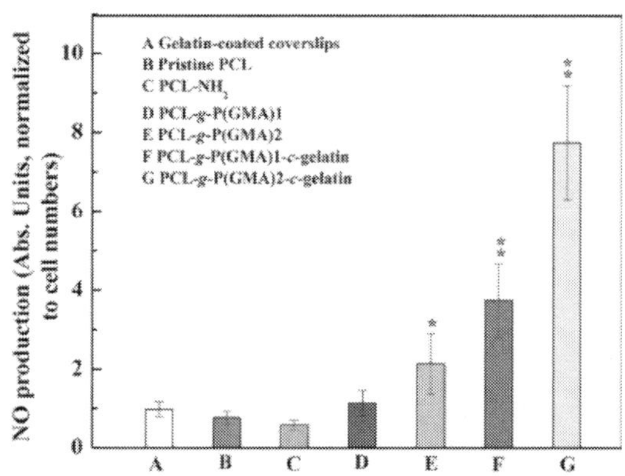

Figure 12. The amount of NO production for the ECs seeded on the pristine and functionalized PCL substrates. Data presented as means ± SD, n=3. **p*<0.05 and ***p*<0.01 corresponds to statistically significant difference as compared to the pristine PCL.

3.5.4. Platelet activation on the bare and endothelialized PCL substrates

Apart from coagulation pathways, platelet activation is considered another important criterion in assessing blood compatibility of the biomaterial surface. The activation of attached platelet results in platelet aggregation and the formation of a thrombus [54]. The subendothelial collagens (i.e., types I - IV) have been found to interact directly with platelets to trigger their activation for thrombogenic initiation [55]. As gelatin is a derivative of collagen, one downside of using gelatin-immobilized surfaces could be the detrimental adhesion and activation of platelets, which could initiate clotting. Therefore, the extent of platelet activation by the different biomaterial surfaces was studied.

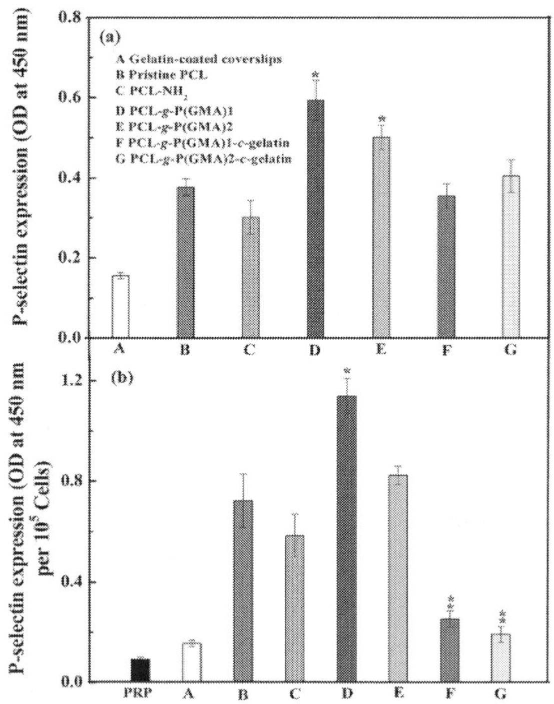

Figure 13. P-selectin expression for the (a) bare and (b) endothelialized surfaces of the pristine and functionalized PCL substrates. Plasma rich platelet (PRP) was used as positive control. *p<0.05 and **p<0.01 corresponds to statistically significant difference compared with the pristine PCL surface.

In this study, platelet activation on the bare and endothelialized PCL substrates was determined using the P-selectin assay, since P-selectin is one of intracellular granular molecules released upon platelet activation and stimulation [5], For the bare PCL substrates, the amount of activated platelets on the gelatin-immobilized PCL surfaces was comparable to those of the pristine PCL surfaces (Fig. 13a). In contrast, platelet activation was found to be significantly higher on the P(GMA)-grafted PCL surfaces, indicative of an increased risk for thrombogenesis. The results suggest that the immobilization of gelatin on the P(GMA)-grafted PCL surfaces not only rendered them with adhesion-promoting properties, but also decreased the risk of inducing the activation of platelets. It has been reported previously that the biological motifs immobilized on the polymers could not bind effectively with the platelet integrin unless their separation distances was between 1.48-2.2 nm [56], and that polymers with small spacer, such as lauric acid-conjugated GRGDS, exhibited no increase in activation [57]. Here, the gelatin was directly conjugated onto the grafted P(GMA) brushes (small spacer), and thus the gelatin motif could not gain access to the binding sites of the platelet integrin (such as $\alpha2\beta1$). In addition, it is probable that the other surface properties (e.g. enhanced hydrophilicity) were also responsible for reduced platelet activation activity.

In the case of the endothelialized PCL substrates, the amount of activated platelets on the gelatin-immobilized PCL surfaces was significantly reduced with respect to the other endothelialized PCL substrates (Fig. 13b), indicative of good anti-thrombogenic behavior of the EC confluent layer. This result is consistent with the high level of NO production by ECs on the gelatinized PCL surface. The P-selectin expression on the endothelialized pristine PCL and PCL-NH_2 surfaces was higher than those on the bare substrates (Fig. 13b), which is line with the well-established fact that subconfluent EC layers were more thrombogenic in nature [13]. The platelet activation was significantly enhanced on the endothelialized P(GMA)-grafted PCL surfaces, as observed by the high levels of P-selectin expression, which is an indication of an increased risk of thrombogenesis for those surfaces. Overall, the above results showed that the thrombogenicity of a biomaterial is influenced by both EC confluency and the surface properties of the biomaterial. Consequently, it can be concluded that the immobilization of gelatin on the P(GMA) brushes prevented platelet activation, but that the presence of the P(GMA) brushes alone led to a pro-thrombogenic surface.

3.5.5. Expression and activity of vWF and MMP-2 of EC on the surface-functionalized PCL samples

In order to further investigate the thrombogenicity of endothelialized surfaces, we performed real-time PCR tests and protein immunoblotting on factors that mediate platelet adhesion to ECs. The expression of von Willebrand factor (vWF) on endothelial cells promotes platelet adhesion and high or abnormal expression of vWF has been implicated in pathological conditions such as thrombosis [58]. Matrix metalloproteinase-2 (MMP-2) is another factor known to be involved in platelet aggregation, as well as to have important roles in the degradation and remodeling of the endothelial extracellular matrix [59, 60]. 7 days after the seeding of ECs, the relative expression of both vWF and MMP-2 mRNA in the EC on both PCL-*g*-P(GMA)1-*c*-gelatin and PCL-*g*-P(GMA)2-*c*-gelatin surfaces were downregulated when compared to the ECs on PCL-*g*-P(GMA) surfaces and gelatin-coated coverslips (Fig. 14a and 14c). Immunoblotting of the vWF protein produced in ECs, also, suggest that it is lower in ECs on PCL-*g*-P(GMA)- *c*-gelatin surfaces in comparison to PCL-*g*-P(GMA) surfaces but differed where the vWF expression of ECs on the pristine PCL surface and gelatin-coated coverslip could not be detected (Fig. 14b). Nevertheless, these results suggest that pro-thrombogenic factors could be increased on bare PCL-*g*-P(GMA) surfaces and the conjugation of gelatin could help to reduce thrombogenicity.

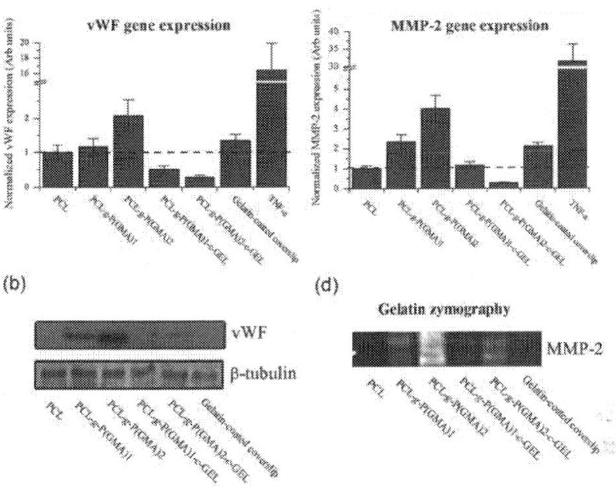

Figure 14. Real-time qPCR revealed that the expression of (a) vWF and (c) MMP-2 in ECs on PCL-*g*-P(GMA)-*c*-gelatin surfaces is lower when compared to PCL-*g*-P(GMA) surfaces and gelatin-coated coverslips. Expression levels are normalized using the housekeeping gene rpl27 and taken relative to ECs on pristine PCL (dotted line) (mean ± SD, n=3). Endothelial cells treated with TNF-αfor 6 h were used as a positive control for the expression of both genes. (c) Immunoblotting of vWF protein expressed in ECs on PCL-*g*-P(GMA)-*c*-gelatin is also reduced when compared to PCL-*g*-P(GMA) surfaces, thus corroborating the real-time PCR results.

4. Conclusion

This study described the successful biofunctionalization of PCL substrates with tunable surface densities of covalently-immobilized gelatin by surface-initiated ATRP of glycidyl methacrylate (GMA). Kinetics studies revealed that the grafting yield of the functional P(GMA) brushes increased linearly with polymerization time, and the amount of immobilized gelatin increased with the concentration of epoxide groups on the P(GMA) brushes. The significant improvement in the adhesion and proliferation of ECs on the gelatin-immobilized PCL substrates were found to be positively correlated to the amount of covalently immobilized gelatin. Blood compatibility tests demonstrated that the ECs cultured on the gelatin-immobilized-P(GMA) surfaces exhibited low platelet activation and significantly increased nitric oxide (NO) production, although the coagulation pathways were not affected both before and after EC coverage. Overall, the high surface-density of immobilized gelatin obtained by surface-initiated ATRP on the PCL surfaces is favorable for EC attachment and proliferation. The attached ECs maintained an unactivated and non-thrombogenic phenotype that mimics the EC lining of a healthy blood vessel. Hence, such a surface may have huge potential for vascular graft applications.

Acknowledgements

This research is supported by the Singapore National Research Foundation under CREATE programme: The Regenerative Medicine Initiative in Cardiac Restoration Therapy (NRF-Technion).

Author details

Shaojun Yuan[1,2*], Gordon Xiong[2], Ariel Roguin[3], Swee Hin Teoh[4] and Cleo Choong[2*]

*Address all correspondence to: yuanshaojun@gmail.com

*Address all correspondence to: cleochoong@ntu.edu.sg

1 Multi-phases Mass Transfer & Reaction Engineering Lab, College of Chemical Engineering, Sichuan University, Chengdu, China

2 School of Materials Science and Engineering, Nanyang Technological University, Singapore

3 Department of Cardiology, Rambam Medical Center, B. Rappaport Faculty of Medicine, Technion – Israel Institute of Technology, Israel

4 School of Chemical and Biomedical Engineering, Nanyang Technological University, Singapore

References

[1] Chlupac J, Filova E, Bacakova L. Blood vessel replacement: 50 years of development and tissue engineering paradigms in vascular surgery. *Physiol Res.* 2009;58:S119-S139.

[2] Mitchell SL, Niklason LE. Requirements for growing tissue-engineered vascular grafts. *Cardiovasc Pathol* 2003;12(2):59-64.

[3] Wang XW, Lin P, Yao QH, Chen CY. Development of small-diameter vascular grafts. *World J. Surg.* 2007;31(4):682-689.

[4] Sharkawi T, Darcos V, Vert M. Poly(DL-lactic acid) film surface modification with heparin for improving hemocompatibility of blood-contacting bioresorbable devices. *J. Biomed. Mater. Res. A* 2011;98A(1):80-87.

[5] Liu HF, Li XM, Niu XF, Zhou G, Li P, Fan YB. Improved hemocompatibility and endothelialization of vascular grafts by covalent immobilization of sulfated silk fibroin on poly(lactic-co-glycolic acid) scaffolds. *Biomacromolecules* 2011;12(8):2914-2924.

[6] Lee JH, Oh SH. MMA/MPEOMA/VSA copolymer as a novel blood-compatible material: Effect of PEO and negatively charged side chains on protein adsorption and platelet adhesion. *J. Biomed. Mater. Res.* 2002;60(1):44-52.

[7] Lee JH, Ju YM, Kim DM. Platelet adhesion onto segmented polyurethane film surfaces modified by addition and crosslinking of PEO-containing block copolymers. *Biomaterials* 2000;21(7):683-691.

[8] Amiji M, Park K. Surface modification of polymeric biomaterials with poly(ethylene oxide), albumin and heparin for reduced thrombogenicity. *J Biomater Sci-Polym Ed* 1993;4(3):217-234.

[9] Chong MSK, Teoh SH, Teo EY, Zhang ZY, Lee CN, Koh S, Choolani M, Chan J. Beyond cell capture: antibody conjugation improves hemocompatibility for vascular tissue engineering applications. *Tissue Eng. Part A* 2010;16(8):2485-2495.

[10] Ku SH, Park CB. Human endothelial cell growth on mussel-inspired nanofiber scaffold for vascular tissue engineering. *Biomaterials* 2010;31(36):9431-9437.

[11] Wu KK, Thiagarajan P. Role of endothelium in thrombosis and hemostasis. *Annu Rev Med* 1996;47:315-331.

[12] Lijnen HR, Collen D. Endothelium in hemostasis and thrombosis. *Prog. Cardiovasc. Dis* 1997;39(4):343-350.

[13] McGuigan AP, Sefton MV. The influence of biomaterials on endothelial cell thrombogenicity. *Biomaterials* 2007;28(16):2547-2571.

[14] Woodruff MA, Hutmacher DW. The return of a forgotten polymer - polycaprolactone in the 21st century. *Prog. Polym.Sci.* 2010;35(10):1217-1256.

[15] Sung HJ, Meredith C, Johnson C, Galis ZS. The effect of scaffold degradation rate on three-dimensional cell growth and angiogenesis. *Biomaterials* 2004;25(26):5735-5742.

[16] Chong MSK, Chan J, Choolani M, Lee CN, Teoh SH. Development of cell-selective films for layered co-culturing of vascular progenitor cells. *Biomaterials* 2009;30(12): 2241-2251.

[17] Serrano MC, Portoles MT, Vallet-Regi M, Izquierdo I, Galletti L, Comas JV, Pagani R. Vascular endothelial and smooth muscle cell culture on NaOH-treated poly (epsilon-caprolactone) films: A preliminary study for vascular graft development. *Macromol Biosci* 2005;5(5):415-423.

[18] de Valence S, Tille JC, Mugnai D, Mrowczynski W, Gurny R, Moller M, Walpoth BH. Long term performance of polycaprolactone vascular grafts in a rat abdominal aorta replacement model. *Biomaterials* 2012;33(1):38-47.

[19] Tillman BW, Yazdani SK, Lee SJ, Geary RL, Atala A, Yoo JJ. The in vivo stability of electrospun polycaprolactone-collagen scaffolds in vascular reconstruction. *Biomaterials* 2009;30(4):583-588.

[20] Walpoth BH, Moller M. Tissue engineering of vascular prostheses. *Chirurg* 2011;82(4):303-310.

[21] Serrano MC, Pagani R, Pena J, Vallet-Regi M, Comas JV, Portoles MT. Progenitor-derived endothelial cell response, platelet reactivity and haemocompatibility parameters indicate the potential of NaOH-treated polycaprolactone for vascular tissue engineering. *J Tissue Eng Regen Med* 2011;5(3):238-247.

[22] Ma ZW, Gao CY, Gong YH, Shen JC. Chondrocyte behaviors on poly-L-lactic acid (PLLA) membranes containing hydroxyl, amide or carboxyl groups. *Biomaterials* 2003;24(21):3725-3730.

[23] Zhu YB, Gao CY, Liu YX, Shen JC. Endothelial cell functions in vitro cultured on poly(L-lactic acid) membranes modified with different methods. *J. Biomed. Mater. Res. A* 2004;69A(3):436-443.

[24] Shin YM, Kim KS, Lim YM, Nho YC, Shin H. Modulation of spreading, proliferation, and differentiation of human mesenchymal stem cells on gelatin- immobilized poly(L-lactide-co-epsilon-caprolactone) substrates. *Biomacromolecules* 2008;9(7): 1772-1781.

[25] Edmondson S, Osborne VL, Huck WTS. Polymer brushes via surface-initiated polymerizations. *Chem Soc Rev* 2004;33(1):14-22.

[26] Coessens V, Pintauer T, Matyjaszewski K. Functional polymers by atom transfer radical polymerization. *Prog. Polym.Sci.* 2001;26(3):337-377.

[27] Xu FJ, Yang XC, Li CY, Yang WT. Functionalized polylactide film surfaces via surface-initiated ATRP. *Macromolecules* 2011;44(7):2371-2377.

[28] Jiang H, Wang XB, Li CY, Li JS, Xu FJ, Mao C, Yang WT, Shen J. Improvement of hemocompatibility of polycaprolactone film surfaces with zwitterionic polymer brushes. *Langmuir* 2011;27(18):11575-11581.

[29] Tiaw KS, Teoh SH, Chen R, Hong MH. Processing methods of ultrathin poly(epsilon-caprolactone) films for tissue engineering applications. *Biomacromolecules* 2007;8(3): 807-816.

[30] Zhu YB, Gao CY, Liu XY, Shen JC. Surface modification of polycaprolactone membrane via aminolysis and biomacromolecule immobilization for promoting cytocompatibility of human endothelial cells. *Biomacromolecules* 2002;3(6):1312-1319.

[31] Causa F, Battista E, Della Moglie R, Guarnieri D, Iannone M, Netti PA. Surface investigation on biomimetic materials to control cell adhesion: the case of RGD conjugation on PCL. *Langmuir* 2010;26(12):9875-9884.

[32] Yuan SJ, Wan D, Liang B, Pehkonen SO, Ting YP, Neoh KG, Kang ET: Lysozyme-coupled poly(poly(ethylene glycol) methacrylate)-stainless steel hybrids and their antifouling and antibacterial surfaces. *Langmuir* 2011;27(6):2761-2774.

[33] Zhu YB, Gao CY, Shen JC. Surface modification of polycaprolactone with poly(methacrylic acid) and gelatin covalent immobilization for promoting its cytocompatibility. *Biomaterials* 2002;23(24):4889-4895.

[34] Chang KY, Hung LH, Chu IM, Ko CS, Lee YD. The application of type II collagen and chondroitin sulfate grafted PCL porous scaffold in cartilage tissue engineering. *J. Biomed. Mater. Res. A* 2010;92A(2):712-723.

[35] Zhang HN, Hollister S. Comparison of bone marrow stromal cell behaviors on poly(caprolactone) with or without surface modification: studies on cell adhesion, survival and proliferation. *J Biomater Sci-Polym Ed* 2009;20(14):1975-1993.

[36] Gabriel M, Amerongen GV, Van Hinsbergh VWM, Amerongen AVV, Zentner A. Direct grafting of RGD-motif-containing peptide on the surface of polycaprolactone films. *J Biomater Sci-Polym Ed* 2006;17(5):567-577.

[37] von Burkersroda F, Schedl L, Gopferich A. Why degradable polymers undergo surface erosion or bulk erosion. *Biomaterials* 2002;23(21):4221-4231.

[38] Bech L, Meylheuc T, Lepoittevin B, Roger P. Chemical surface modification of poly(ethylene terephthalate) fibers by aminolysis and grafting of carbohydrates. *J. Polym Sci Pol Chem* 2007;45(11):2172-2183.

[39] Moulder J F, Sobol FE, Bomben KD. Handbook of X-ray photoelectron spectroscopy. Eden Prairie, Minn: Perkin-Elmer Corp.; 1992.

[40] Siegwart DJ, Oh JK, Matyjaszewski K. ATRP in the design of functional materials for biomedical applications. *Prog. Polym.Sci.* 2012;37(1):18-37.

[41] Xu FJ, Cai QJ, Li YL, Kang ET, Neoh KG. Covalent immobilization of glucose oxidase on well-defined poly(glycidyl methacrylate)-Si(111) hybrids from surface-initiated atom-transfer radical polymerization. *Biomacromolecules* 2005;6(2):1012-1020.

[42] Chan K, Gleason KK. Photoinitiated chemical vapor deposition of polymeric thin films using a volatile photoinitiator. *Langmuir* 2005;21(25):11773-11779.

[43] Wang T, Kang ET, Neoh KG, Tan KL, Cui CQ, Lim TB. Surface structure and adhesion enhancement of poly(tetrafluooethylene) films after modification by graft copolymerization with glycidyl methacrylate .*J. Adhes. Sci. Technol.* 1997;11(5):679-693.

[44] Eckert AW, Grobe D, Rothe U. Surface-modification of polystyrene-microtitre plates via grafting of glycidylmethacrylate and coating of poly-glycidylmethacrylate. *Biomaterials* 2000;21(5):441-447.

[45] Xia Y, Boey F, Venkatraman SS. Surface modification of poly(L-lactic acid) with biomolecules to promote endothelialization. *Biointerphases* 2010;5(3):FA32-FA40.

[46] Marquard H, Selkirk JK, Sims P, Kuroki T, Heidelbe C, Huberman E, Grover PL. Malignant transformation of cells derived from mouse prostate by epoxide and other derivatives of polycylic hydrocarbons. *Cancer Res* 1972;32(4):716-720.

[47] Cheng ZY, Teoh SH. Surface modification of ultra thin poly(ε-caprolactone) films using acrylic acid and collagen. *Biomaterials* 2004;25(11):1991-2001.

[48] Arima Y, Iwata H. Effect of wettability and surface functional groups on protein adsorption and cell adhesion using well-defined mixed self-assembled monolayers. *Biomaterials* 2007;28(20):3074-3082.

[49] van der Zijpp YJT, Poot AA, Feijen J. ICAM-1 and VCAM-1 expression by endothelial cells grown on fibronectin-coated TCPS and PS. *J. Biomed. Mater. Res. A* 2003; 65A(1):51-59.

[50] Kamal AH, Tefferi A, Pruthi RK. How to interpret and pursue an abnormal prothrombin time, activated partial thromboplastin time, and bleeding time in adults. *Mayo Clin Proc* 2007;82(7):864-873.

[51] Liu YA, Wang W, Wang J, Wang YL, Yuan Z, Tang SM, Liu M, Tang H. Blood compatibility evaluation of poly(D,L-lactide-co-beta-malic acid) modified with the GRGDS sequence. *Colloid Surf B-Biointerfaces* 2010;75(1):370-376.

[52] Hansson KM, Tosatti S, Isaksson J, Wettero J, Textor M, Lindahl TL, Tengvall P. Whole blood coagulation on protein adsorption-resistant PEG and peptide functionalised PEG-coated titanium surfaces. *Biomaterials* 2005;26(8):861-872.

[53] Graves JE, Greenwood IA, Large WA. Tonic regulation of vascular tone by nitric oxide and chloride ions in rat isolated small coronary arteries. *Am J Physiol-Heart Circul Physiol* 2000;279(6):H2604-H2611.

[54] Allen RD, Zacharski LR, Widirstky ST, Rosenstein R, Zaitlin LM, Burgess DR. Transformation and motility of human-platelets - details of the shape change and release reaction observed by optical and electron-microscopy. *J Cell Biol* 1979;83(1):126-142.

[55] Saelman EUM, Nieuwenhuis HK, Hese KM, Degroot PG, Heijnen HFG, Sage EH, Williams S, McKeown L, Gralnick HR, Sixma JJ. Platelet-adhesion to collagen type-I through type-VIII under condtions of stasis and flow is mediated by α2β1-integrin. *Blood* 1994;83(5):1244-1250.

[56] Hu B, Finsinger D, Peter K, Guttenberg Z, Barmann M, Kessler I, Escherich A, Moroder L, Bohm J, Baumeister W, Sui SF. Intervesicle cross-linking with integrin alpha(IIb)beta(3) and cyclic-RGD-lipopeptide. A model of cell-adhesion processes. *Biochemistry* 2000;39(40):12284-12294.

[57] Kidane AG, Punshon G, Salacinski HJ, Ramesh B, Dooley A, Olbrich M, Heitz J, Hamilton G, Seifalian AM. Incorporation of a lauric acid-conjugated GRGDS peptide directly into the matrix of a poly(carbonate-urea)urethane polymer for use in cardiovascular bypass graft applications. *J. Biomed. Mater. Res. A* 2006;79A(3):606-617.

[58] Brill A, Fuchs TA, Chauhan AK, Yang JJ, De Meyer SF, Köllnberger M, et al. von Willebrand factor–mediated platelet adhesion is critical for deep vein thrombosis in mouse models. Blood. 2011; 117(4):1400-1407.

[59] Kazes I, Elalamy I, Sraer J-D, Hatmi M, Nguyen G. Platelet release of trimolecular complex components MT1-MMP/TIMP2/MMP2: involvement in MMP2 activation and platelet aggregation. Blood. 2000; 96(9):3064 –3069.

[60] Ben-Yosef Y, Lahat N, Shapiro S, Bitterman H, Miller A. Regulation of endothelial matrix metalloproteinase-2 by hypoxia/reoxygenation. Circulation Research. 2002; 90(7):784 –791.

Chapter 8

Cell Adhesion to Biomaterials: Concept of Biocompatibility

M. Lotfi, M. Nejib and M. Naceur

Additional information is available at the end of the chapter

http://dx.doi.org/10.5772/53542

1. Introduction

Cell adhesion is a dynamic process that results from specific interactions between cell surface molecules and their appropriate ligands. Adhesion can be found between adjacent cells (cell-cell adhesion) as well as between cells and the extracellular matrix (ECM) (cell-matrix adhesion). Adhesion is an extremely important concept in both practical and theoretical terms. Unfortunately, there is no completely satisfactory definition of the term that fulfils the needs of both the theoretical surface chemist and the practicing technologist. It is assumed as a state in which two bodies (usually, but not necessarily dissimilar) are held together by intimate interfacial contact in such a way that mechanical force or work can be applied across the interface without causing the two bodies to separate.

Cell membrane are crucial to the adhesion of the cell and therefore to its life. Indeed, plasma membrane encloses the cell, defines its boundaries, and maintains the essential differences between the cytosol and the extracellular environment. In all cells the plasma membrane also contains proteins that act as sensors of external signals, allowing the cell to change its behavior in response to environmental cues; these receptors transfer information rather than ions or molecules across the membrane. Plasma membrane has the structure of a thin film of lipid and protein molecules linked together mostly through non covalent interactions. These lipid molecules are arranged as a continuous bilayer and are responsible for the basic structure of the membrane and the protein molecules embedded into it control most of the functions of the membrane. In the plasma membrane some proteins serve as structural links that connect the membrane to the cytoskeleton and/or to either the extracellular matrix (ECM) or an adjacent cell, while others serve as receptors to detect and transducer chemical signals in the cell's environment [1].

Besides keeping a multicellular organism together, cell adhesion is also a source of specific signals to adherent cells; their phenotype can thus be regulated by their adhesive interactions. In fact, most of the cell adhesion receptors were found to be involved in signal transduction. By interacting with growth factor receptors they are able to modulate their signaling efficiency. Therefore, gene expression, cytoskeletal dynamics and growth regulation all depend, at least partially, on cell adhesive interactions [2].

In this chapter, I tried to find a possible correlation between polyelectrolyte multilayer films and human gingival fibroblasts to test these biomaterials biocompatibility. This represents a fundamental step needed to know about a possible use in a biological field (i.e. as implant). For that purpose, I characterized each solid surface used as a surface on which fibroblasts were cultured; by calculating their surface free energy and evaluating their chemical heterogeneity, roughness and wettability using contact angle measurement. Thereafter, I followed the adhesion of fibroblasts, their proliferation and their morphology.

2. Polyelectrolyte multilayer film

2.1. Biomaterials: Generality and interest

During a consensus conference in 1986, a definition was given for biomaterials. Indeed, a biomaterial is «a non-living material used and designed to be integrated with biological systems». Biomaterials are defined according to their domain of use and regroup metals and alloys, ceramics, polymers (i.e. collagen)[3].

Biomaterials were used since the pharaoh's time to replace injured and affected organs. Pharaoh had used pure natural materials but presenting integration's problems. Since that, researches had grown up rapidly in this field in order to design the "ideal" material which will be more accepted by the human body. The designed material was referred to as "biomaterial" afterwards and will recover a lot of biomedical applications for implants and tissues injuries covering.

Biomaterials' design must take into account the purpose and the place of its use. This biomaterial must have a well defined shape depending on his position within the body. Indeed, for orthopedic usage, a biomaterial must conform to some criteria and regulations such as: a good mechanical structure, a good resistance to corrosion and metal fatigue. For vascular surgery, a biomaterial must not induce thrombosis, in odontology a biomaterial must withstand changes that can occur to temperature (coffee, cool drinks), to pH (alcohol, lemon…) and to the buccal cavity [4].

Making reliable and cheap biomaterials is being a new challenge for researchers and industries. In fact, the infallibility of every biomaterial depends on the materials from which it's made of. Consequently, there's a great demand in developing new suitable biomaterials (or making the existing ones better) used in multidisciplinary fields and involving physics, chemistry and biology.

http://dx.doi.org/10.5772/53542

In this study, the biomaterials used for fibroblasts adhesion are made of polyelectrolytes using the layer-by-layer technique based on alternating oppositely charged polyelectrolytes on glass probes (more details are shown in paragraph III.2).

2.2. Polyelectrolytes

Polyelectrolytes are highly charged nanoscopic objects or macromolecules. Their electric charge density appears as more or less continuous, when it is seen from distances to the macromolecule equal to several times to the intercharge distance, giving them the polyelectrolytic character. Obviously, their properties will be extremely different according to their geometry. Massive spherical objects will behave like colloids, whereas linear flexible objects will keep some of the macromolecular polymeric character [5]. They are defined as materials for which the solution's properties in dissolvent presenting a high permittivity are governed by electrostatic interactions for distances superior to the molecular dimensions [6]. Polyelectrolytes are by no way a mere superposition of electrolytes and polymers properties. New and rather unexpected behaviours are observed:

- Whereas polymers exhibit only excluded volume effects, the long ranged coulomb interactions, which are present in polyelectrolytes, give rise to new critical exponents.
- The main difference with electrolytes is that one kind of ions, the counterions are stuck together along a chain, and the collective contribution of the charged monomers causes a strong field in the vicinity of the chain, even at very low dilution.

These materials are widely used in industries as dispersive substances in aqueous medium, flocculants to aggregate sludge and industrial waste. Recently, they were used to make films by alternating thin layers of polymers of medical use such as dental prosthesis, fabrication of transplantable organs etc…

Polymers differ by their structure, their surface composition and their biological properties:

2.2.1. Biological properties

The biological properties reflect the origin of polymers. Indeed, one can distinguish three different origins for polymers [7]:

- Natural polymers coming from animal, vegetal and mineral origins
- Artificial polymers with natural basic components and chemically transformed functions in their units (monomers)
- Synthetic polymers presenting synthetic basic components which are often very similar to those of natural polymers

2.2.2. Physico-chemical properties

According to Oudet [7], polymers have different physical properties. The most important are their thermal conductivity reflecting polymers' behaviour under temperature changes.

The second interesting physical property is their optical reactions towards light (refraction, reflection angle, polarization, absorption…). Moreover, polymers are characterized by their ability as electrical conductors or insulators.

From the chemical point of view, Fowkes [8] presumed the existence of different polymers surface structure: polymers with polar surfaces (polyethylene), polymers with acid (polyimide) or basic (polystyrene) sites dominance and others are regrouping both acid and basic characters (polyamide). These surfaces are governed by specific (dispersive forces attraction) and non-specific interactions (acid-base interaction).

Polymers properties are strongly influenced by molecular interactions such as Van der Waals interactions (low energy bonds), hydrogen interactions (low energy bonds having an electrostatic origin) and ionic interactions due to electrostatic attractions and repulsions between ions or ionized groups.

2.3. Polyelectrolyte multilayer film

2.3.1. Generality

In recent years, polyelectrolyte multilayer film has been widely developed in different fields and for a variety of purposes. This kind of ultrathin film can be fabricated from oppositely charged polyelectrolytes using a method called self-assembly discovered by Decher and co-workers in 1992 and allows surface modification and therefore controlling their properties at the molecular (or even the atomic) level.

These films are of a great interest for covering biomaterials used as implants [9, 10] and therefore they will be in contact with cells [11]. Layer-by-layer assembly of polyelectrolytes is a simple and suitable method for coating different substrates such as glass, silicon, thermoplastic and even curved surfaces [12, 13].

It's known that biomaterials must present two main conditions to be admitted for integration in the biological system: to be biocompatible with this system and to have definite mechanical and electrical properties depending on their use [14]. The next implants generation has a tendency to be bioactive, besides its biocompatibility, thanks to substrate coating with bioactive substances.

2.3.2. Fabrication method and application fields

Multilayer polyelectrolyte films are made by alternating oppositely charged polyelectrolytes (polyanions and polycations) on glass slides (Figure 1).

Film's formation is based on charge overcompensation of the newly adsorbed polyions. Indeed, a polyanion (negative charge) added to a polycation (positive charge), previously deposited on the substrate, will neutralise the excess of positive charges and therefore create a new negatively charged polyelectrolyte layer. This step can be repeated as many times as the needed number of layers is reached [15].

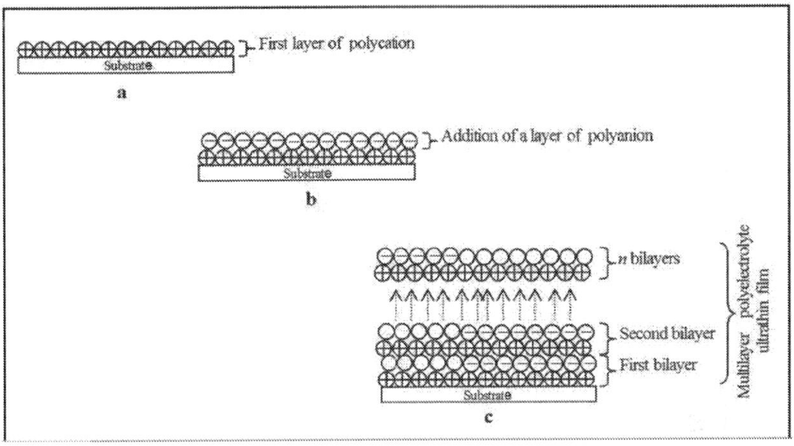

Figure 1. Layer-by-layer polyelectrolyte film's fabrication. This assembly method is based on alternating oppositely charge polyanion (positive charge) and polycations (negative charge) on a solid substrate. One bilayer consists in one polycation associated with one polyanion and the film is a set of n bilayers.

This adsorption mechanism is governed by electrostatic interactions which represent, besides other secondary interactions (hydrogen bond or dispersive force), a paramount parameter for the final structure of the formed film [16].

Polyelectrolyte multilayer films are used in different fields: orthopedic surgery (hip prosthesis...), cardiovascular (artificial heart, vascular prosthesis...), odontology (dental restoration...), ophthalmology (contact lenses...), urology (catheters, artificial kidney...), endocrinology (artificial pancreas, biosensors...), aesthetic surgery and other domains [17].

2.4. Polyelectrolyte film surface characterization

This study is possible by investigating surface wettability and calculating surface free energy. Indeed, wettability is the aptitude of a substrate to be coated by a thin liquid film while dipped in a liquid solution. This method is used to follow the substrate behaviour in relation to its environment and can be done thanks to the contact angle measurement. In this paper we are interested in the dynamic contact angle method using Wilhelmy plate method, treated later. This method, besides giving information about substrate surface hydrophilicity and hydrophobicity, allows us to evaluate the surface roughness and chemical heterogeneity. Moreover, with the results found, we measured the polyelectrolyte film's surface free energy according to Van Oss theory.

2.4.1. Contact angle measurement

There are a variety of simple and inexpensive techniques for measuring contact angles, most of which are described in detail in various texts and publications and will be mentioned only briefly here. The most common direct methods (Figure 2) include the sessile drop (a), the captive bubble (b) and the tilting plate (c). Indirect methods include tensiometry and geometric analysis of the shape of a meniscus. For solids for which the above methods are not applicable, such as powders and porous materials, methods based on capillary pressures, sedimentation rates, wetting times, imbibition rates, and other properties, have been developed [18].

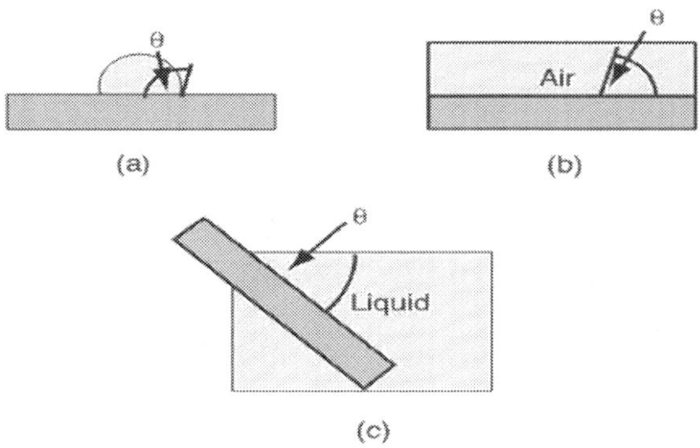

Figure 2. The more common systems of contact angle measurement showing the sessile drop (a), the captive bubble (b) and the tilting plate (c). θ is the contact angle to be measured.

2.4.1.1. *The sessile drop method*

It's a static contact angle measurement method which consists in putting down a liquid drop on the solid plate we want to characterize its surface by measuring the contact angle made by the drop on this surface. Indeed, when a drop of a liquid is putted down on a solid surface; three phases system occurs: solid, liquid and gas (Figure 3).

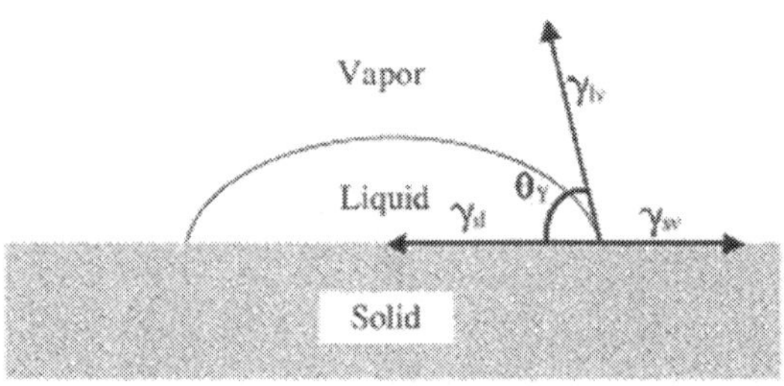

Figure 3. Static contact angle measurement with the sessile drop method

The drop's profile is being changed depending on the physico-chemical characters of the solid surface, on the adhesion forces newly created at the interface solid/liquid and on the cohesion forces of the liquid. This change will affect the contact angle value revealing the surface state (hydrophobic or hydrophilic, rough or smooth, homogeneous or heterogeneous...) and the different forces occurred are linked together according to Young's equation [19]:

$$\gamma_{sv} = \gamma_{sl} + \gamma_{lv} cos\theta,$$

Where γ_{sv}, γ_{sl} and γ_{lv} represent the "surface tensions" of the interface solid/gas, solid/liquid and liquid/gas, respectively, and θ represents the contact angle.

2.4.1.2. The captive bubble method

It's a derivative of the sessile drop method and consists in making an air bubble (or a bubble from a less dense and non miscible liquid such as dodecane, octane and octadecane) on a solid surface immersed in pure water or in other liquid with a well known physico-chemical characters. So, it's possible to measure the contact angle made by this bubble with the immersed solid surface (see Figure 2).

2.4.1.3. The tilting plate method

The tilting plate method is to slowly tilt a contact angle sample until the sessile drop on it begins to move in the downhill direction. At that time, the downhill contact angle is the advancing angle and the uphill angle the receding contact angle [20].

The principal alternative to the tilting plate method is having the dispense needle remain immersed in the sessile drop and pumping in until the drop expands in base area and pumping out until the drop contracts in base area. Often the tilting plate measurement is carried out on an instrument with a mechanical platform that tilts the stage and the camera together.

It has been shown that these methods are a subject of controversy. However, the dynamic contact angle measurement using the Wilhelmy plate method has been shown to be easier for use and gives more information about the surface characterized.

2.4.1.4. The dynamic contact angle method: The tensiometer

In our study, we used the Wilhelmy plate method (Tensiometer 3S, GBX, France) which allows a dynamic measurement of the contact angle hysteresis. Indeed, the tensiometer used for the measurement will measure the force applied to the substrate while immerged in a liquid thanks to a balance where the substrate was hanged (Figure 4)

In each case, the polyelectrolyte film coated glass slide was immersed into and then drawn out of the measurement liquid. Therefore, the tensiometer will evaluate the advancing angle (θ_a) when the liquid moves forward the substrate surface and thereafter the receding angle (θ_r) when the liquid resorbs from the substrate. The difference between θ_a and θ_r is called contact angle hysteresis H ($H = \theta_a - \theta_r$) and is useful for understanding the wettability of the film. It gives us information about the surface film mobility, its reorganization and roughness [21].

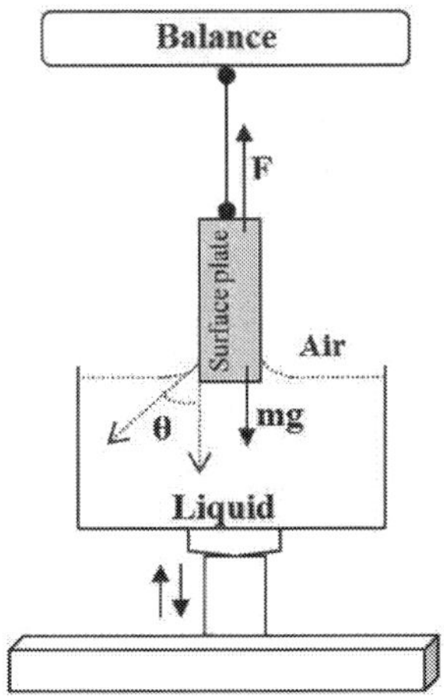

Figure 4. The Wilhelmy plate method for dynamic contact angle measurement. The surface plate is partially immerged in the up down moving liquid container. Curves (Loops) are automatically drawn by a software associated to the Tensiometer according to F = f (Immersion depth)

When a substrate is immersed in a liquid, three forces occur (see Figure 4): the gravity force, the upthrust buoyancy and the capillary forces. Therefore, by measuring the applied force according to the immersion depth and as we previously know the dimension of the substrate; one can calculate the wetting forces according to the equation [22]:

$$F = mg + p * \gamma_{LV} * \cos\theta - F_b \qquad (1)$$

Where F represents the force measured (mN/m), m is the substratum mass, g is the acceleration constant induced by the gravity, p is the substratum perimeter (cm^2), γ_{LV} is the surface free energy (mN/m) of the liquid used for measurement (constant), θ: the contact angle between the liquid and the substratum (°) and F_b is the force related to the upthrust buoyancy.

Usually, we make several immersion/emersion cycles for the substratum we are investigating and the different loops (one loop corresponds to one immersion/emersion cycle) are drawn by a software associated to the Tensiometer according to Force = f (immersion depth). Moreover, the substratum weight is assumed to be nil by a direct correction fixing the pre-immersion force to the value of zero. Therefore, the previous equation ([Eq. 1]) becomes:

$$F_{(zero\ immersion)} = p * \gamma_{LV} * \cos\theta$$

As the surface energy of the liquid of measurement is previously known, therefore the contact angle could be deduced.

It has been shown that the contact angle changes depending on the nature of the film and on its charges and thickness. The nature of liquid of measurement, the speed and temperature of measurement are also involved in this change [23]. Indeed, the thickness of the film can affect its elasticity which will induce a difference in the liquid diffusion into this film and therefore the film's swelling level changes affecting the contact angle. A previous study made by Elbert et al.[24] has shown a clear effect of the film layers' number on the wettability of the film.

The liquid used for measurement can affect the surface wettability by the mean of its pH which varies from a liquid to another and controls the acid or base character as well as the liquid polarity. These parameters are responsible for the rearrangement of the biomaterial's groups at its contact. This reorganization is also depending on the liquid diffusion into the polymer and on the effect of solubilization induced by the liquid to this polymer. This phenomenon represents an interesting mechanism for explaining contact angle hysteresis especially when the liquid concerned is water. Indeed, water has small molecules which allow it to diffuse easily. Therefore, after diffusion into a polymer, water will confer its hydrophilic character to this polymer which is being to have some kind of elasticity responsible for the reorganization of its polar groups as a reaction to the high surface energy level of water which is responsible for the high energy level at the interface [25]. Concerning the dynamic contact angle measurement speed, it affects the contact period between the biomaterial and the liquid and therefore it will change the period of time needed for the rearrangement of the surface polar groups during contact with the liquid. As each film has its own defined reorganization time, therefore different contact angles were found for the same surface at different measurement speeds. Moreover, every polymer has a defined glass transition temperature (T_g) able to induce a change on the surface wettability depending on the temperature of measurement [26].

2.4.2. Surface free energy calculation

It's interesting to know the value of surface free energy of a biomaterial because it has an effect on wettability as shown by Van Oss [27]. While the contact between the biomaterial and the liquid generates an interface solid/liquid which will consume, during its formation, a defined energy called the interface energy. The reversible adhesion force represents, therefore, the difference in the energy level between the initial state characterized by two surfaces [28, 29]: solid surface with the energy (γ_s) and liquid surface with the energy (γ_l); and the final state (γ_{sl}).

The surface free energy is a kind of an attraction force of the surface which cannot be measure directly but calculated after contact angle measurement in different measurement liquids (with different surface free energies) according to Owends et Wendt or Van Oss' approaches. Their theories are complementary but Van Oss' approach has been shown to give more information. It consists in the following equation [27]:

$$\gamma_S = \gamma_S^{LW} + 2\,(\gamma_S^{+}.\gamma_S^{-})^{½}$$

where γ_S represents the surface free energy of the biomaterial surface, γ_S^{WL} : the dispersive component and γ_S^{+}, γ_S^{-} represent the polar components (acid- base).

The different components of the solid and the liquid surface free energies as well as the contact angle are related by this equation:

$$\gamma_L\,(1 + cos\theta) = 2\,((\gamma_S^{LW}.\gamma_L^{LW})^{½} + (\gamma_S^{+}.\gamma_L^{-})^{½} + (\gamma_L^{+}.\gamma_S^{-})^{½}),$$

This equation contains three unknown parameters: γ^{LW}, γ^{+} and γ^{-}; the contact angle measurement must be done with three different measurement liquids in order to solve this equation and calculate the surface free energy of our polyelectrolyte film. For this purpose, we used three different liquids: water, diiodomethane and formamide.

2.4.3. Evaluation of the surface roughness and heterogeneity

Theses parameters are deduced from the shapes of the curves drawn (loops). Indeed, the more the surface is rough; the more the curve is deformed (non linear curve). However, the more the surface is smooth; the more the curve presents a linear shape (no deformations observed). Otherwise, a roughness of about 100 nm has been shown to induce contact angle hysteresis. As for surface heterogeneity, it can be concluded from the different contact angle hysteresis values measured in the case of a negligible roughness.

Concerning the different polyelectrolyte films used in this study, a previous investigation was made by Picart and coworkers [30]. They measured the roughness by the AFM technique, refractive index and thickness are estimated by optical waveguide light mode spectroscopy, and zeta potential is measured by streaming potential measurements. Indeed, these parameters give us information about the chemical heterogeneity of the polyelectrolyte used.

Many studies had observed an important dependence of the contact angle hysteresis on the surface composition and topography (roughness) [31, 32]. Therefore, the more the surface is rough; the more it's hydrophilic and vise versa and the more this surface is composed of small molecules, the less the liquid diffusion in the biomaterial surface is disturbed leading to a low contact angle value. According to Morra et al.[33], this is may be due to existence of two different effects while studying the wettability of rough and homogeneous biomaterials: the barrier effect, where hysteresis increases with increasing the surface roughness due to an important rigidity of the substrate, and the capillary attraction at the surface which can affect Young's concept. Indeed, the capillary effect induces an increase of both the advancing and receding contact angles in the case of a surface presenting a contact angle superior to 90° at the equilibrium state. In the opposite case (contact angle inferior to 90° at the equilibrium state), the inverse situation will happen and the contact angle variations will be less important than those corresponding to the barrier effect. Only in the case of a contact angle equals to 90°, the capillary effect is negligible.

2.5. Conclusion

When a drop of liquid is placed on a solid surface, the liquid will either spread across the surface to form a thin, approximately uniform film or spread to a limited extent but remain as a discrete drop on the surface. The final condition of the applied liquid to the surface is taken as an indication of the wettability of the surface by the liquid or the wetting ability of the liquid on the surface. The quantitative measure of the wetting process is taken to be the contact angle, which the drop makes with the solid as measured through the liquid in question.

The wetting of a surface by a liquid and the ultimate extent of spreading of that liquid are very important aspects of practical surface chemistry. Many of the phenomenological aspects of the wetting processes have been recognized and quantified since early in the history of observation of such processes. However, the microscopic details of what is occurring at the various interfaces and lines of contact among phases has been more a subject of conjecture and theory than of known facts until the latter part of this century when quantum electrodynamics and elegant analytical procedures began to provide a great deal of new insight into events at the molecular level. Even with all the new information of the last 20 years, however, there still remains a great deal to learn about the mechanisms of movement of a liquid across a surface.

3. Fibroblast cells

3.1. Human gingival fibroblasts

3.1.1. Generality

Fibroblasts are spindle-shaped connective-tissue cells of mesenchymal origin that secretes proteins and especially molecular collagen from which the extracellular fibrillar matrix of connective tissue forms. They have oval or circular nucleus and a little developed cytoplasm giving rise to long prolongation forms [34]. These cells do not have a basal lamina and their surfaces are often in contact with the fibers of the collagen. Their cytoplasm contains a rough endoplasmic reticulum, an important Golgi apparatus, few mitochondria and a little bit quantity of cytoplasmic filaments. Fibroblasts synthesize enormous quantities of the extracellular matrix constituents. Indeed, the majority part of the extracellular matrix components consists of collagen made in the intracellular space where fibroblasts sustain structural modifications.

3.1.2. Gingival tissue

It's the tissue that surrounds the necks of teeth and covers the alveolar parts of the jaws; broadly: the alveolar portion of a jaw with its enveloping soft tissues [35]. It consists in a pink connective tissue with fibrous collagen surrounded by an epithelial tissue. Its pink color changes from one person to another, depending on pigmentation, epithelium thickness, its keratinization level and on the underlying vascularization [36]. Fibroblasts are the basic component of the gingival chorion whose intercellular matrix is essentially formed by collagen and elastin.

3.2. Cell-Biomaterial: Interface and interactions

3.2.1. Biocompatibility concept

While a cell is in contact with a biomaterial, many reactions can occur and a sensing phenomenon will launch between this cell and the biomaterial [37]. Indeed, the cell has a signal network reached as a result of the surface exploration and sensing made in order to verify whether the new environment (biomaterial) is in accordance with its expected physiological conditions necessary for a normal biological activity [38]. Thus, before putting a new material in contact with a cell it's of a great importance to choose the corresponding material in such a way that this material obey the cell's norm by not being toxic or injurious and not causing immunological rejection. In one word, this material must be biocompatible.

The biological tolerance of a biomaterial led scientists to regroup the different parameters and mechanisms controlling the interface biomaterial/cell (or tissue) so that they can deduce a concrete and a common definition for biocompatibility concept. Indeed, biocompatibility includes the understanding of the interactive mechanisms relating the biomaterial with its biological environment. Generally, biocompatibility represents the ability of a material to be accepted by a living organism.

In 1987, Williams D.F suggested the following definition «biocompatibility is the ability of a material to be used with an appropriate and suitable reaction of the host for a specific application».

According to Exbrayat [39] « biocompatibility is a set of the different interrelations between a biomaterial and its environment, and their biological local or general consequences, immediate or delayed, reversible or definitive».

Indeed, biocompatibility is a group of networks that liaises between the biomaterial and its environment and takes into account the possible effect of this biomaterial on its environment and vice versa. Interactions existing in the interface biomaterial/biological environment differ by their intensity and their duration period depending both on the biomaterial and on the tissue in contact.

Characterizing the surface properties of a biomaterial before putting it in contact with a cell seems to be an obligation. This step allows us to know about different parameters and characters of this biomaterial (topography, roughness, surface energy etc.) in order to find a correlation with the cell behavior and therefore we can adjust these physico-chemical properties, when making the biomaterial, so that we have a normal and physiological cell behavior in contact with that biomaterial.

3.2.2. *Cell adhesion*

It is well known that during the contact between a cell and a material, information will be transferred from the material surface to the cell and this contact will induce, in return, an alteration to the material. This situation may cause material remodelling [40,22].

Cells adhere to surfaces through adhesion proteins (i.e. fibronectin, collagen, laminin, vitronectin) using specific cell receptors, called integrins, attached to the cell membrane. Indeed, when fibroblasts grow on a substrate, most of their cell surface is separated from the substratum by a gap of more than 50 nm; but at focal contacts, this gap is reduced to 10 to 15 nm. The main transmembrane linker proteins of focal contacts belong to the integrin family and the cytoplasmic domain of the integrin binds to the protein talin, which in turn binds to vinculin, a protein found also in other actin-containing cell junction. Vinculin associates with α-actinin and is thereby linked to an actin filament [1].

Besides their role as anchors, focal contacts can also relay signals from the extracellular matrix (ECM) to the cytoskeleton. Several protein kinases are localized to focal contacts and seems to change their activity with the type of the substratum on which the rest. These kinases can regulate the survival, growth, morphology, movement, and differentiation of cells in response to new environment. Figure 5 shows a possible arrangement of these different proteins during a focal contact.

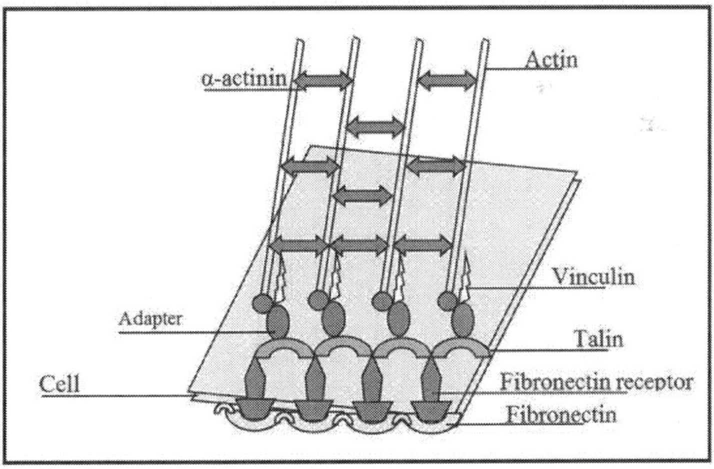

Figure 5. Adhesion proteins involved in focal contacts

The formation of focal contacts occurs when the binding of matrix glycoprotein, such as fibronectin, on the outside of the cell causes the integrin molecules to cluster at the contact site. Fibronectins are associated together by proteoglycans and constitute thins fibers of the extracellular matrix (ECM).

3.2.2.1. Extracellular matrix

The extracellular matrix (ECM) represents an important element in the processes of cell adhesion. Indeed, at this level, cell adhesion is under the control of a well defined zone in the cytoplasmic membrane called focal contact. At this zone, filaments of actin are linked to fibronectin through an intracellular complex of proteins, the adherence complex. The extracellular matrix (ECM) is made of different proteins such as fibronectins, collagen, laminin, vitronectin [41] and represents the mediator of cell adhesion thanks to its integrins.

Although the extracellular matrix generally provides mechanical support to tissues, it serves several other functions as well. Different combinations of ECM components tailor the extracellular matrix for specific purposes: strength in a tendon, tooth, or bone; cushioning in cartilage; and adhesion in most tissues. In addition, the composition of the matrix, which can vary, depending on the anatomical site and physiological status of a tissue, can let a cell know where it is and what it should do (environmental cues). Changes in ECM components, which are constantly being remodeled, degraded, and resynthesized locally, can modulate the interactions of a cell with its environment. The matrix also serves as a reservoir for many extracellular signalling molecules that control cell growth and differentiation. In addition, the matrix provides a lattice through or on which cells can move, particularly in the early stages of tissue assembly [42].

Many functions of the matrix require transmembrane adhesion receptors that bind directly to ECM components and that also interact, through adapter proteins, with the cytoskeleton. The principal class of adhesion receptors that mediate cell–matrix adhesion are integrins, a large family of αβ heterodimeric cell surface proteins that mediate both cell–cell and cell–matrix adhesions and inside-out and outside-in signalling in numerous tissues.

3.2.2.2. Adhesion proteins and receptors in fibroblast cells

Different proteins and their receptors are involved in fibroblast cells adhesion process. The most important and known are fibronectins and their receptors; integrins:

- Fibronectins

  Fibronectins are dimers of two similar polypeptides linked at their C-termini by two disulfide bonds; each chain is about 60–70 nm long and 2–3 nm thick. The combination of different repeats composing the regions, another example of combinatorial diversity, confers on fibronectin its ability to bind multiple ligands [40].

  Fibronectins help attach cells to the extracellular matrix by binding to other ECM components, particularly fibrous collagens and heparan sulfate proteoglycans, and to cell surface adhesion receptors such as integrins. Through their interactions with adhesion receptors (e.g., α5β1 integrin), fibronectins influence the shape and movement of cells and the organization of the cytoskeleton. Conversely, by regulating their receptor-mediated attachments to fibronectin and other ECM components, cells can sculpt the immediate ECM environment to suit their needs.

- Integrins

Integrins are the principle adhesion receptors; a large family of αβ heterodimeric cell surface proteins that mediate both cell–cell and cell–matrix. They are transmembrane proteins that mediate interactions between adhesion molecules on adjacent cells and/or the extracellular matrix (ECM). They have diverse roles in several biological processes including cell migration during development and wound healing, cell differentiation, and apoptosis. Their activities can also regulate the metastatic and invasive potential of tumor cells. They exist as heterodimers consisting of alpha and beta subunits. Some alpha and beta subunits exhibit specificity for one another, and heterodimers often preferentially bind certain cell adhesion molecules, or constituents of the ECM.

Although they themselves have no catalytic activity, integrins can be part of multimolecular signalling complexes known focal adhesions. The two subunits, designated as alpha and beta, both participate in binding.

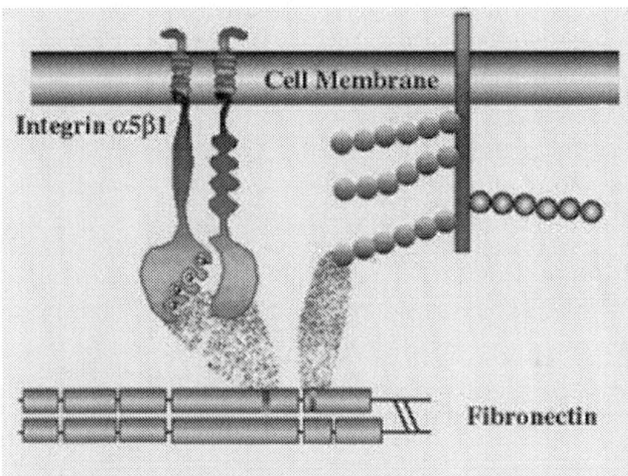

Figure 6. Fibronectin binding to its Integrin receptor (adapted from internet)

Integrins participate in cell-cell adhesion and are of great importance in binding and interactions of cells with components of the extracellular matrix such as fibronectin. Importantly, integrins facilitate "communication" between the cytoskeleton and extracellular matrix; allow each to influence the orientation and structure of the other. It is clear that interactions of integrins with the extracellular matrix can have profound effects on cell function, and events such as clustering of integrins activates a number of intracellular signally pathways.

3.2.3. *Cell adhesion: The physical process*

Biological systems exhibit electromagnetic activity in a wide frequency range from the static or quasistatic electric field to optical bands. Fröhlich [43] presumed that biological matter has anomalous polarization properties (e.g. induction of great electric dipole after electric field application). Static charge distribution of dipole and/or multipole nature exists (e.g. in protein molecules). Vibrations in biological molecules, therefore, generate an electromagnetic field [44]. Pokorny et al.[45], assume that the Fröhlich electromagnetic field can be a fundamental factor of cell adherence.

Surface topography is of an important interest in cell adhesion as well as its chemical composition. Indeed, it has been shown that cells adhere and proliferate depending on the surface roughness and the more the surface is rough the more cell adhesion and proliferation is better [46]. This effect depends on the cell type. For fibroblasts, they line up along the biomaterial surface microstructures and may adapt their shape with uneven surfaces.

Moreover, recent studies had shown that a weak change in the surface roughness may induce different cell reactions such as change in their shape and their way of adhesion [47, 48].

3.2.3.1. Forces involved in cell adhesion

According to Richards [49], cell adhesion to biomaterials is done thanks to focal adhesion sites which represent strict contact sites with the substrate in a so limited space. For fibroblasts, it has been shown the existence of a force called cohesion force responsible for keeping contact between cells themselves. However, this force is weaker than the adhesion force involved while a cell adheres to a biomaterial. This difference in force level depends on the cell type and on the nature of the biomaterial used for adhesion, and may explain the different ways of cell adhesion and spreading on different surface structures.

3.2.3.2. Surface free energy

Surface free energy is a thermodynamic measurement which contributes to the interpretation of the phenomena occurring in interfaces. It has an important effect on cell adhesion in the way that every change in its value induces the modification of the surface wettability, and therefore cell behaviour will be affected too [50, 51, 52].

Cell-biomaterial interface depends on the physico-chemical properties of the biomaterial and every change in the chemical composition or in the electric charge of the surface will affect its surface free energy.

3.2.4. Parameters involved in cell adhesion

3.2.4.1. Surface roughness

Surface roughness has been the subject of many studies as a deciding factor in the process of cell adhesion to biomaterials. Ponsonnet et al.[53] had studied the behaviour of fibroblast cells while adhering to titanium surface with different roughness; they found that cells had adhered to the surface using thin cytoplasmic structures. Indeed, these cells presented a flattened shape spreading practically over the substrate surface after adhesion to smooth surfaces. However, on rough surfaces, cell morphology was affected by the surface grooves and they were reoriented by the surface structure.

According to Richards [48], smooth titanium surfaces always increase fibroblasts adhesion and proliferation better than rough surfaces. They suggested that this kind of surfaces should be a better candidate for biological implant thanks to their ability to resist to bacterial infections. Indeed, their weak roughness is unfavourable to the adhesion of bacteria.

3.2.4.2. The electric charge effect

In the majority of the studies carried out about biomaterials made from polyelectrolyte film, as in our case, the electric charge effect is in proportion with the thickness of the film built and depends on the charged functional group of the polyelectrolyte used [54].

For Andrade [25], the notion of the nature of an electric charge is important to be mentioned but its effect is not significant and doesn't induce an efficient change on surface wettability. However, it has been shown that a better adhesion of cells was observed on negatively charged polyelectrolyte [55]. In reality, most of the existed cells and their corresponding adhesion proteins are negatively charged. Nevertheless, this charge can be without any effect in the case when functional groups become able to control cell adhesion mechanism by their hydrophilic or hydrophobic character as it will be shown later in this text. Dubois [56] presumed that an electric charge trapped within an insulating biomaterial, none associated to a particular chemical group, is able to affect its biological environment. Moreover, Maroudas [57] revealed the dependence of cell adhesion and spreading on a solid surface on the surface charge of the substrate.

3.2.4.3. Chemical composition

The different chemical components of a biomaterial must be studied and known before to start investigating cell adhesion to that biomaterial. Therefore, this step is fundamental for concluding about the biocompatibility of a given biomaterial and its effect on cell adhesion [58].

The wettability of a surface depends on the chemical composition of the material and each change than can occur at this level will disturb cell adhesion process [59]. Besides the effect of the biomaterial, the adhered cell type plays an important role in adhesion. Indeed, for the same biomaterial surface, different cell reactions were observed for two types of cells [60]; this kind of biomaterial seems to be biocompatible with one cell type but not tolerated by the other cell type.

According to Marmur [61], most of the materials in the nature are rough and heterogeneous and contact angle may change along the contact line with a value depending on the roughness and heterogeneity level.

3.2.4.4. Surface hydrophilicity and hydrophobicity

Contact angle measurement allows us to calculate surface free energy [62]. It also allows knowing about the polar or non polar nature of the interactions at the interface liquid/solid. Moreover, one can deduce from it the hydrophilic or the hydrophobic character of a surface [63].

A study about polyelectrolyte films found that hydrophobic interactions on a surface induce the adsorption of proteins and stabilise the complex formed [64]. Indeed, it has been proved that myoglobin or lysozymes are able to adhere to polystyrene sulfonate (PSS) and form many layers. However, this adhesion was not possible when using another surface having the same electric charge as PSS but with a hydrophilic character. The electrostatic interactions between the protein complex and this hydrophilic surface were easily destructed after water rinsing. Thus, surface hydrophilicity and hydrophobicity are a determinant parameter for substrate wettability on account of the rearrangement of the functional groups at the surface of a biomaterial in contact with a cell [65, 66, 67]. Indeed, it has been shown that fibroblast cells adhere and proliferate better on biomaterials with a moderate hydrophilicity [68, 69].

Andrade [66] presumed that, in the case of deformable materials, an elasticity model of 3.5 10^5 dyn/cm^2 is necessary for avoiding contact angle change. A roughness below 0.1 μm has a negligible effect on contact angle. Most of the materials holding over than 20 to 30 % of water present a receding contact angle (θ_r), in water, near zero because of the hydrophilic character which dominates the interface in these conditions. The same author estimated that the majority of polymers have a changeable volume which can be the reason for contact angle change: this change is depending on the duration of the contact with water, on the nature of the liquid and on the temperature of measurement. Non existent contact angle hysteresis may be due to the duration of contact between the material and the liquid which is shorter or longer than the measurement time needed for recording contact angle change. Therefore, surface hydrophilicity and hydrophobicity depends on the volume blowing of the material, on the diffusion phenomenon and on the mobility and reorientation of the molecules on the material surface.

Some materials are able to go out of shape in contact with a liquid depending on their mechanical properties and on their relaxation time and temperature. So, what characterizes a polymer is its chemical composition, roughness, mobility, wettability, surface free energy and its electric charge [70].

3.2.4.5. Surface wettability: Contact angle hysteresis

Contact angle hysteresis is the result of contact angle change between the surface we are characterizing and another ideal surface physico-chemically homogeneous. It's the direct result of a different sensitivity to the wettability process of heterogeneous surfaces. According to Rupp et al.[71], the receding contact angle value (θ_r) is under the control of the small hydrophilic particles of the surface which are able to disturb or to delay the non wettability process. Indeed, when the hysteresis remains constant after many immersion and emersion cycles it's called thermodynamic (or true) hysteresis. However, in the opposite case, it's called kinetic hysteresis (see Figure 7).

Thermodynamic hysteresis is due to the surface roughness and heterogeneity. Nevertheless, kinetic hysteresis is caused by the adsorption mechanisms (due to the liquid phase), surface polar group's reorientation and surface deformation [24].

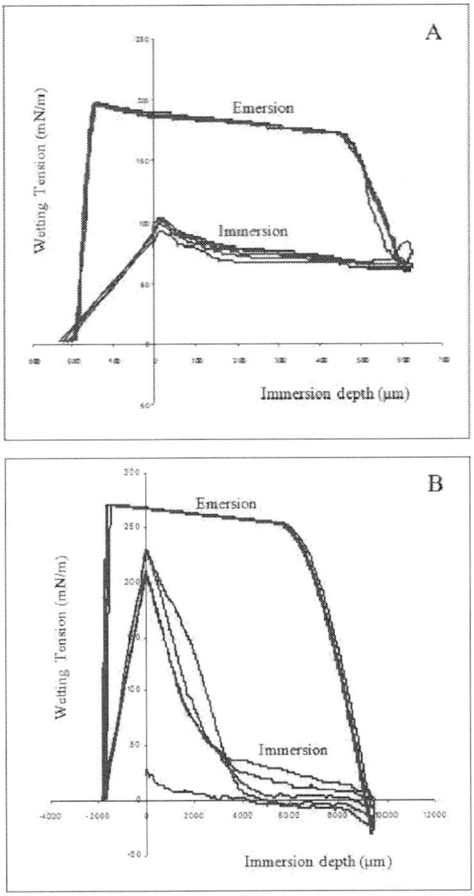

Figure 7. Immersion and emersion loops showing the two types of hysteresis: (A): thermodynamic hysteresis and (B): kinetic hysteresis. The sample is repeatedly immersed in the liquid leading to typical hysteresis loops. From each loop, wettability parameters (advancing and receding contact angle or wetting tension) can be calculated

Contact angle hysteresis is often assigned to the surface roughness and heterogeneity. Actually, a study made by Lam et al. [26], have shown that hysteresis is related to the molecules' mobility, the liquid diffusion and the surface swelling. These authors had observed a close dependence between the liquid molecules size and the liquid/material contact duration. Liquid resorption and retention are the direct causes of hysteresis. However, as the liquid surface free energy is higher that that of the material; therefore the liquid retention into the material will increase the material surface free energy and thus reduces the receding contact angle (θ_r). Indeed, liquids having smaller molecular chains (or smaller molecular weight) diffuse faster into the polymer surface leading to an important decrease in contact angle.

According to Shananan et al.[72], contact angle hysteresis is related to the polymer polarity. Indeed, when a polymer gets in touch with a polar liquid (water), it orients its mobile polar groups on the surface in order to increase the interfacial water/polymer energy and therefore decreasing the system surface free energy. In the other hand, when the polymer is contact with a non polar liquid, its functional groups conserve their state and will not reorient. These authors assumed the existence of two parameters behind hysteresis: the intrinsic polarity of the material and the mobility of its polar groups on the surface. Nishioka et al.[73], had observed that the advancing contact angle hysteresis is under the control of surface sites more hydrophobic than those controlling the receding contact angle hysteresis.

The contact angle hysteresis observed on hydrophilic and hydrated polymers is due to the polar groups' orientation on the interfaces polymer/liquid and polymer/air. This reorientation represents the polymer reaction to every environmental change (air, liquid). The receding contact angle (θ_r) depends on the contact duration with water, the environment temperature and on the glass transition temperature (T_g) of the material itself. Each material has its own glass transition temperature (T_g) allowing a defined molecular mobility sufficient for an important rearrangement [74].

3.3. Conclusion

The concepts of solid surfaces assumed that the surfaces in question were effectively rigid and immobile. Such assumptions allow one to develop certain models and mathematical relationships useful for estimating and understanding surface energies, surface stresses, and specific interactions, such as adsorption, wetting, and contact angles. It is assumed that the surfaces themselves do not change or respond in any specific way to the presence of a contacting liquid phase, thereby altering their specific surface energy [75]. Although such assumptions are (or may be) valid for truly rigid crystalline or amorphous solids, they more often than not do not apply strictly to polymeric surfaces.

In contact with condensed phases, especially liquids, surface relaxations and transitions can become quite important leading to a possible dramatically change in the interfacial characteristics of a polymer with possibly important consequences in a particular application. And since the processes are time-dependent, the changes may not be evident over the short span of a normal experiment. For critical applications in which a polymer surface will be in contact with a liquid phase, such as implant device for biomedical application, it is not only important to know the surface characteristics (e.g., coefficient of friction, adhesion, adsorption)under normal experimental conditions but also to determine the effects of prolonged (equilibrium) exposure to the liquid medium of interest. It is therefore important for biomedical as well as many other applications that the surface characteristics of a material of interest be determined under conditions that mimic as closely as possible the conditions of use and over extended periods of exposure to those conditions, in addition to the usual characterizations.

http://dx.doi.org/10.5772/53542

4. Experimental and results

4.1. Polyelectrolyte multilayer film preparation

Before use, glass slides were cleaned in 0.01 M SDS and then in 0.1 N HCl, both for 10 min in a boiling water bath, followed by a pure water rinse. Polyelectrolyte solutions were prepared by dissolution of the polyelectrolyte powders in 0.15 M NaCl (using ultrapure water filtered with a MilliQ system, Millipore) at a concentration of 1mg/l for PLL, PGA and HA and 5 mg/l for PEI, PSS and PAH. For all the films, the precursor layer was always PEI (polycation), followed by the alternate adsorption of polyanions/polycations for 12 min adsorption times and two rinses in the 0.15 M NaCl solution [76]. The glass slides held in a slide holder were dipped into the different polyelectrolyte baths for the preparation of three different types of film, ending either by the polycation or polyanion: $(PSS/PAH)_{10}$, $(PSS/PAH)_{10}$–PSS; $(PGA/PLL)_5$, $(PGA/PLL)_5$–PGA; and $(HA/PLL)_5$, $(HA/PLL)_5$–HA. Cleaning was made before film characterization. The films were all prepared at the same pH before being in contact with culture medium. Poly(styrene-4-sulfonate) (PSS,MW=70 kDa), Poly(allylamine hydrochloride) (PAH,MW=70 kDa) and Poly(ethyleneimine) (PEI,MW=70 kDa) are purchased from Aldrich. Poly(l-lysine) (PLL, MW=32 KDa) Poly(l-glutamic acid) (PGA, MW=72 KDa) were obtained from Sigma and Hyaluronan (HA,MW=400 kDa) from Bioiberica. Sodium dodecyl sulfate (SDS) was purchased from Sigma and sodium chloride (NaCl, purity ~ 99%) from Aldrich, glass slides (18x18 cm^2 square and 14x14 cm^2 disk), respectively, were obtained from CML, France.

4.2. Contact angle measurement and Surface Free Energy (SFE) calculation

The measurements were performed with a Wilhelmy balance for the characterization of solids using the 3S tensiometer and the corresponding software (GBX, France). For these experiments, the glass slides were coated with polyelectrolyte multilayer films on both sides. Before beginning the measurements, the films were washed in 18.2 MΩ Millipore water for 30–45 min in order to eliminate the NaCl traces that could modify the results. Samples were then dried at 30 °C for 2 h. The dynamic contact angle hysteresis was determined at 20°C for each film and five wetting/dewetting cycles were carried out at a 50 μm/s speed.

Three liquids were used as a probe for surface free energy calculations: diiodomethane, formamide (Sigma Chemical CO, St Louis, MO, USA) and distilled water. The final contact angle used for this calculation was the average of the 2nd to 5th cycle advancing contact angle (θ_a) and the surface free energies of the different films were calculated using the Van Oss (VO) approach, as usual with sessile drop method contact angles:

$$\gamma_S = \gamma_S{}^d + 2\,(\gamma_S{}^+ . \gamma_S{}^-)^{1/2}$$

This method produces the dispersive ($\gamma_S{}^d$) and the polar acid–base ($\gamma_S{}^+$, $\gamma_S{}^-$) components. Solid and liquid SFE components and contact angle are related according to the equation below:

$$\gamma_L\,(1 + \cos\theta) = 2\,((\gamma_S{}^d . \gamma_L{}^d)^{1/2} + (\gamma_S{}^+ . \gamma_L{}^-)^{1/2} + (\gamma_L{}^+ . \gamma_S{}^-)^{1/2})$$

Were γ_L is the SFE of the liquid and γ_s the SFE of the surface.

4.3. Cell adhesion, viability and morphology study

For adhered cell counting, image analysis was performed on a Quantimet 570 (Leica, UK) fitted to an epifluorescence microscope (Axioplan, Zeiss, DE) and a black-and-white charge-coupled device (CCD) camera (LH51XX-SPU, Lhesa Electronique, FR). The scanning was carried out using a ten times lens (NA=0.3) and a filter set adapted for propidium iodide fluorescence observation (BP 546/12 nm, DM 580 nm, LP 590 nm). Microscope focus and stage were motorized and software controlled.

The cell viability was determined with the MTT colorimetric assay. It was measured at 570 nm with a 96-well microplate reader (Becton Dinkinson, Lincoln Park, USA) on a spectrophotometer (Bio-Tek Instruments, Winooski, USA). The blank reference was taken for wells containing only the MTT solution.

The morphology of the cells was analyzed after 120 min (day 0), 2 and 7 days of culture using a scanning electron microscopy (Philips, EDAX XL-20) and phase contrast microscopy.

4.4. Results

4.4.1. Contact angle measurement

The different contact angle values found are shown in Table 1. Experiments were performed at 20 °C at a speed of 50 μm/s. One can observe that contact angle depends on the film's nature (physico-chemical composition) which differs from a polymer to another.

	Water	**Formamide**	**Diiodomethane**
Glass	43 ± 2	23 ± 3	43 ± 3.1
$(HA/PLL)_{10}$	12 ± 2	00	00
$(HA/PLL)_5$	81.9 ± 1.8	49.6 ± 2	43.5 ± 3
$(HA/PLL)_5$-HA	87.8 ± 1.2	00	45 ± 2.9
$(PGA/PLL)_5$	55.2 ± 3	14.7 ± 2.5	39.1 ± 1.2
$(PGA/PLL)_5$-PGA	44.1 ± 3.1	00	40.7 ± 1.6
$(PSS/PAH)_{10}$	49.2 ± 1.8	23.6 ± 3	00
$(PSS/PAH)_{10}$-PSS	53 ± 1.9	12.1 ± 3.3	00

Table 1. Dynamic contact angle

4.4.2. SFE values

SFE and its component's values are summarized in Table 2. (HA/PLL) films have the lowest SFE value and (PSS/PAH) films have the highest value. The outermost layer of the film does not have a great influence.

	$(HA/PLL)_5$	$(HA/PLL)_5$-HA	$(PGA/PLL)_5$	$(PGA/PLL)_5$-PGA	$(PSS/PAH)_{10}$	$(PSS/PAH)_{10}$-PSS	Glass	Thermanox
Surface Free Energy (mN/m)	42.4	44.95	53.3	48.26	57.1	58.9	50	35
Dispersive component (mN/m)	38.8	37.1	40.23	39.5	50.8	50.8	35	36
Acid component (mN/m)	1.2	11.85	3.56	9.46	0.46	1.7	3	3
Basic component (mN/m)	2.86	1.3	14.6	2.4	23.76	7	25	0
Acid-basic component (mN/m)	3.56	7.8	13.9	8.86	6.3	8.1	17	1

Table 2. Surface Free Energy (SFE) and its components for the different films used. The SFE of PSS/PAH is higher compared to the other films.

4.4.3. Cell adhesion

Figure 8 shows the percentage of fibroblasts that have adhered after 2 h in culture. The highest adhesion is found with $(PGA/PLL)_5$ film (95%) and the lowest on $(HA/PLL)_5$ film (49%).

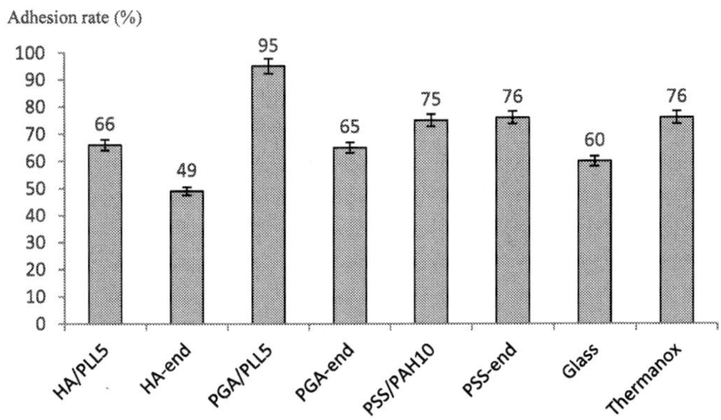

Figure 8. Fibroblast adhesion rate after 2 h in culture onto different films. The percentage represents the number of the adhered cells compared to the initial number of seeded cells.

4.4.4. Cell viability and proliferation rate

Cell viability was evaluated on the different types of film at different time intervals (0, 2 and 7 days) with the MTT assay (Figure 9A). The $(PGA/PLL)_5$–PGA films exhibited a good proliferation rate (Figure 9B) and the $(PSS/PAH)_{10}$ films were the most favorable to cell proliferation.

4.4.5. Cell morphology

Good adhesion is observed on $(PGA/PLL)_5$ film (Figure 10A) whereas bad adhesion was found on $(HA/PLL)_5$-HA film (Figure 10B). Typical morphology at day 2 on a $(PGA/PLL)_5$–PGA film is presented in Figure 10C. After seven days in culture, the difference in morphology for the cells that had adhered to the different films was even more striking. Cells in contact with $(HA/PLL)_5$–HA exhibit necroses (Figure 11A) whereas the cells exhibit elongated and spread morphologies on the highly proliferative $(PSS/PAH)_{10}$ films (Figure 11B).

http://dx.doi.org/10.5772/53542

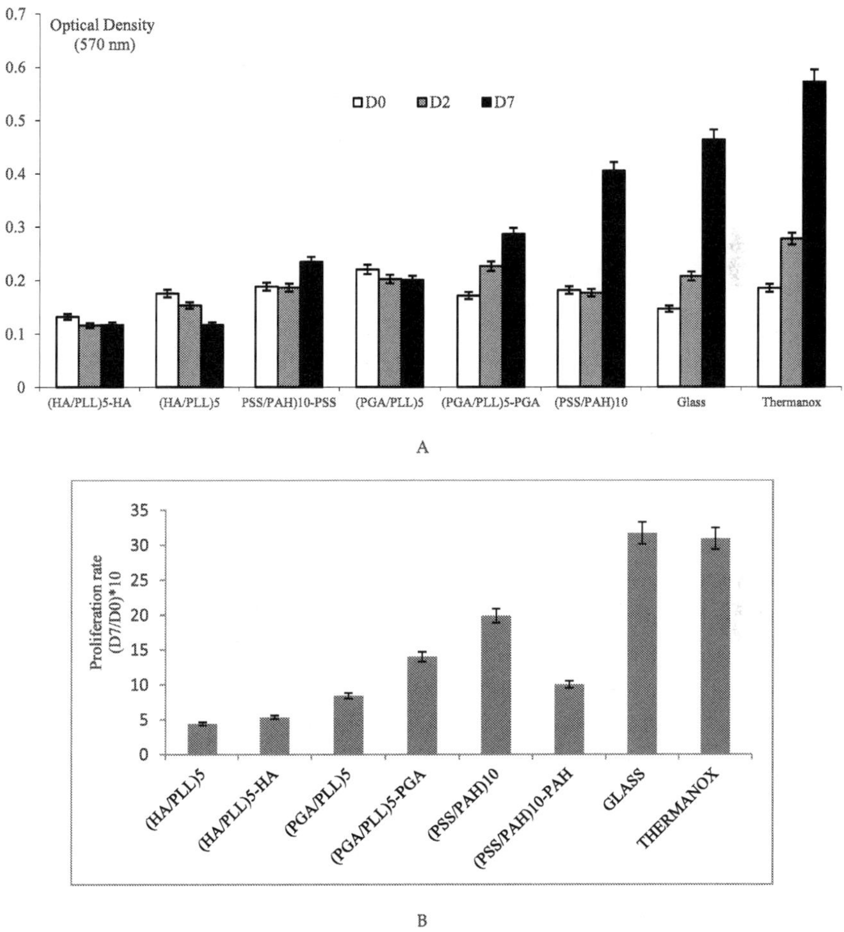

Figure 9. A. Cell viability (MTT test) on each film type followed over a seven day period at: day 0 (D0), day 2 (D2), and day 7 (D7), B. Proliferation rate on the different films as estimated by the ratio (D7/D0)

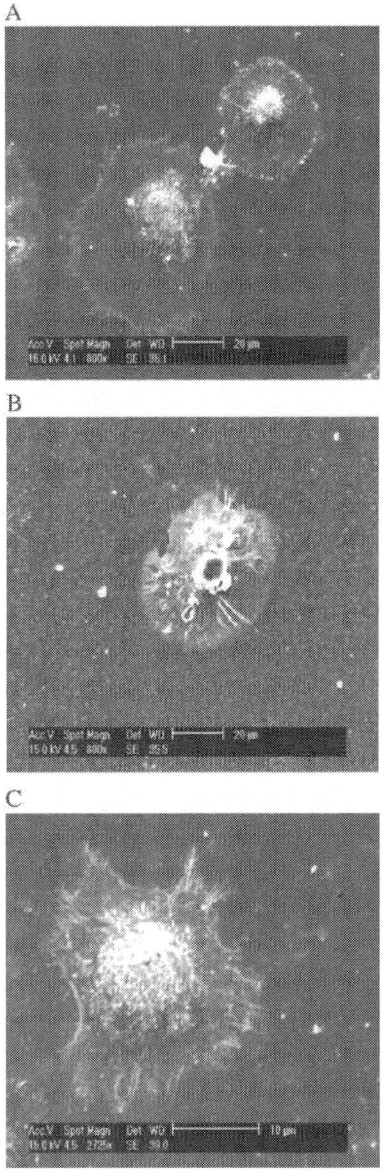

Figure 10. SEM images of cells adhering to different polyelectrolyte multilayer films. (A) (PGA/PLL)5(x800) on the first day, (B) HA/$(PLL)_5$–HA (x800) on the first day, (C) $(PGA/PLL)_5$–PGA film observed on the second day (x2725).

http://dx.doi.org/10.5772/53542

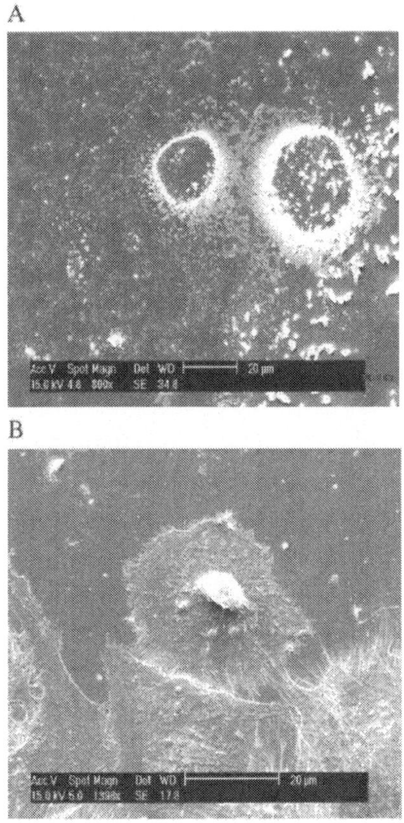

Figure 11. SEM images of cell morphology after seven days of culture. (A) $(HA/PLL)_5$–HA (x800) film, a $(PSS/PAH)_{10}$ film observed at different magnifications (B) (x1398): some areas are at confluency.

4.4.6. Correlation between cell adhesion and films SFE

No correlation was found between the wettability parameters or the SFE parameters and the fibroblast proliferation ratio. However, the adhesion rate at 2 h was correlated to both SFE basic component and the SFE acid component (Figure 12). For the adhesion rate, the SFE basic component is optimum at 15 mN/m (Figure 12A) whereas the acid one is optimum at about 5 mN/m (Figure 12B).

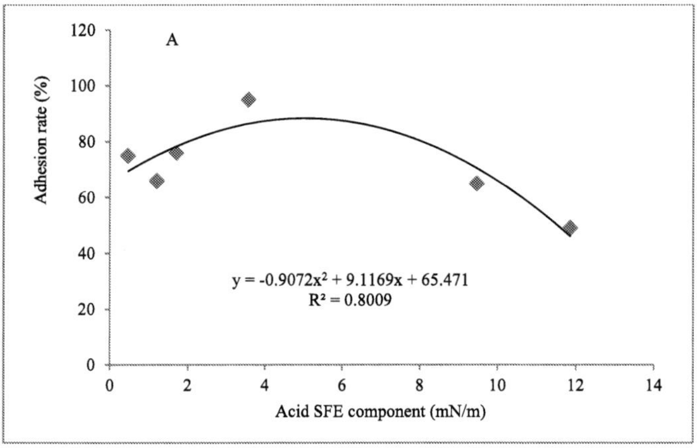

A

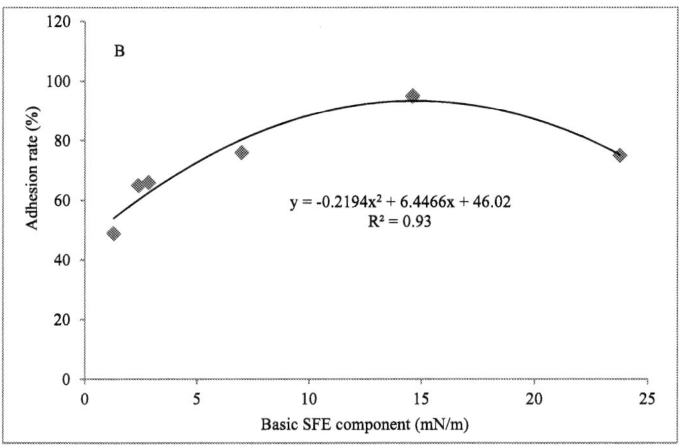

B

Figure 12. SEM images of cell morphology after seven days of culture. (A) $(HA/PLL)_5$–HA (x800) film, a $(PSS/PAH)_{10}$ film observed at different magnifications (B) (x1398): some areas are at confluency. Figure 12B. Correlation between cell adhesion rate and Basic SFE component. An optimum is found for 15 mN/m with good polynomial correlation (R^2=0.93)

http://dx.doi.org/10.5772/53542

4.5. General conclusion

Cell adhesion is a paramount parameter for the biomaterial tissue. These biomaterials, by their surface properties (chemical composition, topography, roughness, surface energy) hold the key of the control of the cell adhesion, proliferation and orientation. Thus, the concept of biocompatibility is seen imposed, it is primarily focused on the interface, sites of the interactions between cells and biomaterials.

The influence of different polyelectrolyte multilayer films (PEM) on gingival fibroblast cell response was studied. Roughness and hydrophobicity/hydrophilicity of the PEM were characterized by contact angle measurement. Polar (acid-basic) components of the surface free energy (SFE) were determined. Surface advancing and receding angles were measured and hysteresis was determined. Cell adhesion, viability and morphology were analyzed.

This work pointed out that cell adherence is a complex process modulated by numerous parameters. Usually, in cell adherence studies and particularly in biomaterial approaches, surface physico-chemical properties are analysed (chemistry, roughness, motility, wettability…).

In our work we tackled the subject of the cellular behavior in contact with a biomaterial by the characterization of the surface of this material. We were interested in physical (topography) and chemical (composition) properties of various polyelectrolyte multilayer films deposited on glass slides, with different charge densities scale and thickness. We have evaluated the wettability of theses biomaterials by measuring the contact angle hysteresis using the Wilhelmy balance tensiometry to study their physico-chemical characteristics in order to understand the effects of surface roughness and chemistry on the fibroblasts behavior. Epifluorescence microscopy, SEM, phase contrast microscopy and MTT test were used to study cell adhesion, proliferation and morphology in order to correlate the film's properties and the cultivated cells response.

Surface hydrophobicity and roughness were found to be unfavourable for both adhesion and proliferation. Adhesion and proliferation were found not to be correlated.

Author details

M. Lotfi[1], M. Nejib[2] and M. Naceur[3]

*Address all correspondence to: lotfi.mhamdi@gmail.com

1 Faculty of Medicine of Monastir, Tunisia, Laboratory of biophysics, Avicenne Avenue, Tunisia

2 Higher Institute of Agronomy of Chott Meriem, Laboratory of animal sciences, Sousse, Tunisia

3 Ecole Polytechnique of Montreal, Canada

References

[1] Aberts B, Bray D, Lewis J, Raff M, Roberts K, D. Watson J. Membrane structure, Molecular biology of the cell, 3rd edition, chapter 10, Garland publishing, Inc., 1994, 477.

[2] Rusciano D. Encyclopedic reference of cancer, p. 24, Manfred Schwab Ed. 2001

[3] Passuti N, Baqey C, Guillot F, Reach G. 2000. In www.frm.org

[4] Shabalovskaya S.A. Biocompatibility aspects of Nitinol as an implant material, Biomedical Materialsand Engeneering, 2002

[5] Durand-Vidal S, Simonin J.-P, Turq P. Polyelectrolytes, Electrolytes at interface, Vol. 1,chapter 6, Kluwer Academic Publishers, 2002, 331

[6] Mini-Encyclopedia of paper making. Wet-End Chemistry, Part Two: Definitions and Concepts

[7] Oudet Ch. Polymères, Structure et propriétés, Introduction, Masson, Paris, Milan, barcelone. 1994

[8] Fowkes F.M. Interface Acid-base/charge-transfer properties, Polymer surface dynamics, Vol.1, Joseph D. Andrade, Plenum Press, New York and London, 1983, 336-371

[9] De Groot C. J, Van Luyn M J. A, Van Dijk-Wolthuis W. N. E, Cadée J. A, Plantinga J. A, Den Otter W,Hennink W. E. In vitro biocompatibility of biodegradable dextran-based hydrogels tested with humanfibroblasts, Biomaterials, Vol. 22, Issue 11, 2001, 1197-1203

[10] Bell J.P, Schmidt R.G, Malofsky A, Mancini D. Controlling factors in chemical coupling of polymers to metals. J. adhesion Sci. Technol, Vol.5, N°10, 1991, 927-944

[11] Serizawa T, Kamimura S, Kawanishi N, Akashi M. Layer -by-layer assembly of Poly(vinyl alcohol) and hydrophobic polymers based on their physical adsorption on surfaces. Langmuir 2002

[12] Caruso F, furlong D N, Ariga K, Ichinose I, Kunitak T. Characterization of polyelectrolyte-protein multilayer films by atomic force microscopy, scanning electron microscopy and Fourrier transform infrared reflection adsorption spectroscopy. Langmuir 1998

[13] Kasemo B.N.L. J, Mattsson H. Surface spectroscopic characterization of titanium implant materials. Appl. Surf. Sci., 1990

[14] Rodil S.E, Olivares R, Arzate H, Muhl S. Properties of carbon films and their biocompatibility using in vitro tests, Diamond and Related Materials, Vol.12, Issues 3-7, 2003, 931-937

[15] Schaaf P, Voegel J. C. Films bioactifs destinés au recouvrement de biomatériaux et ciblant une régénération tissulaire : de nouvelles voies originales? Pathologie Biologie, Vol. 50, 2002, 189-193

[16] Hammond P.T. Recent explorations in electrostatic multilayer thin film assembly. Current Opinion in Colloid & Interface Science, Vol.4, 1999, 430-442

[17] Sedel L, Janot C, Biomatériaux, document found in internet, 1996

[18] Meyers D. Wetting and spreading, Surfaces, Interfaces, and Colloids: Principles and Applications, chapter 17, Second Edition. John Wiley & Sons, Inc., 1999, 419

[19] Young T. Philos. Trans. R. Soc. 95 (1805)65.

[20] www.firsttenangstroms.com. Tilting plate example, 3 February 2004

[21] Chen J, Luo G, Cao W. The study of layer-by-layer ultrathin films by the dynamic contact angle method. Journal of Colloid and Interface Science, Vol.238, 2001, 62-69

[22] Defay P, Peter G. Dynamic surface tension, Surface Colloid Sci., 1969; 3: 27-82

[23] Okano T, Yamada N, Okuhara M, Sakai H, Sakurai Y. Mechanism of cell detachment from temperature- modulated, hydrophilic-hydrophobic polymer surfaces. Biomaterials 16, 1995, 297-303

[24] Elbert D.L, Hubbel J.A. Self-assembly and steric stabilization at heterogeneous biological surfaces using adsorbing block copolymers, Chem. Biol, 5, 83-177

[25] Andrade J. The contact angle and interface energetics, surface and interfacial aspects of biomedical polymers, Vol.1, Joseph D .Andrade, Plenum Press, New York and London, 1983, 249-289

[26] Lam C. N. C, Wu R, Li D, Hair M.L, Neumann A.W. Study of the advancing and receding contact angles: liquid sorption as a cause of contact angle hysteresis. Advances in Colloid and Interface Science, Vol.96, 2002, 169-191

[27] Van Oss C.J. Hydrophobicity and hydrophilicity of biosurfaces. Current opinion in colloid and interface science, 1997, 503-512

[28] Takahara A. Block copolymers and hydrophilic city, Modern approaches to wettability: theory and applications, Malcolm E. Schrader and George Loeb. Plenum Press, New York, 1992, 206-210

[29] Kamusewitz H, Possart W, Paul D. The relation between Young's equilibrium contact angle and the hysteresis on rough paraffin wax surfaces. Colloids and Surfaces A: Physicochemical and Engineering Aspects, Vol.156, Issues 1-3, 15, 1999, 271-279

[30] Picart C, Ph. Lavalle Ph, Hubert P, Cuisinier F.J.G, Decher g, Schaaf p, Voegel J.C. Build-up mechanism for poly-L-lysine/Hyaluronine acid films onto a solid surface. Langmuir 2002

[31] Otsuka H, Nagasaki Y, Kataoka K. Dynamic wettability study on the functionalized PEGylated layer on a polylactide surface constructed by the coating of aldehyde-ended poly(ethylene glycol) (PEG)/ polylactide (PLA) block copolymer. Science and technology of advanced materials, 2000, 9-21

[32] Estarlich F.F, Lewey S.A, Nevell T.G, Torpe A.A, Tsibouklis J, Upton A.C. The surface properties of silicone and fluorosilicone coating materials immersed in seawater. Biofluing, Vol. 16, 263-275

[33] Morra M, Occhiello E, Garbassi F. Knowledges about polymer surfaces from contact angle measurements. Adv. Colloid Interfaces Sci.32, 1990, 79-116

[34] Vogler E.A. Wettability, Interfacial chemistry in biomaterials science, Ed. John C. Berg, Marcel Dekker, Inc, New York, Basel, Hong Kong, 1993

[35] Merriam-Webster's Collegiate® Dictionary. 10th edition with Merriam-Webster's Collegiate® Thesaurus. Version 3.0.327

[36] Comte V. PhD thesis, Faculty of Medicine of Lyon, March 18th 2003

[37] Castner D.G, Ratner B.D. Biomedical surface science: Foundations to frontiers. Surface Science, Vol.500, Issues 1-3, 2002, 28-60

[38] Jones F. H. Teeth and bones: applications of surface science to dental materials and related biomaterials. Surface Science Reports, Vol.42, Issues 3-5, 2001, 75-205

[39] Exbrayat P. Question C.E.S: La biocompatibilité des matériaux dentaires. Journal de Biomatériaux Dentaires, Vol.13. 1998

[40] Hubbel J.A. In situ material transformations in tissue engineering, Mat.Res.Soc.Bull, 21, 1996, 5-33

[41] El-Amin S.F, Lu H, Khan Y, Burems J, Mitchell J, S.Tuan R, Laurencin C.T. Extracellular matrix production by human osteoblasts cultured on biodegradable polymers applicable for tissue engineering. Biomaterials, 24, 2003

[42] Berk L, Kaiser M, Scott K, Darnell Z. Integrating cells into tissues, Molecular cell biology, 5th edition, Chapter 6, p.201

[43] Frölch H. Coherent electric vibrations in biological systems and cancer problem, IEEE Trans, 1978; MTT-26:613-617, 1978

[44] Pokorny J, Jandova. A, Sorfova J, Kobilkova J, Trojan S, Costato M, Milani M. Leukocytes of cancer patients give evidence of fundamental physical forces: a pathway for a new view. Laser & Technology, Vol.6, N° 1-2, 15-23, 1996.

[45] Pokorny J, Fiala J, Jandova A, Kobilkova J. Fröhlich electromagnetic field and adherence of leukocytes. 2nd Int. Meeting Microwaves in Med, 11-14 Oct. 1993, Rome

[46] Eisenbarth E, Meyle J, Nachtigall W, Breme J. Influence of the surface structure of titanium materials on the adhesion of fibroblasts, Biomaterials, 17. 1996

[47] Eisenbarth E, Linez P, Biehl V, Velten D, Breme J, Hildebrand H. F. Cell orientation and cytoskeleton organisation on ground titanium surfaces, Biomolecular Engineering, Vol.19, 2002, 233-237

[48] Schwartz J, Avaltroni M.J, Danahy M.P, Silverman B.M, Hanson E.L, Schwarzbauer J.E, Midwood K.S, Gawalt E.S. Cell attachment and spreading on metal implant materials, Materials science and engineering C, Vol.23, Issue 3, 2003, 395-400

[49] Richards R.G. The effect of surface roughness on fibroblast adhesion in vitro, Injury, Volume 27. Supplement 3, 1996, S

[50] Shelton R.M, Rasmussen A.C, Davies J.E. Protein adsorption at the interface between charged polymer substrata and migration osteoblasts. Biomaterials, 9, 1988, 24-29.

[51] Drummond C, Israelachvili J. Surface forces and wettability. Petroleum science and engineering, 2002, 123-133

[52] Satriano C, Carnazza S, Guglielmino S, Marletta G. Surface free energy and cell attachment onto ion- beam irradiated polymer surfaces. Article in press. Beam interactions with materials and atoms. 2003

[53] Ponsonnet L, Comte V, Othmane A, Lagneau C, Charbonnier M, Lissac M, Jaffrezic N. Effect of surface topography and chemistry on adhesion, orientation and growth of fibroblasts on nickel-titanium substrates. Materials Science & Engineering. 2002

[54] Elbert D, Herbert C.B, Hubbel J.A. Thin polymer layers formed by polyelectrolyte multilayer technique on biological surfaces. Langmuir, 1999

[55] Forest P, Master Diploma report, 2002

[56] Dubois J.C , PhD thesis, Faculty of Medicine of Lyon, 2002

[57] Maroudas N.G. Adhesion and spreading of cells on charged surfaces. J theor Biol, 49, 417-424, 1975

[58] Fischer D, Li Y, Ahlemeyer B, Krieglstein J, Kissel T. In vitro cytotoxicity testing of polycations: influence of polymer structure on cell viability and haemolysis. Biomaterials, 2003, 1121-1131

[59] Good R.J. Contact angle, wetting, and adhesion: a critical review. In: K.L. Mittal Editor, Contact Angle, Wetting and Adhesion VSP, Utrecht, the Netherlands (1993), 3–36.

[60] Troyen-Toth P, Vantier D, Haikel Y,Voegel J.C, Schaaf P, Chluba J, Ogier J. Viability, adhesion and bone phenotype of osteoblast-like cells on polyelectrolyte multilayer films, Wiley Periodical, Inc. 2001

[61] Marmur A. Equilibrium contact angles: theory and measurement. Colloids and Surfaces, Vol.116, 1996, 55-61

[62] Kwok D. Y, Neumann A. W. Contact angle interpretation in terms of solid surface tension. Colloids and Surfaces A: Physicochemical and Engineering Aspects, Vol.61, Issue 1, 2000, 31-48

[63] Lam C.N.C, Kim N, Hui D, Kwok D.Y, Hair M.L, Neumann A.W. The effect of liquid properties to contact angle hysteresis. Colloids and Surfaces A: Physicochemical and Engineering Aspects, Vol. 189, Issues 1-3, 2001, 265-278

[64] Kotov N.A. Layer-by-layer self-assembly: the contribution of hydrophobic interactions, Nanostructured Materials, Acta Metallurgica Inc., Vol.12, 1999, 789-796

[65] Warocquier-Cleront M.R, Legris C, Degrange M, Sigot-Luizard M.F. Correlation between substratum roughness and wettability, cell adhesion and cell migration. J. Biomed. Mater. Res., 36 (1), 1997, 99-108

[66] Andrade J, Gregonis D.E, Smith L.M. Polymer surface dynamics; surface and interfacial aspects of biomedical polymers, Vol.1, Joseph D. Andrade, Plenum Press, New York and London, 1983, 14-39

[67] Voegler .E. Structure and reactivity of water biomaterial surfaces, Advances in colloid and interface science. 1998

[68] Shelton R.M, Rasmussen A.C, Davies J.E. Protein adsorption at the interface between charged polymer substrata and migration osteoblasts, Biomaterials, 9, 1988, 24-29

[69] Janssen M.I,. Van Leeuwen M.B.M, Scholtmeijer K, van Kooten T.G, Dijkhuizen L, Wösten H.A.B. Coating with genetic engineered hydrophobin promotes growth of fibroblasts on a hydrophobic solid. Biomaterials, Vol.23, 2002, 4847-4854

[70] Hunter A, Archer C.W, Walker P.S, Blunn G.W. Attachment and proliferation of osteoblasts and fibroblasts on biomaterials for orthopedic use. Biomaterials, Vol.16, 4, 1995

[71] Rupp F, Scheideler L, G-Gerstorfer J. Effect of heterogenic surfaces on contact angle hysteresis: Dynamic contact angle analysis in material sciences, Chem. Eng. Technol. 25. 2002

[72] Shanahan M.E.R, Carre A, Moll S, Schultz J. Une nouvelle interprétation de l'hystérèse de mouillage des polymères. J. Chim. Phys. Chim. Biol.1996

[73] Nishioka G.M, Wesson S.P. A computer model for wetting hysteresis. 3. Wetting behavior of spatially encoded heterogeneous surfaces, Colloids and Surfaces, Vol.118, 1996, 247-256

[74] Orkoula M.G, Koutsoukos P.G, Robin M, Vizika O, Cuiec L. Wettability of CaCo3 surfaces, Colloids and surfaces. 1999

[75] Vold M.J, Vold R.D. Colloid and Interface Chemistry, Addison-Wesley, Reading, MA, Chapter 3, 198

[76] L. Mhamdi, C. Picart, C. Lagneau, A. Othmane, B. Grosgogeat, N. Jaffrezic-Renault and L. Ponsonnet. Study of the polyelectrolyte multilayer thin films properties and correlation with the behavior of the human gingival fibroblasts. Material sciences and engineering C 26, 2006, 273-281

Section 3

Drug and Gene Delivery

Chapter 9

Nanoparticles Based on Chitosan Derivatives

Ylenia Zambito

Additional information is available at the end of the chapter

http://dx.doi.org/10.5772/54944

1. Introduction

A tremendous effort has been and is currently being devoted to the research in the field of pharmaceutical nanotechnology. Several peculiar properties of gelled polymeric nanosize (<1 μm) particulate systems have been reported, among which the ability to encapsulate either small molecular weight or macromolecular active principles in mild conditions and protect them from degradation by the harsh pH conditions or enzymes they may encounter in the organism, promote transport of actives across mucosal barriers, undergo internalization by cells thereby carrying actives into them. Chitosan, a copolymer of glucosamine and *N*-acetylglucosamine, obtained by deacetylation of the naturally-occurring chitin, has been studied as a basic biomaterial for preparing pharmaceutical nanoparticles, because it is biodegradable and has a very low toxicity [1-5] besides an ability to promote transport of drugs, peptides and proteins across mucosal barriers [6-10]. The preparation procedures of chitosan nanoparticles, their characterization for drug encapsulation efficiency, physical and biopharmaceutical properties, and toxicity have been covered by recent reviews (11,12). Chitosan has been subjected to derivatization, taking advantage of the reactivity of the primary amino group in position 2, or the hydroxyl group in position 6 of its repeating unit, glucosamine. The derivatization changed the physicochemical, biopharmaceutical and biological properties of the parent chitosan and each derivative type lent itself to preparing nanoparticles with their own physical properties (size, shape, surface charge), drug encapsulation and release capability, biopharmaceutical and biological properties (mucoadhesivity, ability to promote drug transport across biological barriers, aptitude for internalization within cells, citotoxicity).

In the following sections the nanoparticles obtained from chitosan derivatives will be surveyed in respect to preparation procedures, interactions with cells and tissues, factors influencing biological properties, pharmaceutical applications.

2. Preparation procedures

2.1. Ionotropic gelation

Ionotropic gelation, that is by far the most used technique for preparation of nanoparticles from chitosan derivatives, was first reported in 1997 by Calvo et al. [13]. The basic concept is that a polycationic polymer in aqueous solution passes, in appropriate conditions, from sol to dispersed gel following electrostatic crosslinking with an adequate anionic substance. This technique has been used with several quaternized chitosans carrying fixed, pH-independent positive charges, the most known of which is *N*-trimethyl chitosan (TMC). Sodium tripolyphosphate (TPP) has widely been employed as the ionotropic crosslinker [14-19].

The nanoparticles prepared by ionotropic gelation of quaternized chitosans with TPP were generally 200-300 nm in size, i.e., smaller than those obtained by the same method starting from plain chitosan which, by the way, showed lesser stability and tended to re-dissolve after some time from formation. The zeta potential was always positive, in the 10-20 mV range. The solution of the chitosan derivative into which the TPP solution was dripped would often contain a surfactant, usually Tween 80, to hinder nanoparticle aggregation and facilitate their re-dispersion after centrifugation. In fact, centrifugation was necessary to clear the particles of non-encapsulated drug.

The technique under discussion has also been used to prepare nanoparticles from thiolated derivatives of chitosan [20-23].

These polymers have shown mucoadhesive properties due to the ability of their thiol groups to form covalent disulfide bridges by reacting with the thiol residues present on the glycoproteins of mucus. For this reason the nanoparticles derived from these chitosan derivatives were themselves endowed with mucoadhesivity. The thiolated nanoparticles formed by ionotropic gelation with TPP were stabilized via oxidation of thiols with H_2O_2 which formed interchain disulfide bonds. These would bestow gastroresistance on the particles, which would be particularly appropriate in case of oral administration of the nanoparticle formulation. However the presence of some non-oxidized thiols on the nanoparticle surface was needed to confer enhanced mucoadhesivity on such a surface. This goal was actually achieved by Bernkop-Schnurch et al. [21]. These authors also studied the crosslinker effect on nanoparticle size. Under similar preparation conditions, sizes in the 200-300 nm range were obtained with TPP as the crosslinker, whereas sizes beyond the micron resulted using Na_2SO_4.

2.2. Gelation from polyelectrolyte complex (PEC) formation

This method involved ionotropic gelation, just as described in the preceding section, only in the present case the crosslinker was a polyanionic polymer with charges opposite to those of the chitosan derivative, with which it formed a PEC. To this purpose *N*-carboxymethyl chitosan, poly(γ-glutamic acid), poly(aspartic acid) and hyaluronic acid were used as polyanions, while TMC and glycidyl trimethyl ammonium chitosan were the polycations [24-27]. In a case both the polyanion (hyaluronic acid) and the polycation (TMC) were thiolated and the nanoparticles were stabilized by the formation of interchain disulfide bonds [28].

2.3. Polymer-drug complexes

Some negatively charged active principles, such as insulin or gene drugs, when mixed with cationic chitosan derivatives in adequate proportions spontaneously formed nanoparticulate dispersions of insoluble complexes [29-34]. TMC nanoparticles obtained by ionotropic gelation with TPP in the presence of insulin were compared with nanoparticles obtained by PEC formation between TMC and insulin. In the latter instance higher encapsulation efficiency and zeta potential (positive), and smaller particle size were observed, which is particularly appropriate for particle internalization into cells. In addition, a higher stability in simulated intestinal fluid (pH 6.8) of the nanocomplex compared to the nanoparticles prepared with TPP resulted [31,32].

2.4. Self-assembly

Amphiphilic derivatives of chitosan in aqueous solution were found, at a critical aggregation concentration (CAC), to spontaneously arrange into nanoparticles of sizes in between 100-400 nm. Such derivatives were prepared by connecting hydrophobic structures to the chitosan or glycol chitosan backbone via the amino group of the chitosan repeating unit. Examples of the above amphiphilic derivatives are the following: glycol chitosan-5β-cholanic acid conjugate [35-40]; palmitoyl chitosan [41]; palmitoyl glycol chitosan [42]; oleoyl chitosan [43]. Other amphiphiles were prepared from chitosans bearing fixed positive charges, in the case of quaternary ammonium palmitoyl glycol chitosan [42], or negative fixed charges, as in the case of linoleic acid-modified *O*-carboxymethyl chitosan [44], or deoxycholic acid-modified *N,O*-carboxymethyl chitosan [45]. Usually, after suspending the polymer in an aqueous medium, probe sonication was applied to limit particle size. The formation of nanoparticles was monitored and the CAC determined fluorometrically, or through UV absorption spectra, or measuring the enthalpy change by a microcalorimeter [42,44,45].

The CAC for the hydrophobically modified chitosan derivatives is usually in the μM range, whereas the CMC of small-molecular weight surfactants is in the mM range. This is one of the most important characteristics of amphiphilic polymers, pointing to stability of the self-aggregates in dilute conditions, such as those the nanoparticles are supposed to encounter after administration to the organism. The CAC values of these polymers have been found to decrease with increasing hydrophobic content of derivatives [44]. In fact, the nanoparticles formed from these chitosan derivatives are characterized by a core-shell structure, i.e., a hydrophobic core in a hydrophilic shell. The drug incapsulation method was chosen on the basis of the hydrophilic or hydrophobic nature of the drug. With hydrophobic drugs the solution of polymer and drug in a water-miscible organic solvent was mixed with an aqueous medium and the organic solvent was cleared away by dialysis or evaporation [36,39,40]. Hydrophobic drugs having a fair water solubility and polar drugs have been loaded into nanoparticles via direct addition to the aqueous polymer dispersion [41,42,44,45]. The non-encapsulated drug has been separated by ultracentrifugation, filtration or dialysis.

3. Interactions with cells and tissues

3.1. Quaternized derivatives

Chitosan has been found to open the tight junctions connecting epithelial cells, through an interaction of its positively charged amino groups with negatively charged sites in the tight junctions, thereby promoting paracellular transepithelial absorption of drugs, peptides and proteins [10,46-57]. The major drawback of unmodified chitosan as an absorption promoter is its insolubility at physiological neutral pH. Therefore the primary amino groups of its repeating units have been quaternized to bestow fixed, pH-independent positive charges on the polymer, thus making it soluble and active as an absorption promoter at physiological pH. In fact, TMC was found to act as an enhancer of drug, peptide and protein permeability across intestinal, nasal, buccal, ocular epithelium [47, 58-66]. TMC was shown to promote not only paracellular but also transcellular drug absorption [66]. Other quaternized chitosan derivatives, namely, *N*-triethylchitosan (TEC) [68], *N,N*-dimethyl *N*-ethylchitosan (DMEC) [69] and *N,N*-diethyl *N*-methylchitosan (DEMC) [70] have been synthesized. Positively charged chitosan and its quaternized derivatives have also exhibited mucoadhesive properties, determined by ionic interactions with the negatively charged sialic acid residues of mucins at neutral or slightly alkaline pH [71].

Particles in the nanosize range have resulted from the interaction of quaternized chitosans with polyanions. Proteins or macromolecular drug models have been encapsulated in these nanoparticles. Ovalbumin (OVA) was encapsulated in nanoparticles, obtained by ionotropic gelation of TMC with TPP, and studied as a nasal delivery system for proteins [19]. No cytotoxicity of nanoparticles on Calu-3 cells, a model of human respiratory function, was evidenced, whereas a partially reversible cilio-inhibiting effect on the ciliary beat frequency of chiken trachea was observed. Confocal laser scanning microscopy (CLSM) of nasal epithelia and nasal associated lymphoid tissue (NALT), incubated with nanoparticles loaded with fluorescein-labelled albumin, showed the presence of fluorescent nanoparticles throughout the cytoplasm of these cells, indicating the transport of albumin-associated TMC/TPP nanoparticles across the nasal mucosa. These findings led the authors to point to these nanoparticles as a potential delivery system for transport of proteins through the nasal mucosa

Other authors studied similar TMC/TPP nanoparticles, loaded with fluorescein isothiocyanate dextran, molecular weight 4400 Da (FD4), as a model of macromolecular drugs [17]. In analogy with the free TMC, the TMC/TPP nanoparticles exhibited the property of opening the tight junctions between cells in the Caco-2 monolayer in vitro and the rat intestinal epithelium ex vivo, thus promoting the permeation of FD4 across the two epithelium models. The nanoparticles also shared, with the free TMC, the property of adhering to the intestinal mucosa. Using CLSM, Sandri et al. [17] showed internalization of their nanoparticles into Caco-2 cells and excised rat jejunum tissue.

Nanoparticles encapsulating fluorescein-labelled bovine serum albumin (BSA) were obtained by ionotropic gelation of alginate-modified TMC with TPP [16]. According to the authors the transport of alginate-modified TMC nanoparticles across the Caco-2 cell in vitro model of gastrointestinal (GI) epithelium was more efficient than that produced by non-modified TMC nanoparticles. However, alginate modification barely had any effect on the trans-epithelial electrical resistance or on paracellular protein transport. Then the hypothesis was made that alginate modification facilitated nanoparticle transport across the Caco-2 monolayer by the transcellular route (transcytosis) by virtue of a reduction of particle size to 100-200 nm (16). The supposedly permeated nanoparticles were assayed by measuring the fluorescence of fluorescein-labelled BSA, which was assumed to be completely associated with the particles. Similar nanoparticles as the above were loaded with urease, a vaccine protein against *Helicobacter pylori* infection. Immunization studies in mice showed that oral administration of urease-loaded TMC nanoparticles generated high titers of both IgG and S-IgA antibodies. The immunostimulating effect was caused by nanoparticle mucoadhesivity and transcytosis by M cells in gut associated lymphoid tissue [16].

OVA-loaded nanoparticles have been prepared from TMC using unmethylated CpG DNA as adjuvant and crosslinker, in place of TPP, for nasal vaccination in mice [15]. TMC/CpG/OVA showed similar physical properties as TMC/TPP/OVA in terms of particle size, zeta-potential and antigen release characteristics, but TMC/CpG/OVA induced a 10-fold higher IgG2a response than TMC/TPP/OVA, and a strong humoral and Th1 type cellular immune responses after nasal vaccination [15].

Nanoparticles derived from the polyelectrolytic complexation of TMC by the polyanionic mono-*N*-carboxymethyl chitosan (MCC), and loaded with fluorescein-labelled BSA were taken up into mouse Balb/c monocyte macrophages. Mice were nasally immunized with tetanus toxoid-loaded TMC/MCC complex nanoparticles. These were shown to induce both mucosal and systemic immune response [24].

Insulin was formulated into nanoparticles formed from quaternized chitosans such as TMC or DEMC via either ionotropic gelation with TPP, or polyelectrolyte complexation by the polyanionic insulin. The PEC method resulted in higher insulin loading efficiency and nanoparticle zeta-potential [31].

Similar nanoparticulate systems loaded with insulin were prepared from other quaternized chitosans, namely, *N*-triethyl chitosan (TEC) and *N*-dimethylethyl chitosan (DMEC), by the PEC method [30]. Insulin was transported ex vivo across the colon membrane of rats when it was formulated into nanoparticles made of quaternized derivatives, better than into those made of plain chitosan. In vivo colon absorption of insulin was enhanced by using insulin-loaded nanoparticles compared to free insulin. Insulin absorption from rat colon was evaluated by its hypoglycemic effect [30].

Poly(γ-glutamic acid) was used by Mi et al. [25] as the anionic polyelectrolyte complexing agent to prepare nanoparticles from TMC by the PEC method, for the oral delivery of insulin. According to the authors insulin was transported across the Caco-2 cell in vitro model of GI epithelium via the paracellular route. In fact, CLSM confirmed the opening of the tight junctions between cells caused by the nanoparticles. The authors propose a mechanism whereby the orally administered nanoparticles with mucoadhesive TMC on their surfaces may adhere and infiltrate into the intestinal mucus, mediate the opening of tight junctions between enterocytes, undergo disintegration, and release insulin, which would permeate through the paracellular pathway to the bloodstream. This hypothesis is contrasting with that, proposed by Chen et al. [16], of protein being carried by TMC/alginate/TPP nanoparticles across the Caco-2 monolayer by transcytosis.

TMC was modified with the specific ligand CSKSSDYQC peptide (CSK) to prepare ionotropically crosslinked TMC-CSK/TPP nanoparticles, loaded with fluorescein isothiocyanate (FITC)-labelled insulin, targeted to the mucus-producing goblet cells [45]. In transport studies across Caco-2/HT29-MTX co-cultured cell monolayer, simulating mucus-producing intestinal epithelium, the CSK modification showed enhanced drug transport ability, even if the target recognition was partially affected by mucus. In pharmacological and pharmacokinetic studies in diabetic rats, the orally administered CSK-modified nanoparticles produced a stronger hypoglycemic effect than the unmodified ones, prompting the authors to state that the former were sufficiently effective as goblet cell-targeting nanocarriers for oral delivery of insulin.

An oral delivery system for paclitaxel, a mitotic inhibitor used in cancer chemotherapy, was devised by encapsulating the drug in *N*-(2-hydroxy-3-trimethylammonium) propyl chitosan chloride (HTCC) nanoparticles prepared by the O/W/O double emulsion temperature-programmed solidification method [72]. CLSM studies suggested that the HTCC nanoparticles could be transported across Caco-2 monolayers via the opening of tight junctions between cells. Also the in vivo absorption of these nanoparticles by the small intestine of rats was shown. These transport properties of nanoparticles were ascribed to their positive surface charge, which was also considered responsible for an enhanced nanoparticle uptake by carcinoma cells. Biodistribution studies after oral administration in subcutaneous LLC tumor-bearing mice showed accumulation of paclitaxel-loaded HTCC nanoparticles in liver, spleen, lung, and kidney tissues, which was ascribed to the uptake of nanoparticles by the reticuloendothelial system, and in tumour tissue through the enhanced permeability and retention (EPR) effect. These results are particularly intriguing as they open the prospect of a targeted oral treatment of cancer by nanomedicine.

3.2. Thiolated derivatives

The thiol groups immobilized on these polymers are supposed to give exchange reactions with disulfide bonds within the mucus, or oxidation reactions with cysteine-rich subdomains of mucus glycoproteins [73, 74], both resulting in the formation of disulfide bonds between thiolated chitosan derivatives and the mucus, which improve the polymer mucoadhesivity. Nanoparticles prepared from this type of chitosan derivatives were supposed to be themselves mucoadhesive, and hence, apt to make nanocarriers for oral drug delivery. In fact, enhanced mucoadhesive properties of nanoparticles prepared by gelation of chitosan-*N*-acetyl cysteine conjugate (chitosan-NAC) with TPP, compared with unmodified chitosan nanoparticles, were found by Wang et al. [23]. Enhanced insulin in vivo absorption via nasal mucosa was found by these authors when insulin-loaded chitosan-NAC/TPP nanoparticles were administered intranasally to rats.

Another thiolated chitosan derivative, chitosan-4-thiobutylamidine (chitosan-TBA) was used by Bernkop-Schnürch et al. [22] to develop a mucoadhesive nanoparticulate delivery system.

Polymer	Gelling agent	Drug	Diameter (nm)	Zeta potential (mV)	Reference
TMC-CSK	TPP	insulin	200-350	3-10	[14]
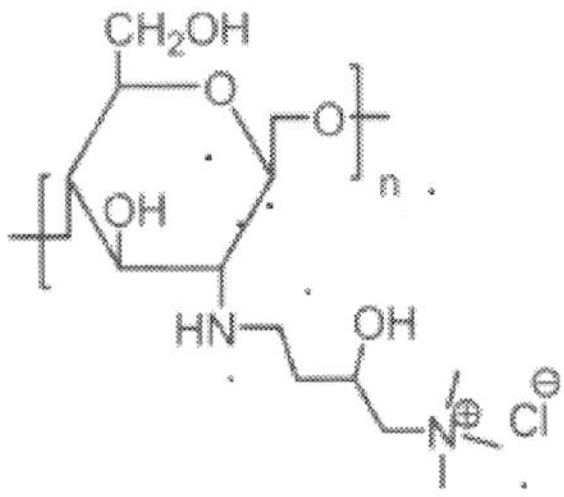 HTCC	o/w/o double emulsion	paclitaxel	130	21	[72]
TMC	cisplatin-hyaluronate	cisplatin	450	45	[27]
TMC	TPP	OVA	300	20	[15]
TMC	MCC	tetanus toxoid, FITC-BSA	not reported	not reported	[24]
HTCC	poly(aspartic acid)	BSA	200-300	55	[26]
TMC	poly(γ-glutamic acid)	insulin	100	30	[25]
TEC, DMEC	insulin	insulin	200	25	[30]
TMC, DEMC	TPP, insulin	insulin	250	25	[31]
TMC	TPP	FITC-BSA	300	14	[16]
TMC	TPP	OVA	254-300	20-61	[19]
TMC	TPP	FD4	200	-	[17]

Table 1. Main characteristics of nanoparticles based on quaternized chitosans

The polymer was first crosslinked ionotropically by TPP, followed by stabilization of the resulting nanoparticles via formation of inter- and intrachain disulfide bonds by thiol oxidation with H_2O_2. Subsequently, TPP was removed by dialysis. The covalently crosslinked particles would not disintegrate in the acidic medium of the stomach. The adhesion to porcine intestinal mucosa was studied after incorporation of fluorescein diacetate into nanoparticles. The more thiol groups were oxidized, the lower was the nanoparticle mucoadhesivity, nevertheless, even when as much as 90% of all thiols were oxidized the mucoadhesivity of chitosan-TBA nanoparticles was twice as high as that of unmodified chitosan nanoparticles.

Polymer	Gelling agent	Drug	Diameter (nm)	Zeta potential (mV)	Reference
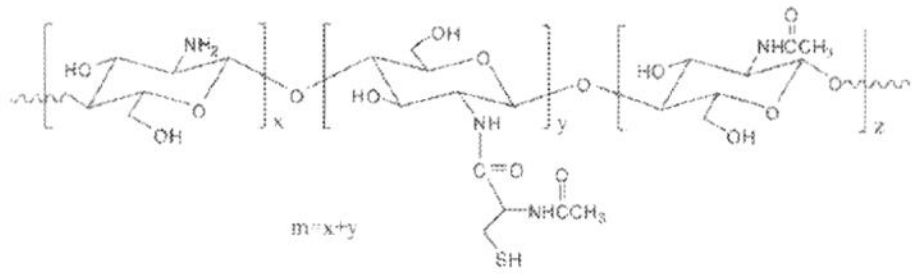 chitosan-NAC	TPP	insulin	140-210	19-31	[23]
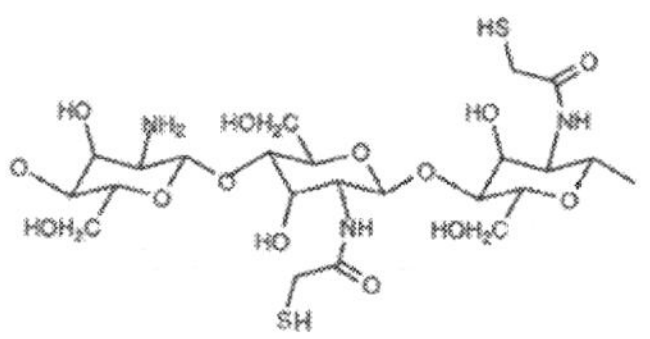 chitosan-TGA	DNA	DNA	75-120	2-20	[34]
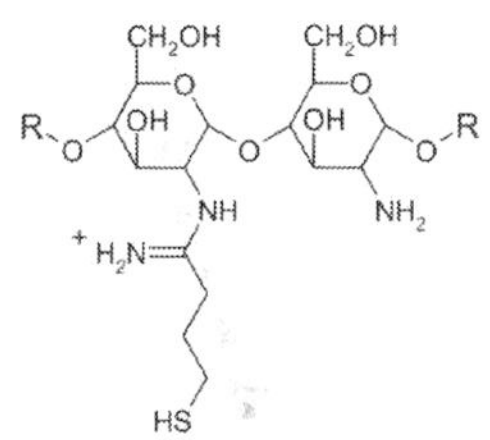 chitosan-TBA	TPP	none	268	4-19	[22]
chitosan-TBA	TPP	none	240	5-11	[21]
thiolated TMC	thiolated HA	OVA	250-350	10-20	[28]
chitosan-TGA	none	pSEAP	212-113	4-8	[33]
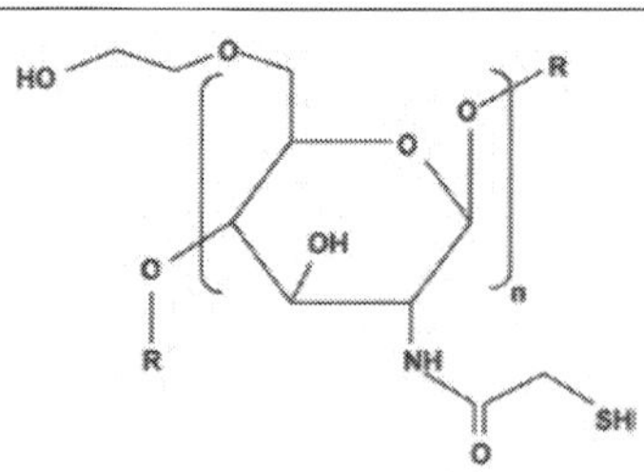 glycol chitosan-TGA	TPP	calcitonin	230-330	21-27	[20]

Table 2. Main characteristics of nanoparticles based on thiolated chitosan derivatives

Glycol chitosan coupled with thioglycolic acid (TGA) was ionotropically gelled with TPP to yield nanoparticles, which showed a twofold increase in mucoadhesion to lung tissue after intra-tracheal administration to rats as compared to non-thiolated nanoparticles. Biocompatibility of nanoparticle formulations with lung tissue was demonstrated. Calcitonin-loaded glycol chitosan and glycol chitosan-TGA nanoparticles resulted in a pronounced hypocalcemic effect for at least 12 and 24 h and a bioavailability of 27 and 40%, respectively [20].

Verheul et al. [28] used the thiol groups of thiolated TMC to spontaneously form interchain disulfide crosslinks with the thiols of thiolated hyaluronic acid (HA), after ionic gelation. OVA-loaded stabilized TMC-S-S-HA nanoparticles demonstrated higher immunogenicity than not stabilized particles, indicated by higher IgG titers, in nasal and intradermal vaccination.

Besides showing enhanced mucoadhesivity and cell penetration properties, nanoparticles made of thiolated chitosans have appeared highly effective as gene delivery systems. Thiolated derivatives, prepared from 33-kDa chitosan by coupling with TGA, formed nanocomplexes with plasmid DNA encoding green fluorescent protein (GFP), that were able to bind and protect plasmid DNA from Dnase I digestion. Thiolated chitosan/DNA nanocomplexes induced higher GFP expression in HEK293, MDCK and Hep-2 cell lines than unmodified chitosan. Nanocomplexes of disulfide-crosslinked thiolated chitosan/DNA showed a sustained DNA release and continuous expression in cultured cells lasting up to 60 h post transfection. Intranasal administration of crosslinked thiolated chitosan/DNA nanocomplexes to mice yielded gene expression that lasted at least 14 days [34].

Nanoparticles containing the gene reporter pSEAP (recombinant Secreted Alkaline Phosphatase) were generated, based on a thiolated chitosan conjugate, chitosan-TGA, crosslinked by thiol oxidation with H_2O_2 to form disulfide crosslinks. Transfection of nanoparticles in Caco-2 cells led to increased protein expression compared to unmodified chitosan nanoparticles. Red blood cells lysis tests provided evidence for no cytotoxicity of nanoparticles. On the basis of their experimental results the authors stated that their crosslinked thiolated chitosan nanoparticles showed the potential for being used as a non-viral vector system for gene therapy [33].

3.3. Amphiphilic derivatives

Amphiphilic derivatives resulted when hydrophobic structures were attached to the hydrophilic chitosan backbone. In aqueous milieu these derivatives would self-assemble into nanoparticles to attain thermodynamic stability. Nanoparticles derived from the self-assembly of amphiphilic derivatives were often intended for cancer therapy. Glycol chitosan (hydrophilic)-cholanic acid (hydrophobic) conjugates self-assembled to form nanoparticles, the in vivo tissue distribution, time-dependent excretion and tumor accumulation of which were monitored in tumor-bearing mice by Park et al. [37]. The particles exhibited prolonged blood circulation time, decreased time-dependent excretion from the body, and increased tumor accumulation with increasing polymer molecular weight. The enhanced tumor targeting by nanoparticles made of high molecular weight glycol chitosan-cholanic acid was ascribed to a better in vivo stability, related to an improvement in blood circulation time [37].

Similar nanoparticles as the above, formed from glycol chitosan-cholanic acid conjugate, loaded with the anticancer drug camptothecin, exhibited significant antitumor effects and high tumor targeting ability towards MDA-MB231 human breast cancer xenografts subcutaneously implanted in nude mice. The significant antitumor efficacy of nanoparticles was ascribed to both their prolonged blood circulation and high accumulation in tumors through the EPR effect [39].

The cellular uptake mechanism and the intracellular fate of nanoparticles formed from glycol chitosan hydrophobically modified with cholanic acid have been reported [40]. These particles showed an enhanced distribution in the whole cells, compared to the parent hydrophilic glycol chitosan polymer. In vitro experiments with endocytic inhibitors suggested that the cellular uptake of these nanoparticles involved several distinct pathways, e.g., clathrin-mediated endocytosis, caveolae-mediated endocytosis, and macropinocytosis. Such a property, along with low toxicity and biocompatibility suggested these hydrophobically modified glycol chitosan nanoparticles as a versatile carrier for the intracellular delivery of therapeutic agents [40].

A further hydrophobically modified chitosan derivative from which self-assembled nanoparticles were obtained was oleoyl chitosan. The toxicity profile of the relevant nanoparticles, evaluated in vitro via hemolysis test and MTT assay, was within acceptable limits. When loaded with the antitumor drug doxorubicin, oleoyl chitosan nanoparticles exhibited inhibitory rates on different human cancer cells (A549, Bel-7402, HeLa, and SGC-7901) significantly higher than the drug solution [43].

Folic acid was conjugated with *O*-carboxymethyl chitosan via the bifunctional 2,2′-(ethylenedioxy)-*bis*-(ethylamine) to obtain an amphiphilic chitosan derivative that would self-assemble into nanoparticles. Folate-mediated endocytosis significantly enhanced the cellular targeting of nanoparticles, thus facilitating apoptosis of cancer cells (HeLa, B16F1). Doxorubicin could be loaded into the nanoparticles. It was observed that survival in cancer cells treated with doxorubicin-loaded nanoparticles was lower than that of normal cells in similar concentrations [75].

The ability of nanoparticles prepared by self-assembly of chitosan amphiphiles to promote oral absorption of hydrophobic and hydrophilic drugs in rats was recently investigated by Siew et al. [42], using quaternary ammonium palmitoyl glycol chitosan as the basic material. The nanoparticles were found to enhance the oral absorption (Cmax) of griseofulvin and cyclosporine A (hydrophobic) and, to a lesser extent, of ranitidine (hydrophilic). Hydrophobic drug absorption was facilitated by the nanomedicine by: (a) increasing the drug dissolution rate, (b) adhering to and penetrating the mucus layer, thus allowing intimate contact between the drug and the GI epithelium absorptive cells, and (c) enhancing transcellular drug transport. As for the absorption of the hydrophilic ranitidine, despite an 80% increase of Cmax there was no appreciable opening of tight junctions by the nanoparticles. No uptake of this type of nanoparticles by epithelial cells is reported [42].

http://dx.doi.org/10.5772/54944

Polymer	Drug	Diameter (nm)	Zeta potential (mV)	Reference
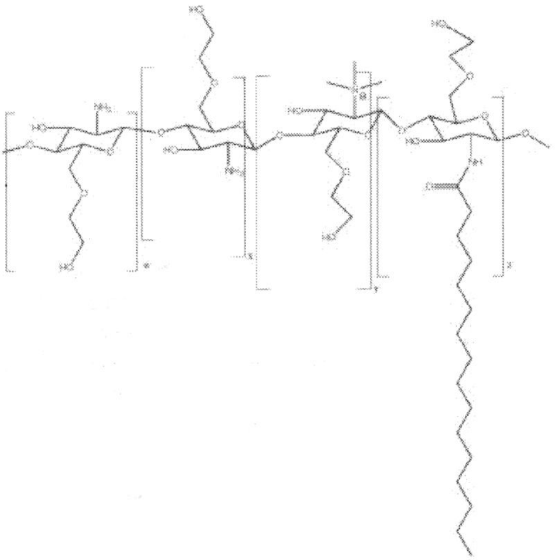 quaternary ammonium palmitoyl glycol chitosan	griseofulvin cyclosporin A ranitidine	100-500	not report-ed	[42]
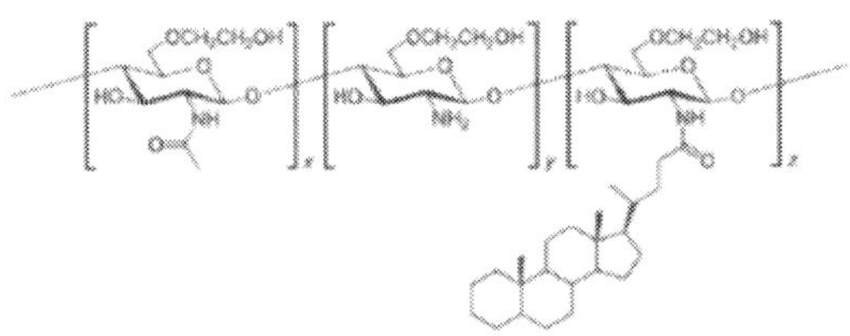 glycol chitosan-5β-cholanic acid	none	300-400	10	[40]
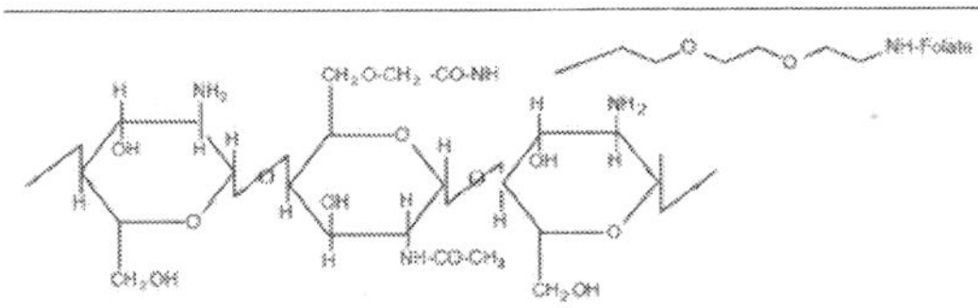 *O*-carboxymethyl chitosan-2,2',(ethylene dioxy)-bis-(ethylamine)-folic acid	doxorubicin	150-200	10-20	[75]
glycol chitosan-5β-cholanic acid	none	230-310	10-11	[37]
glycol chitosan-5β-cholanic acid	camptothe-cin	250-350	not report-ed	[39]

Polymer	Drug	Diameter (nm)	Zeta potential (mV)	Reference
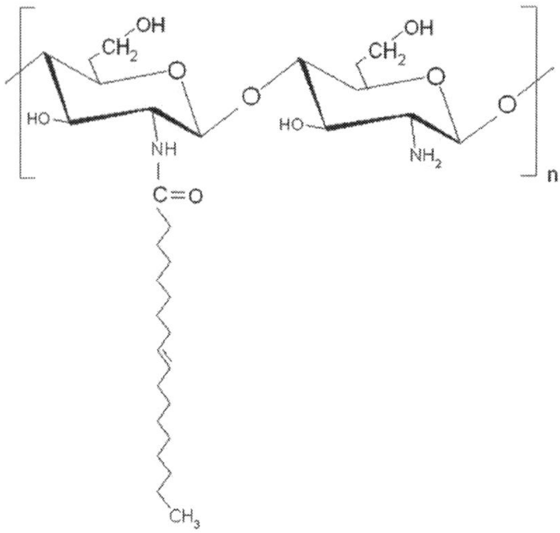 oleoyl-chitosan	doxorubicin	250-350	not reported	[43]

Table 3. Main characteristics of nanoparticles based on anphiphilic chitosans

4. Concluding remarks

Three families of chitosan derivatives have been synthesized and used to prepare nanoparticles for pharmaceutical application, namely, polycations obtained by introducing quaternary ammonium groups on the polymer backbone; thiolated derivatives, and amphiphilic derivatives obtained by attaching hydrophobic structures to the chitosan or glycol chitosan backbone. The nanoparticles prepared from the quaternary ammonium-chitosan derivatives, especially via the PEC formation method, have shown improved stability and physical properties (smaller size, higher zeta potential) compared to nanoparticles from unmodified chitosan. The thiolated derivatives offered the opportunity to stabilize the nanoparticles by covalent crosslinks formed from interchain thiol oxidation to disulfide, which made the particles stable in the GI environment. The critical aggregation concentration of the amphiphilic hydrophobically modified chitosan derivatives is usually very low, which implies stability of the self aggregates in dilute conditions, such as those encountered by the nanoparticles in the organism. The nanoparticulate systems prepared from chitosan derivatives have generally shown acceptable cytotoxicity. In accord with the known behavior of particles of a size smaller than 500 nm, they have shown endocytic uptake by cells. Smaller particles with higher zeta potential have shown more aptitude to endocytosis. Ionotropically crosslinked TMC nanoparticles are a potential vehicle for transport of proteins across mucosal epithelia, as they have been found to open the tight junctions between epithelial cells. Indeed, nanoparticles based on quaternized chitosan are a promising vehicle for the oral administration of insulin, especially if the chitosan derivative is conjugated with the specific ligand CSKSSDYQC peptide. Also interesting is the nanosystem based on the quaternary ammonium-chitosan conjugate HTCC, which was orally absorbed by the rat small intestine and subsequently accumulated in carcinoma tissue by the EPR effect. These results are particularly intriguing as they open the prospect of a targeted oral treatment of cancer by nanomedicine. Nanoparticles prepared from thiolated chitosan derivatives have shown a particular mucoadhesivity implying a suitability for making nanocarriers for transmucosal protein delivery. Also this type of nanoparticles have appeared highly effective as gene delivery systems and have shown the potential for being used as a non-viral vector system for gene therapy. Nanoparticles derived from the self-assembly of amphiphilic chitosan derivatives were often intended for cancer therapy. Glycol chitosan hydrophobically modified with cholanic acid yielded nanoparticles with comparatively high in vivo stability, responsible for a prolonged blood circulation time, which led to high accumulation in tumors through the EPR effect. This type of nanoparticles can be taken up by cells through distinct pathways, which points to this system as a versatile carrier for the intracellular delivery of therapeutic agents. Folic acid, conjugated with *O*-carboxymethyl chitosan to obtain doxorubicin-loaded self-assembled nanoparticles, could mediate particle endocytosis by cancer cells with consequent cell apoptosis. In conclusion the present survey has endorsed the concept that chitosan derivatization can lead to new basic materials for nanosystems with unique pharmaceutical performances.

Abbreviations

BSA Bovine serum albumin

CAC Critical aggregation concentration

CLSM Confocal laser scanning microscopy

CSK CSKSSDYQC peptide

DEMC *N,N*-diethyl *N*-methyl chitosan

DMEC *N,N*-dimethyl *N*-ethyl chitosan

EPR Enhanced permeability and retention effect

FD4 Fluorescein isothiocyanate dextran, molecular weight 4400 Da

FITC Fluorescein isothiocyanate

GFP Green fluorescent protein

GI Gastrointestinal

HA Hyaluronic acid

HTCC *N*-(2-hydroxy-3-trimethylammonium) propyl chitosan chloride

LLC Lewis lung carcinoma

MCC Mono-*N*-carboxymethyl chitosan

NAC *N*-acetyl cysteine

NALT Nasal associated lymphoid tissue

OVA Ovalbumin

PEC Polyelectrolyte complex

pSEAP Recombinant secreted alkaline phosphatase

TBA Thiobutyl amidine

TEC *N*-triethyl chitosan

TGA Thioglycolic acid

TMC N-trimethyl chitosan

TPP Sodium tripolyphosphate

Author details

Ylenia Zambito*

Address all correspondence to: zambito@farm.unipi.it

Dipartimento di Farmacia, Università di Pisa, Italy

References

[1] Lee, K. Y, Ha, W. S, & Park, W. H. Blood compatibility and biodegradability of partially N-acylated chitosan derivatives. Biomaterials (1995). , 16-1211.

[2] Muzzarelli RAAHuman enzymatic activities related to the therapeutic administration of chitin derivative. Cellular and Molecular Life Scince (1997). , 53-131.

[3] Onishi, H, & Machida, Y. Biodegradation and distribution of water-soluble chitosan in mice. Biomaterials (1999). , 20-175.

[4] Chandy, T, & Sharma, C. P. Chitosan as a biomaterial. Artificial Cells and Artificial Organs (1990). , 18-1.

[5] Aspeden, T. J, Mason, J. D, Jones, N. S, Lowe, J, Skaugrud, O, & Illum, L. Chitosan solutions on in vitro and in vivo mucociliary transport rates inhuman turbinates and volunteers. Journal of Pharmaceutical Sciences (1997). , 86-509.

[6] Illum, L, Farraj, N. F, & Davis, S. S. Chitosan as a novel nasal delivery system for peptide drugs. Pharmaceutical Research (1994). , 11-1186.

[7] Illum, L, Jabbal-gill, I, Hinchcliffe, M, Fisher, A. N, & Davis, S. S. Chitosan as a novel nasal delivery system for vaccines. Advanced Drug Delivery Reviews (2001). , 51-81.

[8] Junginger, H. E, & Verhoef, J. C. Macromolecules as safe penetration enhancers for hydrophilic drugs-a fiction? Pharmaceutical Sciences Technology Today (1998). , 1-370.

[9] Borchard, G. Lueßen HL, de Boer AG, Verhoef JC, Lehr CM, Junginger HE. The potential of mucoadhesive polymers in enhancing intestinal peptide drug absorption: III. Effects of chitosan glutamate and carbomer on epithelial tight junctions in vitro. Journal of Controlled Release (1996). , 39-131.

[10] Di Colo GZambito Y, Burgalassi S, Nardini I, Saettone MF. Effect of chitosan and of N-carboximethylchitosan on intraocular penetration of topically applied ofloxacin. International Journal of Pharmaceutics (2004). , 273-37.

[11] Peniche, H, & Peniche, C. Chitosan nanoparticles: a contribution to nanomedicine. Polymer Intenational (2001). , 60-883.

[12] Wang, J. J, Zeng, Z. W, Xiao, R. Z, Xie, T, Zhou, G. L, Zhan, X. R, & Wang, S. L. Recent advances of chitosan nanoparticles as drug carriers. International Journal of Nanomedicine (2011). , 6-765.

[13] Calvo, P, Remunan-lopez, C, Vila-jato, J. L, & Alonso, M. J. Novel hydrophilic chitosan-polyethylene oxide nanoparticles as protein carriers Journal of Polymer Scince (1997). , 63-125.

[14] Jin, Y, Song, Y, Zhu, X, Zhou, D, Chen, C, Zhang, Z, & Huang, Y. Goblet cell-targeting nanoparticles for oral insulin delivery and the influence of mucus on insulin transport. Biomaterials (2012). , 33-1573.

[15] Slütter, B, & Jiskoot, W. Dual role of CpG as immune modulator and physical crosslinker in ovalbumin loaded N-trimethyl chitosan (TMC) nanoparticles for nasal vaccination. Journal of Controlled Release (2010). , 148-117.

[16] Chen, F, Zhang, Z. R, Yuan, F, Qin, X, Wang, M, & Huang, Y. In vitro and in vivo study of N-trimethyl chitosan nanoparticles for oral protein delivery. Intenational Journal of Pharmaceutics (2008). , 349-226.

[17] Sandri, G, Bonferoni, M. C, Rossi, S, Ferrari, F, Gibin, S, & Zambito, Y. Di Colo G, Caramella C. Nanoparticles based on N-trimethylchitosan: Evaluation of absorption

properties using in vitro (Caco-2 cells) and ex vivo (excised rat jejunum) models. European Journal of Pharmaceutics and Biopharmaceutics (2007). , 65-68.

[18] Dehousse, V, Garbacki, N, Jaspart, S, Castagne, D, Piel, G, Colige, A, & Evrard, B. Comparison of chitosan/siRNA and trimethylchitosan/siRNA complexes behaviour in vitro. International Journal of Biological Macromolecules (2010). , 46-342.

[19] Amidi, M, Romeijn, S. G, Borchard, G, Junginger, H. E, Hennink, W. E, & Jiskoot, W. Preparation and characterization of protein-loaded N-trimethyl chitosan nanoparticles as nasal delivery system. Journal of Controlled Release (2006). , 111-107.

[20] Makhlof, A, Werle, M, Tozuka, Y, & Takeuchi, H. Nanoparticles of glycol chitosan and its thiolated derivative significantly improved the pulmonary delivery of calcitonin. International Journal of Pharmaceutics (2010). , 397-92.

[21] Bernkop-schnurch, A, Heinrich, A, & Greimel, A. Development of a novel method for the preparation of submicron particles based on thiolated chitosan. European Journal of Pharmaceutics and Biopharmaceutics (2006). , 63-166.

[22] Bernkop-schnürch, A, Weithaler, A, Albrecht, K, & Greimel, A. Thiomers: Preparation and in vitro evaluation of a mucoadhesive nanoparticulate drug delivery system. Intenational Journal of Pharmaceutics (2006). , 317-76.

[23] Wang, X, Zheng, C, Wu, Z, Teng, D, Zhang, X, Wang, Z, & Li, C. Chitosan-NAC nanoparticles as a vehicle for nasal absorption enhancement of insulin. Journal of Biomedical Materials Research Part B: Applied Biomaterials (2009). , 88-150.

[24] Sayin, B, Somavarapu, S, Li, X. W, Sesardic, D, Senel, S, & Alpar, O. H. TMC-MCC (N-trimethyl chitosan-mono-N-carboxymethyl chitosan) nanocomplexes for mucosal delivery of vaccines. European Journal of Pharmaceutical Sciences (2009). , 38-362.

[25] Mi, F. L, Wu, Y. Y, Lin, Y. H, Sonaje, K, Ho, Y. C, Chen, C. T, Juang, J. H, & Sung, H. W. Oral delivery of peptide drugs using nanoparticles self-assembled by poly(γ-glutamic acid) and a chitosan derivative functionalized by trimethylation. Bioconjugate Chemistry (2008). , 19-1248.

[26] Wang, T. W, Xu, Q, Wu, Y, Zeng, A. J, Li, M, & Gao, X. Quaternized chitosan (QCS/poly (aspartic acid) nanoparticles as a protein drug-delivery system. Carbohydrate Research (2009). , 344-908.

[27] Cafaggi, S, Russo, E, Stefani, R, Parodi, B, Caviglioli, G, Sillo, G, Bisio, A, Aiello, C, & Viale, M. Preparation, characterisation and preliminary antitumour activity evaluation of a novel nanoparticulate system basedon a cisplatin-hyaluronate complex and N-trimethyl chitosan. Invest New Drugs (2011). , 29-443.

[28] Verheul, R. J, Slütter, B, Bal, S. M, Bouwstra, J. A, Jiskoot, W, & Hennink, W. E. Covalently stabilized trimethyl chitosan-hyaluronic acid nanoparticles for nasal and intradermal vaccination. Journal of Controlled Release (2011). , 156-46.

[29] Bayat, A, Larijani, B, Ahmadian, S, Junginger, H. E, & Rafiee-tehrani, M. Preparation and characterization of insulin nanoparticles using chitosan and its quaternized derivatives. Nanomedicine (2008). , 4-115.

[30] Bayat, A, Dorkoosh, F. A, Dehpour, A. R, Moezi, L, Larijani, B, Junginger, H. E, & Rafiee-tehrani, M. Nanoparticles of quaternized chitosan derivatives as a carrier for colon delivery of insulin: Ex vivo and in vivo studies. Intenational Journal of Pharmaceutics (2008). , 356-259.

[31] Sadeghi AMMDorkoosh FA, Avadi MR, Saadat P, Rafiee-Tehrani M, Junginger HE. Preparation, characterization and antibacterial activities of chitosan, N-trimethyl chitosan (TMC) and N-diethylmethyl chitosan (DEMC) nanoparticles loaded with insulin using both the ionotropic gelation and polyelectrolyte complexation methods. International Journal of Parmaceutics (2008). , 355-299.

[32] Jintapattanakit, A, Junyaprasert, V. B, Mao, S, Sitterberg, J, Bakowsky, U, & Kissel, T. Peroral delivery of insulin using chitosan derivatives: A comparative study of polyelectrolyte nanocomplexes and nanoparticles. International Journal of Pharmaceutics (2009). , 342-240.

[33] Martien, R, Loretz, B, Sandbichler, A. M, & Bernkop-schnürch, A. Thiolated chitosan nanoparticles: transfection study in the Caco-2 differentiated cell culture. Nanotechnology (2008).

[34] Lee, D, Zhang, W, Shirley, S. A, Kong, X, Hellermann, G. R, Lockey, R. F, & Mohapatra, S. S. Thiolated chitosan/DNA nanocomplexes exhibit enhanced and sustained gene delivery. Pharmaceutical Research (2007). , 24-157.

[35] Yoo, H. S, Lee, J. E, Chung, H, Kwon, I. C, & Jeong, S. Y. Self-assembled nanoparticles nanoparticles containing hydrophobically modified glycol chitosan for gene delivery. Journal of Controlled Release (2005). , 103-235.

[36] Kim, J. H, Kim, Y. S, Kim, S, Park, J. H, Kim, K, Choi, K, Chung, H, Jeong, S. Y, Park, R. W, Kim, I. S, & Kown, I. C. Hydrophobically modified glycol chitosan nanoparticles as carriers for paclitaxel. Journal of Controlled Release (2006). , 111-228.

[37] Park, K, Kim, J. H, Nam, Y. S, Lee, S, Nam, H. Y, Kim, K, Park, J. H, Kim, I. S, Choi, K, Kim, S. Y, & Kwon, I. C. Effect of polymer molecular weight on the tumor targeting characteristics of self-assembled glycol chitosan nanoparticles. Journal of Controlled Release (2007). , 122-305.

[38] Kim, J. H, Kim, Y. S, Park, K, Kang, E, Lee, S, Nam, H. Y, Kim, K, Park, J. H, Chi, D. Y, Park, R. W, Kim, I. S, Choi, K, & Kwon, I. C. Self-assembled glycol chitosan nanoparticles for the sustained and prolonged delivery of antiangiogenic small peptide drugs in cancer terapy. Biomaterials (2008). , 29-1920.

[39] Min, K. H, Park, K, Kim, Y. S, Bae, S. M, Lee, S, Jo, H. G, Park, R. W, Kim, I. S, Jeong, S. Y, Kim, K, & Kwon, I. C. Hydrophobically modified glycol chitosan nanoparticles-

encapsulated camptothecin enhance the drug stability and tumor targeting in cancer therapy. Journal of Controled Release (2008). , 127-208.

[40] Nam, H. Y, Kwon, S. M, Chung, H, Lee, S. Y, Kwon, S. H, Jeon, H, Kim, Y, Park, J. H, Kim, J, Her, S, Oh, Y. K, Kwon, I. C, Kim, K, & Jeong, S. Y. Cellular uptake mechanism and intracellular fate of hydrophobically modified glycol chitosan nanoparticles. Journal of Controlled Release (2009). , 135-259.

[41] Chen, K. J, Chiu, Y. L, Chen, Y. M, Ho, Y. C, & Sung, H. W. Intracellularly monitoring/imaging the release of doxorubicin from pH-responsive nanoparticles using Förster resonance energy transfer. Biomaterial (2011). , 32-2586.

[42] Siew, A, Le, H, Thiovolet, M, Gellert, P, Schatzlein, A, & Uchegbu, I. Enhanced oral absorption of hydrophobic and hydrophilic drugs using quaternary ammonium palmitoyl glycol chitosan nanoparticles. Molecular Pharmaceutics (2012). , 9-14.

[43] Zhang, J, Chen, X. G, Li, Y. Y, & Liu, C. S. Self-assembled nanoparticles based on hydrophobically modified chitosan as carriers for doxorubicin. Nanomedicine: Nanotechnology, Biology, and Medicine (2007). , 3-258.

[44] Tan, Y. L, & Liu, C. G. Self-aggregated nanoparticles from linoleic acid modified carboxymethyl chitosan: Synthesis, characterization and application in vitro. Colloids and Surfaces B: Biointerface (2009). , 69-178.

[45] Jin, Y, Song, Y, Zhu, X, Zhou, D, Chen, C, Zhang, Z, & Huang, Y. Goblet cell-targeting nanoparticles for oral insulin delivery and the influence of mucus on insulin transport. Biomaterials (2012). , 33-1573.

[46] Schipper, N. G, Varum, K. M, & Artusson, P. Chitosans as absorption enhancers for poorly absorbable drugs. 1. Influence of molecular weight and degree of acetylation on drug transport across human intestinal epithelial (Caco-2) cells. Pharmaceutical Research (1996). , 13-1686.

[47] Kotzé, A. F, & De Leeuw, B. J. Lueßen HL, de Boer AG, Verhoef JC, Junginger HE. Chitosans for enhanced delivery of therapeutic peptides across intestinal epithelia: in vitro evaluation in Caco-2 cell monolayers. Internationa Journal of Pharmaceutics (1997). , 159-243.

[48] Lueßen HLRentel CO, Kotzé AF, Lehr CM, de Boer AG, Verhoef JC, Junginger HE. Mucoadhesive polymers in peroral peptide drug delivery. IV. Polycarbophil and chitosan are potent enhancers of peptide transport across intestinal mucosae in vitro. Journal of Controlled Release (1997). , 45-15.

[49] Lueßen HLde Leew BJ, Langmeÿer MWE, de Boer AG, Verhoef JC, Junginger HE. Mucoadhesive polymers in peroral peptide drug delivery. VI. Carbomer and chitosan improve the intestinal absorption of the peptide drug buserelin in vivo. Pharmaceutical Research (1996). , 13-1668.

[50] Illum, L, Farraj, N. F, & Davis, J. J. Chitosan as novel nasal delivery system for peptide drugs. Pharmaceutical Research (1994). , 11-1186.

[51] Tengamnuay, P, Sahamethapat, A, Sailasuta, A, & Mitra, A. K. Chitosans as nasal absorption enhancers of peptides: comparison between free amine chitosans and soluble salts. International Journal of Pharmaceutics (2000). , 197-53.

[52] Sinswat, P, & Tengamnuay, P. Enhancing effect of chitosan on nasal absorption of salmon calcitonin in rats: comparison with hydroxypropyl- and dimethyl-β-cyclodextrins. Intenational Journal of Pharm (2003). , 257-15.

[53] Hinchcliffe, M, Jabbal-gill, I, & Smith, A. Effect of chitosan on the intranasal absorption of salmon calcitonin in sheep. Journal of Pharmacy and Pharmacology (2005). , 57-681.

[54] Zhang, Y. J, Ma, C. H, Lu, W. L, Zhang, X, Wang, X. L, Sun, J. N, & Zhang, Q. Permeation-enhancing effects of chitosan formulations on recombinant hirudin-2 by nasal delivery in vitro and in vivo. Acta Pharmacologica Sinica (2005). , 26-1402.

[55] Di Colo GBurgalassi S, Chetoni P, Fiaschi MP, Zambito Y, Saettone MF. Gel-forming erodible inserts for ocular controlled delivery of ofloxacin. International Journal of Pharmaceutics (2001). , 215-101.

[56] Schipper NGMVarum KM, Stemberg P, Ocklind G, Lennernäs H. Chitosans as absorption enhancers of poorly absorbable drugs. 3: Influence of mucus on absorption enhancement. European Journal of Pharmaceutical Sciences (1999). , 8-335.

[57] Florea, B. I, Thanou, M, Junginger, H. E, & Borchard, G. Enhancement of bronchial octreotide absorption by chitosan and N-trimethyl chitosan shows linear in vitro/in vivo correlation. Journal of Controlled Release (2006). , 110-353.

[58] Sandri, G, Rossi, S, Bonferoni, M. C, Ferrari, F, & Zambito, Y. Di Colo G, Caramella C. Buccal penetration enhancement properties of N-trimethyl chitosan: Influence of quaternization degree on absorption of a high molecular weight molecule. International Journal of Pharmaceutics (2005). , 297-146.

[59] Kotzé, A. F, Luessen, H. L, De Leew, B. J, De Boer, B. G, & Verhoef, J. C. Junginger HE. N-trimethyl chitosan chloride as a potential absorption enhancer across mucosal surfaces: in vitro evaluation in intestinal epithelial cells (Caco-2). Pharmaceutical Research (1997). , 14-1197.

[60] Kotzé, A. F, Thanou, M, Luessen, H. L, De Boer, B. G, Verhoef, J. C, & Junginger, H. E. Enhancement of paracellular drug transport with highly quaternized N-trimethyl chitosan chloride in neutral environments: in vitro evaluation in intestinal epithelial cells (Caco-2). Journal of Pharmaceutical Sciences (1999). , 88-253.

[61] Thanou, M, Verhoef, J. C, Marbach, P, & Junginger, H. E. Intestinal absorption of octreotide: N-trimethyl chitosan chloride (TMC) ameliorates the permeability and ab-

sorption properties of the somatostatin analogue in vitro and in vivo. Journal of Pharmaceutical Sciences (2000). , 89-951.

[62] Thanou, M, Florea, B. I, Langemeyer, M. W, & Verhoef, J. C. Junginger HE. N-trimethyl chitosan chloride (TMC) improves the intestinal permeation of the peptide drug buserelin in vitro (Caco-2 cells) and in vivo (rats), Pharmaceutical Research (2000). , 17-27.

[63] Hamman, J. H, Stander, M, & Kotzé, A. F. Effect of the degree of quaternization of N-trimethyl chitosan chloride on absorption enhancement: in vivo evaluation in rat nasal epithelia. Internationa Journal of Pharmaceutics (2002). , 232-235.

[64] Florea, B. I, Thanou, M, Junginger, H. E, & Borchard, G. Enhancement of bronchial octreotide absorption by chitosan and N-trimethyl chitosan shows linear in vitro/in vivo correlation. Journal of Controlled Release (2006). , 110-353.

[65] Di Colo G,Burgalassi S,. Zambito Y, Monti D, Chetoni P. Effects of different N-trimethyl chitosans on in vitro/in vivo ofloxacin transcorneal permeation. Journal of Pharmaceutical Sciences (2004). , 93-2851.

[66] Zambito, Y, & Zaino, C. Di Colo G. Effects of N-trimethylchitosan on transcellular and paracellular transcorneal drug transport. European Journal of Pharmaceutics and Biopharmaceutics (2006). , 64-16.

[67] Hamman, J. H, & Schultz, C. M. Kotzé AF. N-trimethyl chitosan chloride: optimum degree of quaternization for drug absorption enhancement across epithelial cells. Drug Development and Industrial Pharmacy (2003). , 29-161.

[68] Avadi, M. R, Zohuriaan-mehr, M. J, Younessi, P, Amini, M, Rafiee-tehrani, M, & Shafiee, A. Optimized synthesis and characterization of N-triethylchitosan. Journal of Bioactive Compatible Polymers (2003). , 18-469.

[69] Bayat, A. Sadeghi AMM, Avadi MR, Amini M, Rafiee-Tehrani M, Shafiee A, Majlesi R, Junginger HE. Synthesis of N,N-dimethyl N-ethylchitosan as a carrier for oral delivery of peptide drugs. Journal of Bioactive Compatible Polymers (2006). , 21-433.

[70] Avadi, M. R. Sadeghi AMM, Tahzibi A, Bayati KH, Pouladzadeh M, Zohuriaan-mehr MJ, Rafiee-Tehrani M. Diethylmethyl chitosan as antimicrobial agent: synthesis, characterization and antibacterial effects. European Journal of Polymers (2004). , 40-1355.

[71] Ludwig, A. The use of mucoadhesive polymers in ocular drug delivery. Advanced Drug Delivery Review (2005). , 57-1595.

[72] Lv, P. P, Wei, W, Yue, H, Yang, T. Y, Wang, L. Y, & Ma, G. H. Porous quaternized chitosan nanoparticles containing paclitaxel nanocrystals improved therapeutic efficacy in non-small-cell lung cancer after oral administration. Biomacromolecules (2011). , 12-4230.

[73] Kast, C. E, & Bernkop-schnurch, A. Thiolated polymers- thiomers: development and in vitro evaluation of chitosan-thioglycolic acid conjugates. Biomaterials (2001). , 22-2345.

[74] Leitner, V. M, Walker, G. F, & Bernkop- Schnürch, A. Thiolated polymers: evidence for the formation of disulfide bonds with mucus glycoproteins. European Journal of Pharmaceutics and Biopharmaceutics (2003). , 56-207.

[75] Sahu, S, Mallick, S. K, Santra, S, Maiti, T. K, Ghosh, S. K, & Pramanik, P. In vitro evaluation of folic acid modified carboxymethyl chitosan nanoparticles loaded with doxorubicin for targeted delivery. Journal of Materials Science: Materials in Medicine (2010). , 21-1587.

Chapter 10

pH-Sensitive Nanocrystals of Carbonate Apatite- a Powerful and Versatile Tool for Efficient Delivery of Genetic Materials to Mammalian Cells

Ezharul Hoque Chowdhury

Additional information is available at the end of the chapter

http://dx.doi.org/10.5772/ 53107

1. Introduction

Delivery of a functional DNA to mammalian cells is an attractive approach for genetic manipulation of the cells in biomedical research as well as in gene therapy for treating critical human diseases. Following delivery to the cytoplasm, a foreign gene enters the nucleus and is transcribed to the corresponding mRNA, which is subsequently transported to the cytoplasm for translation into a specific protein. However, a gene-silencing element, such as an antisense onligonucleotide or a small interfering RNA blocks the transcription of a target mRNA. Thus, nucleic acid delivery has been an essential tool either to turn on or off the expression of a particular gene in basic research laboratories.

Intensive research in the last three decades led to the development of a number of viral and non-viral vectors. However, an ideal vector in terms of safety and efficacy is still lacking. Synthetic non-viral vectors, such as cationic polymers, lipids and peptides, are relatively safe, but extremely inefficient. On the other hand, viral systems are by far the most effective means of DNA delivery to mammalian cells, but some major limitations including toxicity, immunogenicity, restricted targeting of specific cell types, restricted DNA carrying capacity, production and packaging problems, recombination and high cost, limit their successful applications in basic research and clinical medicine. The effectiveness of a viral particle is the result of its highly evolved and specialized structure basically composed of a protein coat surrounding a nucleic acid core. Such a highly organized structure can prevent viral particles from unwanted interactions with serum components, while promoting subsequent internalization by cells, escape from endosomes, and release of genetic material from the particle either before or after entering the nucleus. Development of a non-viral approach having the beneficial virus-like properties and lacking the disadvantageous aspects would emerge as the most attractive one for implementation in research laboratories and gene therapy.

A major barrier to the non-viral delivery is low uptake of DNA across the plasma membrane of a cell owing to the inappropriate and ineffective interactions of the DNA delivery vehicle with the cell membrane. Negatively charged DNA molecules are usually condensed with cationic reagents to allow formation of the complexes carrying net positive charges. The resulting complexes can interact electrostatically with anionic heparan sulfate proteoglycans (syndecans) on cell surface and reach the cytoplasmic side in the form of endosomes through endocytosis [1]. The extremely low pH and enzymes within the late endosomes usually bring about degradation of entrapped DNA and associated complexes. Finally, DNA that survives both endocytic processing and cytoplasmic nucleases must dissociate from the condensed complexes either before or after nuclear translocation through nuclear pore or during cell division.

Many therapeutic applications demand a vehicle with capability of delivering transgene(s) to a selective cell type in order to increase the expression efficacy and alleviate any side effect. A common strategy in non-viral case involves attachment of a targeting moiety onto a polycation (lipid or polymer) backbone which finally condenses the DNA through ionic interactions. Targeting moiety can enable the resulting DNA carrier to bind to a receptor, lectin, antigen or cell-adhesion molecule on plasma membrane prior to internalization via endocytosis or phagocytosis. Polylysine, the first backbone used for gene delivery has been conjugated to a diverse set of cell-specific ligands, such as asialoorosomucoid [2], transferrin [3], epidermal growth factor (EGF) [4], mannose [5], fibroblast growth factor (FGF) [6] and antibodies [7] for targeting, respectively, hepatocytes via asialoglycoprotein receptors, transferrin receptor-positive cells, EGF receptor-carrying cells, macrophages through membrane lectins, FGF receptor-bearing cells and lymphocytes via surface-bound antigens. In the similar fashions, polymers like polyethylenimine and liposomes have been coupled to other cell surface receptor-specific ligands in addition to those described above, such as integrin-binding peptide conjugated onto PEI to target integrins on cell surfaces [8] and vitamin folate conjugated onto liposomes through a polyethylene spacer to target folate receptor-bearing cells [9].

Cell adhesion molecules (integrin, syndecan, cadherin, selectin) which are a diverse group of cell surface proteins mediating interactions between cells, and between cells and the extracellular matrix, are valuable targets for precise gene delivery to haematopoietic cells, airway epithelial cells, tumor cells and vascular endothelial cells using synthetically designed non-viral vectors [10].

Recently, we have reported on the development of a safe, efficient nano-carrier system of carbonate apatite which can assist both intracellular delivery and release of DNA leading to very high level of trans-gene expression in cancer and primary cells [11-13]. We have also revealed a new approach of organic-inorganic hybrid carrier devised by complexing fibronectin and E-cadherin-Fc chimera electrostatically with nano-particles of carbonate apatite [14, 15]. Specific recognition to cell surface integrin and E-cadherin molecules through double ligand-coated nano-particles, resulted in synergistic acceleration of transgene delivery and consequential expression into embryonic stem cells. Instead of simultaneous mixing of DNA and cell-adhesive molecules in particle-preparation medium and subsequent incubation, step-

http://dx.doi.org/10.5772/ 53107

wise addition and incubation of DNA and the protein molecules, results in improved DNA loading and decreased particle diameter with ability of recognizing stem cell surface for more efficient transgene delivery. Activation of PKC which might up-regulate both integrin and E-cadherin, enhances transgene expression in mouse embryonic carcinoma cells.

2. Materials and Methods

2.1. Reagents

Plasmids, pGL3 (Promega) containing a luciferase gene under SV40 promoter and pEGFP-N2 (CLONTECH Laboratories, Inc.) having a green fluorescence protein gene under CMV promoter were propagated in the bacterial strain XL-1 Blue and purified by QIAGEN plasmid kits. Lipofectamine 2000 and DMEM were purchased from Invitrogen and Gibco BRL, respectively. Fibronectin was bought from Sigma and expression as well as purification of E-cad-Fc fusion proteins was done according to the previously described report [16].

2.2. Cell Culture

HeLa cells were cultured in 75-cm^2 flasks in Dulbecco's modified Eagle's medium (DMEM, Gibco BRL) supplemented with 10% fetal bovine serum (FBS), 50 μg penicillin ml^{-1}, 50μg streptomycin ml^{-1} and 100μg neomycin ml^{-1} at 37°C in a humidified 5% CO_2-containing atmosphere. F9, a mouse teratocarcinoma stem cell line and EB3, a mouse embryonic stem cell line were cultured in gelatin-coated 25-cm^2 flasks. F9 cells were maintained in Dulbecco's modified Eagle's medium (DMEM, Gibco BRL) supplemented with 10% fetal bovine serum (FBS) at 37°C in a humidified 5% CO_2-containing atmosphere. Feeder-free murine ES cells were maintained in KNOCKOUT-DMEM (Invitrogen), supplemented with 1 mM L-glutamine, 1% nonessential amino acids (Invitrogen), 0.1 mM β-mercaptoethanol (Sigma Chemical), 10% FBS and 1,000 units/ml leukemia inhibitory factor (LIF) (Chemicon). All media contained 50 μg/ml penicillin, 50 μg/ml streptomycin, and 100 μg/ml neomycin.

2.3. Transfection of cells

Cells from the exponentially growth phase were seeded at 50,000 cells per well into 24-well plates the day before transfection. 3 to 6 μl of 1 M $CaCl_2$ was mixed with 2 μg of plasmid DNA in 1 ml of fresh serum-free $HCO3^-$ - buffered (pH 7.5) medium (DMEM) and incubated for 30 min at 37^{0}c for complete generation of DNA/carbonate apatite particles. For generation of ECM protein-embedded carbonate apatite particles, fibronectin and E-cad-Fc proteins were added either alone or together to a final concentration of 5 μg/ml, to Ca^{2+} and DNA-containing DMEM followed by incubation at 37^{0}c for 30 min. Medium with generated ECM protein-associated or non-associated, DNA-containing particles was added with 10% FBS to the rinsed cells. After 4 hr incubation, the medium was replaced with serum supplemented medium and the cells were cultured for 1 day. Luciferase gene expression was monitored by using a commercial kit (Promega) and photon counting (TD-20/20 Luminometer, USA). Each transfection experiment was done in triplicate and transfection efficiency was expressed as mean light units per mg of cell protein. For lipofectamine-mediated transfection, protocol provided by Invitrogen was followed in a 24-well plate. Cells were incubated with DNA/lipofactamine complexes in serum-free media for 4 hr and like above, grown for 1 day after replacement with fresh serum media.

For transfection with calcium phosphate-DNA co-precipitation, briefly, 12 μg of plasmid DNA was added to 300 μl of a solution containing 250 mM $CaCl_2$. This solution was added to 300μl of a 2×HBS (50 mM Hepes, 140 mM NaCl, 1.5 mM $Na_2HPO_4.2H_2O$, pH 7.05) and mixed rapidly by gentle pipetting twice. The DNA/CaPi mixture was incubated at room temperature for the period of time indicated. After addition of 100 μl of the incubated mixture drop-wise to 1 ml serum supplemented media of each well, cells were incubated for 4 hr and like above, after replacement with fresh serum media, grown for 1 day.

2.4. MTT assay

HeLa cells were transfected and cultured for 1 day as described above. 30 μl of MTT solution (5mg/ml) was added to each well and incubated for 4 hrs. 0.5 ml of DMSO was added after removal of media. After dissolving crystals and incubating for 5 min at 37^{0}C, absorbance was measured in a microplate reader at 570 nm with a reference wavelength of 630 nm.

2.5. Chemical analysis

Following generation of carbonate apatite as described above, using 6 mM Ca^{2+} and no DNA, precipitated particles were lyophilized after centrifugation and washing with distilled deionized water. Other apatite particles generated as described above, were also similarly lyphilized. Calcium and phosphorus contents were determined using SPS1500VR Atomic Absorption Spectrophotometer. Carbon and fluorine were estimated by CHNS-932 (Leco, USA) and SX-elements micro analyser, YS-10 (Yanaco, Japan), respectively.

2.6. Infrared spectroscopy

Fourier transform-infrared spectroscopy of apatite particles prepared as described above, was performed using FT/IR-230, JASCO. The samples were ground in a mortar and approximately 1 mg was thoroughly mixed with 300 mg of ground spectroscopic grade KBr. Transparent pellets were prepared in a KBr die with an applied load of 8000 kg, under a vacuum of 0.5 torr.

2.7. X-ray diffraction

The x-ray diffraction powder reflections of the particles prepared as described above, were recorded using M18XHF-SRA diffractometer system.

2.8. Particle size measurements

For visualization by a scanning electron microscope (SEM), a drop of DNA-carbonate apatite suspension prepared according to the instructions in transfection protocol, was added to a carbon-coated SEM stage and dried, followed by observation by a high resolution SEM (S-800, Hitachi, Japan). Dynamic light scattering (DLS) measurement for particle suspension was carried out with a Super-dynamic Light Scattering Spectrophotometer, 'Photal' (Otsuka Electronics) at 75 mW Ar laser.

2.9. Confocal laser scanning microscopy

pGL3 vector was labeled with PI at a PI/DNA ratio of 1:1 and particles generated with this labeled plasmid (described in transfection protocol), were incubated with HeLa cells for 6 hours. Acidic compartments were labeled with 5μM LysoSensor, according to the instructions provided by Molecular Probes, and membrane-bound precipitates were removed by 5 mM EDTA in PBS before observation by LEICA TCS-NT.

2.10. SDS-PAGE and Western blotting

Following generation of carbonate apatite as described above using 3 mM Ca^{2+} and required amount of fibronectin or E-cad-Fc chimera and no DNA and centrifugation at 15000 rpm for 5 min at 4⁰C, precipitated particles were washed with water with several centrifugation steps to remove unbound proteins and dissolved with 50 mM EDTA in PBS for subsequent analysis by 7.5% SDS-PAGE in reducing condition. In order to see particle-bound fibronectin, after SDS-PAGE, the gel was stained with Coumassie blue, washed and dried. For detection of particle-associated E-cad-Fc, proteins after being run by SDS-PAGE were transferred to PVDF membrane (Immobilon, Millipore) and 80 mA current was applied for 90 min to complete transfer of the proteins. The PVDF membrane was washed with PBS (-)-containing 0.1% Tween 20 and then blocked for 1 hr at room temperature by "Blocking One" (Nacalai Tesque, Japan). The membrane was incubated with horseradish peroxidase (HRP)-conjugated anti-mouse IgG for 1 hr and washed with PBS-T three to four times to completely remove non-specific interactions. Enhanced chemiluminescence system (Amersham Bioscience) was used for visualization.

2.11. DNA labelling, fluorescence microscopy and flow cytometry

Plasmid DNA was labelled non-covalently with propidium iodide (PI) using 1:1 weight ratio of DNA to PI in the particle preparation medium. Labelled DNA inside the cells was observed by a fluorescence microscope (Olympus-IX71), following 4 hr incubation of differentially formulated particle suspensions with F9 cells and removing extracellularly bound particles by 5 mM EDTA in PBS. For flow cytometric analysis using FACS Calibur (Becton, Dickinson and Company), 1 day after transfection with pEGFP plasmid DNA, F9 cells were collected in a sorter buffer following treatment with trypsin-EDTA and repeated centrifugation and washing of the resulting cell pellet with PBS (-) (2 times).

3. Results and Discussion

3.1. Generation of carbonate apatite particles

Addition of only 3 mM Ca^{2+} to the HCO_3^- - buffered cell culture medium (DMEM or Willium E, pH 7.5) and incubation at 37⁰C for 30 min, resulted in microscopically visible particles. Generation of these particles only in HCO_3^--, but not in Hepes-buffered media or solution (pH 7.5) containing the same amount of exogenous Ca^{2+} and phosphate (0.9 mM), indicates the possible involvement of carbonate along with phosphate and Ca^{2+} in particle formation. Elemental analysis proved the existence of C (3%), P (17%) and Ca^{2+} (32%) and FT-IR spectra (Fig. 1a) identified carbonate, as evident from the peaks between 1410 and 1540 cm^{-1} and at approximately 880 cm^{-1}, along with phosphate in the particles, as shown by the peaks at 1000-1100 cm^{-1} and 550-650 cm^{-1}. X-ray diffraction patterns (Fig. 1b) indicated less crystalline nature, represented by broad diffraction peaks of the particles, compared to that of hydroxyaptite (Fig. 1c) – an intrinsic property of carbonate apatite [12].

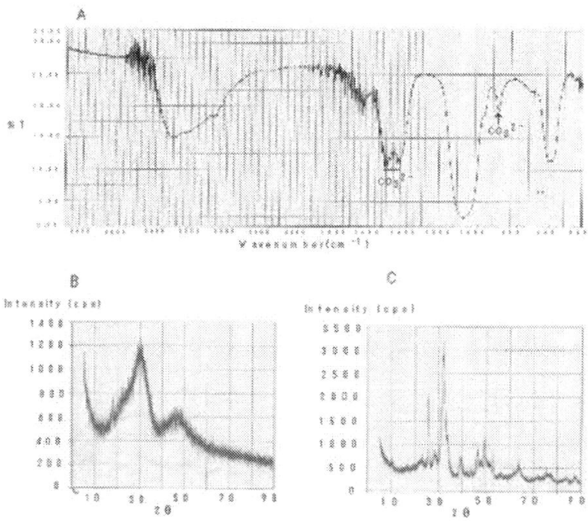

Figure 1. Infrared spectra of generated carbonate apatite (A) and X-ray diffraction patterns of carbonate apatite (B) and hydroxyapatite (C).

3.2. Influences of pH and temperature on generation of effective particles of carbonate apatite

We have investigated a long range of pH (7.0 to 7.9) of the HCO_3^-–buffered medium as well as incubation temperatures (25 °C to 65 °C) in order to make particles by exogenously added Ca^{2+} and subsequently transfect HeLa cells using the generated particles. Interestingly, the optimal Ca^{2+} concentrations required for generation of effective number of DNA/carbonate apatite particles leading to the high transfection efficiency, were inversely related to the pHs of the media (Fig. 2-a) and the incubation temperatures (Fig. 2-b). Thus, while 4 mM Ca^{2+} was sufficient to induce particle formation at pH 7.4 by incubating the Ca^{2+}-supplemented buffered medium for 30 min at 37 °C, only 1 mM^{2+} was enough to stimulate particle generation to the similar level at pH 7.9. Like pH, incubation temperatures have also profound and

http://dx.doi.org/10.5772/53107

sensitive effects on particle formulation and subsequent trans-gene delivery. Thus, at the incubation temperature of 37 °C, 3 mM Ca^{2+} was able to induce the proper "supersaturation" whereas at 65 °C, only 1 mM Ca^{2+}could stimulate "supersaturation" development to a similar extent – a prior need for generation of the particles. The decline below the high efficiency level of transfection was due to the formation of too few particles (microscopically observed) since increase in pH or temperature contributed to the development of "supersaturation" by increasing the ionization of phosphate and carbonate in the solution. The new system of particle synthesis is, therefore, very flexible since it allows us to make particles at a wide range of pH and temperatures. The analysis also indicates that induction of "supersaturation" as required for particle formation, can be delicately controlled by manipulating the parameters.

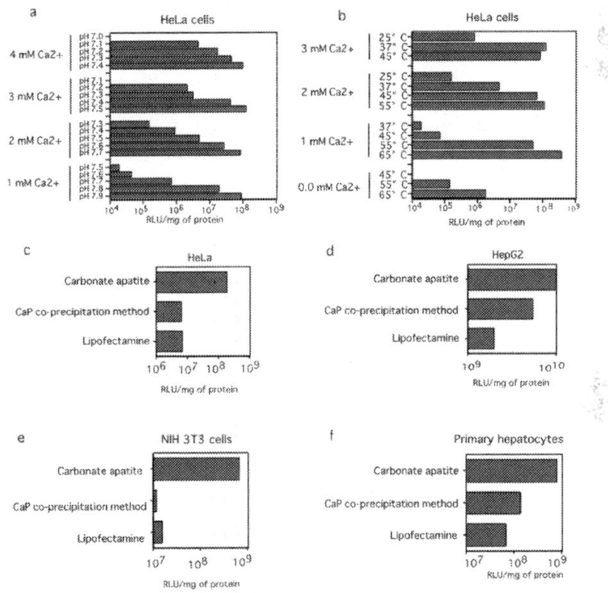

Figure 2. Regulation of trans-gene delivery and expression facilitated by carbonate apatite particles. Regulation of trans-gene expression by the nanoparticles of carbonate apatite generated at a wide range of pH and temperature. DNA/carbonate apatite particles were generated by addition of 1 to 4 mM Ca^{2+} and 2 µg plasmid DNA to 1 ml HCO_3^- (40 mM)-buffered DMEM medium with a pH range from 7.0 to 7.9, followed by incubation for 30 min either at 37 °C (a) or at 25 °C to 65 °C (b). Transfection of HeLa cells, HepG2, NIH3T3 and primary hepatocytes was performed in the same manner as mentioned in 'Materials and methods' section.

3.3. Tranfection efficiency and cell viability assessment

To evaluate the role of carbonate apatite as a powerful carrier of genetic material, we compared transfection efficiency of different techniques including two frequently used ones- CaP co-precipitation method and lipofection. In HeLa cell, for example, luciferase expression level for carbonate apatite-mediated transfection was over 25-fold higher than for lipofection and CaP co-precipitation method (Fig. 2c). Nano gram level of DNA was even sufficient for efficient transgene expression (Fig. 2c). Transfection efficiency was also significantly high in HepG2 (Fig. 2d), NIH 3T3 cells (Fig. 2e) and mouse primary hepatocytes (Fig. 2f). We performed MTT assay in HeLa cells (not shown here) to clarify that high transfection efficiency was accompanied by high viability of the cells [12].

3.4. Estimation of particle sizes and cellular uptake of particle-associated plasmid DNA

To explore why carbonate apatite is so efficient as a vector for gene delivery, we investigated two basic properties of carbonate apatite [12]. Carbonate, when present in the apatite structure, limits the size of the growing apatite crystals and increases the dissolution rate [12]. We carried out scanning electron microscopic observation of generated carbonate apatite (Fig. 3A) which revealed reduced growth of the crystals, most of which had diameters of 50 to 300 nm. We verified this size limiting effect of carbonate by observing cellular uptake of the PI (propidium iodide)-labeled plasmid DNA adsorbed to the apatites, since large particles are phagocytosed less efficiently than small ones [12]. DNA was carried into the cells by carbonate apatite (Fig. 3B-c) at least 10 times more efficiently than hydroxyapatite, generated by 1 min incubation (Fig. 3B-d). Longer period (30 min) incubation resulted in large hydroxyapatite particles [12], showing significantly reduced transfection efficiency [12] (Fig. 2A) due to extremely low cellular uptake of DNA [12]. Our findings, therefore, clearly suggest that carbonate apatite is superior over hydroxyapatite for its intrinsic property of preventing crystal growth, leading to high efficiency cellular uptake of DNA.

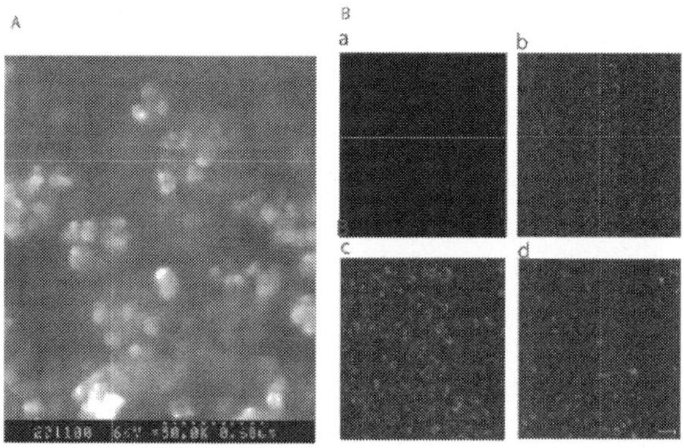

Figure 3. A, scanning electron microscopy, showing limited growth of generated carbonate apatite cryatals. Scale bar, 600 nm. B, cellular uptake of PI-labeled plasmid DNA associated with carbonate apatite and hydroxyapatite. a, no uptake of DNA (control), since endocytosis was blocked by energy depletion (50 mM 2-deoxy glucose and 1 mM Na-azide). DNA/carbonate apatite particles were prepared in 1 ml serum-free media (described in legend to Fig. 3) using 6 mM Ca^{2+} and 2 µg DNA. 40 ng (b) and 200 ng (c) of DNA in 20 µl and 100 µl of 1ml suspension respectively, were allowed for cellular uptake for 4 hr. d, 2 µg of DNA adsorbed to hydroxyapatite (described in experimental protocol) was allowed for uptake for the same period of time. Bar indicates 50 µM.

http://dx.doi.org/10.5772/53107

3.5. Endosomal escape of plasmid DNA carried by nanoparticles

To evaluate the role of endosomal escape of DNA in transgene expression, following endocytosis of PI-labeled plasmid DNA, we labeled endosomes with LysoSensor (a fluorescence probe for endosomes). Following 6 hr of DNA uptake by cells, a significant portion of DNA (red colour) appeared to be released from endosomes (green colour) after colocalization of plasmid DNA with endosomes (Fig. 4).

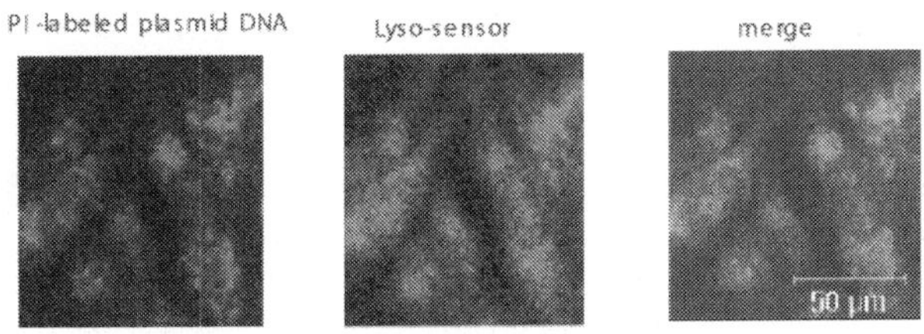

Figure 4. Endosomal escape of endocytosed PI–labeled DNA, as evident after colocalization with a fluorescence probe (Lyso-Sensor) for endosomes.

3.6. Relationship of endosomal pH and crystalline properties of particles affecting transfection

Treatment with bafilomycin A1, a specific inhibitor of v-ATPase (a proton pump for acidification of endocytic vesicles) resulted in drastic reduction of transfection efficiency in HeLa cells (Fig. 5A), which indicated that acidic environment might be necessary for solubilization of carbonate apatite to release DNA from the apatite. To establish this notion, we generated fluoridated carbonate apatite to see the effect of solubility of the particles on transfection efficiency, since incorporation of fluoride reduces the solubility of the apatite [12]. Surprisingly, transfection efficiency was reduced gradually to a significant extent (100 fold) with increasing fluoride level in carbonate apatite (Fig. 5B).

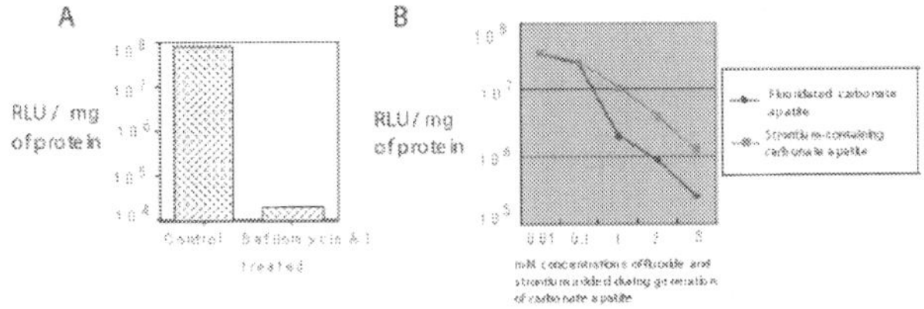

Figure 5. A, Effect of bafilomycin A1 (an inhibitor of v-ATPase) on transfection. HeLa cells were incubated with DNA/carbonate apatite particles and 200 nM bafilomycin A1 for 6 hr. After washing with 5 mM EDTA in PBS, cells were grown for 1 day and luciferase expression was detected. B, Changes in luciferase expression for increasing concentrations of F^- (0.01 to 3 mM) and strontium (0.01 to 3 mM) added during generation of DNA/carbonate apatite particles.

To establish a relationship between transfection efficiency and dissolution rates of the apatites, turbidity (320 nm) measurement was done as an indicator of their solubilization, following an acid load in solution of generated apatites. Carbonate apatite generated in presence of increasing concentrations of NaF, showed gradual decrease in dissolution rates, as evident from changes in turbidity, following adjustment of pH from 7.5 to 7.0 with 1 N HCl (Fig. 6 A, B), which is consistent with gradually reduced transfection efficiency of fluoridated carbonate apatites (Fig. 5 B). With decreasing pH from 7.0 to 6.8, carbonate apatite was completely solubilized within 1 min, whereas fluoridated carbonate apatite was partially dissolved (Fig. 6 B).

To examine whether dissolution rates of apatites are correlated with their degree of crystallization, we studied x-ray diffraction of the apatites (Fig. 7), which clearly indicates that apatite with higher degree of crystallization, had lower solubility (Fig. 6A). In other words, apatites with higher crystallinity (Fig. 7) showed lower transfection efficiency (Fig. 5 B). The gradual increase in crystallinity owing to increased level of incorporated fluoride in carbonate apatite (Fig. 7) resulted in gradual decrease in transfection efficiency (Fig. 5B).

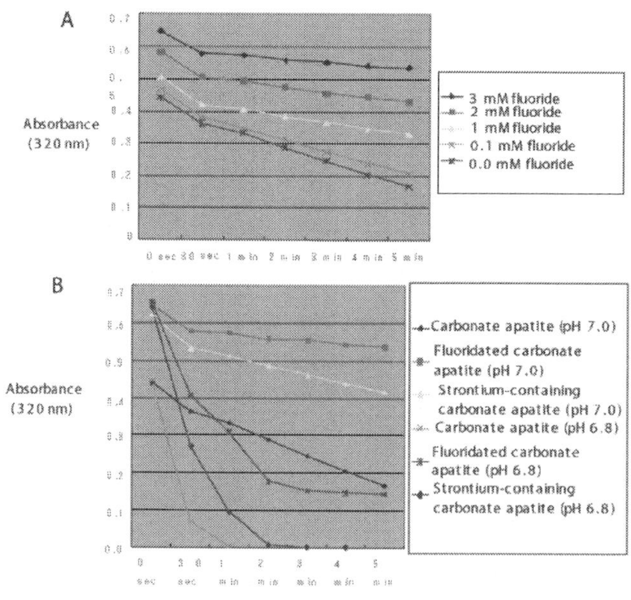

Figure 6. A, Dissolution rates (at pH 7.0) of fluoridated carbonate apatites prepared by addition of 0-3 mM F^- during generation of carbonate apatite at pH 7.5 (described in experimental protocol), were studied by turbidity measurement at 320 nm of apatite suspensions just after being adjusted to the pH 7.0 with 1 N HCl. B, Dissolution rates of carbonate apatite, fluoridated carbonate apatite and strontium-containing carbonate apatite at pHs of 7.0 and 6.8.

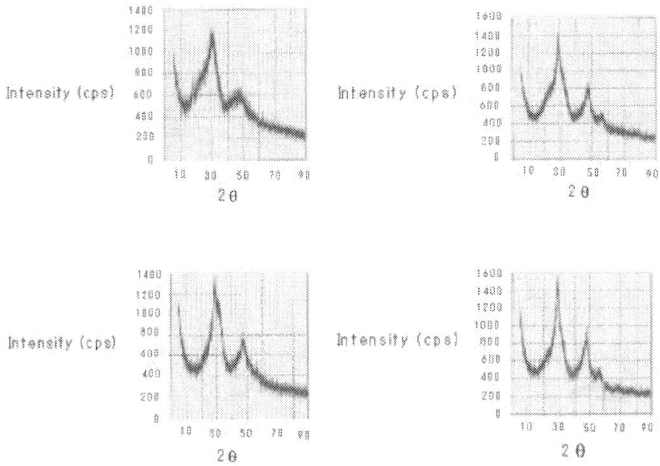

Figure 7. X-ray diffraction patterns of carbonate apatite (A) and fluoridated carbonate apatites, containing 0.65% (B), 1.43% (C) and 2.5% (D). Carbonate apatite was generated by addition of 6 mM Ca^{2+} and fluoridated carbonate apatites by addition of 1 mM (B), 2 mM (C) and 3 mM (D) NaF along with 6 mM Ca^{2+} to HCO_3^- -buffered medium (pH7.5), followed by incubation at 37°C

To establish that decreased transfection efficiency was only due to decreased solubility of fluoridated carbonate apatite, but not by any other fluoride-mediated effects, we examined the effects of strontium which, when incorporated into carbonate apatite, is known to improve the crystallinity and reduce the solubility of the apatite, but to a lesser extent than fluoride [12]. As expected, addition of strontium chloride during preparation of carbonate apatite reduced its dissolution rate but to a level less than that observed for fluoride (Fig 6B). Moreover, transfection efficiency was gradually decreased with increasing concentrations of strontium chloride during generation of DNA/carbonate apatite particles (Fig. 5B). Taken together, our findings suggest that intracellular release of DNA through dissolution of apatite should play a major role in carbonate apatite-mediated transfection.

3.7. Immobilization of cell-adhesive molecules on nano-particle surface

Since embryonic stem cells produce substantial amount of fibronectin-specific integrins as well as E-cadherin as transmembrane proteins [17], we hypothesized that if the nano-particles of carbonate apatite could be complexed with fibronectin and E-cadherin, either individually or together, they might recognize in the immobilized state the corresponding receptors on cell surface in order to facilitate quick internalization of the composite particles across the plasma membrane through endocytosis. These nano-apatite particles possess anion- and cation-binding domains which enable them to bind to both acidic and basic amino acids of protein molecules [18, 19]. On the other hand, fibronectin as well as E-cadherin are rich in acidic amino acid residues [19, 20] which make them excellent candidates for possible binding with the apatite particles. We have examined whether these "cell adhesive molecules" could, in deed, bind to the particles, by SDS-PAGE and Western blot analysis, following generation of apatite-protein composites and decomplexation through EDTA-mediated particle dissolution. Whereas binding affinity of the particles for fibronectin was relatively lower requiring higher amount of initially added fibronectin as observed by SDS-PAGE, almost all E-cad-Fc was found to be associated with the particles as verified by Western blot analysis (Fig 8). Very high affinity for E-cadherin could be interpreted by the previous report that E-cadhein has many exposed acidic residues in several loop structures responsible for binding divalent cation Ca^{2+} [20].

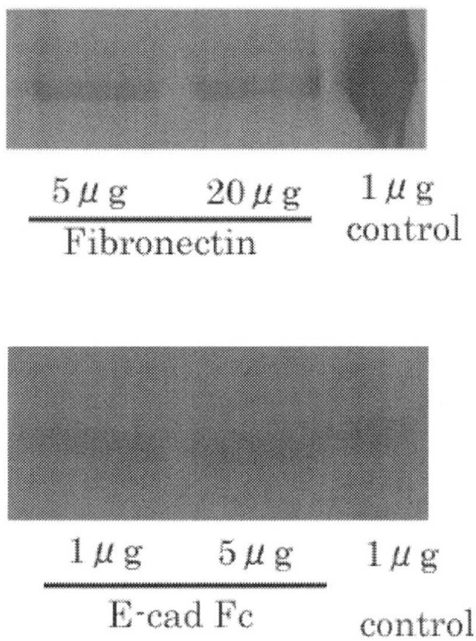

Figure 8. Analysis of the binding of cell-adhesive proteins to nano-particles. Particles were prepared by addition of 3 μl of 1 M $CaCl_2$ and 5 to 20 μg of fibronectin or 1 to 5 μg of E-cad-Fc to 1 ml bicarbonate-buffered DMEM and incubation for 30 min at 37°C. Generated particles were centrifuged at 15000 rpm for 5 min and washed 2 times with H_2O to remove the unbound proteins, followed by EDTA treatment to dissolve the particles. SDS-PAGE and Western blot analysis were performed in order to see, respectively, particle-associated fibronectin (A) or E-cad-Fc (B).

3.8. Enhanced cellular uptake of DNA by immobilized cell-adhesive molecules

In order to explore whether apatite particles functionalized with fibronectin and E-cad-Fc can facilitate enhanced delivery of apatite-associated plasmid DNA across the plasma membrane, we examined cellular uptake of the DNA labeled with propidium iodide (PI) [19], following 4 hr incubation of F9 cells with various particle formulations. As shown in Fig. 9, while with only apatite particles, delivery of PI-labeled DNA into the cells was extremely low, complexation of the particles with either fibronectin or E-cad-Fc resulted in significantly improved DNA delivery, suggesting that immobilized fibonectin or E-cad-Fc retained their functionalities in order to recognize specific cell surface integrin or E-cadherin, respectively for enabling subsequent internalization of the whole particle composite through endocytosis [19, 21]. Moreover, the apatite particles when complexed with both fibronectin and E-cad-Fc, demonstrated more pronounced DNA delivering activity compared to the particles embedded with either fibronectin or E-cad-Fc, indicating a synergistic effect of the multifunctional particles on endocytosis through simultaneous recognition of extracellular domains of specific integrin as well as E-cadherin molecules.

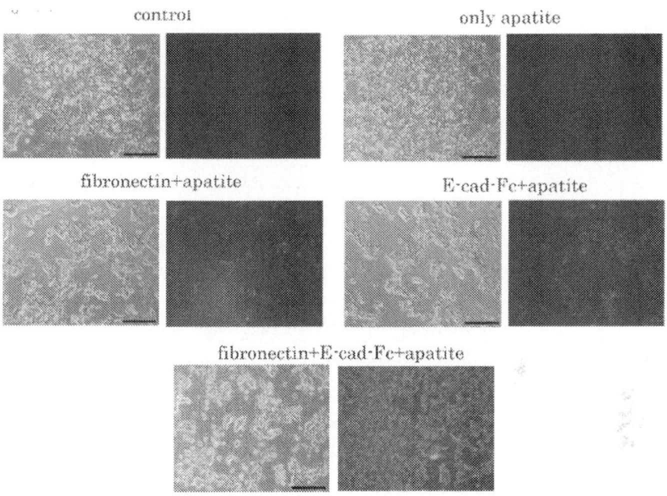

Figure 9. Effects of particle-immobilized proteins on cellular internalization of plasmid DNA. Particles were prepared by addition of 3 µl of 1 M $CaCl_2$, 2 µg of PI-labelled plasmid DNA and 5 µg of fibronectin and/or 5 µg of E-cad-Fc to 1 ml bicarbonate-buffered DMEM and incubation for 30 min at 37°C. F9 cells were incubated with the generated particles for 4 hr, washed with 5 mM EDTA in PBS and visualized by a fluorescence microscope (scale bar, 50 µm).

3.9. Quantitation and validation of trans-gene expression facilitated by cell-adhesive molecules

Since expression of a trans-gene is the result of overcoming a number of barriers including entry into the cells, release from the particle and endosomes and finally nuclear translocation [17, 18], we have investigated whether improved DNA delivery as a result of integrin- and E-cadherin-mediated endocytosis of composite particles (Fig. 9) contributed to the similar extent to final protein expression. Quantitative luciferase expression analysis indicated that particles complexed with fibronectin or E-cad-Fc promoted trans-gene expression with a value which was almost 20 times higher than that achieved for the particles only (Fig 10). A prior optimization study demonstrated that 1 to 5 μg/ml of fibronectin as well as E-cad-Fc conferred the best transfection efficiency and was, therefore, maintained for all subsequent experiments. Finally, synergistic activity of fibronectin and E-cad-Fc which caused huge cellular uptake of DNA (Fig. 9), further accelerated gene expression efficiency with a value almost 3 times higher than that observed for commercially available lipofectamine (Fig. 10). With increasing the total amount of initially added DNA up to 4 μg, a further increase in trasfection efficiency was observed (data not shown here) possibly due to the higher loading of DNA into the crystals with the consequence of more DNA getting inside the cells. The high level of expression could directly be observed by fluorescence microscopy which demonstrated many GFP-expressing F9 cells (Fig. 11). Fluorescence Activated Cell Sorting (FACS) analysis demonstrated that almost 60% cells were GFP-positive following transfection with the particles carrying, in addition to pEGFP plasmid DNA, both fibronectin and E-cad-Fc (Fig 11). MTT assay was performed in F9 cells to clarify that high transfection efficiency was not accompanied by significant toxicity of the cells (data not shown here). In order to establish that such organic-inorganic hybrid particles promote trans-gene delivery and expression through specific interactions with cell-surface molecules (integrin or E-cadherin), we added increasingly high amounts of free fibronectin to the preformed particle suspension carrying both fibronectin and E-cadherin and incubated with the cells for the same period of time (4 hr) as followed in usual transfection procedure. Transfection efficiency decreased as the concentration of free fibronectin increased from 5 to 100μg/ml, indicating the involvement of specific interactions between immobilized fibronectin and the corresponding specific integrin receptors (Fig 12). At a sufficiently high concentration (300 μg/ml), free fibronectin drastically reduced luciferase expression suggesting that high amount of fibronectin molecules not only saturate their specific integrins and block binding of immobilized fibronectin needed for particle internalization, but also shield cell-surface E-cadherin and prevent specific binding of particle surface-embedded E-cad-Fc chimera leading to very low cellular uptake of particle-associated DNA and diminished luciferase expression. Since embryonic stem cells are the final target for genetic modification in regenerative medicine, we applied the new transfection approach to mouse embryonic stem cells. As shown in Fig. 13, only apatite particles were extremely inefficient in transfecting the cells, whereas fibronectin-bound particles to some extent promoted GFP expression and fibronectin and E-cad-Fc-bound particles to a significant extent accelerated trans-gene expression, thus proposing that the synergistic effect is a universal way of accelerating trans-gene delivery and expression using inorganic nano-particle-associated cell recognizable proteins. Quantitative luciferase expression in embryonic stem cells indicated that particles complexed with fibronectin and E-cad-Fc individually, promoted trans-gene expression with efficiency approximately 9 and 7 times higher, respectively, than that achieved with the particles only (Fig. 14). However, when the particles were associated with both fibronectin and E-cadherin-Fc, a synergistic effect resulted in remarkable level of transgene expression leading to almost 40 and 28 times higher efficiency than that obtained by apatite particles and widely used lipofectamine 2000 system [14].

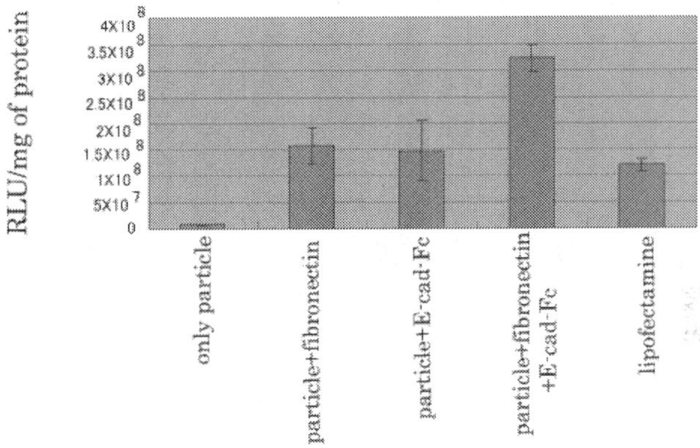

Figure 10. Comparison of luciferase expression for differentially formulated particles. Particles were prepared by addition of 3 µl of 1 M $CaCl_2$, 2 µg of luciferase plasmid DNA and 5 µg of either fibronectin, E-cad-Fc or both to 1 ml bicarbonate-buffered DMEM and incubation for 30 min at 37°C. F9 cells were incubated with the generated particles for 4 hr and after replacement of particle-containing media with fresh media, further incubated for 1 day in order to quantitate luciferase expression. Transfection efficiency was normalized after estimation of total proteins in cell lysate.

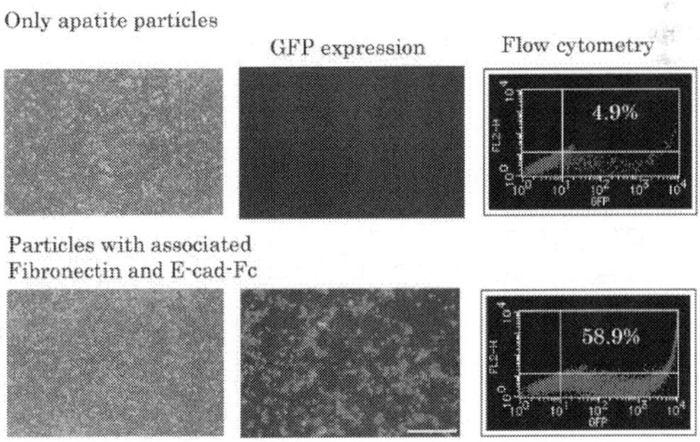

Figure 11. Comparison of GFP expression between only particles and fibronectin/E-cad-Fc-embedded-particles. Particles were prepared by addition of 3 µl of 1 M $CaCl_2$, 2 µg of pEGFP plasmid DNA and 5 µg of fibronectin and 5 µg of E-cad-Fc to 1 ml bicarbonate-buffered DMEM and incubation for 30 min at 37°C. F9 cells were incubated with the generated particles for 4 hr and after replacement of particle-containing media with fresh media, further incubated for 1 day in order to both observe and quantitate GFP expression by fluorescence microscopy and flow cytometry, respectively (scale bar, 50 µm).

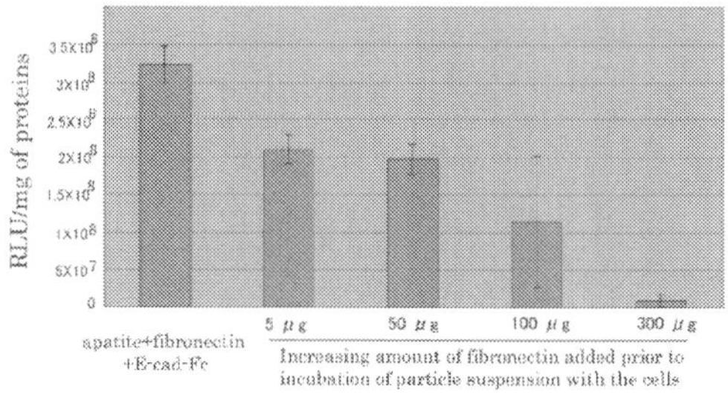

Figure 12. Blocking of integrin-mediated trans-gene delivery by excess free fibronectin. Particles were prepared by addition of 3 µl of 1 M $CaCl_2$, 2 µg of luciferase plasmid DNA and 5 µg of fibronectin and 5 µg of E-cad-Fc to 1 ml bicarbonate-buffered DMEM and incubation for 30 min at 37°C. F9 cells were incubated with the generated particles in presence or absence of increasingly high concentrations of free fibronectin for 4 hr and after replacement of particle-containing media with fresh media, further incubated for 1 day in order to quantitate luciferase expression. Transfection efficiency was normalized after estimation of total proteins in cell lysate.

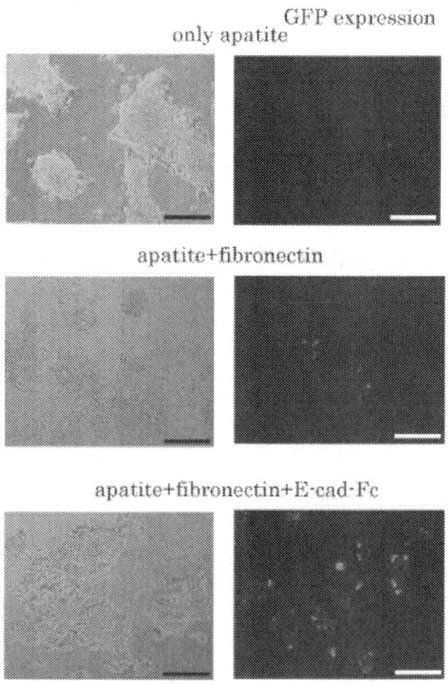

Figure 13. Enhancement of GFP expression in mouse embryonic stem cells with fibronectin/E-cad-Fc-embedded-particles. Particles were prepared by addition of 3 µl of 1 M $CaCl_2$, 2 µg of pEGFP plasmid DNA and 5 µg of fibronectin and 5 µg of E-cad-Fc to 1 ml bicarbonate-buffered DMEM and incubation for 30 min at 37°C. Embryonic stem cells were incubated with the generated particles for 4 hr and after replacement of particle-containing media with fresh media, further incubated for 1 day in order to see GFP expression by a fluorescence microscope (scale bar, 50 µm).

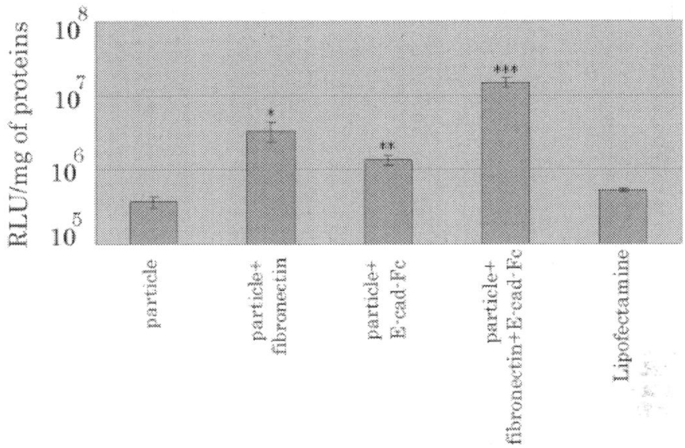

Figure 14. Comparison of luciferase expression for differentially formulated apatite particles and liposomes. Particles were prepared by addition of 3 µl of 1 M $CaCl_2$, 2 µg of luciferase plasmid DNA and 2 µg of either fibronectin, E-cad-Fc or both to 1 ml bicarbonate-buffered DMEM and incubation for 30 min at 37°C. Embryonic stem cells were incubated with the generated particles for 4 hr and after replacement of particle-containing media with fresh media, further incubated for 1 day in order to quantitate luciferase expression. Transfection efficiency was normalized after estimation of total proteins in cell lysate. Transfection by lipofectamine was performed according to the instructions provided by Invitrogen.

3.10. DNA binding with differentially formulated cell adhesive protein-embedded particles

Since direct mixing of DNA and cell-adhesive proteins in Ca^{2+} and PO_4^{3-}-containing medium prior to induction of particle formation by incubation at 37°C for 30 min, could interfere with maximum DNA loading due to the competitive binding of the proteins to the growing crystals, we investigated DNA binding efficiency by first adding DNA to the particle-preparation medium prior to time-dependent addition of the proteins [22]. As shown in Fig. 15, in the direct mixing process (control), DNA binding is much higher for E-cadherin-Fc compared to fibronectin, indicating that E-cadherin-Fc facilitates DNA loading probably by accelerating particle growth because turbidity of particle suspension was higher for E-cadherin-Fc than for fibronectin (not shown). It is worth mentioning that only particles have also higher affinity towards DNA (almost 40%) that the particles associated with fibronectin which showed lower turbidity than the particles (mentioned before), suggesting again that particle growth has a significant role in the observed DNA binding efficiency. When cell adhesive proteins were added after 5, 10 and 20 min from the start of incubation of DNA-containing particle-preparation medium, followed by incubation for an additional 25, 20 and 10 min respectively, DNA binding to the particles was enhanced to a significant extent for fibronectin, E-cadherin and fibronectin/E-cadherin-Fc compared to the control, suggesting than a competitive inhibition of DNA binding happens in the direct mixing procedure while delaying addition of the proteins to the growing crystals and DNA favors optimal DNA binding to the particles. Decreased DNA binding to the particles with which E-cadherin and fibronectin/E-cadherin-Fc were incubated for only 1 min, could be due to the reduced growth of the particles for too long time absence of E-cadherin-Fc in particle-preparation medium.

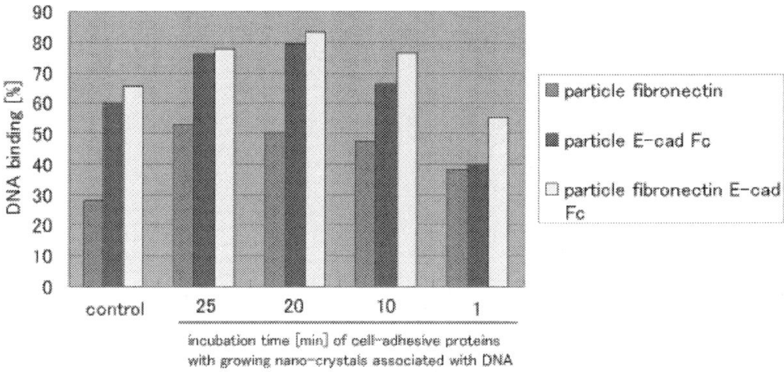

Figure 15. Binding affinities of DNA to differentially formulated cell adhesive protein-embedded particles. Particles in the control samples were prepared by addition of 3 µl of 1 M $CaCl_2$, 2 µg of luciferase plasmid DNA and 2 µg of either fibronectin, E-cad-Fc or both to 1 ml bicarbonate-buffered DMEM and incubation at 37°C for 30 min. Formation of the particles in experimental samples was done by addition of fibronectin, E-cadherin-Fc or both after 5, 10, 20 and 29 min from the start of incubation of DNA-containing particle preparation medium, followed by incubation for an additional 25, 20, 10 and 1 min respectively. F9 cells were incubated with the generated particles for 4 hr and after replacement of particle-containing media with fresh media, further incubated up to 1 day for quantitation of luciferase expression. Transfection efficiency was normalized after estimation of total proteins in cell lysate.

3.11. Size determination for differentially formulated cell adhesive protein-embedded particles

Particle growth kinetics is correlated to the size of the finally formed particles and excessive growth lead to big size particles being inefficient for intracellular DNA delivery [11]. Since E-cadherin-Fc favors particle growth by making bridges among the neighboring E-cadherin-anchored crystals [14], prolonged incubation together with DNA for generation of functional particles might lead to large complex particles. As shown in Fig. 16, fibronectin association maintained the average particle diameter close to 300 nm whereas E-cadherin-Fc or fibronectin/E-cadherin-Fc induced the particle growth with an average diameter of approximately 900 nm. However, addition of E-cadherin-Fc or fibronectin/E-cadherin-F after 5, 10, 20 and 29 min from the start of incubation of DNA-containing particle-preparation medium, followed by incubation for an additional 25, 20, 10 and 1 min respectively, resulted in the particles of decreasing sizes with a minimum average value of approximately 300 nm. On the other hand, time-dependent association of fibronectin having no role in particle growth induction, demonstrated no significant change in overall particle diameter, suggesting that particle growth is the size-determining factor for cell-adhesive protein-embedded particles.

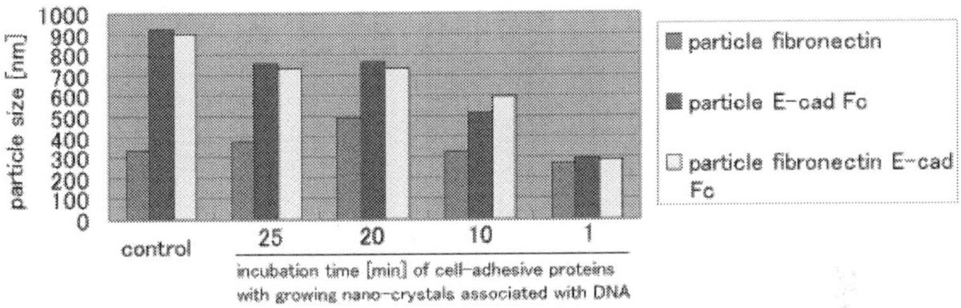

Figure 16. Estimation of sizes for differentially formulated cell adhesive protein-embedded particles. Following preparation of different particles as mentioned in the legend to Figure 4, dynamic light scattering (DLS) measurement was performed with a Super-dynamic Light Scattering Spectrophotometer.

3.12. Cellular delivery of DNA in association with cell adhesive protein-embedded particles

Both DNA binding to the particles and particle size contribute to the overall uptake of DNA by cells. As shown in Fig. 17, only particles were very inefficient in delivering propidium (PI)-labeled plasmid DNA into F9 cells whereas particles being associated with fibronectin or E-cadherin-Fc significantly increased cellular delivery of labeled DNA in a 4 hr uptake study. Moreover, particles when complexed with both fibronectin and E-cadherin-Fc in direct mixing with DNA, synergistically accelerated delivery of PI-labeled DNA into the cells. Particles prepared by addition of fibronectin or fibronectin/E-cadherin-Fc after 5, 10 and 20 min from the start of incubation of labeled DNA-containing particle preparation medium and incubation for an additional 25, 20 and 10 min respectively, mediated increased cellular delivery of labeled DNA, indicating that transgene delivery is well-controlled by the sizes as well as the DNA-loading efficiency of cell adhesive protein-embedded particles. Reduced DNA uptake level for the small size particles with which cell-adhesive proteins were incubated for a very short time (1 min) could be accounted for their inefficient binding with the cell-recognition molecules. The reason for low DNA uptake for the particles to which only E-cadherin-Fc was adsorbed in a time-dependent manner, is still not clear and might be related to the serum instability of the complex particles at the time of transgene delivery.

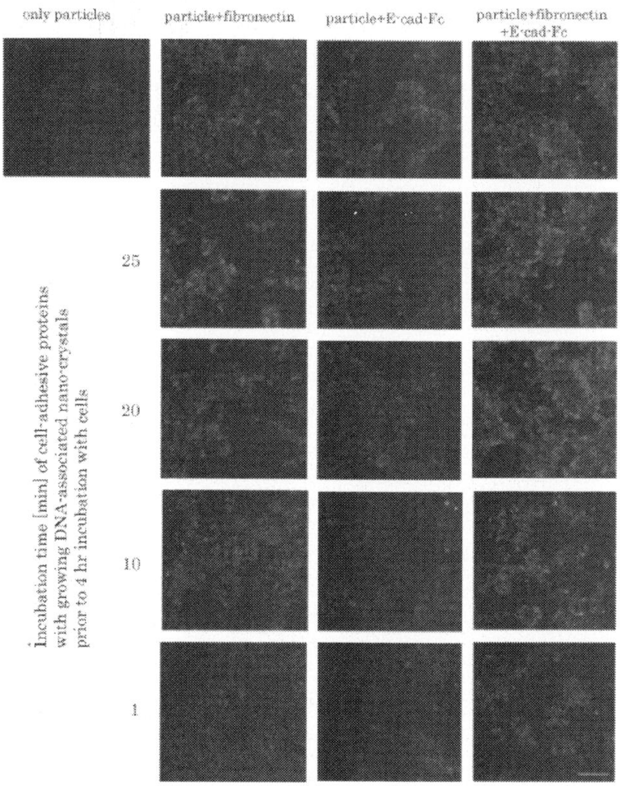

Figure 17. Differentially formulated cell adhesive protein-embedded particles for cellular delivery of DNA. Particles in the control samples were prepared by addition of 3 µl of 1 M $CaCl_2$, 2 µg of PI-labelled plasmid DNA and 2 µg of either fibronectin, E-cad-Fc or both to 1 ml bicarbonate-buffered DMEM and incubation at 37°C for 30 min. Formation of the particles in experimental samples was done by addition of fibronectin, E-cadherin-Fc or both after 5, 10, 20 and 29 min from the start of incubation of DNA-containing particle preparation medium, followed by incubation for an additional 25, 20, 10 and 1 min respectively. F9 cells were incubated with the generated particles for 4 hr, washed with 5 mM EDTA in PBS and visualized by a fluorescence microscope (scale bar, 100 µm).

3.13. Transfection efficiency achieved with cell adhesive protein-embedded particles

Since transgene expression is the result of overcoming a number of barriers including entry into the cells, release from the particles and endosomes, and finally nuclear translocation [11], we checked whether accelerated DNA delivery owing to the improved DNA loading capacity and smaller sizes of fibronectin and E-cadherin-Fc-anchored carbonate apatite particles, contributed to the similar extent to final protein expression (Fig. 18). Quantitative luciferase expression demonstrated that particles generated by addition of fibronectin and fibronectin/E-cadherin-Fc after 5 min from the start of incubation of DNA-containing medium and incubation for an additional 25 min, enhanced 2 and 3-fold higher transgene expression than the control samples prepared by direct mixing with DNA. This is a significant achievement considering the high expression level already achieved with control samples [15]. The decline in luciferase expression for other samples is consistent with the low efficiency of DNA delivery as described before.

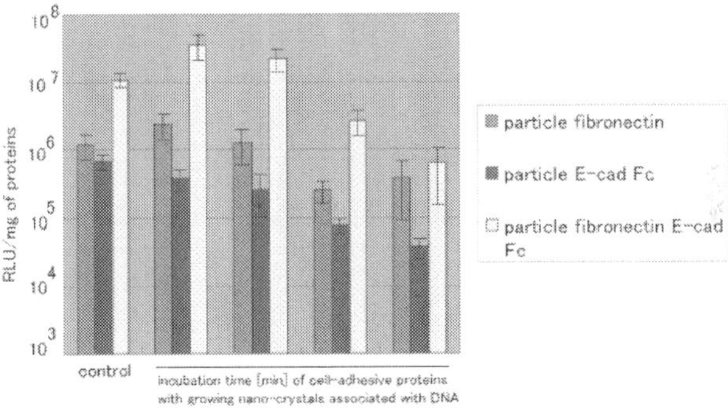

Figure 18. Intracellular expression of luciferase gene delivered by differentially formulated cell adhesive protein-embedded particles. Particles in the control samples were prepared by addition of 3 µl of 1 M $CaCl_2$, 2 µg of PI-labelled plasmid DNA and 2 µg of either fibronectin, E-cad-Fc or both to 1 ml bicarbonate-buffered DMEM and incubation at 37°C for 30 min. Formation of the particles in experimental samples was done by addition of fibronectin, E-cadherin-Fc or both after 5, 10, 20 and 29 min from the start of incubation of DNA-containing particle preparation medium, followed by incubation for an additional 25, 20, 10 and 1 min respectively. F9 cells were incubated with the generated particles for 4 hr and after replacement of particle-containing media with fresh media, further incubated up to 1 day in order to quantitate luciferase expression. Transfection efficiency was normalized after estimation of total proteins in cell lysate.

3.14. Role of protein kinase C on immobilized fibronectin and E-cad-Fc-mediated gene delivery

Since protein kinase C (PKC) in "inside-out" signaling cascade enhances integrin affinity towards ECM proteins promoting cell adhesion and spreading [23, 24] and up regulates endocytosis and recycling of E-cadherin [21], we have investigated the effect of Phorbol 12-myristate 13-acetate (PMA), a specific activator of PKC on trans-gene delivery mediated by particle-immobilized fibronectin and E-cadherin-Fc. As shown in Fig. 19, while only carbonate apatite particles are very inefficient in transfecting F9 cells even in presence of increasing doses of PMA (0 to 100 nM), fibronectin- or E-cad-Fc-embedded particles showed significant increment in luciferase gene expression (2 to 10 times) depending on PMA concentrations. Surprisingly, particles when associated with both of the "cell adhesive molecules" remarkably enhanced trans-gene expression resulting in almost 8, 14, 20 and 92-fold higher efficiency due to the presence of PMA at 1, 10, 50 and 100 nM concentrations, respectively. Immobilization of either fibronectin or E-cad-Fc on the particles also showed a dramatic increment in transgene expression, indicating clearly that both of the transmembrane proteins integrin and E-cadherin are up-regulated in response to PKC activation to promote efficient internalization of the bio-functional particles across the plasma membrane (data not shown here) and subsequent expression of the particle-associated DNA in cytoplasm [27] (Fig 19).

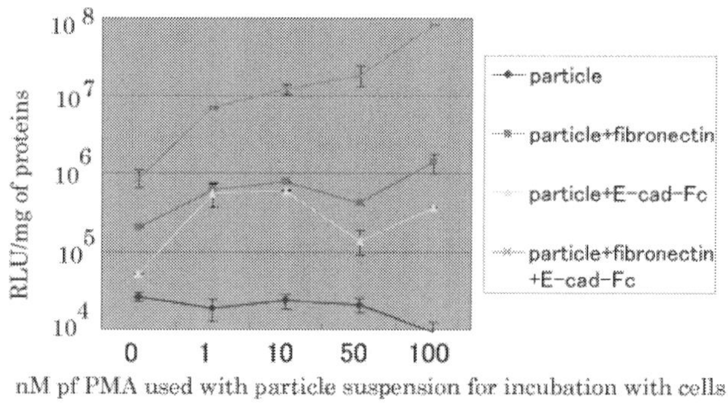

Figure 19. Effects of PMA on trans-gene expression mediated by fibronectin/E-cad-Fc-embedded-particles. Particles were prepared by addition of 3 μl of 1 M $CaCl_2$, 2 μg of luciferase plasmid DNA and 2 μg of either fibronectin, E-cad-Fc or both to 1 ml bicarbonate-buffered DMEM and incubation for 30 min at 37°C. F9 cells were incubated with the generated particles in presence of increasingly high concentrations of PMA (o to 100 nM) for 1 hr and after replacement of particle- and PMA-containing media with fresh media, further incubated for 1 day in order to quantitate luciferase expression.

3.15. Transfection efficiency achieved in leukemia cells with cell adhesive protein-embedded particles

T cell expresses on its membrane α4β1 and α5β1 integrins which can bind fibronectin during lymphocyte adhesion and migration from vascular compartment to the injured tissues [22]. Moreover, $\alpha_E\beta7$ integrin on some T cells can interact with epithelial E-cadherin for tissue-specific retention of lymphocytes [22]. We, therefore, aimed to functionalize the surface of DNA-associated nanocrystals with fibronectin and E-cadherin-Fc for transgene delivery through integrin-mediated endocytosis [22].

As shown in Fig. 20, luciferase expression in Jurkat cells was significantly low after delivery of luciferase gene-containing plasmid DNA with the help of carbonate apatite particles. A 3-fold enhancement in transgene expression was observed following delivery with fibronectin-embedded particles. Transgene expression could be further increased to the level (up to 6 times) equivalent to that of lipofection with the particles complexed with both fibronectin and E-cadherin-Fc. Since lymphocytes posses 2 different types of integrins (α4β1 and α5β1) being able to bind fibronectin [22], particles with electrostatically associated fibronectin could recognize any of the two receptors for efficient endocytosis in Jurkat cells leading to high transgene expression. However, particles with adsorbed E-cadherin-Fc reduced transfection efficiency below the level obtained with particles only, indicating that binding of E-cadherin-Fc probably neutralizes the positive charges of the particles as required for their subsequent interaction with anionic cell surface and additionally, E-cadherin-Fc on the particle surface might have low affinity interaction with the cell membrane integrin ($\alpha_E\beta7$). On the other hand, the highest gene expression obtained with the particles complexed with both fibronectin and E-cadherin-Fc could be interpreted by the strong affinity of the composite particles towards the cell membrane due to the specific and synchronized recognition of the two different ligands on particle surface to their corresponding integrin receptors on plasma membrane, resulting in fast endocytosis of the particles along with DNA.

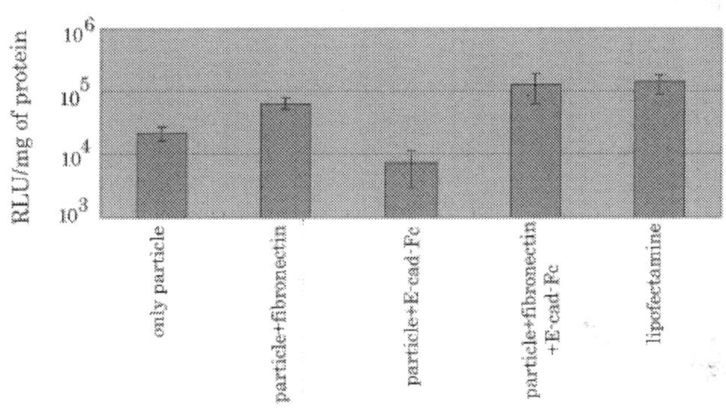

Figure 20. Comparison of luciferase expression among differentially formulated apatite particles. Particles were prepared by addition of 3 µl of 1 M $CaCl_2$, 2 µg of luciferase plasmid DNA and 2 µg of either fibronectin, E-cad-Fc or both to 1 ml bicarbonate-buffered DMEM and incubation for 30 min at 37°C. Jurkat cells were incubated with the generated particles for 1 day followed by quantitation of luciferase expression. Transfection efficiency was normalized after estimation of total proteins in cell lysate. Transfection by lipofectamine was performed according to the instructions provided by Invitrogen. Reproducibility of the result was established by performing same the experiment in another day.

4. Conclusions

Stem cells possessing the inherent capability of transforming into many cell types, have been shown tremendous potential for cell-based therapies in regenerative medicine for neurological disease or injury [28], diabetes [29] and myocardial infarct [30]. The *in vitro* differentiated derivatives of stem cells are thought to be able to repair or replace damaged cells, tissues or organs. However, compared to embryonic stem cells, adult stem cells are likely more difficult to be implemented into useful therapies considering their limited pluripotency. Transgene delivery could be a powerful strategy for specific differentiation of embryonic stem cells since several transcription factors have been demonstrated to regulate stem cell differentiation to specific cell types of heart, pancreas, liver and neurons [31-36]. On the other hand, tumor cells such as leukemia and lymphoma cells are obvious and attractive targets for gene therapy. Gene transfer and expression for cytokine and immunomodulatory molecules in various kinds of tumor cells have been shown to mediate tumor regression and anti-metastatic effects [37-40]. Moreover, genetically modified leukemia cells expressing co-stimulatory molecules or cytokines are likely to have significant therapeutic roles for patients with leukemia [41, 42].

Among the existing approaches for transgene delivery, viral systems suffer from their potential life-threatening effects of immunogenicity and carcinogenicity whereas non-viral ones, although safe, possess significant limitation in terms of efficacy [43]. Development of a safe as well as an efficient carrier is, therefore, an urgent requirement for effective implementation of the stem cells in regenerative medicine and the leukemia (or lymphocytes) in cancer treatment.

We have established a novel type of non-viral gene delivery systems based on pH-sensitive inorganic nanoparticles and revealed an innovative strategy for surface-functionalization of these biodegradable nanoparticles through their ionic interactions with "cell-adhesive molecules". Moreover, the new approach has directly been applied for highly efficient delivery and expression of a trans-gene into "hard-to-transfect" embryonic stem cells- a success with tremendous future for stem-cell based therapeutic development. The involvement of E-cadherin and fibronectin in intercellular and extracellular interactions of cultured undifferentiated embryonic stem cells may exclude the possibility of stem cell differentiation following transfection with the new nano-apatite carriers associated with E-cad-Fc and fibronectin. More the same approach has successfully used to transfect the leukemia cells having potential application in cancer therapy.

Acknowledgements

This work has financially been supported by a research grant (Project ID 02-02-09-SF0013) of the Ministry of Science, Technology and Innovation (MOSTI), Malaysia.

Author details

Ezharul Hoque Chowdhury*

Address all correspondence to: md.ezharul.hoque@med.monash.edu.my

Jeffrey Cheah School of Medicine and Health Sciences, Faculty of Medicine, Nursing and Health Sciences, Monash University (Sunway Campus), Australia

http://dx.doi.org/10.5772/53107

References

[1] Kopatz, I., Remy, J. S., & Behr, J. P. (2004). A model for non-viral gene delivery: through syndecan adhesion molecules and powered by actin. *Journal Gene Medicine*, 6, 769-776.

[2] Wu, G. Y., & Wu, C. H. (1997). Receptor-mediated in vitro gene transformation by a soluble DNA carrier system. *Journal Biological Chemistry*, 262, 4432-4439.

[3] Cotton, M., Längle-Rouault, F., Kirlappos, H., Wagner, E., Mechtler, K., Zenke, M., Beug, H., & Birnstiel, M. L. (1990). Transferrin-polycation-mediated introduction of DNA into human leukemic cells: stimulation by agents that affect the survival of transfected DNA or modulate transferrin receptor levels. *Proceedings of the National Academy of Sciences*, 87, 4033-4037.

[4] Schaffer, D. V., & Lauffenburger, D. A. (1998). Optimization of cell surface binding enhances efficiency and specificity of molecular conjugate gene delivery. *Journal Biological Chemistry*, 273, 28004-28009.

[5] Erbacher, P., Bousser, M. T., Raimond, J., Monsigny, M., Midoux, P., & Roche, A. C. (1996). Gene transfer by DNA/glycosylated polylysine complexes into human blood monocyte-derived macrophages. *Human Gene Therapy*, 7, 721-729.

[6] Hoganson, D. K., Chandler, L. A., Fleurbaaij, G. A., Ying, W., Black, ME, Doukas, J., Pierce, G. F., Baird, A., & Sosnowski, BA. (1998). Targeted delivery of DNA encoding cytotoxic proteins through high-affinity fibroblast growth factor receptors. *Human Gene Therapy*, 9, 2565-2575.

[7] Buschle, M., Cotton, M., Kirlappos, H., Mechtler, K., Schaffner, G., Zauner, W., & Birnstiel, M. L. (1995). Receptor-mediated gene transfer into human T lymphocytes via binding of DNA/CD3 antibody particles to the CD3 T cell receptor complex. *Human Gene Therapy*, 6, 753-761.

[8] Erbacher, P., Remy, J. S., & Behr, J. P. (1999). Gene transfer with synthetic virus-like particles via the integrin-mediated endocytosis pathway. *Gene Therapy*, 6, 138-145.

[9] Wang, S., Lee, R. J., Cauchon, G., Gorenstein, D. G., & Low, P. S. (1995). Delivery of antisense oligodeoxyribonucleotides against the human epidermal growth factor receptor into cultured KB cells with liposomes conjugated to folate via polyethylene glycol. *Proceedings of the National Academy of Sciences*, 92, 3318-3322.

[10] Parkes, R. J., & Hart, S. L. (2000). Adhesion molecules and gene transfer. *Advanced Drug Delivery Reviews*, 44, 135-152.

[11] Chowdhury, E. H. (2007). pH-sensitive nano-crystals of carbonate apatite for smart and cell-specific transgene delivery. *Expert Opinion Drug Delivery*, 4, 193-196.

[12] Chowdhury, E. H., Maruyama, A., Kano, A., Nagaoka, M., Kotaka, M., Hirose, S., Kunou, M., & Akaike, A. (2006). pH-sensing nano-crystals of carbonate apatite: ef-

fects on intracellular delivery and release of DNA for efficient expression into mammalian cells. *Gene*, 376, 87-94.

[13] Chowdhury, E. H., Kutsuzawa, K., & Akaike, T. Designing Smart Nano-apatite Composites: The Emerging era of non-viral gene delivery. *Gene Therapy and Molecular Biology*, 9, 301-316.

[14] Kutsuzawa, K., Maruyama, K., Akiyama, Y., Akaike, T., & Chowdhury, E. H. (2008). Efficient transfection of mouse embryonic stem cells with cell-adhesive protein-embedded inorganic nanocarrier. *Analytical Biochemistry*, 372, 122-124.

[15] Kutsuzawa, K., Chowdhury, E. H., Nagaoka, M., Maruyama, K., Akiyama, Y., & Akaike, T. (2006). Surface functionalization of inorganic nano-crystals with fibronectin and E-cadherin chimera synergistically accelerates trans-gene delivery into embryonic stem cells. *Biochemical and Biophysical Research Communications*, 350, 514-520.

[16] Shirayoshi, Y., Okada, T. S., & Takeichi, M. (1983). The calcium-dependent cell-cell adhesion system regulates inner cell mass formation and cell surface polarization in early mouse development. *Cell*, 35, 631-638.

[17] Oka, M., Tagoku, K., Russell, T. L., Nakano, Y., Hamazaki, T., Meyer, E. M., Yokota, T., & Terada, N. (2002). CD9 is associated with leukemia inhibitory factor-mediated maintenance of embryonic stem cells. *Molecular Biology of the Cell*, 13, 1274-1281.

[18] Chowdhury, E. H., Zohra, F. T., Tada, S., Kitamura, C., & Akaike, T. (2004). Fibronectin in collaboration with Mg^{2+} enhances transgene expression by calcium phosphate coprecipitates. *Analytical Biochemistry*, 335, 162-164.

[19] Chowdhury, E. H., Nagaoka, M., Ogiwara, K., Zohra, F. T., Kutsuzawa, K., Tada, S., Kitamura, C., & Akaike, T. (2005). Integrin-supported fast rate intracellular delivery of plasmid DNA by extracellular matrix protein embedded calcium phosphate complexes. *Biochemistry*, 44, 12273-12278.

[20] Overduin, M., Harvey, T. S., Bagby, S., Tong, K. I., Yau, P., Takeichi, M., & Ikura, M. (1995). Solution structure of the epithelial cadherin domain responsible for selective cell adhesion. *Science*, 267, 386-389.

[21] Le Joseph, T. L., Yap, S. R., , A. S., & Stow, J. L. (2002). Protein kinase C regulates endocytosis and recycling of E-cadherin. *American Journal of Physiology- Cell Physiology*, 283, C489-C499.

[22] Kutsuzawa, K., Maruyama, K., Akiyama, Y., Akaike, T., & Chowdhury, E. H. (2008). The influence of the cell-adhesive proteins E-cadherin and fibronectin embedded in carbonate-apatite DNA carrier on transgene delivery and expression in a mouse embryonic stem cell line. *Biomaterials*, 29, 370-376.

[23] Kolanus, W., & Seed, B. (1997). Integrins and inside-out signal transduction: converging signals from PKC and PIP3. *Current Opinion in Cell Biology*, 9, 725-731.

[24] Besson, A., Wilson, T. L., & Yong, V. W. (2002). The anchoring protein RACK1 links protein kinase Cepsilon to integrin beta chains. Requirements for adhesion and motility. *Journal Biological Chemistry*, 277, 22073-22084.

[25] Stephens, L. E., Sonne, J. E., Fitzgerald, M. L., & Damsky, C. H. (1993). Targeted deletion of beta 1 integrins in F9 embryonal carcinoma cells affects morphological differentiation but not tissue-specific gene expression. *Journal of Cell Biology*, 123, 1607-1620.

[26] Maeno, Y., Moroi, S., Nagashima, H., Noda, T., Shiozaki, H., Monden, M., Tsukita, S., & Nagafuchi, A. (1999). alpha-catenin-deficient F9 cells differentiate into signet ring cells. *American Journal of Pathology*, 154, 1323-1328.

[27] Kutsuzawa, K., Maruyama, K., Akiyama, Y., Akaike, T., & Chowdhury, E. H. (2007). Protein kinase C activation enhances transfection efficacy of cell-adhesive protein-anchored carbonate apatite nanocrystals. *Analytical Biochemistry*, 37, 116-117.

[28] Webber, D. J., & Minger, S. L. (2004). Therapeutic potential of stem cells in central nervous system regeneration. *Current Opinion in Investigational Drugs*, 5, 714-719.

[29] Hussain, M. A., & Theise, N. D. (2004). Stem-cell therapy for diabetes mellitus. *Lancet*, 364, 203-205.

[30] Mathur, A., & Martin, J. F. (2004). Stem cells and repair of the heart. *Lancet*, 364, 183-192.

[31] Duncan, S. A., Navas, MA, Dufort, D., Rossant, J., & Stoffel, M. (1998). Regulation of a transcription factor network required for differentiation and metabolism. *Science*, 281, 692-695.

[32] Li, J., Ning, G., & Duncan, S. A. (2000). Mammalian hepatocyte differentiation requires the transcription factor HNF-4alpha. *Genes and Development*, 14, 464-474.

[33] Ishizaka, S., Shiroi, A., Kanda, S., Yoshikawa, M., Tsujinoue, H., Kuriyama, S., Hasuma, T., Nakatani, K., & Takahashi, K. (2002). Development of hepatocytes from ES cells after transfection with the HNF-3beta gene. *FASEB Journal*, 16, 1444-1446.

[34] Dohrmann, C., Gruss, P., & Lemaire, L. (2000). Pax genes and the differentiation of hormone-producing endocrine cells in the pancreas. *Mechanisms of Development*, 92, 47-54.

[35] Kim, S. K., & Hebrok, M. (2001). Intercellular signals regulating pancreas development and function. *Genes and Development*, 15, 111-127.

[36] Xian, H. Q., & Gottlieb, D. I. (2001). Peering into early neurogenesis with embryonic stem cells. *Trends in Neurosciences*, 24, 685-686.

[37] Adams, S. W., & Emerson, S. G. (1998). Gene therapy for leukemia and lymphoma. *Hematology/Oncology Clinics of North America*, 12(3), 631-48.

[38] Schakowski, F., Buttgereit, P., Mazur, M., Märten, A., Schöttker, B., Gorschlüter, M., & Schmidt-Wolf, I. G. (2004). Novel non-viral method for transfection of primary leukemia cells and cell lines. *Genetic Vaccines and Therapy*, 2(1), 1.

[39] Szeps, M., Erickson, S., Gruber, A., Castro, J., Einhorn, S., & Grandér, D. (2003). Effects of interferon-alpha on cell cycle regulatory proteins in leukemic cells. *Leukemia and Lymphoma*, 44(6), 1019-25.

[40] Finke, S., Trojaneck, B., Lefterova, P., Csipai, M., Wagner, E., Kircheis, R., Neubauer, A., Huhn, D., Wittig, B., & Schmidt-Wolf, I. G. (1998). Increase of proliferation rate and enhancement of antitumor cytotoxicity of expanded human CD3+ CD56+ immunologic effector cells by receptor-mediated transfection with the interleukin-7 gene. *Gene Therapy*, 5(1), 31-9.

[41] Notter, M., Willinger, T., Erben, U., & Thiel, E. (2001). Targeting of a B7-1 (CD80) immunoglobulin G fusion protein to acute myeloid leukemia blasts increases their costimulatory activity for autologous remission T cells. *Blood*, 97(10), 3138-45.

[42] Kato, K., Cantwell, MJ, Sharma, S., & Kipps, T. J. (1998). Gene transfer of CD40-ligand induces autologous immune recognition of chronic lymphocytic leukemia B cells. *Journal of Clinical Investigation*, 101(5), 1133-1141.

[43] Chowdhury, E. H., & Akaike, T. (2005). Bio-functional inorganic materials: an attractive branch of gene-based nano-medicine delivery for 21st century. *Current Gene Therapy*, 5, 669-676.

Section 4

Biomaterials for Tissue Engineering and Regeneration

Chapter 11

Innovative Strategies for Tissue Engineering

Juliana Lott Carvalho, Pablo Herthel de Carvalho,
Dawidson Assis Gomes and
Alfredo Miranda de Goes

Additional information is available at the end of the chapter

http://dx.doi.org/10.5772/53337

1. Introduction

Unmet need. It is beyond dispute that human population is ageing. For the first time in history, people age 65 and over will outnumber children under age 5. This trend is emerging around the globe, and will bring several challenges for health technologies. For instance, in a few decades, the loss of health and life worldwide will be greater from chronic diseases than from infectious diseases and accidents [4].

The report made by the National Institute of Health and the National Institute of Aging (NIH/NIA) underscores the unmet needs lying ahead for regenerative medicine. Chronic diseases, in opposition to infectious diseases, are mainly treated by regenerative approaches, instead of immunization and antibiotics.

Actually, many of those challenges are already present in our daily living: malformations [9,14], accidents [16], chronic infections [9,15], and end-organ failure [17] (usually occurs during the final stages of degenerative and other diseases), may, in some cases, only be treated by organ replacement. In fact, end-organ failure alone already affects millions of Americans. More specifically, nearly six million Americans suffer from heart failure with approximately 550,000 new cases diagnosed annually, 530,000 Americans suffer from end-stage renal disease and nearly 25 million Americans suffer from chronic obstructive pulmonary disease with an estimated 12 million new annual diagnoses [5]. As already stated, for many of those patients, organ transplantation is their only treatment option. Currently, organ transplantation is considered the best option for some patients and achieves up to 98,5% of patient 1-year survival rates [18]. Unfortunately, though, current organ shortage/recovering engenders waiting lines of up to three or more years [5]. During 2008, for instance, the number of heart transplantations decreased 2,67%, even though waiting lines increased during the same year. Such decrease occurred mainly due to a reduction in number of recovered organs [18].

It is clear that alternatives to organ transplantation need to be developed as soon as possible. That`s where tissue engineering comes into picture.

Tissue engineering. Tissue engineering refers to an "interdisciplinary field that applies the principles of engineering and the life sciences toward the development of biological substitutes that restore, maintain, or improve tissue function" [3]. The term was first coined by Dr. Fung, from California University, which suggested this name during the National Science Foundation Meeting, in 1987 [2]. The first official definition dates to 1988, though, when Skalak and Fox published it after the "Tissue engineering Meeting" held in Lake Tahoe, USA during that year [1].

In 1993, Langer and Vacanti described three strategies for the creation of new tissue in vitro [3].

1. *Isolated cells or substitutes.* The concept of treating injured tissues with isolated cells is currently regarded as cell therapy. Infusion of cells, e.g. stem cells, has presented several promising results, and have already been approved for human use for specific applications [21] but in some cases, is hindered by the lack of fixation of cells in the site of lesion. When injected systemically, stem cells are attracted to injured tissues, but are also found in several organs such as lungs, liver and spleen [56].

2. *Tissue inducing substances.* At the time, tissue inducing substances included growth factors, small molecules, and other classes of molecules which, if delivered in the organism, would promote several effects on cells, such as growth [58], survival [58], migration [57] and neo tissue formation.

3. *Cells placed on or within matrices.* Associated cells and substrates provide the injured tissue with continuity, and promotes cell attachment and fixation. In this context, scaffolds may be associated with inducing substances, providing means to combine all the aforementioned strategies. The combination of cells and matrices, in addition to inducing substances or not, is currently the main strategy for tissue engineering, as depicted in Figure 1.

Currently, tissue engineering focuses mainly of associating cells with supports (also called biomaterials or scaffolds), in order to: i. promote cell attachment and restrict their distribution in the tissue, ii. direct cell distribution and differentiation, iii. sustain large tissue losses while new tissue is formed, and ultimately, to iv. lead to new tissue formation.

Since its early days, tissue engineering has significantly evolved in each of its pillars – Cells, signaling molecules and scaffolds. This evolution covered both conceptual aspects - as evidenced above – as well as practical aspects, mainly reflected in the achievements of the field (for more information, go to conclusion section). Unfortunately, even though cells and signaling molecules platforms have evolved during the past decades, leading to major field evolution, the degree of success of tissue engineering methods is still highly dependent on the properties of the scaffold. Therefore, this study focuses on the main Achille`s Hill of tissue engineering: production of scaffolds for biological applications.

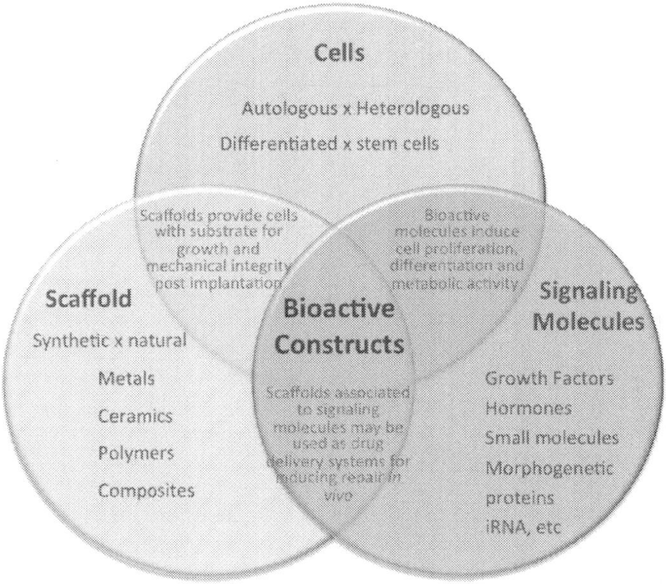

Figure 1. Tissue engineering strategies. Based on [59]. Tissue engineering may be performed by several different approaches, as proposed by Langer and Vancanti. In order to obtain tissue regeneration, cells, scaffolds and signaling molecules may be introduced into the body alone or in association. Currently, the association of all three elements, composing bioactive constructs, is proposed to be the best option for tissue engineering.

In the present chapter, we present the current status of tissue engineering. First, we present a comprehensive picture of classical tissue engineering approaches, as well as an analytic view of its main achievements and limitations. Secondly, we present innovative and paradigm breaking strategies for successful tissue engineering, accompanied by the history and rationale behind each of them. Finally, we analyze the next steps of tissue engineering translation into the clinic.

2. Classical tissue engineering approaches

Association of cells to classic biodegradable solid/porous biomaterial represents a dominating conceptual framework in tissue engineering. Actually, men have used biomaterials (alone, not associated to cells) in order to substitute eventual tissue loss since ancient civilizations [11]. In the early days, all kinds of materials derived from natural and manufactured sources were used as biomaterials. Natural materials included wood and shells, and manufactured materials comprised metals such as iron, gold and zinc. The host responses to these materials were extremely varied, and only after the concept of sterility, biomaterial implants began to achieve consistent safety. During the last 30 years, further progress has been made in understanding the interactions between the tissues and the materials.

Since then, tissue reconstitution evolved to a more sophisticated approach, in which the regeneration of the tissue/organ was clearly viewed as the ideal way of treating injuries, compared to biomaterial science, which simply reconstitutes tissue structure, without restoring tissue function in most situations. This paradigm evolution led to the emergence of tissue engineering.

Tissue engineering based on biomaterials relies on four main classes of materials: i. Polymers, ii. Ceramics, iii. Metals and iv. Composites (blends and combinations of the aforementioned materials). Biomaterials may derive from natural or synthetic sources [12].

The association of cells to biomaterials is called construct and is the base of current tissue engineering, as already stated. Construct-based classic tissue engineering platform derives from several basic assumptions, as described by Mironov et al., 2009 [31]: "1) cell growth is substrate attachment-dependent; cells need a solid substrate for attachment and proliferation; 2) tissue constructs must have an organo-specific shape; a solid scaffold is essential to keep the desired shape; a tissue construct could not maintain its shape without a solid rigid scaffold; 3) the scaffold serves not only as an attachment substrate, but also as a source of inductive and instructive signals for cell differentiation, migration, proliferation and orientation; 4) the porous structure of a solid scaffold will allow optimal cell seeding, tissue construct viability, and vascularization; and 5) mechanical properties initially provided by the rigid solid scaffold after its biodegradation will be maintained by controlled neomorphogenesis of parenchymal and stromal tissue synthesized *in vitro* or *in vivo* in the tissue construct".

Considering these basic assumptions, currently, classic tissue engineering is made taking several aspects into account, which will be highlighted below.

1. *Construct design.* Currently, scaffold design is a complex science, in which several aspects are carefully addressed in order to produce successful constructs. Those aspects include, but are not restricted to:

i. *Choosing the most suitable biomaterial for the envisioned application.* Organs and tissues in the human body present different characteristics, therefore, the ideal biomaterial must reproduce as many as those features as possible. Such aspects include tissue resistance, elasticity, resilience, and chemical composition, among others, in addition to biocompatibility. For instance, a biomaterial designed to be used in a skin construct must be thin and elastic. A biomaterial for application in corneas, on the other hand, must also be thin, but most importantly, it must be transparent.

ii. *Customizing the material in order to promote cell colonization.* The ideal construct must promote rapid and equal cell adhesion and colonization, therefore scaffolds are usually porous and present a surface which is recognized directly or indirectly (by promoting protein adsorption) by the cell as a substrate for attachment. Usually, materials such as chitosan and collagen are used as biomaterials, due to their resemblance to the extracellular matrix and their efficacy in promoting cell adhesion [26,27]. In case of other materials, such as metals and ceramics, biomaterial surface may not be easily recognized by cells. In order to improve cell contact and adhesion by cells, many strategies have been developed, such as blending biomaterials, recovering them with other substances or still covering them with protein residues recognized by cells, such as isoleucine-lysine-valine-alanine-valine (IKVAV) residue, derived from Laminin [28,29].

iii. *Studying material tribology and surface topography* [19]. For several tissues, such as muscles and tendons, cell organization is paramount for optimal tissue function. The heart, for instance, works as a pumping organ, and needs to contract in a specific, synchronized way in order to actually eject blood into body and lungs. If muscle fibers are not synchronized, no pumping force is generated, and great deficit of function is witnessed by the patient [60]. Tendon is a specific connective tissue composed of parallel collagen fibers. Along with the heart, tendon constitutes another of several tissues which depend on specific cell organization for proper function. It has been thoroughly described in literature that tendon strength is directly linked to cell orientation, and, following injuries, the lack of orientation of scar tissue promotes tendon weakness, leading to repetitive lesion [61]. Tissue engineering approaches for tendon must promote cell alignment in order to achieve significant benefit for patients. cell alignment has been shown to be achievable and effective in promoting tissue organization and maturation [30], as shown in Figure 2.

iv. *Optimizing biomaterial degradation rate.* Tissue engineering has been envisioned to promote tissue regeneration, therefore in this context, biomaterials should be biodegradable. Ideally, biomaterial should gradually degrade, at the same rate as neotissue is formed.

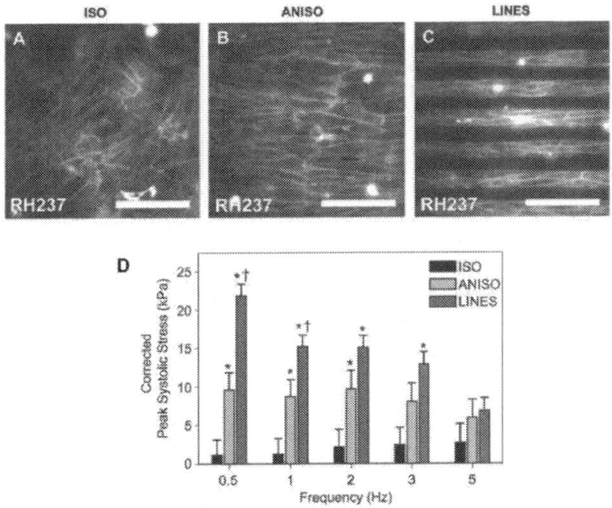

Figure 2. Cell orientation promotes major contractile strength of construct. In order to assess cell orientation importance for construct function, Dr. Parker`s group engineered bidimensional cardiac muscles with different micropatterned surfaces in order to promote degrees of cell orientation. Confluent unaligned isotropic (A), aligned anisotropic (B) and non confluent, 20μm spaced, parallel arrays of myocardial fibers (C) were build and studied *in vitro*. The cited work showed that contractile force increases with major sarcomere alignment, as measured by peak systolic stress in kPa (D). This panel was based on [60] and was kindly provided by Dr. Kevin Kit Parker, from Harvard University.

2. *Cell seeding.* Cell seeding is paramount for construct optimization, and must be carefully planned. Usually, most cells have low capacity of invasion, therefore, cell seeding must be optimized to promote an equal distribution of cells along construct surface and interior [62].

3. *Construct maintenance in vitro.* The maintenance of small constructs may be achievable in static cultures, but when it comes to larger constructs, static culture is hindered by the limitation of nutrient diffusion. Several strategies have been developed for large construct maintenance *in vitro,* represented mainly by bioreactors, as depicted in Figure 3.

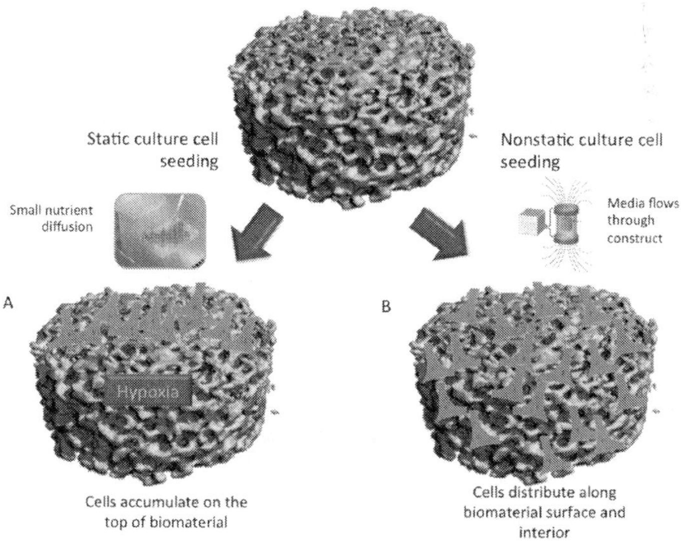

Figure 3. Importance of non-static cell seeding and culture for construct equal distribution of cell elements and viability of construct. As depicted, static culture seeding promotes cell accumulation in specific regions of the construct. As no media is perfused through the construct, nutrient distribution along the biomaterial depends on diffusion only, which leads to normoxic construct edges, colonized with cells and hypoxic center, which is not feasible to be colonized by cells. Nonstatic culture strategies, on the other hand, promote equal cell seeding and colonization of the construct, as it enhances nutrient diffusion. Even though nonstatic culture is not sufficient to promote viability of large constructs, it increases maximum size of constructs and is more adequate for tissue engineering purposes.

4. *Construct implant.* The ultimate function of a construct built *in vitro* is to be implanted and substitute/regenerate an injured organ or tissue. Construct implant must be performed in order to promote construct integration and viability. Several techniques have already been applied to promote construct long term viability, such as designing VEGF (vascular endothelial growth factor) releasing constructs; previously implanting the construct in an ectopic site, in order to promote *in vivo* vascularization, prior to implantation; and reducing the size of implants, among others.

In spite of its major advances, scaffold-based tissue engineering suffers from several limitations, as explored by Mironov et al., 2009 [31], and further explained here:

1. Vascularization of thick tissue constructs

Cell survival requires continuous supply of nutrients and oxygen, as well as the removal of metabolites, which, if accumulated, might be toxic. Such demand is addressed mainly by osmosis, therefore, to facilitate nutrient and metabolite flow, cells must be kept near vessels and capilars. Actually, few cells are able to survive at more than 200um distance from the nearest blood vessel [8]. Cells cultured in tridimensional scaffolds also need to be maintained in homeostasis in order to survive. *In vitro*, several strategies have been developed in order to maintain construct viability prior to implantation, mainly through the development of bioreactors, as reviewed by Rauh et. *al.* [10]. *In vivo*, however, none of those strategies are applicable, and only through vascularization cells are kept alive, especially for modular organs, such as heart, kidney and liver, which are organized in functioning units and require their own vascular supply [9]. Studies indicate that vessels grow at a rate of <1mm a day [6], and although the effectiveness of an enhanced angiogenic response using various growth factors has been demonstrated in many tissue systems, the rate of angiogenesis hasn`t been accelerated so far [7]. Considering the relatively large sizes of constructs for humans, it is clear the urgent need to promote faster vascularization of tissue constructs, or to improve cell survival in scaffolds.

Actually, several attempts of increasing tissue vascularization are underway.

As previously mentioned, increasing cell survival is also an interesting strategy. Actually, recently, oxygen generating scaffolds have been developed and tested with encouraging results, even though no *in vivo* tests were performed [7].

2. Precise placing of different multiple cell types inside 3D porous scaffolds is technologically challenging.

Modular organs, such as the heart, liver, kidneys and others, are complex structures of several types of cells, including stromal and parenchymal cells. They function as working units, such as muscle fibers, liver lobules and kidney nephrons. Modular organs also count on intrinsic vascular system for cell survival, constituting incredibly difficult organs to build *in vitro*, even though many papers have shown a significant capacity of self-organization of several cells [22,23,24]. For such organs, whole organ approaches are more suitable then partial reconstitution of those structures.

On the other hand, non-modular organs have witnessed several successful strategies, such as the construction of bladders [44] and the recellularization of tracheas [15], both of which have already been translated to the clinic.

3. Achieving organo-specific level of cell density in tissue constructs remains a big challenge

Currently, porous matrices are paramount to allow cell invasion and colonization of the matrix. Porous present in the matrices are usually optimized to have specific sizes and to be interconnected, in order to permit cell invasion. The size of the produced porous is usually large, though, and cells within the scaffold are not able to fully fill it and achieve cell density similar to natural tissues. Therefore, it is almost as if cells were still in two-dimensional surfaces [31]. Actually, extracellular matrix molecules can be washed out from 3D porous scaffolds in the same way as in 2D cultures, and may not provide means for real tridimensional tissue formation.

4. Recent reports on the effect of matrix rigidity on (stem) cell differentiation can undermine the value of solid rigid biodegradable scaffolds at least for certain tissue applications [35-37]

Stem cell differentiation has traditionally employed cocktails of various growth factors, but recently, mechanobiological concepts have been described as important to cell fate decision. The mechanism underlying cellular response to tension comprises the force generated by myosin bundles sliding along actin filaments and transmission to the ECM. Transduction of these signals link the extracellular and intracellular worlds, ultimately affected by proteins such as Rho GTPases, which not only regulate contraction of stress fibers, but also regulate gene expression by acting over their effector target proteins, [45].

Actually, matrix rigidity has been involved in embryonic development, as well as adult stem cell differentiation. As expected, rigid surfaces facilitates adult stem cell differentiation into bone, and soft surfaces lead to differentiation of adult stem cells into soft tissues, such as fat or central nervous system (brain) [45].

5. Biodegradability of constructs

Even though it makes sense that biomaterials should be absorbed by the body in order to give space to neotissue formation, the same is not true when whole tissue engineering is planned. There is no use in spending efforts in order to build a construct, which will be invaded by inflammatory cells and vessels and disorganized, previously to being substituted by neotissue.

Therefore, even though current tissue engineering techniques are fairly successful in treating bone, skin and cartilage loss, they are extremely limited in treating large tissue loss, as well as in regenerating complex tissues, such as heart annd kidneys, among several other tissues and organs.

3. Innovative tissue engineering approaches

3.1. Decellularized matrices

The Extracellular Matrix (ECM) represents the three-dimensional fibrilar protein scaffold, produced by cells of each tissue and organ, which surrounds and anchors them. It is kept in a state of dynamic reciprocity with those cells, in response to changes in the microenvironment. ECM has been shown to provide cues that affect cell migration, proliferation, gene expression and differentiation [32, 37].

The ECM is obviously the optimal support for tissue engineering, as it provides the perfect chemical composition, surface topology and physical properties experienced by cells *in vivo in their niche* [32]. Even though sometimes that`s exactly what is needed to be avoided (e.g. Central Nervous System ECM has been shown to contain molecules which inhibit axonal growth and hinders tissue regeneration [33,34]), ECM has been considered a great option for tissue engineering.

Recently, it has been shown that cell sensibility towards ECM chemical composition is higher than previously expected. For instance, Tsai et al. showed that MG63, an osteoblast like cell lineage, behaves differently when grown in collagen or gelatin electrospun matrices. When grown in electrospun collagen, MG63 did not show variation on cell attachment or proliferation rates. On the other hand, cells seeded on electrospun collagen showed increased expression of osteogenic genes such as Osteopontin and alkaline phosphatase. Collagen and gelatin present high chemical composition similarity, varying mainly in secondary and tertiary structure. Such fact underscores the strikingly cell sensitivity to all aspects of ECM chemical and physical composition [62]. It also underscores the potential of decellularized matrices on tissue engineering.

Decellularized tissues have been used in regenerative medicine approaches since the early eighties [38], specially focused on treating cardiovascular diseases by engineering vascular grafts. Most of the grafts produced, derived from synthetic and natural sources suffered from several limitations. When the issue of natural graft calcification and immunological recognition were related to residual cellular components of unmodified biological materials, decellularization techniques began to be developed [38,39].

Initially, decellularization was considered for tissue grafts. Developed techniques are continuously evolving, as every cell removal agent and method currently available alters ECM composition and cause some degree of ultrastructure disruption. Decellularization agents include chemical, biological and physical agents, each of them with different mechanisms of action.

More specifically: acids and bases promote hydrolytic degradation of biomolecules; hypotonic solutions lyses cells through osmosis with minimal changes in matrix molecules and tissue architecture; hypertonic solutions dissociates DNA from proteins; ionic, non-ionic, and zwitterionic detergents solubilize cell membranes leading to effective removal of cellular material from tissue; solvents, such as alcohol and acetone, promote either cell lysis by dehydration or solubilization and removal of lipids and biological agents, such as enzymes, and chelating agents act through protein cleavage and disrupting cell adhesion to ECM. Finally, physical agents promote cell lysis through freezing and thawing cycles, electroporation or pressure [32].

The most effective agents for decellularization of each tissue and organ will depend upon many factors, including the tissue's cellularity, density, lipid content, and thickness [32].

Lately, whole organ decellularization began to be performed, offering an interesting option for modular organs such as the heart, lung and kidneys. In 2008, Ott et al. not only performed whole heart decellularization, but also recellularized the organ with neonatal cardiomyocytes and obtained organ function [17]. This groundbreaking work highlighted the possibilities of decellularized matrix-based whole organ tissue engineering.

The major breakthrough of organ decellularization is to obtain scaffolds with perfect (or very similar) chemical composition and tridimensional structure, compared to natural organs. In addition, the vascular bed is completely preserved, facilitating *in vitro* maintenance of the construct via perfusion bioreactors, as well as *in vivo* viability of the construct, which may be reconnected to the circulatory system of host, also shown by Ott et al [17]. As decellularization is performed making use of vascular system of organs, virtually any vascularized organ may be decellularized, disregarding its size, as depicted in Figure 4. Acellular organs, such as tracheas, may also be decellularized through different protocols [15].

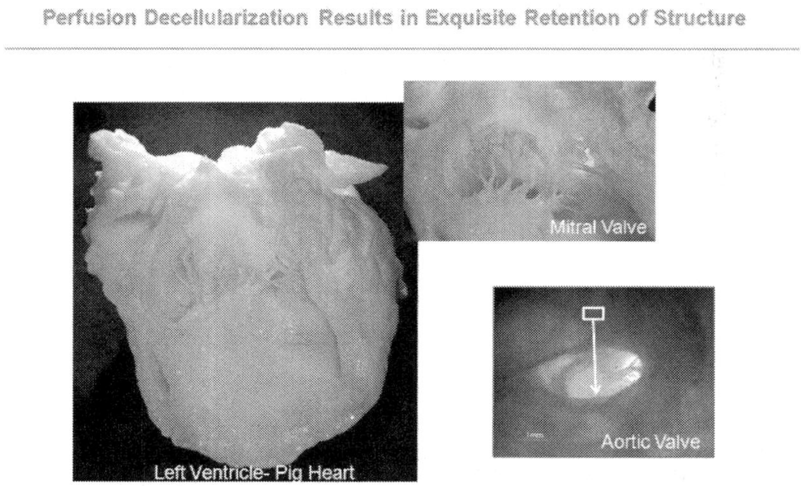

Figure 4. Decellularized pig heart. As published by Ott et al., perfusion decellularization is feasible in small rat organs, but also in bigger size organs, such as the pig heart. Illustration owned by Miromatrix Medical Inc., available at http://miromatrix.com/technology/perfusion-decellularization-recellularization/, accessed on September 2012.

3.2. Biomimetic scaffolds

As already mentioned, an optimal scaffold attempts to mimic the function of the natural extracellular matrix [52]. The functionality of most tissues is related to their complex architecture, therefore mimicking and recapitulating this complexity *in vitro* is paramount for successful tissue engineering. In the *in vivo* condition, cells are surrounded by other cells and by the extracellular matrix (ECM), whose components, such as collagen, elastin, and laminin, are organized in nanostructures (i.e., fibers, triple helixes, etc) with specific bioactive motifs that regulate cell homeostasis [42].

Following the initial concept of mimicking the ECM chemical composition, it has been shown that the structure of the cell-surrounding niche is also paramount for optimal cell function. In accordance, modulating the scaffold microarchitecture is one of the most potent ways of achieving biomimetic tissues. Advances in microfabrication technologies have been exploited by an increasing number of research groups. Often technologies from other engineering disciplines have been translated and used in creating microfeatures in engineered scaffolds in a controlled manner. These include photolithographic approach of the electrical engineering, electrospinning tools of the textile industry, emulsification and fluid dynamics principles of chemical engineering and rapid prototyping methods of mechanical engineering. The latter will be covered in further detail in the next section [40].

Vacanti and coworkers pioneered the concept of engineering a vasculature using photolithographic techniques (which use light, e.g. UV, to selectively remove parts of a thin film or the bulk of a substrate), literally generating channels within biomaterials [43].

Electrospining techniques have also been considered and tested for tissue engineering application, due to their potential of producing polymer fibers with nano to micrometer diameter scale that are physically and topographically comparable to the collagen fibers, commonly found in the natural ECM, as shown in Figure 5. It has been extensively employed in tissue engineering strategies, including vascularization strategies [41].

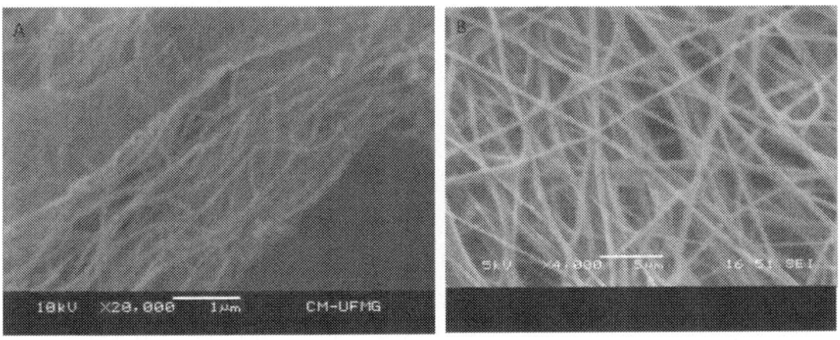

Figure 5. Similarity between ultrastructure of natural collagen fibers (A) and electrospun biomaterials (B). Source: A – author`s unpublished data; B - http://en.wikipedia.org/wiki/Electrospinning, accessed on September 2012.

Electrospun biomaterials must also be carefully fabricated, as even fiber diameter variation results in different cell behavior, as observed in endothelial cells cultured on electrospun poly(l-lactide-co-ε-caprolactone) with different fiber diameters. In contrast to cells cultured on fibers of 0.3 or 1.2μm, cells cultured on 7μm presented lower cell adhesion, spreading and proliferation [63].

Even though electrospining has presented promising results, it also suffers from some limitations, such as poor cell invasion, as usually the electrospun biomaterials are highly compacted, impeding cell migration towards the inner side of the scaffold. The search for different solvents and electrospining conditions may solve this issue, promoting less fiber compactation.

Still, as cited, many excellent constructs have been built using those techniques, and in association with the other tissue engineering strategies presented in the present chapter, strongly contribute to novel advances in the field.

Another exciting tissue engineering strategy, which have gained growing interest since the nineties, has been the hydrogel approach. Hydrogels are 3D cross-linked insoluble, hydrophilic networks of polymers that partially resemble the physical characteristics of native ECM. The biocompatibility of various hydrogels (e.g., collagen, agarose and polyethylene glycol) is well characterized, and the possibility of optimizing their physico-chemical and mechanical properties to levels that are desirable for tissue scaffolds, in order to achieve cell encapsulation, immobilization, and drug delivery turn hydrogels into an extremely promising technique [64].

Hydrogels have been successfully used in mimicking ECM of simple tissues, composed of one cell type. In many cases, hydrogels provide means for nutrient diffusion, facilitating cell maintenance *in vitro* and *in vivo*. Still, maintaining the viability of high cell density constructs remains a challenge as well as promoting cell organization within the scaffold. In many cases, construct implantation near host rich vascularization sites may be an effective strategy to promote construct viability *in vivo*.

In the cell`s point of view, hydrogels possess the advantage of completely surround encapsulated cells, and providing tridimensional substrate for cell interaction. This strategy prevents cell polarization which is common to regular scaffolds. Usually, as already stated, even though most scaffolds may be tridimensional macroscopically, they are commonly seen as bidimensional surfaces by cells.

As expected, truly tridimensional environments promote several effects over cultured cells. Some of them include, but are not limited to: cell morphology/spreading [65], cell motility and proliferation [66], and metabolic rate [67]. Obviously, all of those cell behaviors reflect differential gene expression.

Even though cell gene expression is a paramount factor to be evaluated in tissue engineering, it must be clearly noted that an exact gene expression profile is not essential for tissue engineering effectiveness. Cells will always interact with their microenvironment. What is important for tissue engineering is to maintain cell plasticity in a manner that, once in vivo, implanted cells start to behave as host cells would.

Finally, a third highly modern and innovative approach for tissue engineering, which shares the truly tridimensional environment for cells as provided by hydrogels, but which doesn`t suffer from vascularization and cell organization limitations presented by the latter, is the organ printing tissue engineering technique, which will be more thoroughly described in the next section.

3.3. Organ printing

As listed before, the main limitations of the solid scaffold approach include the low level of precision in cell placement, especially when engineering multicellular constructs, considering the intrinsic problem of vascularization of thick tissue constructs [53]. Ideally, a possible way to solve many of those aforementioned problems would be to assemble cells and ECM elements at the same time and in an organized way, in order to obtain the most similar structure found in a functional organ as possible. Over the years, technology evolution turned tissue engineers unreachable dream into a feasible objective to be fulfilled in the next years or decades. This strategy is known as organ printing and/or robotic biofabrication, and offer interesting alternatives to solid scaffold-based tissue engineering.

According to the First International Workshop on Bioprinting and Biopatterning, organ printing was defined as "The use of material transfer processes for patterning and assembling biologically relevant materials (molecules, cells, tissues, and biodegradable biomaterials) with a prescribed organization to accomplish one or more biological functions" [45]. In fact, this technology could be defined as computer-aided, layer-by-layer deposition of biologically relevant materials [53].

The ultimate goal of organ-printing technology is to fabricate 3D vascularized functional living human organs suitable for clinical implantation in reasonable time scales. Other applications of this technology are in histogenesis and organogenesis, pharmacological tests and disease research [45, 46, 47, 48].

Wilson and Boland (2003) showed protein and cell printing using a commercial ink-jet device can be possible. In this technology, either individual cells or small clusters are printed over ECM hydrogels, designed for involve printed cells and to provide them with desired signals. Therefore, organ printing has derived from hydrogel classical approach, described in the previous section. The method is rapid, versatile and cheap. Its disadvantage is that it is difficult to assure high cell density needed for the fabrication of solid organ structures. Furthermore, due to the high speed of cell deposition, considerable damage is caused to cells, although the latest developments in the field have led to considerable improvement in cell survival [54, 55]. In the other approach, mechanical extruders are used to place 'bio-ink' particles, multicellular aggregates of definite composition into a supporting environment, the 'bio-paper', according to computer-generated templates consistent with the topology of the desired biological structure. Organoids are formed by postprinting fusion of the bio-ink particles and the sorting of cells within the bio-ink particles. The advantage of this technology is that the bio-ink particles represent small 3D tissue fragments. Thus, cells in them are in a more physiologically relevant arrangement, with adhesive contacts with their neighbors, which may assure the transmission of vital molecular signals. Both inkjet and extruder bioprinting are compatible with rapid prototyping [55].

In this biomanufacturing a precise layer-by-layer placement of self-assembled tissue spheroids in sprayed tissue fusion permissive hydrogels is used to obtain an organ or tissue. The hydrogels work as "biopapers" and cell blocks or tissue spheroids work as a "bioink". Both biopaper and bioink must be optimized in order to obtain viable tissues. For instance, biopapers vary according to each "printed" organ, and cell spheroids vary in properties according to their composition. Cell viability during and after printing is an obvious goal for bioprinting [45]. Preliminary studies of both ink-jet and laser forward transfer indicated that cells can survive deposition condition forces. Problems associated with ink-jet delivery of cell suspensions may also come about from the high shear stresses observed during ejection and impact of a fluid drop [50, 51, 52].

In many cases, the bioprinting process requires that before and during printing, cells and molecules must be carried in a fluid vehicle that shortly after printing requires consolidation and should consequently behave as a viscoelastic solid. This phase change must occur without damage to the biochemical, cells, or more complex units within the fluid, which presents a considerable challenge. Concurrently, tissue printed mustn't be too solid, or cell spheroids won`t interact and form a continued tissue.

Organ printing is a technology that promises to transform tissue engineering into a commercially successful biomedical industry. Unlike other tissue engineered approaches, organ printing involves the high throughput generation of organs, relying on automated cell sorters, cell and organ bioreactors and robotic bioprinters, most of them which are already commercially available [46]. However, much research is necessary to turn this technology into reality of clinical application.

4. Conclusion: Tissue engineering — From the bench to the bedside

It is known that any technology takes about 20 years to reach the market, and despite progress in many fields, this timeframe has yet to shorten [20]. Accordingly, tissue engineering, which has officially given its first steps during the late eighties, hasn`t brought many products to the bedside [20].

In contrast to biomaterials - which are readily available as hip implants, contact lenses, silicon breast prosthesis, among others –, and cell therapy – which is also available for bone marrow transplants, as well as its first allogeneic stem cell therapy products [21] -, constructs have been successfully produced for only few applications, largely limited to non-modular organs such as skin epidermis, corneal epithelium and cartilage [40]. Indeed, Apligraf - a bilayered skin substitute - was the first allogeneic cell based therapy to be approved by the US Food and Drug Administration (FDA), receiving permission for sale as a treatment for venous leg ulcers13.

Apligraf is constructed by culturing human foreskin-derived neonatal fibroblasts in a bovine type I collagen matrix over which human foreskin-derived neonatal epidermal keratinocytes are then cultured and allowed to stratify [68]. Even though it is considered one of the first tissue engineering products ever approved for commercialization, Apligraf doesn`t directly restore skin, but transiently protects and provides injured skin with scaffold and signaling molecules (produced by the cells within the construct) which fosters and accelerate skin regeneration.

Engineered bladders and airways have also been built and implanted *in vivo*, but as they require highly customized and complex approaches, they are available to a small number of patients, and are not considered products to be sold, such as Apligraf and other similar products.

Therefore, it is clear that, in spite of recent advances, tissue engineering has much to deliver. Innovative strategies, such as the presented in this chapter, present out of the box solutions

http://dx.doi.org/10.5772/53337

for some of the present challenges in the field, and may one day constitute the major breakthroughs to finally catalyze translating tissue engineering from bench to bedside.

Author details

Juliana Lott Carvalho[1], Pablo Herthel de Carvalho[2], Dawidson Assis Gomes[1] and Alfredo Miranda de Goes[1]

*Address all correspondence to: julianalott@gmail.com

1 Department of Biochemistry and Immunology, Federal University of Minas Gerais, Belo Horizonte, Brazil

2 Department of Veterinary Clinicals and Surgery of the Federal University of Minas Gerais, Belo Horizonte, Brazil

References

[1] Skalak R, Fox CF. Tissue engineering. New York: Liss; 1988.

[2] Lanza RP, Langer R, Vacanti J. Principles of tissue engineering. 3rd edition. Elsevier Inc; 2007

[3] Langer RL, Vacanti JP. Tissue Engineering. Science 1993; 260(5110) 920-926.

[4] National Institutes of Aging. Why population aging matters – A global perspective. http://www.nia.nih.gov/research/publication/why-population-aging-matters-global-perspective (accessed 18 july 2012).

[5] Song JJ, Ott HC. Organ engineering based on decellularized matrix scaffolds. Trends in Molecular Medicine 2011. Article in press. doi:10.1016/j.molmed.2011.03.005.

[6] Mikos AG, Sarakinos G, Ingber DE, Vacanti J, Langer R. Prevascularization of porous biodegradable polymers. Biotech Bioengineering 1993; 42(6) 716-23.

[7] Oh SH, Ward CL, Atala A, Yoo JJ, Harrison BS. Oxygen generating scaffolds for enhancing engineered tissue survival. Biomaterials 2009; 30(5) 757-762).

[8] Kaully T, Kaufman-Francis K, Lesman A, Levenberg S. Vascularization--the conduit to viable engineered tissues. Tissue Eng Part B Rev. 2009; 15(2) 159-69.

[9] Alberti C. Hollow organ tissue engineering: short updating about current approaches and forecast for major research advances. Giornal e Chirurgia 2011; 32(8-9) 345-351.

[10] Rauh J, Milan F, Günther KP, Stiehler M. Bioreactor systems for bone tissue engineering. Tissue Eng Part B Rev. 2011; 17(4) 263-80.

[11] Williams DF, Cunningham J. Materials in Clinical Dentistry. Oxford; Oxford University Press, 1979.

[12] Ratner BD. Biomaterials Science: An Introduction to Materials in Medicine. London: Elsevier Academic Press, 2004

[13] Pham C, Greenwood J, Cleland H, Woodruff P, Maddern G. Bioengineered skin substitutes for the management of burns: a systematic review. Burns 2007; 33(8) 946-57

[14] Burchill LJ, Ross HJ. Heart transplantation in adults with end-stage congenital heart disease. Future Cardiol. 2012; 8(2) 329-42.

[15] Macchiarini P, Jungebluth P, Go T, Asnaghi MA, Rees LE, Cogan TA, Dodson A, Martorell J, Bellini S, Parnigotto PP, Dickinson SC, Hollander AP, Mantero S, Conconi MT, Birchall MA. Clinical transplantation of a tissue-engineered airway. The Lancet 2008; 372: 2023–30.

[16] Molzahn AE, Starzomski R, McCormick J. The supply of organs for transplantation: issues and challenges. Nephrol Nurs J. 2003;30(1) 17-26.

[17] Ott HC, Matthiesen TS, Goh SK, Black LD, Kren SM, Netoff TI, Taylor DA. Perfusion-decellularized matrix: using nature's platform to engineer a bioartificial heart. Nat Med. 2008; 14(2) 213-21.

[18] U.S. Organ Procurement and Transplantation Network and the Scientific Registry of Transplant Recipients. 2009 OPTN / SRTR Annual Report: Transplant Data 1999-2008. http://www.ustransplant.org/annual_Reports/current/default.html (accessed 2 august 2012).

[19] Chaureya V, Blocka F, Sua YH, Chiang PC, Botchweyc E, Choub CF, Swamia NS. Nanofiber size-dependent sensitivity of fibroblast directionality to the method of alignment of the scaffold. Acta Biomaterialia 2012. Epub ahead of print.

[20] Future Directions in Regenerative Medicine. Business Insights, February 2012.

[21] Osiris therapeutics, Press release: Osiris Therapeutics Reports Second Quarter 2012 Financial Results. http://investor.osiris.com/ (accessed 6 august 2012).

[22] Wei C, Larsen M, Hoffman MP, Yamada KM. Self-organization and branching morphogenesis of primary salivary epithelial cells. Tissue Eng. 2007;13(4) 721-35.

[23] Takebe T, Sekine K, Suzuki Y, Enomura M, Tanaka S, Ueno Y, Zheng YW, Taniguchi H. Self-organization of human hepatic organoid by recapitulating organogenesis *in vitro*. Transplant Proc. 2012; 44(4) 1018-20.

[24] Takebe T, Koike N, Sekine K, Enomura M, Chiba Y, Ueno Y, Zheng YW, Taniguchi H. Generation of functional human vascular network. Transplant Proc. 2012; 44(4) 1130-3.

[25] Green JB, Dominguez I, Davidson LA. Self-organization of vertebrate mesoderm based on simple boundary conditions. Dev Dyn. 2004; 231(3) 576-81.

[26] Miranda SC, Silva GA, Mendes RM, Abreu FA, Caliari MV, Alves JB, Goes AM. Mesenchymal stem cells associated with porous chitosan-gelatin scaffold: A potential strategy for alveolar bone regeneration. J Biomed Mater Res A. 2012. doi: 10.1002/jbm.a. 34214. [Epub ahead of print]

[27] Miranda SC, Silva GA, Hell RC, Martins MD, Alves JB, Goes AM. Three-dimensional culture of rat BMMSCs in a porous chitosan-gelatin scaffold: A promising association for bone tissue engineering in oral reconstruction. Arch Oral Biol. 2011; 56(1) 1-15.

[28] Ding T, Luo ZJ, Zheng Y, Hu XY, Ye ZX. Rapid repair and regeneration of damaged rabbit sciatic nerves by tissue-engineered scaffold made from nano-silver and collagen type I. Injury. 2010; 41(5) 522-7.

[29] X. Lin, K. Takahashi, Y. Liu, P.O. Zamora. Enhancement of cell attachment and tissue integration by a IKVAV containing multi-domain peptide. Biochimica et Biophysica Acta 2006; 1760(9) 1403–1410.

[30] Choi JS, Lee SJ, Christ GJ, Atala A, Yoo JY. The influence of electrospun aligned poly(ε-caprolactone)/collagen nanofiber meshes on the formation of self-aligned skeletal muscle myotubes. Biomaterials 2008; 29(19), 2899–2906.

[31] Mironov V, Visconti RP, Kasyanov V, Forgacs G, Drake CJ, Markwald RR. Organ printing: tissue spheroids as building blocks. Biomaterials 2009; 30(12) 2164-74.

[32] Crapo PM, Gilbert TW, Badylak SF. An overview of tissue and whole organ decellularization processes. Biomaterials 2011; 32(12) 3233–3243.

[33] Giger RJ, Hollis ER, Tuszynski MH. Guidance molecules in axon regeneration. Cold Spring Harb Perspect Biol. 2010; 2(7) a001867.

[34] Sharma K, Selzer ME, Li S. Scar-mediated inhibition and CSPG receptors in the CNS. Exp Neurol. 2012. [Epub ahead of print]

[35] Parka JS, Chua JS, Tsou AD, Diopa R, Tanga Z, Wang A, Li S. The effect of matrix stiffness on the differentiation of mesenchymal stem cells in response to TGF-b. Biomaterials 2011; 32 3921-3930.

[36] Chaudhuri O, David J. Mooney. Stem-cell differentiation: Anchoring cell-fate cues. Nature Materials 2012; 11 568–569.

[37] Reilly GC, Engler AJ, Intrinsic extracellular matrix properties regulate stem cell differentiation. Journal of Biomechanics 2010; 43(1) 55-62.

[38] Malone JM, Brendel K, Duhamel RC, Reinert RL. Detergent-extracted small-diameter vascular prostheses. Journal of Vascular Surgery 1984. 1(1) 181-194.

[39] Schmidt CE, Baier JM. Acellular vascular tissues: natural biomaterials for tissue repair and tissue engineering. Biomaterials 2000; 21(22) 2215-31.

[40] Zorlutuna P, Annabi N, Camci-Unal G, Nikkhah M, Cha JM, Nichol JW, Manbachi A, Bae H, Chen S, Khademhosseini A. Microfabricated Biomaterials for Engineering 3D Tissues. Advanced Biomaterials 2012; 4 1782–1804.

[41] Zonari A, Novikoff S, Electo NRP, Breyner NM, Gomes DA, et al. Endothelial Differentiation of Human Stem Cells Seeded onto Electrospun Polyhydroxybutyrate/ Polyhydroxybutyrate-Co-Hydroxyvalerate Fiber Mesh. PLoS ONE 2012; 7(4) e35422. doi:10.1371/journal.pone.0035422

[42] Perán M, García MA, López-Ruiz E, Bustamante M, Jiménez G, Madeddu R, Marchal JA. Functionalized Nanostructures with Application in Regenerative Medicine. Int. J. Mol. Sci. 2012; 13 3847-3886.

[43] Kaully T, Kaufman-Francis K, Lesman A, Levenberg S. Vascularization: The Conduit to Viable Engineered Tissues. Tissue Eng Part B Rev. 2009; 15(2) 159-169.

[44] Atala A, Bauer SB, Soker S, Yoo JJ, Retik AB. Tissue-engineered autologous bladders for patients needing cystoplasty. Lancet 2006; 367(9518) 1241-1246.

[45] Mironov V, Reis N, Derby B. Bioprinting: A Beginning. Tissue Engineering 2006; 12(4) 631-634.

[46] Mironov V, Kasyanov, V, Markwald RR. Organ printing: from bioprinter to organ biofabrication line. Current Opinion in Biotechnology 2011; 22 667-673.

[47] Mironov V, Trusk T, Kasyanov, Little S, Swaja R, Markwald RR. Biofabrication: a 21st century manufacturing paradigm. Biofabrication 2009; 1 1-16.

[48] Partridge R, Conlisk N, Davies JA. In-lab three-dimensional printing. An inexpensive tool for experimentation and visualization for the field of organogenesis. Organogenesis 2012; 8(1) 22-27.

[49] Lalan S, Pomerantseva I, Vacanti JP. Tissue engineering and its potential impact on surgery. World Journal of Surgery 2001; 25 1458-1466.

[50] Mironov V, Boland T, Trusk T, Forgacs G, Markwald RR. Organ printing: computer-aided jet-based 3D tissue engineering. TRENDS in Biotechnology 2003; 21(4) 157-161.

[51] Mironov V, Visconti RP, Kasyanov V, Forgacs G, Drake CJ, Markwald RR. Organ printing: tissue spheroids as building blocks. Biomaterials 2009; 30 2164-2174.

[52] Yeong WY, Chua CK, Leong KF, Chandrasekaran M. Rapid prototyping in tissue engineering: challenges and potential. TRENDS in Biotechnology, 22(12) 663–662.

[53] Mirononv V, Kasyanov V, Drake C, Markwald RR. Organ printing: promises and challenges. Regenerative Medicine 2008; 3(1) 93-103.

[54] Wilson WC, Boland T. Cell and organ printing 1: protein and cell printers. The Anatomical Record Part A 2003; 272 491-496.

[55] Jakab K, Norotte C, Marga F, Murphy K, Vunjak-Novakovic G, Forgacs G. Tissue engineering by self-assembly and bio-printing of living cells. Biofabrications 2010; 2 1-14.

[56] Assis AC, Carvalho JL, Jacoby BA, Ferreira RL, Castanheira P, Diniz SO, Cardoso VN, Goes AM, Ferreira AJ. Time-dependent migration of systemically delivered bone marrow mesenchymal stem cells to the infarcted heart. Cell Transplantation 2010;19(2) 219-30.

[57] Kumar S, Ponnazhagan S. Mobilization of bone marrow mesenchymal stem cells *in vivo* augments bone healing in a mouse model of segmental bone defect. Bone 2012; 50(4) 1012-8.

[58] Liu Z, Yang D, Xie P, Ren G, Sun G, Zeng X, Sun X. MiR-106b and MiR-15b modulate apoptosis and angiogenesis in myocardial infarction. Cell Physiol Biochem 2012; 29(5-6) 851-62.

[59] Daher JD, Chahine NO, Greenberg AS, Sgaglione NA, Grande DA. New methods to diagnose and treat cartilage degeneration. Nature Reviews Rheumatology 2009; 5 599-607.

[60] Feinberg AW, Alford PW, Jin H, Ripplinger CM, Werdich AA, Sheehy SP, Grosberg A, Parker KK. Controlling the contractile strength of engineered cardiac muscle by hierarchal tissue architecture. Biomaterials 2009; 33(23) 5732-5741.

[61] Wang JH, Jia F, Gilbert TW, Woo SL. Cell orientation determines the alignment of cell-produced collagenous matrix. J Biomech. 2003; 36(1) 97-102.

[62] Tsai SW, Liou HM, Lin CJ, Kuo KL, Hung YS, Weng RC, Hsu FY. MG63 osteoblast-like cells exhibit different behavior when grown on electrospun collagen matrix versus electrospun gelatin matrix. PLoS One 2012; 7(2) e31200.

[63] Kwon IK, Kidoaki S, Matsuda T. Electrospun nano- to microfiber fabrics made of biodegradable copolyesters: structural characteristics, mechanical properties and cell adhesion potential. Biomaterials 2005; 26(18) 3929-39.

[64] Geckil H, Xu F, Zhang X, Moon S, Demirci U. Engineering hydrogels as extracellular matrix mimics. Nanomedicine 2010; 5(3) 469–484.

[65] Burdick JA, Vunjak-Novakovic G. Engineered microenvironments for controlled stem cell differentiation. Tissue Eng Part A. 2009; 15(2) 205-19.

[66] Tibbitt MW, Anseth KS. Hydrogels as extracellular matrix mimics for 3D cell culture. Biotechnol Bioeng. 2009; 103(4) 655-63.

[67] Zahir N, Weaver VM. Death in the third dimension: apoptosis regulation and tissue architecture. Current Opinion in Genetics and Development 2004; 14(1) 71-80.

[68] Zaulyanov L, Kirsner RS. A review of a bi-layered living cell treatment (Apligraf) in the treatment of venous leg ulcers and diabetic foot ulcers. Clin Interv Aging 2007; 2(1) 93-8.

Chapter 12

Biofabrication of Tissue Scaffolds

Ning Zhu and Xiongbiao Chen

Additional information is available at the end of the chapter

http://dx.doi.org/10.5772/54125

1. Introduction

Tissue engineering is an emerging interdisciplinary field that applies the principles of life science and engineering to produce engineered tissues for the repair and replacement of damaged tissues or organs [1]. In tissue engineering, tissue scaffolds play a crucial role. A tissue scaffold is a three-dimensional (3D) structure made from biological materials and biomaterials, which is used to facilitate cell/tissue growth and the transport of nutrients and wastes while degrading gradually itself. To fabricate such tissue scaffolds, a number of fabrication techniques have been developed and reported in the literature and these techniques can generally be classified into two categories: conventional and advanced. Conventional techniques [2], including solvent-casting, particulate-leaching, and freeze drying, can build scaffolds with interconnected porous structures. However, they offer little capacity to precisely control pore size, pore geometry, pore interconnectivity, and spatial distribution of pores or allow for the construction of internal channels within the scaffolds. Ideally, scaffolds should not only provide a supporting structure but also the chemical, mechanical, and biological signals required to respond to environmental stimuli. As an alternative to conventional scaffold fabrication methods, advanced fabrication techniques have recently been developed in tissue engineering, such as electrospinning [3], a nanotechnology-based fabrication technique, and rapid prototyping [4], a class of techniques by which a 3D scaffold is fabricated by laying down multiple, precisely formed layers in succession. With the development of such advanced tissue engineering fabrication techniques, the new concept of "biofabrication" has emerged. Biofabrication is defined as the production of complex living and non-living biological products from raw materials, such as living cells, molecules, extracellular matrices, and biomaterials. It has the potential to be the manufacturing paradigm of the 21st century and makes a significant contribution to the development of tissue engineering strategies [5].

2. Overview of scaffold-fabrication techniques

2.1. Conventional fabrication techniques

Many techniques are available to process synthetic and natural biomaterials into various scaffolds. These include conventional techniques, such as solvent-casting and particulate-leaching [6], gas foaming [7], phase separation [8], melt molding [9], and freeze drying [10], among others. An overview of these different techniques follows.

1. Solvent-casting and particulate-leaching (Figure 1a):

Solvent-casting and particulate-leaching techniques involve using a polymer solution uniformly mixed with salt particles of a specific diameter. The solvent then evaporates leaving behind a polymer matrix with salt particles embedded throughout. The composite is immersed in water, where the salt leaches out to produce a porous structure [11]. Highly porous scaffolds with porosity values up to 93% and average pore diameters up to 500 μm can be formed using this technique. A disadvantage of this technique is that it can only be used to produce thin membranes up to 3 mm thick [12].

2. Gas foaming (Figure 1b):

During the gas foaming process, molded biodegradable polymers are pressurized at high pressures with gas-foaming agents, such as CO_2 and nitrogen [13], water [14], or fluoroform [15], until the polymers are saturated. This results in nucleation and growth of gas bubbles with sizes ranging between 100 and 500 μm in the polymer. This technique has the advantage of being an organic solvent-free process; the major drawback is that the process may yield a structure with largely unconnected pores and a non-porous external surface [16].

3. Phase separation (Figure 1c):

During the phase separation process, a polymer solution is quenched and undergoes a liquid-liquid phase separation to form two phases; a polymer-rich phase and a polymer-poor phase. The polymer-rich phase solidifies and the polymer poor phase is removed, leaving a highly porous polymer network [17]. The micro- and macro-structure of the resulting scaffolds are controlled by varying process parameters such as polymer concentration, quenching temperature, and quenching rate. The process is conducted at low temperatures, which is beneficial for the incorporation of bioactive molecules in the structure. Using phase separation techniques, nano-scale fibrous structure enables to be formed, which mimics natural extracellular matrix architecture and provides a better environment for cell attachment and function [18].

4. Melt molding (Figure 1d):

Melt molding involves filling a mold with polymer powder and a porogen component and then heating to above the glass-transition temperature of the polymer while applying pressure to the mixture [19]. During the fabrication process, the raw materials will bind together to form a scaffold with designed specified external shape. Once the mold is removed, the

http://dx.doi.org/10.5772/54125

porogen is leached out and the porous scaffold is then dried. Melt-molding with porogen-leaching is a non-solvent fabrication process that allows independent control of morphology and shape. Drawbacks include the possibility of residual porogen and high processing temperatures that preclude the ability to incorporate bioactive molecules.

5. Freeze drying (Figure 1e):

Polymeric porous scaffolds can be prepared by freeze drying. In the freezing stage, the polymer solution is cooled down to a certain temperature at which all materials are in a frozen state and the solvent forms ice crystals, forcing the polymer molecules to aggregate into the interstitial spaces. In the second phase, the solvent is removed by applying a pressure lower than the equilibrium vapor pressure of the frozen solvent. When the solvent is completely sublimated, a dry polymer scaffold with an interconnected porous microstructure remains [20, 21]. The porosity of the scaffolds depends on the concentration of the polymer solution; pore size distribution is affected by the freezing temperatures. Apart from fabricating porous scaffolds, this technique is also used to dry biological samples to protect their bioactivities [22].

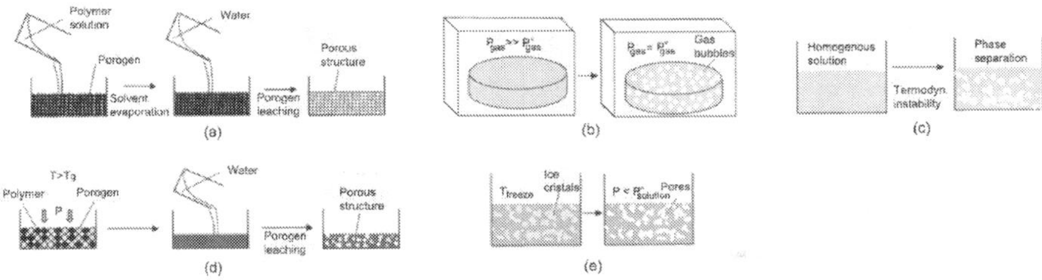

Figure 1. Schematic of conventional scaffold fabrication techniques: (a) solvent-casting and particulate-leaching process: A polymer solution is cast into a mold filled with porogen particles, then the solvent is allowed to evaporate and the porogen is leached out; (b) gas foaming process: Polymer samples are exposed to high pressure allowing saturation of the gas into the polymer; the subsequent gas pressure reduction causes the nucleation of bubbles; (c) phase separation process: A thermodynamical instability is established in a homogeneous polymer solution that separates into a polymer-rich and a polymer-poor phase; (d) melt molding process. A mold filled with polymer powder and porogen component is heated to above the polymer glass-transition temperature (T_g) and a pressure (P) is applied to the mixture. The porogen is then leached out, leaving a porous structure; (e) freeze drying process: A polymer solution is cooled down, leading to the formation of solvent ice crystals. Then the solvent is removed by using a pressure lower than the equilibrium vapor pressure of the solvent (P° solution), leaving a porous structure. (Modified from [23])

2.2. Advanced biofabrication techniques

1. Electrospinning

Electrospinning is a fabrication technique utilizing electrical charges to draw fine fibers up to the nanometer scale. The technique was invented by Cooley and Morton in 1902. The fiber electrospinning can also be traced back to the 1930s [24]. In the past decade, significant developments in electrospinning have allowed for creation of scaffolds with different materials and, hence, this technique has gained a high popularity in tissue engineering research.

Nanofibrous architectures are known to modulate effects on a wide variety of cell behaviors. Nanofibrous architectures can positively affect cell binding and spreading compared to micropore and microfibrous architectures (Figure 2). Nanofibrous scaffold architectures have larger surface areas to adsorb proteins than micro-architectures, presenting more binding sites to cell membrane receptors [25]. The exposure of additional cryptic binding sites may also be affected by adsorbed proteins. Furthermore, cells growing in a 3D nanofibrous structural environment are able to exchange nutrients and utilize receptors throughout their surface, while cells in flat culture conditions are limited to nutrient exchange on only one side. Electrospinning techniques have been widely employed to fabricate porous scaffolds with nanofibrous architectures that can mimic the structure and biological functions of the natural extracellular matrix [26]. This technique is able to generate fibers with diameters ranging from 2 nm to several micrometers using solutions of both natural and synthetic polymers, with small pore sizes and high surface area to volume ratios. A typical electrospinning setup includes three parts: a syringe pump containing the polymeric materials, a high voltage source to generate high electric field for spinning, and a collector to collect the fibers [27] (Figure 3). During scaffold fabrication, the following electrospinning parameters are very important with respect to the fiber morphology: polymer solution parameters (viscosity, molecular weight of polymer, polymer conductivity, surface tension), processing parameters (applied voltage, distance between tip and collector, flow rate), and environment parameters (humidity, temperature). Nanofibers with high surface area to volume ratios are most suitable for tissue engineering applications [28].

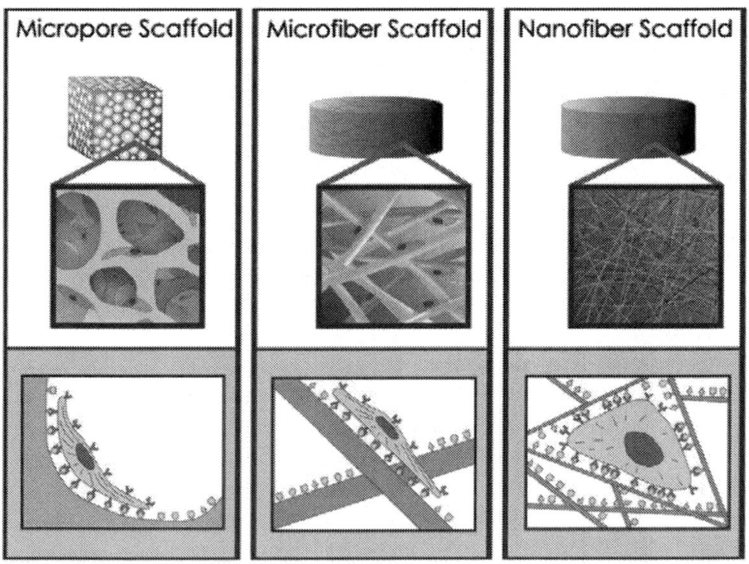

Figure 2. Scaffold architecture affects cell binding and spreading. (Modified from [25])

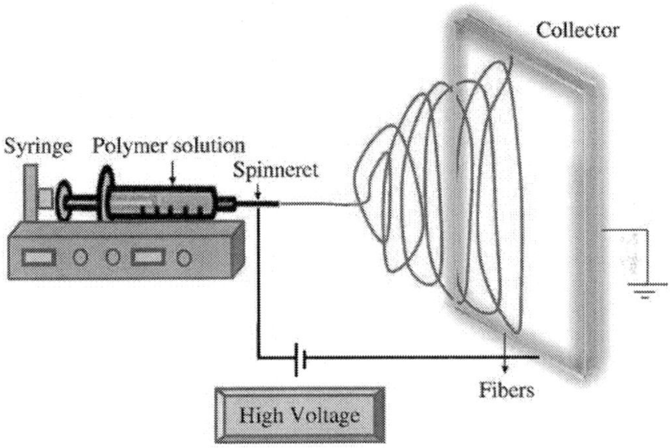

Figure 3. Schematic of electrospinning apparatus. (Modified from [29])

2. Rapid prototyping

As an alternative to conventional scaffold fabrication methods, a group of techniques based on rapid prototyping (RP) has recently been introduced within the tissue engineering field. RP techniques, based on computer assisted design (CAD) and manufacturing (CAM) techniques, allow for better control of scaffold internal microstructure and external macroshape compared to conventional fabrication techniques [4, 30]. Three basic RP system types: liquid-based, solid-based, and powder-based can be selected based on the properties of different scaffold biomaterials. The primary RP processes applied to tissue scaffold fabrication include stereolithography (SLA) [31], selective laser sintering (SLS) [32], fused deposition modeling (FDM) [33], three dimensional (3D) printing [34], and 3D plotting [35]. The choice of materials for the RP techniques includes various polymers, ceramics, and metals. Recently, RP techniques have also demonstrated their capacity for embedding living cells [36, 37] and growth factors [38] into scaffolds during the fabrication process and thus their utility for creating biomimetic tissue scaffolds.

Technology	Types	Materials	Advantages	Disadvantages	Tissue Engineering Applications
SLA	Liquid-based	Polymers, wax or wax compounds	Good mechanical strength; easy to remove support materials; easy to achieve small features	Limited to reactive resins (mostly toxic)	Bone [39], Heart valves [40].
SLS	Powder-based	Metals, ceramics, bulk polymers	Good mechanical strength; high accuracy; broad range of materials	Elevated temperatures; local high energy input; uncontrolled porosity	Bone [41, 42], Cartilage [43].
FDM	Solid-based	Some thermoplastic polymers/ ceramics	Low costs; good mechanical strength; versatile in lay-down pattern design	Elevated temperatures; small range of bulk materials	Bone [44], Adipose [45], Cartilage [46].
3D printing	Powder-based	powder of bulk polymers; ceramics	Fast processing; low costs; no toxic components; water used as binder	Material must be in powder form; weak bonding between powder particles; rough surface; trapped powder issue; might require post-processing	Bone [47],
3D plotting	Liquid-based or solid-based	Swollen polymers (hydrogels); thermoplastic polymers; reactive resins; ceramics	Broad range of materials and conditions; incorporation of cells and proteins	Slow processing; no standard condition; time consuming adjustment to new materials; low mechanical strength	Bone [48], Cartilage [49].

Table 1. RP techniques for tissue engineering

3. Applications of biofabrication to tissue engineering

3.1. Biofabrication and architectural design of scaffolds

The microstructure of scaffolds is increasingly believed to contribute significantly to the diffusion of nutrients and metabolic wastes, spatial organization of cell growth and the development of specific biological functions in tissues. A scaffold with high porosity is desirable for the easy diffusion of nutrients and metabolic wastes and is also beneficial for cell migration and neo-vascularization. A high surface area to volume ratio favors cell attachment and growth. The effect of scaffold pore size on tissue regeneration is also emphasized by experiments demonstrating that (1) an optimum pore size of 5 μm is good for neo-vascularization; (2) 5 - 15 μm pores are beneficial for fiberblast ingrowth; (3) 20 - 125 μm pores can affect the regeneration of adult mammalian skin; and (4) fibrovascular tissues require pores sizes greater than 500 μm [38, 50]. Advanced biofabrication techniques are able to design and precisely control the architecture of scaffolds. They can build scaffolds with reproducible morphology and microstructure that varies across the scaffold matrix to resemble natural tissues with complex hierarchical structures.

Conventional lyophilization can only form porous structures with random orientation. An improved technique for fabricating scaffolds with a linearly oriented architecture is called "freeze casting". Freeze casting facilitates directional solidification of solutions or slurries [51]. During freeze casting, the polymer solution is pipetted into a cylindrical mold fitted with a copper bottom plate and secured onto the temperature-controlled copper cold finger of the freeze casting system. The cold finger temperature is lowered at a constant cooling rate to a final temperature, resulting in the directional solidification of the material dispersion. When the ice is sublimated by freeze drying, the porous microstructure of the resulting scaffold is a negative template of the ice crystals. Freeze casting has been used to produce a wide range of porous, oriented scaffolds from organic and inorganic materials [51, 52]. This technique is also suitable for the fabrication of nerve conduit scaffolds with a featured porous structure that may guide axon growth.

Different scaffold fabrication techniques can be combined to capitalize on their respective positive features for varying applications. The combination of rapid prototyping with lyophilization, in which the polymer solution is dispensed on substrates with a controllable temperature and the strands formed are frozen and lyophilized to remove the solvent, is called "rapid freeze prototyping" technique [53]. This technique has the advantage of fabricating scaffolds with both sub-millimeter and micrometer sized pores [54] (Figure 4). The optimized porous scaffolds can accommodate tissue ingrowth at different scales, from cells to tissues. Scaffolds can also be cold processed so that the polymer can be bio-functionalized without compromising their function during manufacturing.

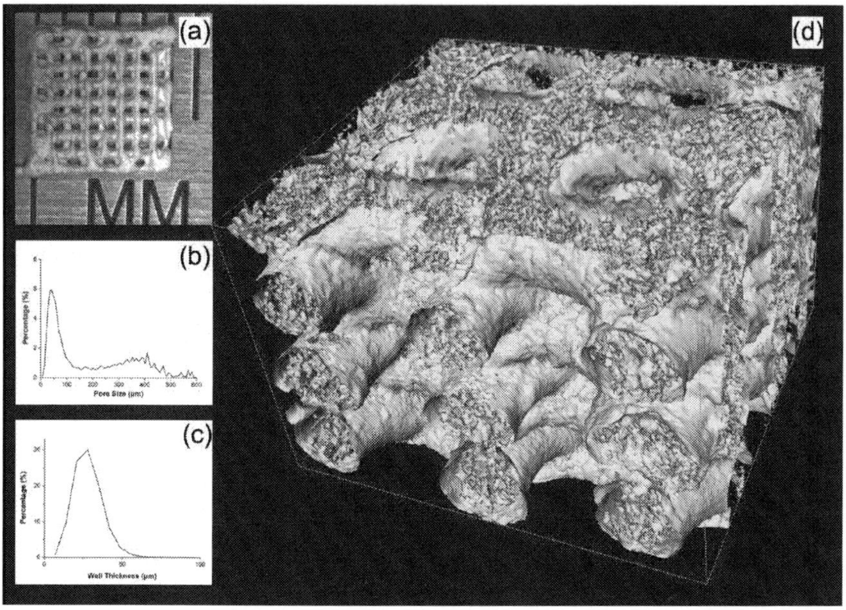

Figure 4. Scaffold fabricated by rapid freeze prototyping: (a) camera image of the scaffold, (b) pore size distribution of the scaffold, (c) wall thickness distribution of the scaffold, and (d) 3D reconstructed model of the scaffold using micro-tomography. (Modified from [54])

3.2. Biofabrication and scaffolds with living cells

Biofabrication of living structures with desired functionality has become a hot topic in tissue engineering in past few years. Conventional cell-seeding methods are inadequate for the development of *in vitro* tissue-test systems because they involve random placement of cells and, therefore, lack the precision necessary for spatial control. Conventional cell-seeding methods are also a type of 2D cell culture. In contrast, cell cultures in 3D structures allow for a more natural cell attachment and focal adhesion in all directions. The most physiologically relevant cell morphology that can be attained on and in three-dimensional scaffolds will provide the best structural cues to regulate cell function [55, 56]. Different methods of fabricating 3D scaffolds with living cells have been developed. One of the methods is to spray living cells into the scaffolds throughout the electrospinning process to produce nanofibrous 3D tissue scaffolds. In this method, cells are periodically sprayed from a pump-action spray bottle onto the developing scaffold during the electrospinning process [56]. The cells can be layered throughout the thickness of the scaffolds, but not incorporated into individual polymer nanofibers.

Living cells also can be directly electrospun, as fine composite threads encapsulating living cells, using a coaxial needle configuration and a biocompatible polymer [57, 58]. The polymer nanofibers accommodate the survival and proliferation of the cells. Advanced rapid

http://dx.doi.org/10.5772/54125

prototyping techniques, such as bioprinting, are more capable of incorporating living cells into scaffolds than other techniques. Introducing cells at almost any arbitrary density and precisely into the desired location of a scaffold is possible by means of rapid prototyping. Hydrogel scaffolds, as delivery vehicles for cells, are suitable for bioprinting processes that seed living cells while constructing scaffolds with specific geometries [36]. A pneumatic dispenser system is used to bioprint the cell-associated scaffolds using polymer solution, such as alginate aqueous solutions. The fabrication parameters including pressure and nozzle velocity can be altered, thus affecting the viability of the cells [59]. Complete biological "scaffold free" tissue substitutes can also be engineered with specific compositions and shapes, by exploiting cell-cell adhesion and the ability of cultured cells to grow their own ECM; such approaches have the advantage of reducing and mediating inflammatory responses to biomaterials [60]. For this concept, extrusion-based bioprinting is an automated deposition method that can generate a fully biological construct which is structurally and functionally close to native tissues. Spherical or cylindrical multicellular units (the bio-ink) are delivered according to a computer-generated template with the hydrogel (the bio-paper) serving as the support material. The cells neither invade nor rearrange within the hydrogel, which keeps its integrity during post-printing fusion and can be easily removed to free the fused multicellular construct (Figure 5). The authentic tissues can be assembled through cell adhesion, cell sorting, and tissue fusion processes [37].

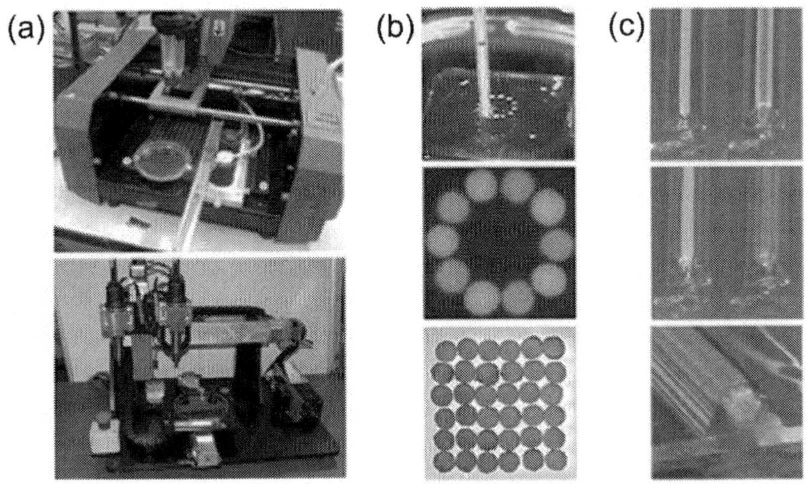

Figure 5. Scaffold-free bioprinting technology: (a) the bio-printer: 3D printing is achieved by means of a three-axis positioning system (stage in *y* and printing heads along *x* and *z* (top: Neatco, Carlisle, Canada; bottom: Organovo-Invetech, San Diego)); (b) spheroids with living cells are delivered one by one into the hydrogel bio-paper according to a computer script; (c) layer-by-layer deposition of cylindrical units of bio-paper (shown in blue) and multicellular cylindrical building blocks. The outcome of printing (spheroids in panel (b), multicellular cylinders in panel (c)) is a set of discrete units, which post-printing fuse to form a continuous structure. (Modified from [60])

4. Summary

Engineered scaffolds are playing an increasingly important role in tissue engineering. Scaffolds should not only have porous structures and provide mechanical support to new tissue regrowth but also have a complex mimetic hierarchical structure and biological features. Conventional scaffold fabrication techniques fail to meet these requirements for tissue regeneration. Biofabrication technologies have demonstrated potentials in this regard and can be used to create regenerative tissues or organs through the combination of state of the art fabrication techniques, materials science, and cell biology.

Author details

Ning Zhu and Xiongbiao Chen

Division of Biomedical Engineering, University of Saskatchewan, Saskatoon, SK, Canada

References

[1] Langer R, Vacanti JP. Tissue Engineering. Science 1993 May;260(5110):920-926.

[2] Liu C, Xia Z, Czernuszka JT. Design and development of three-dimensional scaffolds for tissue engineering. Chemical Engineering Research & Design 2007 Jul;85(A7): 1051-1064.

[3] Prabhakaran MP, Ghasemi-Mobarakeh L, Ramakrishna S. Electrospun composite nanofibers for tissue regeneration. Journal of Nanoscience and Nanotechnology 2011 Apr;11(4):3039-3057.

[4] Peltola SM, Melchels FPW, Grijpma DW, Kellomaki M. A review of rapid prototyping techniques for tissue engineering purposes. Annals of Medicine 2008;40(4): 268-280.

[5] Mironov V, Trusk T, Kasyanov V, Little S, Swaja R, Markwald R. Biofabrication: a 21st century manufacturing paradigm. Biofabrication 2009 Jun;1(2):022001.

[6] Hsu SH, Su CH, Chiu IM. A novel approach to align adult neural stem cells on micropatterned conduits for peripheral nerve regeneration: A feasibility study. Artificial Organs 2009 Jan;33(1):26-35.

[7] Yang Y, De Laporte L, Rives CB, Jang JH, Lin WC, Shull KR, et al. Neurotrophin releasing single and multiple lumen nerve conduits. Journal of Controlled Release 2005 Jun;104(3):433-446.

[8] Khorasani MT, Mirmohammadi SA, Irani S. Polyhydroxybutyrate (PHB) scaffolds as a model for nerve tissue engineering application: Fabrication and in vitro assay. International Journal of Polymeric Materials 2011;60(8):562-575.

[9] Pego AP, Poot AA, Grijpma DW, Feijen J. Biodegradable elastomeric scaffolds for soft tissue engineering. Journal of Controlled Release 2003 Feb;87(1-3):69-79.

[10] Huang YC, Huang YY, Huang CC, Liu HC. Manufacture of porous polymer nerve conduits through a lyophilizing and wire-heating process. Journal of Biomedical Materials Research Part B-Applied Biomaterials 2005 Jul;74B(1):659-664.

[11] Mikos AG, Thorsen AJ, Czerwonka LA, Bao Y, Langer R, Winslow DN, et al. Preparation and characterization of poly(L-lactic acid) foams. Polymer 1994 1994;35(5): 1068-1077.

[12] Mikos AG, Sarakinos G, Vacanti JP, Langer R, Cima LG, inventors. Biocompatible polymer membranes and methods of preparation of three dimensional membrane structures. United States Patent No. 5514378, 1996.

[13] Di Maio E, Mensitieri G, Iannace S, Nicolais L, Li W, Flumerfelt RW. Structure optimization of polycaprolactone foams by using mixtures of CO_2 and N_2 as blowing agents. Polymer Engineering and Science 2005 Mar;45(3):432-441.

[14] Haugen H, Ried V, Brunner M, Will J, Wintermantel E. Water as foaming agent for open cell polyurethane structures. Journal of Materials Science-Materials in Medicine 2004 Apr;15(4):343-346.

[15] Parks KL, Beckman EJ. Generation of microcellular polyurethane foams via polymerization in carbon dioxide .2. Foam formation and characterization. Polymer Engineering and Science 1996 Oct;36(19):2417-2431.

[16] Quirk RA, France RM, Shakesheff KM, Howdle SM. Supercritical fluid technologies and tissue engineering scaffolds. Current Opinion in Solid State & Materials Science 2004 Jun-Aug;8(3-4):313-321.

[17] Lee KWD, Chan PK, Feng XS. Morphology development and characterization of the phase-separated structure resulting from the thermal-induced phase separation phenomenon in polymer solutions under a temperature gradient. Chemical Engineering Science 2004 Apr;59(7):1491-1504.

[18] Ma PX, Zhang RY. Synthetic nano-scale fibrous extracellular matrix. Journal of Biomedical Materials Research 1999 Jul;46(1):60-72.

[19] Thomson RC, Wake MC, Yaszemski MJ, Mikos AG. Biodegradable polymer scaffolds to regenerate organs. Advances in Polymer Science 1995: 245-274.

[20] Pikal MJ, Shah S, Roy ML, Putman R. The secondary drying stage of freeze-dring kinetics as a function of temperature and chamber pressure. International Journal of Pharmaceutics 1990 May;60(3):203-217.

[21] Liapis AI, Bruttini R. A theory for the primary and secondary drying stages of the freeze-drying of pharmaceutical crystalline and amorphous solutes - comparison between experimental - data and theory. Separations Technology 1994 Jul;4(3):144-155.

[22] Bischof JC, He XM. Thermal stability of proteins. In: Lee RC, Despa F, Hamann KJ, editors. Cell Injury: Mechanisms, Responses, and Repair, 2005. p. 12-33.

[23] Puppi D, Chiellini F, Piras AM, Chiellini E. Polymeric materials for bone and cartilage repair. Progress in Polymer Science 2010 Apr;35(4):403-440.

[24] Subbiah T, Bhat GS, Tock RW, Pararneswaran S, Ramkumar SS. Electrospinning of nanofibers. Journal of Applied Polymer Science 2005 Apr 15;96(2):557-569.

[25] Stevens MM, George JH. Exploring and engineering the cell surface interface. Science 2005 Nov 18;310(5751):1135-1138.

[26] Huang ZM, Zhang YZ, Kotaki M, Ramakrishna S. A review on polymer nanofibers by electrospinning and their applications in nanocomposites. Composites Science and Technology 2003 Nov;63(15):2223-2253.

[27] Pham QP, Sharma U, Mikos AG. Electrospinning of polymeric nanofibers for tissue engineering applications: A review. Tissue Engineering 2006 May;12(5):1197-1211.

[28] Jain KK. Role of nanotechnology in developing new therapies for diseases of the nervous system. Nanomedicine 2006 Jun;1(1):9-12.

[29] Bhardwaj N, Kundu SC. Electrospinning: A fascinating fiber fabrication technique. Biotechnology Advances 2010 May-Jun;28(3):325-347.

[30] Yeong WY, Chua CK, Leong KF, Chandrasekaran M. Rapid prototyping in tissue engineering: challenges and potential. Trends in Biotechnology 2004 Dec;22(12):643-652.

[31] Karalekas DE. Study of the mechanical properties of nonwoven fibre mat reinforced photopolymers used in rapid prototyping. Materials & Design 2003 Dec;24(8):665-670.

[32] Wiria FE, Leong KF, Chua CK, Liu Y. Poly-epsilon-caprolactone/hydroxyapatite for tissue engineering scaffold fabrication via selective laser sintering. Acta Biomaterialia 2007 Jan;3(1):1-12.

[33] Hutmacher DW, Schantz T, Zein I, Ng KW, Teoh SH, Tan KC. Mechanical properties and cell cultural response of polycaprolactone scaffolds designed and fabricated via fused deposition modeling. Journal of Biomedical Materials Research 2001 May;55(2):203-216.

[34] Lam CXF, Mo XM, Teoh SH, Hutmacher DW. Scaffold development using 3D printing with a starch-based polymer. Materials Science & Engineering C-Biomimetic and Supramolecular Systems 2002 May 31;20(1-2):49-56.

[35] Landers R, Mulhaupt R. Desktop manufacturing of complex objects, prototypes and biomedical scaffolds by means of computer-assisted design combined with computer-guided 3D plotting of polymers and reactive oligomers. Macromolecular Materials and Engineering 2000 Oct;282(9):17-21.

[36] Khalil S, Sun W. Bioprinting endothelial cells with alginate for 3D tissue constructs. Journal of Biomechanical Engineering-Transactions of the ASME 2009 Nov;131(11).

[37] Mironov V, Kasyanov V, Drake C, Markwald RR. Organ printing: promises and challenges. Regenerative Medicine 2008 Jan;3(1):93-103.

[38] Yu D, Li Q, Mu X, Chang T, Xiong Z. Bone regeneration of critical calvarial defect in goat model by PLGA/TCP/rhBMP-2 scaffolds prepared by low-temperature rapid-prototyping technology. International Journal of Oral and Maxillofacial Surgery 2008 Oct;37(10):929-934.

[39] Chang PSH, Parker TH, Patrick CW, Miller MJ. The accuracy of stereolithography in planning craniofacial bone replacement. Journal of Craniofacial Surgery 2003 Mar; 14(2):164-170.

[40] Sodian R, Loebe M, Hein A, Martin DP, Hoerstrup SP, Potapov EV, et al. Application of stereolithography for scaffold fabrication for tissue engineered heart valves. ASAIO Journal 2002 Jan-Feb;48(1):12-16.

[41] Williams JM, Adewunmi A, Schek RM, Flanagan CL, Krebsbach PH, Feinberg SE, et al. Bone tissue engineering using polycaprolactone scaffolds fabricated via selective laser sintering. Biomaterials 2005 Aug;26(23):4817-4827.

[42] Cruz F, Simoes J, Coole T, Bocking C. Direct manufacture of hydroxyapatite based bone implants by Selective Laser Sintering, 2nd International Conference on Advanced Research and Rapid Prototyping (VRAP 2005), September 28 – October 1, 2005, Leiria, Portugal.

[43] Chen C-H, Chen J-P, Lee M-Y. Eeffects of gelatin modification on rapid prototyping PCL scaffolds for cartilage engineering. Journal of Mechanics in Medicine and Biology 2011 Dec;11(5):993-1002.

[44] Rohner D, Hutmacher DW, Cheng TK, Oberholzer M, Hammer B. In vivo efficacy of bone-marrow-coated polycaprolactone scaffolds for the reconstruction of orbital defects in the pig. Journal of Biomedical Materials Research Part B-Applied Biomaterials 2003 Aug 15;66B(2):574-580.

[45] Wiggenhauser PS, Mueller DF, Melchels FPW, Egana JT, Storck K, Mayer H, et al. Engineering of vascularized adipose constructs. Cell and Tissue Research 2012 Mar; 347(3):747-757.

[46] Koo S, Hargreaves BA, Gold GE, Dragoo JL. Fabrication of custom-shaped grafts for cartilage regeneration. The International Journal of Artificial Organs 2010 Oct;33(10): 731-737.

[47] Seitz H, Rieder W, Irsen S, Leukers B, Tille C. Three-dimensional printing of porous ceramic scaffolds for bone tissue engineering. Journal of Biomedical Materials Research Part B-Applied Biomaterials 2005 Aug;74B(2):782-788.

[48] Yilgor P, Sousa RA, Reis RL, Hasirci N, Hasirci V. 3D plotted PCL scaffolds for stem cell based bone tissue engineering. Macromolecular Symposia 2008 2008;269:92-99.

[49] El-Ayoubi R, Degrandpre C, DiRaddo R, Yousefi AM, Lavigne P. Design and dynamic culture of 3D-scaffolds for cartilage tissue engineering. Journal of Biomaterials Applications 2011 Jan;25(5):429-444.

[50] Whang K, Healy KE, Elenz DR, Nam EK, Tsai DC, Thomas CH, et al. Engineering bone regeneration with bioabsorbable scaffolds with novel microarchitecture. Tissue Engineering 1999 Spr;5(1):35-51.

[51] Wegst UGK, Schecter M, Donius AE, Hunger PM. Biomaterials by freeze casting. Philosophical Transactions of the Royal Society a-Mathematical Physical and Engineering Sciences 2010 Apr 28;368(1917):2099-2121.

[52] Zhang H, Cooper AI. Aligned porous structures by directional freezing. Advanced Materials 2007 Jun 4;19(11):1529-1533.

[53] Pham CB, Leong KF, Lim TC, Chian KS. Rapid freeze prototyping technique in bio-plotters for tissue scaffold fabrication. Rapid Prototyping Journal 2008;14(4):246-253.

[54] Zhu N, Li MG, Cooper D, Chen XB. Development of novel hybrid poly(l-lactide)/chitosan scaffolds using the rapid freeze prototyping technique. Biofabrication 2011;3(3):034105.

[55] Albrecht DR, Underhill GH, Wassermann TB, Sah RL, Bhatia SN. Probing the role of multicellular organization in three-dimensional microenvironments. Nature Methods 2006 May;3(5):369-375.

[56] Seil JT, Webster TJ. Spray deposition of live cells throughout the electrospinning process produces nanofibrous three-dimensional tissue scaffolds. International Journal of Nanomedicine 2011 2011;6:1095-1099.

[57] Townsend-Nicholson A, Jayasinghe SN. Cell electrospinning: a unique biotechnique for encapsulating living organisms for generating active biological microthreads/scaffolds. Biomacromolecules 2006 Dec 11;7(12):3364-3369.

[58] Hamid Q, Sun W, Asme. Coaxial electrospinning biopolymer with living cells, ASME 2010 First Global Congress on NanoEngineering for Medicine and Biology (NEMB2010), February 7-10, 2010, Houston, Texas, USA; Paper no. NEMB2010-13282, pp. 167-172.

[59] Khalil S, Sun W. Biopolyrner deposition for freefonn fabrication of hydrogel tissue constructs. Materials Science & Engineering C-Biomimetic and Supramolecular Systems 2007 Apr;27(3):469-478.

[60] Jakab K, Norotte C, Marga F, Murphy K, Vunjak-Novakovic G, Forgacs G. Tissue engineering by self-assembly and bio-printing of living cells. Biofabrication 2010 Jun; 2(2).

Chapter 13

Biomaterials and Stem Cell Therapies for Injuries Associated to Skeletal Muscular Tissues

Tiago Pereira, Andrea Gärtner, Irina Amorim,
Paulo Armada-da-Silva, Raquel Gomes,
Cátia Pereira, Miguel L. França, Diana M. Morais,
Miguel A. Rodrigues, Maria A. Lopes, José D. Santos,
Ana Lúcia Luís and Ana Colette Maurício

Additional information is available at the end of the chapter

http://dx.doi.org/10.5772/ 53335

1. Introduction

Skeletal muscle injuries are common in humans, particularly in athletes and it is important to develop new methods to improve muscle regeneration. Skeletal muscle has good regenerative ability, but the extent of muscle injury might prevent complete regeneration, especially in terms of functional recovery. Severe lesions, like those originated by trauma associated with loss of healthy muscular tissue and development of fibrous tissue scar and irreversible muscular atrophy after long-term peripheral nervous injuries are examples of those situations where regeneration is limited. An alternative approach for the restoration of the damaged skeletal muscular tissue, considered to be an ultimate treatment of some traumatic or degenerative diseases, is the transplantation of stem cells that limit the fibrosis and the atrophy of the involved muscle masses, and even imply the myocytes regeneration and local revascularization [1]. Stem cells and regenerative medicine is a fast emerging field with rapid strides of progress and focus on human health. Successful clinical use of stem cells in regenerative medicine depends on 3 important features:

i. stem cells can grow and divide indefinitely,

ii. stem cells can differentiate into specialized cell types, like skeletal muscle; and

iii. the stem cells can be delivered to the site of lesion associated to biomaterials, nowadays available with excellent characteristics concerning biocompatibility.

The purpose of this review is to describe the current research lines in the skeletal muscle regeneration field, with special emphasis to the work performed by our research group in testing different biomaterials and cellular therapies, emphasizing the use of mesenchymal stem cells (MSCs) isolated from the Wharton's jelly of the umbilical cord. We also focused our research in developing skeletal muscular lesion models which could be reproducible. It is important to state, that a multidisciplinary team has a crucial role in the development of these biomaterials associated to cellular systems, and in pre-clinical tests. MSCs comprise a rare population of multipotent progenitor cells with a great therapeutic potential since they are capable of self-renewal and multi-lineage differentiation. Due to this ability, MSCs appear to be an attractive tool in the context of Tissue Engineering and cell-based therapy concerning skeletal muscle regeneration. Several biomaterials associated to MSCs from the Wharton's jelly of the umbilical cord have been tested in standard lesions of the rat muscle and the results of these tests will be discussed here. The umbilical cord matrix is an important and safe source of MSCs with positive effects in nerve and skeletal muscle regeneration, with no ethical or technical issues.

2. Skeletal muscle tissue

2.1. Basic structure and terminology

The muscle fibers are the basic contractile units of skeletal muscles. They are individually surrounded by a connective tissue layer and grouped into bundles to form a skeletal muscle [1]. Each muscle is surrounded by a layer of dense connective tissue - the epimysium - which is continuous with the tendon. The muscle is composed of numerous bundles of muscle fibers – fascicles - which are separated from each other by another connective tissue layer named perimysium. The endomysium is the connective tissue that separates individual muscle fibers from each other. Mature muscle cells are termed muscle fibers or myofibers. Each myofiber is a multinucleate syncytium formed by fusion of immature muscle cells termed myoblasts. In the cytoplasm of each myofiber – the sarcoplasm - lays the contractile apparatus of the cell which is composed of sarcomeres arranged in series to form myofibrils, which give myofibers their striated appearance. The sarcomeres contain a number of proteins, including alpha-actinin — which is the major constituent of the Z band — and actin and myosin, which are the major components of the thin and thick filaments, respectively. The sarcoplasm, located between the myofibrils, is called the intermyofibrillar network and contains the mitochondria, lipid, glycogen, T-tubules, and sarcoplasmic reticulum [1-3]. Skeletal muscles are highly vascularized to provide essential nutrients for muscle function. As the myofiber matures, it is contacted by a single motoneuron and expresses characteristic molecules for contractile function, principally different myosin heavy chain isoforms and metabolic enzymes. Both the motoneuron and the myoblast origin have been implicated to play a role in specifying the myofiber contractile properties, although the precise mechanisms remain to be defined [1].

2.1.1. *Fiber type*

Individual adult skeletal muscles are composed of a mixture of myofibers with different physiological properties, ranging from a slow-contracting/fatigue-resistant type to a fast-contracting/non-fatigue-resistant type. The proportion of each fiber type within a muscle determines its overall contractile properties [1]. The slow contracting *soleus* muscle is rich in myofibers expressing the slow type I myosin heavy chain isoform, whereas the fast contracting plantaris muscle is devoid of slow type I myofibers [1-3]. The most informative methods to delineate muscle fiber types are based on specific myosin profiles, specially the myosin heavy chain (MHC) isoform complement. According to the major MHC isoforms found in adult mammalian skeletal muscles, the following pure fiber types exist: slow type I with MHCIb, and three fast types, namely type IIA with MHCIIa, type IID with MHCIId and type IIB with MHCIIb [4]. Despite having different physiological properties, the basic mechanism of muscle contraction is similar in all myofiber types and is the result of a "sliding mechanism" of the myosin-rich thick filament over the actin-rich thin filament after neuronal activation [5]. The connective tissue framework in skeletal muscle combines the contractile myofibers into a functional unit, in which the contraction of myofibers is transformed into movement via myotendinous junctions at their ends, where myofibers attach to the skeleton by tendons. Thus the functional properties of skeletal muscle depend on the maintenance of a complex framework of myofibers, motor neurons, blood vessels, and extracellular connective tissue matrix [1].

2.2. Regeneration of the skeletal muscle

Regeneration is a unique adaptation of skeletal muscle that occurs in response to injury. Following direct trauma or disease, the regeneration of skeletal muscle results in restoration, to some degree, of the original structure and function of the muscle tissue [6]. Skeletal muscle regeneration is a physiological response of the tissue to traumatic or pathological injuries and its progress depends on the type of damaged muscle and the extent of the injury. Under normal conditions, the regenerated muscle is morphologically and functionally indistinguishable from undamaged muscle [1]. Regeneration resembles the process of formation of skeletal muscle during embryogenesis. Skeletal myogenesis begins in the somites where multipotencial mesodermal cells commit to the myogenic lineage. These mononucleated myoblasts then fuse and form multinucleated cells (myotubes) that ultimately develop into mature myofibers [1, 7]. During the course of muscle development, a distinct subpopulation of myoblasts fails to differentiate and remains associated with the surface of the developing myofiber as quiescent muscle satellite cells (SCs) in fully developed mature skeletal tissue [1, 7].

2.2.1. *Satellite Cells and other cells involved in regeneration of skeletal muscle tissue*

During regeneration and muscle repair, SCs fuse together or to the existing fibers to form new muscle fibers [8]. Although the number of SCs is greatly reduced in aged muscle, those remaining maintain an intrinsic capacity to regenerate the muscle tissue as efficiently as in younger muscles. A vital condition for successful regeneration is the presence of SCs in the uninjured portions of the basal membrane of the myofiber, along with its ability for reinnervation and revascularization. After a skeletal muscle injury, myofibers become completely desintegrated via myolysis and the SCs are realeased from the basal membrane. From this point SCs start to divide and are capable of differentiating into muscle fibers, reestablishing myofiber's architecture and restoring the muscle function [9, 10]. In post-natal skeletal muscle, PW1 expression is detected in SCs and a subset of interstitial cells and is markedly upregulated during muscle regeneration [11]. These interstitial multipotent stem cells are extralaminal and exhibit fibroblastic morphology but do not express the same myogenic markers such as Pax7 [10]. $PW1^+/Pax7^-$ interstitial cells (PICs) are myogenic *in vitro* and efficiently contribute to skeletal muscle regeneration *in vivo* as well as generating satellite cells and PICs. PICs show bipotential behavior *in vitro,* generating both smooth and skeletal muscle. Isolated PICs do not express Pax7 or MyoD, but they convert to a $Pax7^+/MyoD^+$ state before forming skeletal muscle *in vitro.* PICs are not derived from a Pax3-expressing parental cell and thus do not share a satellite cell lineage; however, PICs do express Pax3 upon conversion to skeletal muscle. PICs are a key cell population that cannot be recruited into the skeletal muscle lineage in the absence of Pax7 function and is likely to contribute to the Pax7 muscle phenotype during postnatal growth. PICs are as abundant as SCs in muscle tissue and correspond to the only population of PW1+/Pax7– cells *in vivo,* requiring Pax7 for their myogenic capacity [11]. PDGFRα+ mesenchymal progenitor cells located in the muscle *interstitium* were also identified as being distinct from SCs. Of the muscle-derived cell populations, only PDGFRα+ cells show efficient adipogenic differentiation both *in vitro* and *in vivo,* being strongly inhibited by the presence of satellite cell-derived myofibres. These results suggest that PDGFRα+ mesenchymal progenitors are the major contributor to ectopic fat cell formation in skeletal muscle that is more conspicuous in perimysium and particularly in perivascular space. The balance between satellite cell-dependent myogenesis and PDGFRα+ cell-dependent adipogenesis, rather than multipotency of satellite cells, has a considerable impact on muscle homeostasis [12]. Hematopoietic and dendritic cells are also present in the perimysium of the skeletal tissue, as well as some lymphocytes and macrophages [10].

2.2.2. Myogenic differentiation

Cells derived from Pax3-expressing cells are myofibres and SCs [11]. Once activated, SCs express factors involved in the specification of the myogenic program, such as Pax-7, desmin, MNFα, Myf5, MRF4 and MyoD. Activated SCs enter the cell cycle and proliferate as indicated by the expression of factors involved in cell cycle progression, such as PCNA and by the incorporation of BrDU. Recently, miRNAs have also been reported to regulate gene expression in skeletal muscle. Upon activation, SCs generate fusion-competent myoblasts and can self-renew at least to a limited extent. Any interruption in the proliferation or fusion of myoblasts, or any alterations in the extracellular matrix leads to the development of fibrosis, compromising the establishment of the correct muscular function [8, 10]. Proliferative MyoD and/or Myf5 positive myogenic cells are termed myoblasts. Both SCs and myoblasts increase their cytoplasmic-nuclear ratio and can migrate along myofibers. Proliferating myoblasts withdraw from the cell cycle to become terminally differentiated myocytes that express the "late" myogenic regulatory factors (MRFs), Myogenin and MRF4, and subsequently muscle-specific genes such as MHC and muscle creatine kinase (CK_M), and stopping Pax7 expression. Myogenic subpopulations have also been identified by their enriched M-cadherin and CD34 expression. M-cadherin can be considered to be a reliable marker for both quiescent and activated SCs. Once fusion of myogenic cells is completed, newly formed myofibers increase in size and myonuclei move to the periphery of the muscle fiber [1, 10, 13].

2.2.3. *Degeneration*

This scenario changes dramatically when the muscle is damaged, in which muscle degeneration after acute injury is characterized by myofiber necrosis and is followed by inflammation, tissue reconstruction and remodeling [10]. The necrosis is triggered by disruption of the myofiber sarcolemma resulting in increased myofiber permeability. The disruption of myofiber integrity is reflected by increased serum levels of muscle proteins, such as CK (usually restricted to the myofiber cytosol) [1]. It has been hypothesized that increased Ca^{2+} influx after sarcolemmal or sarcoplasmic reticulum damage results in a loss of Ca^{2+} homeostasis and increased Ca^{2+}-dependent proteolysis that drives tissue degeneration resulting in focal or total autolysis depending on the extent of the injury [1].

2.2.4. *Inflammation*

The early regenerative response in skeletal muscle is similar to that in other tissues and requires the coordinated regulation of inflammation, extracellular matrix remodeling, and myofiber growth [14]. The early phase of muscle injury is usually accompanied by the activation of mononucleated cells, mainly inflammatory cells and myogenic cells. Factors released by the injured muscle activate inflammatory cells residing within the muscle, which in turn provide the chemotactic signals to circulating immune cells. Neutrophils are the first immune cells to invade the injured muscle, with a significant increase in their number being observed as early as 1–6h after myotoxin or exercise-induced muscle damage. After neutrophil infiltration and 48h post-injury, macrophages become the predominant inflammatory cell type within the site of injury. Macrophages infiltrate the injured site and through phagocytosis remove cellular debris and may affect other aspects of muscle regeneration by activating myogenic cells [1]. Testosterone has a documented ability to modulate the activity of immune, fibroblast, and myogenic precursor cells, which are all components of regeneration [14].

3. Skeletal muscle injury models

In order to study the process of muscle regeneration in a controlled and reproducible way, it was necessary to develop experimental models of muscle injury [1]. In this sense, a variety of experimental models that compromise skeletal muscle function or destroy this tissue is available. Each of the injury models can potentially have a different effect on the fate of resident cells and circulating cells within the muscle bed after the trauma [10]. A large number of studies, involving a variety of experimental injuries, such as injection of myotoxic agents, crush, ischemia, denervation and muscular dystrophies, demonstrate the unique ability of skeletal muscle for regeneration, irrespectively the precise method used to induce the initial injury [15]. In this review we will focus on chemical and mechanical models of skeletal muscle injury, adding a new model of muscle injury based on surgical myectomy, developed in order to mimic severe losses of skeletal muscle mass. Other models, like exercise and denervation, will also be outlined. The latter is not a model of injury but else of skeletal muscle disuse but that also can be used to investigate skeletal muscle remodeling. There is a variety of other genetic models that are essential in studying diseases like Duchenne muscle dystrophy (Mdx mouse is currently the most widely used in this case) but will not be discussed in this review.

3.1. Chemical methods of skeletal muscle injury

The use of myotoxins, such as bupivacaine (Marcaine), cardiotoxin (CTX), and notexin (NTX) is perhaps the easiest and most reproducible way to induce muscle injury and regeneration. Myotoxins are also widely used to induce skeletal muscle injury because their inoculation by intramuscular injection does not require complex surgery. Several chemical agents are known to produce skeletal muscle damage. Severe muscle fiber damage, like breakdown of sarcolemma and myofibrils, has been described after intramuscular injections of 0.75% bupivacaine, 2% mepivacaine, or 2% lidocaine associated to epinephrine [18]. While lidocaine can cause rapid destruction of skeletal muscle fibers, long-acting anesthetics, like bupivacaine, are more often used to cause skeletal muscle injury in rodents [19].

3.1.1. Bupivacaine

The bupivacaine injection procedure is simple and quick, does not involve extensive surgery, and induces a regeneration process which is qualitatively similar to that observed in other model systems. Doses of 1.5 and 1% wt/vol produce significant levels of muscle injury and subsequent regeneration, but these doses also produce large regions of ischemic muscle tissue. Doses of 0.75 and 0.5% bupivacaine are also effective in inducing regeneration and produce little or no ischemia [16]. Muscle fiber necrosis is extremely rapid after induced bupivacaine injury [14]. Injection of the drug into small skeletal muscles of rat or mouse leads to immediate and massive myonecrosis followed by phagocytosis of necrotic debris and a rapid and apparently complete regeneration of muscle fibers 3-4 wk after injection. The peak isometric twitch and tetanic tensions produced by rat fast-twitch *extensor digitorum longus* muscle injected with bupivacaine returns to normal values by 21 d after injection [17]. Morphological analysis has shown that many indexes of successful regeneration in healthy muscle can be completed within 2–3 weeks of recovery from injury [14]. The sequence of fiber breakdown induced by bupivacaine is similar to that of progressive muscular dystrophy [18] and it is also striking that the same types of muscle fibers are spared by both Duchenne's muscular dystrophy and bupivacaine toxicity. It has been suggested that bupivacaine may disrupt Ca^{2+} homeostasis *in vivo,* triggering Ca^{2+}-activated cellular death pathways that include proteolysis. This suggestion is supported by the findings that

i. bupivacaine affects sarcoplasmic reticulum function *in vitro,*

ii. extracellular Ca^{2+} omission delays the morphological changes and decreases the protein degradation rate that are observed in isolated rat soleus muscle exposed to bupivacaine, and

iii. bupivacaine uncouples isolated rat liver and heart mitochondria and decreases mitochondrial membrane potential and oxygen consumption both in cultured fibroblasts and Ehrlich tumor cells [19].

Extracellular Ca^{2+} plays a part in mediating the muscle damage caused by bupivacaine but other factors must also be involved [20]. For example macrophage invasion is necessary for complete degeneration of myofibrillar components [21]. Saito and Nonaka [22] injected 0,5ml of 0,5% bupivacaine after *soleus* muscle exposure of Wistar rats, and observed that SCs proliferation began at almost the same time as following muscle crush injuries. Bupivacaine is still commonly used for the purpose of studying the mechanisms of skeletal muscle regeneration following injury [7, 14, 15, 22].

3.1.2. Cardiotoxin and Notexin

Snake venom is known for a long time to directly affect the skeletal muscle, producing fibrillation, contractures and depolarization of the sarcolemma. Although initially ascribed to the phospholipase A content of this venom, muscle contracture and depolarization seem to be related to the cardiotoxic action of cobra venom [23]. Notexin (NTX) is a phospholipase A2 neurotoxin peptide extracted from snake venoms that blocks neuromuscular transmission by inhibition of acetylcholine release. Cardiotoxin (CTX) is also a peptide isolated from snake venoms, but it is a protein kinase C-specific inhibitor that appears to induce the depolarization and contraction of muscle cells, disruption of membrane structure, and lysis of various cell types [1]. CTX is postulated to be neurotoxic as its injection destroys neuromuscular junctions [24]. However, CTX might cause direct destruction of muscle tissues [25]. Snake CTX polypeptide is now known to be a potent inducer of muscle contracture with phospholipase A likely acting in accelerating the action of CTX rather than in augmenting it [23, 26]. Dantrolene antagonizes CTX-induced contractures, suggesting a role for Ca^{2+} derived from the sarcoplasmic reticulum in CTX action. CTX rapidly lowers the threshold for Ca^{2+}-induced Ca^{2+} release in heavy sarcoplasmic reticulum fractions. The mechanism of action involved in contractures of skeletal muscle appears to be related to the immediate and specific effect of CTX (Ca^{2+} release by the sarcoplasmic reticulum) [27, 28].

A more recent study by Gutiérrez and Ownby [25] focused on the role of PLA_2 as important myotoxic components in these venoms suggesting that myotoxic PLA_2s binds to acceptors in the plasma membrane leading to its disruption and pronounced Ca^{2+} influx which, in turn, initiates a complex series of degenerative events associated with contracture, activation of calpains and cytosolic Ca^{2+}-dependent PLA_2s, and mitochondrial Ca^{2+} overload. Fourie et al. [30] already had suggested that the biological effects of CTX could be a consequence of inhibition of plasma membrane (Ca^{2+} + Mg^{2+})-ATPases. The local myonecrosis is often associated with other effects, such as hemorrhage, blistering and edema, in a complex pattern of local tissue damage. Apart from membrane-active CTXs, snake venom hemorrhagic metalloproteinases also cause myonecrosis, but the mechanism involved is likely to be an indirect one, probably related to ischemia [25]. CTX is a useful model for muscle regeneration that does not influence muscle architecture like basal lamina or microvasculature, making the regeneration process less complicated than other models like crush, where for example, inadequate blood supply might result in an increase of fibrosis. CTX injection also results in faster and more extensive muscle degeneration, and an earlier start of the reconstruction phase, than muscle crushing [24].

3.2. Mechanical methods of skeletal muscle injury

Crush injuries of the skeletal muscle can occur in considerable numbers following natural disasters or acts of war and terrorism. They can also occur sporadically after industrial accidents or following periods of unconsciousness from drug intoxication, anesthesia, trauma or cerebral events [31]. Crushing as a method of inducing muscle injury and regeneration was first described by Bassaglia and Gautron [32], and has since been used in several published research studies [24]. Muscle damage occurs at three distinct stages: at the time of the initial mechanical crushing force, during the period of ischemia and during the period of reperfusion [31]. It has been hypothesized that ischemia is the primary instigator of local muscle damage following crush injuries [33]. However, studies have shown that although skeletal muscle tissue can survive circulatory ischemia for 4h, the mechanical force sustained in crushing, along with ischemia, causes skeletal muscle death in only 1 h. Studies of enzyme release suggest that most damage to myocytes occurs during the reperfusion stage rather than the ischemic stage [31]. Animal models of muscle injury should closely mimic the clinical situation. Among these models open crush lesion have been used frequently, allowing standardized evaluation of regeneration in a selected muscle. For application of the trauma, either forceps or custom-made devices have been used. There are two types of muscle-crush models described in the literature: the segmental crush and the complete crush, where only 4-6% of the fibers remain intact [34]. There are different forms to accomplish the segmental crush model but most of them include the use of a surgical instrument (hemostatic clamp e.g.) to produce a standardized closing force in a specific area of a muscle causing a compression contusion injury [35]. One of the important steps of this procedure is denervation, which makes the initial steps of regeneration less painful for the animal. Skeletal muscle contusion can also be performed without skin incision by dropping a mass over a selected muscle. This technique was used by Iwata, Fuchioka [36] employing a 640g mass dropped from a 25 cm height onto an impactor (diameter 10 mm) placed on the belly of the rat medial *gastrocnemius*. This procedure damaged around 47% of the entire cross-sectional area of both medial and lateral *gastrocnemius*. At day 2 post-injury, an intense inflammatory response and necrotized myofibers with infiltrated mononuclear cells were observed. No myotubes were found at this stage. However, a number of regenerative myotubes were detected at days 7, 14, and 21 days post-injury. This study also showed that normal locomotion recovers prior to isometric force and complete regeneration of the injured muscle [36]. The main disadvantage of the complete muscle crush is the potential damaging of myoneural junctions which triggers not only regeneration of muscle substance but also initial innervation deficits. These deficits always lead to impaired healing [34]. Histological analysis of muscle regeneration after crush injury shows an initial phase of inflammation followed by SCs activation, myotube regeneration and fibrosis of the muscle. It has been shown that development of fibrotic tissue is one of the main factors affecting the recovery of muscle function after traumatic muscle injury [34]. In a qualitative assessment performed by our group we tested the open crush lesion in the *tibialis anterior* (TA) muscle of adult Sasco Sprague rats. Different standardized force intensities, durations of muscle compression (30 seconds and 1 minute) and time points (3, 8, 15 and 21 days post-surgery) were considered for the histological evaluation of skeletal muscle injury. Hematoxilin-eosin (HE) and Masson's trichrome staining were employed in this preliminary study. At day 3 post-surgery, myofiber damage was evident and the lymph nodes were reactive due to the active inflammatory process. The presence of fibrosis was evident only following 15 days from the initial injury. This evaluation revealed that the crush model was not the most appropriate for *in vivo* evaluation of cellular therapies for skeletal muscle regeneration aid, since the extent of this injury type did not present the magnitude required to accurately appreciate the biological effects of MSCs utilization [38].

3.3. Myectomy and myotomy

The loss of a portion of a skeletal muscle poses a unique challenge for regeneration of muscle tissue and restoration of its normal structure and function [39]. In the event of large-scale soft tissue traumas, extensive loss of full-thickness native tissue architecture renders the wound site unable to support normal regeneration process. In severe tissue injuries the acute inflammatory response is followed by formation of a provisional fibrin matrix derived from trauma-associated blood clotting and this matrix is then infiltrated by type I collagen-producing fibroblasts [40]. In order to mimic those situations, new experimental models have been developed in which a defined portion of the muscle tissue is removed, creating a myectomy defect within the muscle. For example, Merrit et al. [39] removed a 0.5 x 1.0 cm or a 1.0 x 1.0 cm fraction of the *gastrocnemius* muscle of rats, creating a small and large defect respectively. This was accomplished lacerating the lateral side of the muscle with a #9 scalpel blade. We have recently developed a novel experimental muscle injury model in the TA muscle of adult Sasco Sprague rat, by using a biopsy punch to create a standardized myectomy defect. Sasco Sprague male rats with 250-300g were used and after a standardized 5 mm diameter myectomy lesion in the mid-belly of the *tibialis anterior* muscle, the defect was completely filled with different vehicles and/or biomaterials, cellular suspensions containing $1x10^6$ human MSCs isolated from Wharton's jelly and conditioned media (Figure 1). This concentrated media contains trophic factors secreted by MSCs during cell culture. In our research work, the myectomy model proved to be the most appropriate for a comprehensive and standard evaluation of the rat skeletal muscular regeneration ability. The regeneration process in other models of lesion, like simple muscle crush, did not present the magnitude required to accurately appreciate the biological effects of MSCs [38].

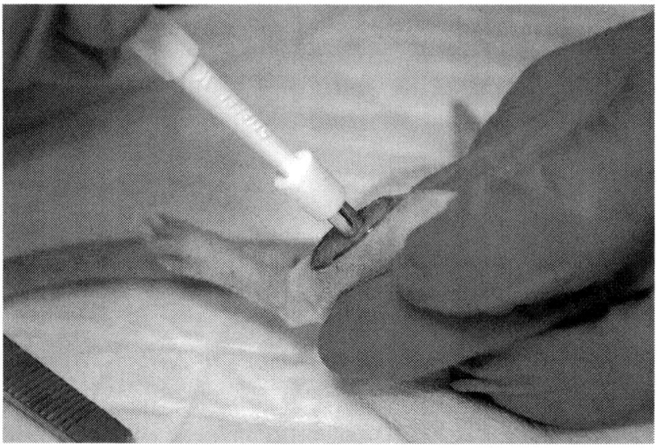

Figure 1. Biopsy punch for myectomy lesion creating a 5 mm Ø defect in the rat *tibialis anterior* (TA) muscle.

Another less invasive model of muscle injury has been used in a number of studies by producing a laceration injury (myotomy) [42-44]. In some cases this was obtained by a partial thickness (50%) cut of gastrocnemius muscles in mice at 60% of their length from their distal insertion, through 75% of their width and then sutured with a modified Kessler stitch and simple sutures using a PDS 7.0 wire (Ethicon, Somerville, New Jersey) [43]. Other studies used a full-thickness (100%) cut though 50% of the *gastrocnemius* muscle width [44]. The advantages of this model are its reproducibility and the ability to apply consistently precise injections into the laceration site [43].

3.4. Denervation (indirect model)

Innervation regulates skeletal muscle mass and muscle phenotype and peripheral nerve injury in the rat is a widely used model to investigate nerve regeneration and can also be employed as a model of muscle inactivity and muscle atrophy. Changes in the muscles may contribute to functional deficit after nerve injury [47]. Denervation induces muscle atrophy and 25 months post denervation muscle fibers cross sectional area of the *extensor digitorium longus* (EDL) muscle diminish to only 2.5% of control animals although their fascicular organization is maintained [47]. The effect of denervation on muscle atrophy is both activity-dependent and activity-independent since the degree of hindlimb muscle atrophy after spinal isolation (activity-independent nerve influence) is less when compared to the atrophy caused by removal of all nerve influences by transecting the sciatic nerve [9]. Two basic mechanisms are responsible for denervation-induced muscle atrophy. First, there is augmented activity of the ubiquitin-proteasome pathway and proteolysis [48]. Second, there is cell death and myonuclei apoptosis conjugated with decreased capacity of satellite cell-dependent reparative myogenesis [49]. Together with atrophy, denervated/reinnervated muscles undergo phenotypical changes and conversion between muscle fiber types [50]. The relative increase in type I or type II muscle fibers following denervation seems to depend on the type of muscle fibers predominant in the muscle, with type II muscle fibers (fast fibers) increasing in proportion in *soleus* (considered a slow muscle) and type I muscle fiber number increasing in *gastrocnemius* and TA muscles [51]. Likewise, the degree of muscle fiber atrophy in short-term denervation (4 weeks) has been noticed to be greater in the muscle fiber type that is more abundant in the affected muscles [52]. Earlier studies suggested that it was possible that denervated muscles could have increased muscle plasticity due to acceleration in the early myoblastic stages of muscle regeneration. Nevertheless McGeachie and Grounds [53] data proved that very few precursors were proliferating in denervated muscle within 30 h after injury, and the onset of myogenesis at 30 h was essentially the same in denervated and innervated muscle. They compared the onset of DNA synthesis in muscle precursors in denervated and innervated muscle of adult BALBc mice regenerating after a simple cut injury. This study concluded that although denervation of skeletal muscles causes an increase in SCs and connective tissue cell turnover, it does not "prime" the general population of muscle precursors to start synthesizing DNA more rapidly after injury than in innervated muscle [53]. After sciatic nerve transection at an adult age, electromyography (EMG) patterns in hindlimb muscles during locomotion remained highly abnormal even after recovery periods lasting 15 or 21 weeks [54]. This may be a limitation when using denervated muscles as a model of muscle injury since regeneration might be affected for a very prolonged period. Like already mentioned, other models of skeletal muscle injury, like complete crush or myectomy, can be accompanied by denervation since these traumatic models may possibly damage peripheral nerves or myoneural junctions. This might also be an undesirable occurrence in the standardization of these models of muscle injury. In fact, our preliminary work using TA myectomy showed that few animals developed severe muscle force deficit after 4 weeks recovery, suggesting that damaged of the supplying nerve occurred in these animals subset.

In our research group, standard peripheral nerve injuries in the rat sciatic nerve model have been performed [57-63] in order to evaluate different therapeutic approaches including several biomaterials and cellular systems to promote sensitive and functional recovery of the nerve. A standard crush injury is performed by a non-serrated clamp (Institute of Industrial Electronic and Material Sciences, University of Technology, Vienna, Austria), exerting a constant force of 54 N for a period of 30s, 10mm above the bifurcation into tibial and common peroneal nerves, inducing a 3mm axonotmesis lesion [57-63]. In order to induce a standard neurotmesis lesion in the rat sciatic nerve model, considered a more serious lesion, under deep anaesthesia, the right sciatic nerve is exposed through a skin incision extending from the greater trochanter to the distal mid-half followed by a muscle splitting incision. After nerve mobilisation, a transection injury is performed (neurotmesis) using straight microsurgical scissors. The nerve is injured at a level as low as possible, in general, immediately above the terminal nerve ramification. To prevent autotomy, a deterrent substance should be daily applied to rat right foot [57-63]. Both experimental injuries induce severe motor deficit and loss of sensory function, evaluated by measuring extensor postural thrust (EPT) and withdrawal reflex latency (WRL), respectively [57-63]. Sensory and motor deficit then progressively decreased along the post-operative, depending on the therapeutic approach used.

Very promising results were obtained with chitosan type III membranes and MSCs isolated from the umbilical cord matrix. In addition, we also perform kinematic analysis of the rat walk which is a more sensitive behavioral test. This analysis is increasingly being used to assess functional recovery in peripheral nerve research because of its higher accuracy and better relationship with histological outcome [57-63]. We should bare in mind that locomotion is also of higher functional relevance since it involves integrated function of both the motor and sensory systems and their respective components, such as skeletal muscles, sensory endings, efferent and afferent nerve fibers and integrative centers within the central nervous system. Muscles innervated by sciatic nerve branches include both dorsiflexors and plantarflexors and, although in our published studies we focused our kinematic analysis only in the stance phase, we now prefer to include analysis of the ankle joint motion also during the swing phase in order to provide additional information [59]. Denervation can be a very useful model of skeletal muscle injury for some experimental studies but some limitations might be pointed out in studies that attempt to focus exclusively on the muscular regeneration process. Nevertheless and as demonstrated by several studies, this muscular regeneration process is highly dependent of the neural supply and the nerve regeneration itself can be influenced by the damaged muscle tissue.

4. Tissue engineering and regenerative medicine

Every day thousands of clinical procedures are performed to replace or repair tissues in the human body that have been damaged through disease or trauma that use tissue engineering technology. The use of constructs for tissue engineering (TE) and regenerative medicine are promising innovative therapies that can address several clinical situations. These constructs are often combination of cells, scaffolds and biological factors. Although there are only a few commercial products currently in the market for cell/drug delivery, probably because each type of cell requires its own specific encapsulating microenvironment with cell-specific material properties and spatially controlled bioactive features, a vast amount of research is being performed worldwide on all aspects of tissue engineering/regenerative medicine exploring polymer materials. To implant cells into defective skeletal muscles, there are two main techniques. The cellular system may be directly injected into the scaffold which is localized in the injury site. It can also be performed by pre-adding the cells to the scaffold via injection or co-culture (in most of the cellular systems, cells are allowed to form a monolayer) and then the biomaterial with the cellular system is implanted in the injured muscle. In case of multiple sites of injury, the systemic administration of cells capable of reaching damaged tissues would be an interesting alternative [64].

4.1. – Scaffolds and Biomaterials

Scaffolds, which are used to deliver cells, drugs, and genes into the body, can take on various forms from porous solid devices to injectable networks, such as a typical three-dimensional porous matrix, a nanofibrous matrix, a hydrogel, and microspheres. Although solid scaffold provide a mechanically strong matrix for seeded cells, hydrogel scaffolds and microspheres are becoming increasingly popular in TE. The spherical nature maximizes the surface area, and the small volume of beads facilitating biomolecular transport. Regarding hydrogels, they have a similar microstructure to the extracellular matrix (ECM) and allow good physical integration into the defect by the use of minimally invasive approaches for material and cell/drug delivery. The biological, chemical, topography features and mechanical properties, as well as the degradation kinetics of hydrogels, can be tailored depending on the application [65-68]. Aligned nanoscale and microscale topographic features in scaffolds have been also reported to influence the alignment of cells. For example, this alignment is an important requirement of functional skeletal muscle since it leads to alignment of myoblasts and cytoskeletal proteins and promote myotube assembly along the nanofibres and microgrooves to mimic the myotube organization in muscle fibres [65, 68-70]. Scaffolds are used successfully in various fields of tissue engineering such as bone formation, periodontal regeneration, cartilage development, as artificial corneas, in tendon repair and in ligament replacement. In addition, the incorporation of drugs (i.e., inflammatory inhibitors and/or antibiotics) into scaffolds or specific molecules to provide adequate signals to the cells is also possible [71] Depending on the medical applications, scaffolds requirements will depend on its function. Hydrogels can be used as a physical barrier to protect the cells from hostile extrinsic factors before delivery, or be used as a matrix to drug controlled release or cell adhesion, growth and differentiation to further improve the secretion of therapeutic proteins from cells. In fact cells are capable of delivering drugs in response to an external stimulus, which is highly advantageous to maintain homeostasis for patients suffering from chronic diseases. For the first application, the scaffold needs:

i. to be biocompatible, by minimizing the patients' immune response, which is detrimental to cell viability, hydrogel stability, and mass transport. Ideally, the scaffold should evoke no or only minimal fibrous tissue reaction, macrophage activation, and cytokine and cytotoxic agent release

ii. to have controllable degradability, being the degradation products not toxic and eliminated easily from the implantation site by the body, and

iii. to have mechanical properties that are sufficient to shield cells from tensile forces without inhibiting biomechanical cues to cells through mechanotransduction pathways that mediate tissue homeostasis, morphogenesis, cell growth, contractility, differentiation, and pathophysiology [70-72].

For the second applications further requirements are needed, mainly:

i. a microstructure that allows for the influx of nutrients and oxygen toward the encapsulated cells and prevents the efflux of therapeutic molecules and cellular wastes away from the scaffold; this is assured through adequate pore size distribution and its interconnectivity. A high surface:volume ratio should be suitable for cell/drug attachment;

ii. adequate drug binding affinity to allow a controllable drug released to be stable when incorporated in the scaffold at a physiological conditions;

iii. bioadhesion, to allows cells and tissues to adhere to scaffolds. Some hydrogels such as fibrin or collagen inherently exhibit bioadhesive properties, but others do not and therefore linker molecules that enable covalent or non-covalent molecular interactions between the scaffold and its surroundings are incorporated;

iv. the mechanical properties of the scaffold, commonly controlled with the polymer concentration and molar ratio between polymers and cross-linking molecules, should match those of the tissue at the implantation site as well as the degradation rate that should match the rate of tissue regeneration [73-77].

4.1.1. Hydrogel scaffolds

Hydrogels, three-dimensional (3D) networks of hydrophilic polymers, are appealing for biological applications because of their high water content, high permeability, biocompatibility, and the ability of be placed into critical defects in a minimally invasive manner [78]. They are being used in a wide range of tissues, including cartilage, bone, muscle, fat, liver, and neurons. For use in drug/cell delivery, hydrogels should be low-viscosity solutions prior to gelling, which is crucial to maintain cell viability during the encapsulation process, and should rapidly gel in the human body. These properties can be fine-tuned through variations in the chemical structure and cross-linking density in hydrogels. Injectable hydrogels can be formed *in situ* by either chemical or physical cross-linking methods [79, 80]. Physical cross-linked hydrogels are capable of phase transition in response to external stimuli such as temperature, pH or both [81]. Chemically cross-linked hydrogels are prepared through photopolymerization, disulfide bond formation, or reaction between thiols and acrylate or sulfone. The latter hydrogels undergo significant volume changes compared to the first ones [81-83]. The pH/temperature-sensitive hydrogels show several advantages over thermo-sensitive ones, such as the absence of clogging during injection and avoidance of pH decreased caused by degradation. The pH/temperature-sensitive copolymer hydrogels can be prepared by combining a pH-sensitive moiety with a temperature-sensitive block. For example, if acidic sulfamethazine oligomers (OSMs) are coupled with thermosensitive poly(e-CL-co-LA)-PEG-poly-(e-CL-co-LA) triblock copolymers a pH/temperature-sensitive hydrogels (OSM-PCLA-PEG-PCLA-OSM) is produced. Photopolymerized hydrogel systems have been reported to provide better temporal and spatial control over the gelation process [79, 81, 84].

4.1.2. Biomaterials for scaffold fabrication

A wide variety of natural and synthetic materials have been used to prepare injectable hydrogels. Natural polymers, which are either components of or have macromolecular properties similar to the natural ECMs, are known to often undergo rapid degradation upon contact with body fluids or medium and show batch-to-batch variation. Synthetic hydrogels offer improved control of the matrix architecture and chemical composition, no immunogenicity, consistent supply of large quantities, but tend to have lower biological activity. Therefore, modification of natural and synthetic derived hydrogels is usually required [79]. A natural biodegradable 3D scaffold can be made of acellular muscle ECM but it's fragile and difficult to handle [85]. Another natural biodegradable scaffold can be created by using fibrin, which leads to a process much similar to wound healing, in which fibrin forms a temporary scaffold to serve tissue regeneration and then is replaced by the physiological ECM. Fibrin has the additional advantage that it binds growth factors [70].

Fibrin hydrogels are made from commercially purified allogeneic fibrinogen and purified thrombin, and have been used in a variety of tissue engineering applications. Its main disadvantages reported to be shrinkage of the gel, low mechanical stiffness and its rapid degradation can be overcome by incorporating other polymers such as gelatin, hyaluronic acid, and chondroitin-6-sulfate. Fibrin glue is clearly distinguished from fibrin hydrogels that are prepared from purified fibrinogen and thrombin. Despite the commercial fibrin glue is available in standardized quality; autologous fibrin glue is cheaper and has no viral transmission and prion infection [86]. Tisseel® VH, is a fibrin glue commercialized by Baxter, and consists of a two-component fibrin biomatrix with highly concentrated human fibrinogen to produce fibrin gel from a blood sample and is safe to be used in TE. FloSeal® is another commercial hemostatic matrix with potential in TE, and consists of a cross-linked bovine-derived Gelatin Matrix component and a human-derived Thrombin component. Literature reports that myoblasts seeded on fibrin gels have been shown to differentiate into contracting muscle fibres and to demonstrate a normal length–tension and force–frequency relationship [87, 88].

Alginate is the designation given to a natural family of biodegradable, biocompatible, hydrophilic and non-toxic polysaccharides extracted from some marine algae and some microorganisms. Alginates are linear block co-polymers composed of two different monomers, β-D-mannuronic acid (M) and α-L-glucuronic acid (G), which are linked by (1-4) glycosidic bonds. The main property of alginate that potentiates its use in different areas, it is its ability to bind some divalent cations such as Ca^{2+} in the carboxylic groups which provides the gelation of the alginate solution. The properties of the gel are dependent of the ratio between M and G monomers (M:G ratio); if the proportion of the G monomer is predominant, a strong brittle gel it obtained, whereas if the proportion of the M monomer is predominant, the formed gel will be weaker, but more flexible, because there are less junction zones between the polymer chains. As alginate is a polyelectrolyte, more specifically a polyanion, it can be ionically associated with a polycation existent in the same solution through hydrogen bonding or electrostatic interactions, forming a polyelectrolyte complex [89, 90]. Cell-encapsulating calcium cross-linked alginate hydrogels have been extensively studied because alginate molecules are anionic polysaccharides and do not associate with many proteins. Since alginate itself is inert for cell attachment and spreading, the cell adhesion properties can be tailored by linking molecules such as RGD peptides to its backbone [91].

Chitosan is a natural and hydrophilic copolymer, and it is composed by two monomeric units, D-glucosamine and N-acetyl-D-glucosamine linked by β(1–4)-glycosidic bond. This linear polysaccharide has been widely studied in medical applications due to its biocompatibility, biodegradability, non-toxicity, fungistaticity, antimicrobial activity, non-carcinogenicity, notable affinity to proteins, promotion of cell adhesion as well as proliferation and differentiation [67]. Chitosan results from the alkaline deacetylation of the chitin and its solubility is mainly influenced by its molecular weight and degree of deacetylation. Some methods have been developed to lower the molecular weight of chitosan by hydrolysis of the polymeric chains, in order to produce chitosan salts which are soluble in water. Chitosan-based hydrogels have been gelled via glutaraldehyde cross-linking, UV irradiation, and thermal variations [81].

Hyaluronic Acid (HA) is a natural, hydrophilic and non-sulfated glycosaminoglycan. This polymer is a linear polysaccharide, in which the repeating unity is a disaccharide composed by two monomers, D-glucuronic acid and N-acetyl-D-glucosamine, linked through alternating β1,3 and β1,4 glycosidic bonds. HA has been used as a biomaterial in various medical applications, due to its biocompatibility, biodegradability, and non-immunogenicity. HA is the main component existent in the extracellular matrix (ECM) of living tissues, namely in the connective, epithelial and neural. This polymer, due to its structural and biological properties, has the ability of mediate the cell signalling and behaviour, and the matrix organization. HA is able of interact with some cell surface receptors, being involved in the tissue hydrodynamics, cell migration and proliferation. Several strategies have been reported to prepare HA-based hydrogels [92, 93].

Among the most widely used synthetic polymers for scaffolds, either alone or copolymerized with synthetic or natural polymers, as biodegradable polymers are polyglycolide, polylactide and its copolymer poly(lactide-co-glycolide), polyphosphazene, polyanhydride, poly(propylene fumarate), polycyanoacrylate, polycaprolactone, polydioxanone, and polyurethanes, and as non-biodegradeable polymers are included polyvinyl alcohol (PVA), polyhydroxyethymethacrylate, and poly(N-isopropylacrylamide) [71, 94].

The majority of natural biomaterials used in clinical applications are derived from animal or human cadavers' sources. In spite of thorough purification methods, these materials bear the inherent risk of transfer viral diseases and may cause immunological body reactions while synthetic biomaterials are not associated with these risks. So, a critical issue in this type of cellular transplants is the search for an optimal vehicle to provide the ideal environment for cell hosting and for the release and conduction of molecules to the site of injury for cell-host interaction. Taking this into account we evaluated different biomaterials as vehicles for the cellular system intended to be tested for skeletal muscle regeneration using our myectomy injury model in the rat. Plasma derived substances, hemostatic matrix solutions and hydrogels (Figure 2 and Figure 3) were tested and the *in vivo* response was compared histologically according to the International Standard ISO 10993-6 (*see* 5.1.3). The following procedure was done under sterile conditions. For preparation of the spherical hydrogel, the polymer solution is prepared by adding in a ratio of 1:1 (V/V), a sodium alginate aqueous solution 7% (m/V) to a sodium hyaluronate aqueous solution 0.5% (m/V), under magnetic stirring. Afterwards the polymer solution is inserted into an insulin syringe and a droplet is released into an excess of cerium nitrate solution 135mM, in order to obtain a cross-linked polymer sphere of approximately 60 μl of volume. Cerium nitrate and sodium hyaluronate solution were sterilized by microfiltration (0.22 μm membrane) and sodium alginate powder is sterilized in an autoclave (120ºC for 15 minutes) previous to the solution preparation (Figure 2 and Figure 3). These tested biomaterials including the spherical hydrogel were not only used as vehicles but their properties were also evaluated and optimized to find a suitable matrix for the cellular implants.

Figure 2. Hydrogel preparation (alginate, hyaluronic acid and cerium).

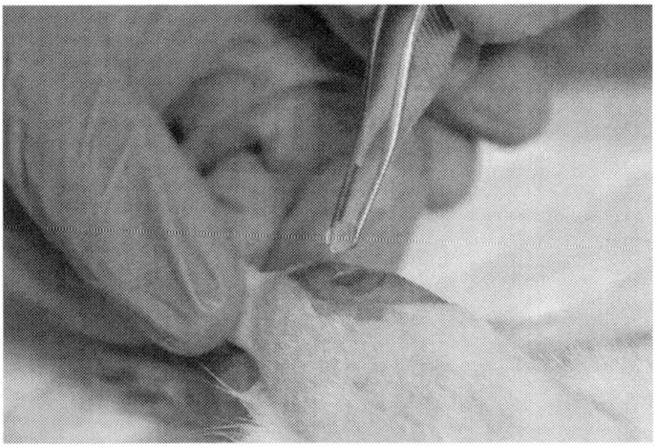

Figure 3. Application of a spherical hydrogel containing $1x10^6$ MSCs from the Wharton's jelly, in the 5 mm Ø TA myectomy defect.

4.2. Cells

There is evidence both from animal studies and clinical investigations that cell therapy involving different types of stem cells application is promising as means to promote regeneration of skeletal muscles following severe injuries. Technical or/and ethical difficulties in obtaining sufficient and appropriate stem cells from the bone marrow or from embryos (obtained from assisted reproduction techniques or somatic nuclear transfer - cloning) have limited the application of this type of therapy. Stem cells are known as an undifferentiated population, with endless self-renewal and sustained proliferation *in vitro* and multilineage differentiation ability [95]. The *in vitro* multilineage differentiation is the concept that gives these cells an extreme priority for use in tissue and cell-based therapies. Stem cells can be loosely classified into 3 categories based on their functional role: hematopoietic stem cells, mesenchymal stem cells (MSCs) and embryonic stem cells [95].

MSCs have become one of the most exciting targets for tissue regeneration due to their high plasticity, proliferative and multilineage differentiation capacity. These cells are capable of differentiating into adipose, bone cartilage and muscle. Among all this notable characteristics, MSCs reveal other properties of great importance, they present low immunogenicity and high immunosuppressive properties due to a decreased or even absence HLA Class II expression [96]. Differentiation potential of MSCs in multilineage end-stage cells has been proven, so as the treatment potential in musculoskeletal disorders [97, 98]. Since their first isolation in 1968, from rat bone marrow [99], MSCs have been isolated with success from almost all tissue sources: skeletal muscle, adipose tissues, synovial membranes, umbilical cord matrix and blood, placental tissue, amniotic fluid among others. Along with differentiation capacity, an increasing amount of data has demonstrated that the MSCs have the capacity of modulating the surrounding environment, by secretion of multiple factors and activation of endogenous progenitor cells [100, 101].

4.2.1. Umbilical cord

From our data and from previously published experimental work, the development of cell therapies associated to biomaterials is a promising tool for increase skeletal muscle regeneration, avoiding the irreversible loss of function and limit the fibrous scar tissue presence [57-59]. Recent years have witnessed an explosion in the number of adult stem cells populations isolated and characterized. While still multipotent, adult stem cells have long been considered restricted, giving rise only to progeny of their resident tissues. Recently, and currently controversial studies have challenged this dogma, suggesting that adult stem cells may be far more plastic than previously appreciated [102, 103]. Extra-embryonic tissues as stem cell reservoirs offer many advantages over both embryonic and adult stem cell sources. The umbilical cord matrix is an important and safe source of MSCs with positive effects in nerve and skeletal muscle regeneration, with no ethical or technical issues. MSC isolated from umbilical cord matrix (Wharton's jelly), as well as embryonic stem cells (ESCs) are originated from inner cell mass of blastocyst [104]. Comparing with ESCs, MSCs have shorter population doubling time; can be easily cultured in plastic flasks, are well tolerated by immune system; therefore transplantation of these cells into non-immune-suppressed animals does not induce acute rejection. Most important, these cells do not originate teratomas [104]. Like bone marrow stromal cells and other MSCs, the MSCs from the Wharton's jelly are plastic adherent, stain positively for markers of the mesenchymal lineage (CD10, CD13, CD29, CD44, CD90, and CD105) and negatively for markers of the hematopoietic lineage. These MSCs are capable of self-renewal with sustained proliferation *in vitro* and can differentiate into multiple mesodermal cells. The high plasticity and low immunogenicity of these cells turn them into a desirable form of cell therapy for the injured musculoskeletal tissue without requiring the use of immunosuppressive drugs during the treatments. Interestingly, these cells, which are HLA class II negative, not only express both an immuno-privileged and immuno-modulatory phenotype, but their HLA complex class I expression levels can also be manipulated, making them a potential cell source for MSC-based therapies. In addition and as previously referred, these cells represent a non-controversial source of primitive mesenchymal progenitor cells that can be harvested after birth, cryogenically stored, thawed, and expanded for therapeutic uses. MSCs from the Wharton's jelly display a high proliferative rate and plasticity, being able to differentiate into adipocytes, osteoblasts, chondrocytes, cardiomyocites, neurons, and glia. More recently, Conconi et al. [105] demonstrated that CD105(+)/CD31(-)/KDR(-) cells are able not only to differentiate *in vivo* towards the myogenic lineage as demonstrated by the co-localization of HLA 1 and sarcomeric tropomyosine antigens, but also to contribute to the muscle regenerative process. These cells were found to differentiate *in vitro* into myoblast-like cells, expressing Myf5 and MyoD after 7 and 11 days of myogenic induction, respectively. The timing of expression of Myf5 and MyoD in CD105(+)/CD31(-)/KDR(-) cells is similar to that described during embryonic development and in myoblast cultures [105].

Using the myectomy model we tested the use of Human MSCs isolated from the Wharton's jelly in order to improve skeletal muscle regeneration. The cells were directly infiltrated into the lesion or delivered by different vehicles including Floseal®, Tisseel®, carboximetilcellulose (Sigma) and spherical hydrogel (own fabrication). MSC from Wharton's jelly were purchased from PromoCell GmbH (C-12971, lot-number: 8082606.7). The MSCs are cultured and maintained in a humidified atmosphere with 5% CO_2 at 37ºC. Mesenchymal Stem Cell Medium, PromoCell (C-28010) is replaced every 48 hours. At 90% confluence, cells are harvested with 0.25% trypsin with EDTA (GIBCO) and passed into a new flask for further expansion. MSCs at a concentration of 2 x 10^5cells are cultured exhibiting a 90% confluence after 3-4 days. The application of human MSCs in rats is possible without inducing any immunossupression in the experimental animals. The MSCs exhibited a normal star-like shape with a flat morphology in culture (Figure 4). A total of 20 Giemsa-stained metaphases of these cells, were analyzed for numerical aberrations. Sporadic, non-clonal aneuploidy was found in 3 cells (41-45 chromosomes) the other 17 metaphases had 46 chromosomes. The karyotype was determined in a completely analyzed G-banding metaphase and no structural alterations were found [57]. The karyotype analysis to the MSCs cell line derived from Human Wharton jelly demonstrated that this cell line hasn't neoplasic characteristics and is stable during the cell culture procedures in terms of number and structure of the somatic and sexual chromosomes. Also, the morphologic characteristics of these cells in culture, observed in an inverted microscope, are normal. These cells presented a star-like shape with a flat morphology, characteristic of the MSCs been adequate to be used in *in vivo* rat experimental model [57]. The MSCs karyotype was studied in order to be sure that these cells did not present any number or structure chromosome abnormalities due to isolation and cell culture procedures before *in vivo* application. This concern was due to the negative effects that some cellular systems, like the ESCs present, inducing the production of teratomas. The cellular systems implanted into the injured skeletal muscle improved the skeletal muscle regeneration since these cells produce growth factors, ECM molecules, and even modulate the inflammatory process.

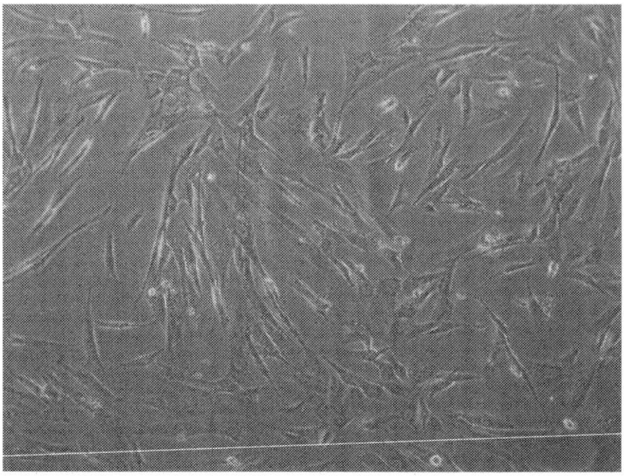

Figure 4. Undifferentiated MSCs from Wharton's jelly, exhibiting a star-like shape with a flat morphology (100x magnification).

5. Evaluation of muscle regeneration

Muscle biopsies should be considered in order to obtain careful clinical assessment or for investigation purposes. After the collection of the muscle samples, they should be immediately equally divided in three. One sample should be placed into formalin (for hematoxilin and eosin - HE), another sample should be fixed into 2.5% purified glutaraldehyde in 0.1M Sorensen phosphate buffer (for electron microscopy - EM) and the other sample should remain unfixed and refrigerated (for histochemistry, biochemistry/genetics analysis).

5.1. Routine histological evaluation

Routine evaluation of the muscle biopsy sample involves the examination of formalin-fixed, paraffin processed sections and unfixed frozen sections with standard histological and enzyme histochemical stains at the light microscopic level. HE is the routine histological stain used for evaluation of basic tissue organization and cellular structure. For HE, the whole piece of tissue should be fixed in a clamp and after the tissue is infiltrated with wax, both longitudinal and cross sections must be cut before embedding. Five levels should be obtained, especially in cases suspected of vasculitis. The parameters that can be evaluated are: the type of inflammatory infiltrate present; examination of the structure of vessels walls (vasculitis and/or fibrinoid necrosis); presence of endomysial and perimysial fibrosis/fatty infiltration; the range of fiber caliber; presence of angulated fibers; increase in number of centrally located nuclei; central capillary migration; split fiber; group atrophy; necrotic/ myopathic (degenerating) fibers; atrophic fibers; regenerating fibers; target fibers; whorl fibers and ring fibers. The rounding of fiber contour and the variation of fiber diameter should also be analyzed, however they are better evaluated with frozen sections (Figure 5).

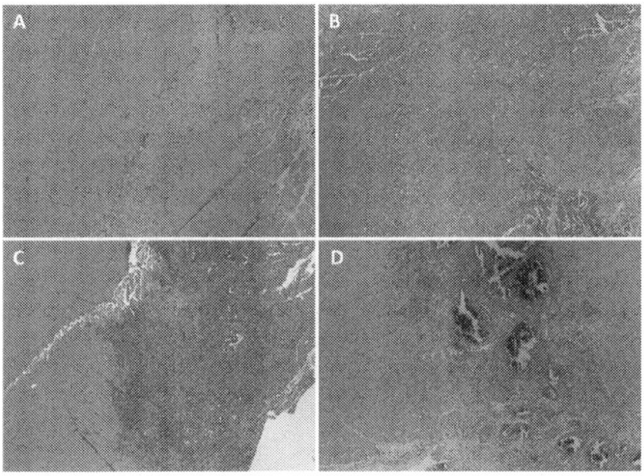

Figure 5. HE staining of TA muscles 15 days post myectomy (A - control) and application of fibrin (B), hydrogel (C), Floseal® (D).

5.1.1. Morphological analysis

Long-standing histological characteristics are still used to identify the mammalian skeletal muscle regeneration process. On muscle cross-sections, these fundamental morphological characteristics are newly formed myofibers of small caliber and with centrally located myonuclei. Newly formed myofibers are often basophilic (reflecting high protein synthesis) and express embryonic/developmental forms of MHC (reflecting *de novo* fiber formation). On muscle longitudinal sections and in isolated single muscle fibers, central myonuclei are observed in discrete portions of regenerating fibers or along the entire new fiber, suggesting that cell fusion is not diffuse during regeneration but rather focal to the site of injury [1]. Cross-sectional area (CSA) analysis is one of the features that can be assessed. This can be achieved with imaging software processing (Scio Image, ImageJ) of HE-stained muscle sections. A predefined number of fibers is traced per sample and should be determined as appropriate by the examination of no additional changes in standard deviation. The classification of small and large fibers can be determined for example by setting three standard deviations from the mean CSA for the uninjured group at different time points [14]. The CSA and number of myotubes can be used to estimate the development degree of muscle regeneration following injury [36].

5.1.2. Collagen quantification

Collagen content in the wound bed can be calculated by image analysis of Masson's Trichrome-stained histological images taken at a predefined image magnification. Color separations must be performed and an analysis threshold must be established for each image series collected using the same brightness and white balance settings. Output images showing only computed blue coverage must be compared to the color images to ensure the representation of truly blue color due to collagen staining. As a control the analysis must be performed on uninjured (control) skeletal muscle tissue sections stained with Masson's Trichrome and collected using the same camera and threshold settings to confirm a collagen content of zero for control tissue. The ratio of blue pixels above the threshold to total pixels in the image is used to calculate the collagen content for each image [40].

5.1.3. International Standard (ISO 10993-6)

The International Standard (ISO 10993-6) specifies test methods for the assessment of the local effects after implantation of biomaterials intended for use in medical devices. These implantation tests are not intended to evaluate or determine the performance of the test specimen in terms of mechanical or functional loading. The local effects are evaluated by a comparison of the tissue response caused by the tested implant to that caused by the control. The objective of the test methods is to characterize the history and evolution of the tissue response after implantation of a medical device/biomaterial including final integration or resorption/degradation of the material. The test sample shall be implanted into the tissues most relevant to the intended clinical use of the material. For short-term testing, animals such as rodents or rabbits are commonly used. During the first two weeks after implantation the reaction due to the surgical procedure itself may be difficult to distinguish from the tissue reaction evoked by the implant and for that reason in our study we collected the muscle samples 15 days after implantation. For degradable/resorbable materials the test period shall be related to the estimated degradation time of the test product. In our case the majority of the vehicles/matrices tested the degradation time is less than 4 days. In the absence of complete degradation, absorption, or restoration to normal tissue structure and function, the overall data collected may be sufficient to allow characterization of the local effects after implantation. A sufficient number of implants shall be inserted to ensure that the final number of specimens to be evaluated will give valid results. The evaluation of the biological response must be accomplished by documenting the macroscopic and histopathological responses as a function of time. The responses to the test sample must be compared to the responses obtained at the control sample or sham operated sites. The scoring system used for the histological evaluation shall take into account the extent of the area affected, either quantitatively (e.g. in micrometres) or semi-quantitatively (Annex E of this Standard) [44, 106]. The biological response parameters, which shall be assessed and recorded, include:

i. the extent of fibrosis/fibrous capsule (layer in μm) and inflammation;

ii. the degeneration as determined by changes in tissue morphology;

iii. the number and distribution as a function of distance from the material/tissue interface of the inflammatory cell types, namely polymorph nuclear neutrophilic leucocytes, lymphocytes, plasma cells, eosinophils, macrophages and multinucleated cells;

iv. the presence, extent and type of necrosis;

v. other tissue alterations such as vascularization, fatty infiltration, granuloma formation and bone formation;

vi. the material parameters such as fragmentation and/or debris presence, form and location of remnants of degraded material;

vii. the quality and quantity of tissue ingrowth, for porous and degradable implant materials [106].

Under the conditions of the study and following the results for the mentioned parameters in the semi-quantitative scoring system (Annex E of this Sandard), the test sample is considered as non-irritant (0,0 up to 2,9), slight irritant (3,0 up to 8,9), moderate irritant (9,0 up to 15,0), severe irritant (> 15) to the tissue as compared to the negative control sample [106]. This test method is used for assessing the biological response of muscle tissue to an implanted material (Annex C of this Standard). As already mentioned, the method compares the biological response to implants of test specimens with the biological response to implants of control specimens. The control materials are those used in medical devices of which the clinical acceptability and biocompatibility characteristics have been established [106]. In our study we developed an adaptation of this Standard by considering the control as the group where the surgical procedure (myectomy) was performed without any biomaterial or cell implantation (Figure 6). Although the surgical technique may profoundly influence the result of any implantation procedure, we assumed that our standardized myectomy lesion could be considered as the *Control* group since we were able to determine the local effects of the different implants by their comparison to the minor effects of the surgical procedure. The surgeries were executed under general anesthesia with a xylazine (1.25 mg/100 g BW im) and ketamine (9 mg/100 g BW im) combination [38].

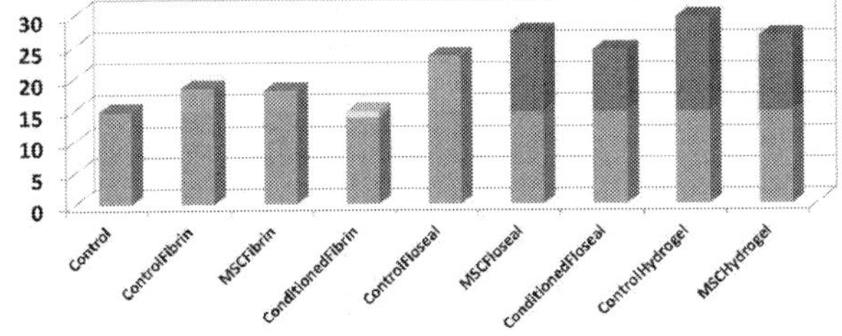

Figure 6. ISO 10993-6 scoring for the groups tested. The *Control* group obtained a score of 14.7 (in blue). Scorings above the *Control* group were considered as non-irritant (in yellow), slight irritant (in orange) and moderate irritant (in red).

5.2. Histochemistry

For histochemistry a basic panel should be performed, preferentially in frozen sections. Depending on the objectives, an extended panel can be done concerning the study of some molecules like the already mentioned Masson's Trichrome-stain or enzyme processes such as ATPase; NADH-TR or Esterase (Bancroft&Stevens). When necessary, other special stains can be performed on paraffin sections.

5.3. Immunohistochemistry

In general, the immunohistochemical stains are utilized for the diagnosis of various muscular dystrophies. They also may help to determine the subtypes of inflammatory cells within an infiltrate or for other investigation purposes. Specific skeletal muscle markers such as myosin heavy chain and desmin can be applied in order to clearly identify this tissue. The distinction of SCs, considered as the reservoir of myogenic precursor cells, from other cells must be made (like plasma cells, which may be occasionally seen under the basal lamina in pathologic conditions). SCs can be easily demonstrated by immunostaining for N-CAM; they also express vimentin. Activated SCs generally express Myo-D and myogenin [107]. The regenerating fibers express N-CAM, MyoD and myogenin, and also embryonic and neonatal isoforms of myosin heavy chain. In contrast to mature fibers, MCH class I histocompatibility complex is expressed in regenerating fibers.

5.4. Immunofluorescence

For the preparation of TA muscles for immunofluorescence they should be embedded in Tissue-Tek OCT compound. Sections are cut at 10 μm using a Leica CM1850 cryostat and placed onto Surgipath microscope slides. Laminin-α2 chain is detected with a 1:500 dilution of rabbit anti-laminin-α2 (2G) polyclonal antibody. The laminin-α1 chain is detected with a rat anti-laminin-α1 monoclonal antibody. Primary rabbit antibodies are detected with a 1:500 dilution of fluorescein isothiocyanate-conjugated anti-rabbit secondary antibody and the rat monoclonal antibody is detected with 1:500 dilution of fluorescein isothiocyanate-conjugated anti-rat secondary antibody. In all immunofluorescence experiments, secondary only antibody controls are included to test for specificity. For mouse monoclonal antibodies, endogenous mouse immunoglobulin is blocked with a mouse-on-mouse (MOM) kit. A 1-μg/ml concentration of tetramethylrhodamine-conjugated wheat germ agglutinin (WGA) is used to define muscle fibers. To examine immune response, cytotoxic T cells are detected with fluorescein isothiocyanate-labeled rat anti-mouse CD8a and macrophages are detected with fluorescein isothiocyanateconjugated anti-mouse F4/80 at 1:1000 (Figure 7 and Figure 8) [50].

http://dx.doi.org/10.5772/53335

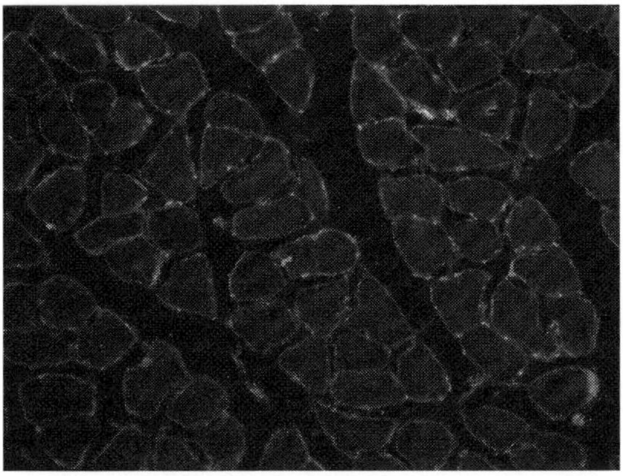

Figure 7. Double imunofluorescence staining for laminin and CD31.

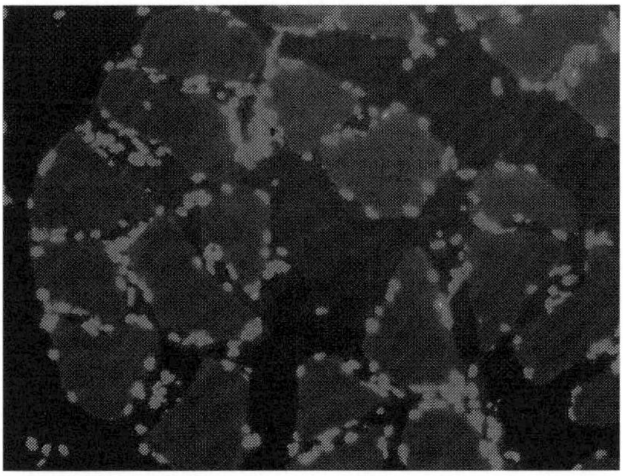

Figure 8. Pax7+ SC counterstained with DAPI.

5.5. Electron microscopy

Electron microscopic (EM) examination of the glutaraldehyde-fixed portion of the biopsy is performed when the light microscopic studies are inconclusive. Thus, it is reserved for selected circumstances in which the pathologist determines that EM has the potential of contributing significantly to determining a specific diagnosis. A specimen placed in glutaraldehyde must be small, approximately 1-2 mm in width and depth, allowing the complete tissue penetration by this fixative. Glutaraldehyde makes tissue brittle and interferes with immunohistochemical studies, so it is not appropriate for the paraffin specimen. With EM, other muscle cell parameters can be analyzed in detail: the myofibril architecture;

the plasma and sarcolemmal membrane; the mitochondria (size, density and shape); T-tubule; amount of lipid; nucleus; phagocytic granules and amount of glycogen. EM is extremely useful in some cases: to identify inclusions primarily found by light microscopy; to help in the characterization of stored material found on light microscopy and define its intracellular localization; to analyze structural abnormalities found by light microscopy; can assist in the diagnosis of mitochondrial myopathy or seeking evidence to support a diagnosis of dermatomyositis (EM can be used to look for tubuloreticular inclusion in endothelial cells when light microscopic fails to reveal it).

5.6. Contraction force measurement

To obtain an estimate of total TA muscle strength reduced by the injury and possibly recovered by the cell/vehicle implants, contractile force due to electrical stimulation can be measured before injury, after injury and at the time of sacrifice (at different time points) for non-implanted and implanted animals. This can be accomplished with the animals under general anesthesia and by anchoring the knee joint using a custom clamping system anchored to the floor of the surgical stereomicroscope stand and attaching a silk ligature to the cleft between digits 1 and 2 that must be anchored to a transducer at the other end. This can also be executed by cutting the TA tendon just before the insertion at the ankle and tying it with a 4.0 nylon suture attached to the isometric transducer. The exposed muscle is stimulated using 2 custom needle electrodes placed at the proximal muscle surface. Electrical stimulation of the TA muscle is applied at 5 volts, 4 ms pulse duration, at 500 ms intervals and the resultant tetanic force recorded (200 points(s) using a BioPac MP-100 (Harvard Apparatus) and accompanying software (Acknowledge^{TM}). The muscle must be kept hydrated during the procedure using sterile saline. Maximum tetanic force is measured by reducing the stimulation interval to 20 ms, generating continuous stimulation simulating tetanus condition. Another method of applying the electrical stimulation can be obtained by exposing the sciatic nerve with an incision in the hamstring region The tibial nerve is cut just after the sciatic nerve splits into the tibial and peroneal nerves to eliminate any contraction from the *gastrocnemius* muscle causing background in the force data. The exposed sciatic nerve is then laid over two electrodes with a small piece of parafilm and should also be kept moist with periodic treatment of mineral oil. Stimulation is made using a supra-maximal square-wave pulse of 0.1 ms duration. Measurements are performed at the length at which maximal extension is obtained during the twitch and the data should be recorded for sub-maximal and maximal isometric force. Specific maximal force should be quantified by correcting for muscle mass [40, 98].

Acknowledgements

The authors would like to thank the support by Dr. José Manuel Correia Costa, from INSRJ, Porto, Portugal; and Biosckin, Molecular and Cell Therapies SA for the umbilical cord units supply and access to the GMP cell culture room (Scientific Protocol between Porto University and Biosckin, Molecular and Cell Therapies SA). This work was supported by Fundação para a Ciência e Tecnologia (FCT), Ministério da Ciência e Ensino Superior (MCES), Portugal, through the financed research project PTDC/DES/104036/2008, and by QREN Nº 1372 para Criação de um Núcleo I&DT para Desenvolvimento de Produtos nas Áreas de Medicina Regenerativa e de Terapias Celulares – Núcleo Biomat & Cell. A Gärtner (SFRH/BD/70211/2010) and I Amorim (SFRH/BD/76237/2011) acknowledge FCT for financial support.

Author details

Tiago Pereira[1,3*], Andrea Gärtner[1,3], Irina Amorim[3], Paulo Armada-da-Silva[2], Raquel Gomes[1,3], Cátia Pereira[2], Miguel L. França[1,3], Diana M. Morais[4], Miguel A. Rodrigues[4], Maria A. Lopes[4], José D. Santos[4], Ana Lúcia Luís[1,3] and Ana Colette Maurício[1,3]

*Address all correspondence to: tiago.vet@gmail.com

1 Centro de Estudos de Ciência Animal (CECA), Instituto de Ciências e Tecnologias Agrárias e Agro - Alimentares (ICETA), Universidade do Porto (UP), Portugal

2 Faculdade de Motricidade Humana (FMH), Universidade Técnica de Lisboa (UTL), Portugal

3 Instituto de Ciências Biomédicas Abel Salazar (ICBAS), Universidade do Porto (UP), Portugal

4 CEMUC, Departamento de Engenharia Metalúrgica e Materiais, Faculdade de Engenharia, Universidade do Porto (FEUP), Portugal

References

[1] Charge, SBP, & Rudnicki, M. A. (2004). Cellular and molecular regulation of muscle regeneration. *Physiological Reviews*, 84(1), 209-38.

[2] Gibson, M. C., & Schultz, E. (1982). The distribution of satellite cells and their relationship to specific fiber types in soleus and extensor digitorum longus muscles. *Anat Rec*, 202(3), 329-37, Epub 1982/03/01.

[3] Snow, M. H. (1983). A quantitative ultrastructural analysis of satellite cells in denervated fast and slow muscles of the mouse. *Anat Rec*, 207(4), 593-604, Epub 1983/12/01.

[4] Pette, D., & Staron, R. S. (2000). Myosin isoforms, muscle fiber types, and transitions. *Microscopy research and technique*, 50(6), 500-9, Epub 2000/09/22.

[5] Huxley, A. F. (2000). Cross-bridge action: present views, prospects, and unknowns. *Journal of biomechanics*, 33(10), 1189-95, Epub 2000/07/19.

[6] Carlson, B. M., & Faulkner, J. A. (1983). The regeneration of skeletal muscle fibers following injury: a review. *Medicine and science in sports and exercise,* 15(3), 187-98.

[7] Yamanouchi, K., Soeta, C., Naito, K., & Tojo, H. (2000). Expression of Myostatin Gene in Regenerating Skeletal Muscle of the Rat and Its Localization. *Biochemical and Biophysical Research Communications,* 270(2), 510-6.

[8] Musaro, A., Giacinti, C., Pelosi, L., Dobrowolny, G., Barberi, L., Nardis, C., et al. (2007). Stem cell-mediated muscle regeneration and repair in aging and neuromuscular diseases. *European journal of histochemistry : EJH,* 51(1), 35-43, Epub 2007/08/21.

[9] Hyatt, J. P., Roy, R. R., Baldwin, K. M., & Edgerton, V. R. (2003). Nerve activity-independent regulation of skeletal muscle atrophy: role of MyoD and myogenin in satellite cells and myonuclei. *American journal of physiology Cell physiology,* 285(5), C1161-73, Epub 2003/07/04.

[10] Gayraud-Morel, B., Chretien, F., & Tajbakhsh, S. (2009). Skeletal muscle as a paradigm for regenerative biology and medicine. *Regenerative medicine,* 4(2), 293-319, Epub 2009/03/26.

[11] Mitchell, K. J., Pannerec, A., Cadot, B., Parlakian, A., Besson, V., Gomes, E. R., et al. (2010). Identification and characterization of a non-satellite cell muscle resident progenitor during postnatal development. *Nat Cell Biol,* 12(3), 257-66, Epub 2010/02/02.

[12] Uezumi, A., Fukada, S-i., Yamamoto, N., Takeda, Si., & Tsuchida, K. (2010). Mesenchymal progenitors distinct from satellite cells contribute to ectopic fat cell formation in skeletal muscle. *Nat Cell Biol,* 12(2), 143-52.

[13] Cooper, RN, Tajbakhsh, S., Mouly, V., Cossu, G., Buckingham, M., & Butler-Browne, G. S. (1999). In vivo satellite cell activation via Myf5 and MyoD in regenerating mouse skeletal muscle. *Journal of cell science,* 112(Pt 17), 289-901, Epub 1999/08/13.

[14] White, J. P., Baltgalvis, K. A., Sato, S., Wilson, L. B., & Carson, J. A. (2009). Effect of nandrolone decanoate administration on recovery from bupivacaine-induced muscle injury. *Journal of Applied Physiology,* 107(5), 1420-30.

[15] Politi, P. K., Havaki, S., Manta, P., & Lyritis, G. (2006). Bupivacaine-induced regeneration of rat soleus muscle: ultrastructural and immunohistochemical aspects. *Ultrastructural Pathology,* 30(6), 461-9.

[16] Jones, GH. (1982). Protein synthesis in bupivacaine (Marcaine)-treated, regenerating skeletal muscle. *Muscle & nerve,* 5(4), 281-90.

[17] Rosenblatt, J., & Woods, R. (1992). Hypertrophy of rat extensor digitorum longus muscle injected with bupivacaine. A sequential histochemical, immunohistochemical, histological and morphometric study. *Journal of Anatomy,* 181(Pt 1), 11.

[18] Nonaka, I., Takagi, A., Ishiura, S., Nakase, H., & Sugita, H. (1983). Pathophysiology of muscle fiber necrosis induced by bupivacaine hydrochloride (Marcaine). *Acta Neuropathologica,* 60(3), 167-74.

[19] Irwin, W., Fontaine, E., Agnolucci, L., Penzo, D., Betto, R., Bortolotto, S., et al. (2002). Bupivacaine Myotoxicity Is Mediated by Mitochondria. *Journal of Biological Chemistry*, 277(14), 12221-7.

[20] Steer, J. H., Mastaglia, F. L., Papadimitriou, J. M., & Van Bruggen, I. (1986). Bupivacaine-induced muscle injury: The role of extracellular calcium. *Journal of the Neurological Sciences*, 73(2), 205-17.

[21] Ishiura, S., Nonaka, I., & Sugita, H. (1986). Biochemical aspects of bupivacaine-induced acute muscle degradation. *Journal of cell science*, 83, 197-212, Epub 1986/07/01.

[22] Saito, Y., & Nonaka, I. (1994). Initiation of satellite cell replication in bupivacaine-induced myonecrosis. *Acta Neuropathologica*, 88(3), 252-7.

[23] Chang, C., Chuang, S. T., Lee, C., & Wei, J. (1972). Role of cardiotoxin and phospholipase A in the blockade of nerve conduction and depolarization of skeletal muscle induced by cobra venom. *British journal of pharmacology*, 44(4), 752.

[24] Czerwinska, A. M., Streminska, W., Ciemerych, M. A, & Grabowska, I. (2012). Mouse gastrocnemius muscle regeneration after mechanical or cardiotoxin injury. *Folia histochemica et cytobiologica/ Polish Academy of Sciences, Polish Histochemical and Cytochemical Society*, 50(1), 144-53, Epub 2012/04/26.

[25] Gutiérrez, J. Ma, & Ownby, C. L. (2003). Skeletal muscle degeneration induced by venom phospholipases A2: insights into the mechanisms of local and systemic myotoxicity. *Toxicon*, 42(8), 915-31.

[26] Lin Shiau, S. Y., Huang, M. C., & Lee, C. Y. (1976). Mechanism of action of cobra cardiotoxin in the skeletal muscle. *The Journal of pharmacology and experimental therapeutics*, 196(3), 758-70, Epub 1976/03/01.

[27] Fletcher, J. E., & Lizzo, F. H. (1987). Contracture induction by snake venom cardiotoxin in skeletal muscle from humans and rats. *Toxicon*, 25(9), 1003-10, Epub 1987/01/01.

[28] Fletcher, J. E., Jiang, S. M., Gong-H, Q., Yudkowsky, M. L., & Wieland, S. J. (1991). Effects of a cardiotoxin from Naja naja kaouthia venom on skeletal muscle: Involvement of calcium-induced calcium release, sodium ion currents and phospholipases A2 and C. *Toxicon*, 29(12), 1489-500.

[29] Ownby, C. L., Fletcher, J. E., & Colberg, T. R. (1993). Cardiotoxin 1 from cobra (Naja naja atra) venom causes necrosis of skeletal muscle in vivo. *Toxicon*, 31(6), 697-709.

[30] Fourie, A. M., Meltzer, S., Berman, M. C., & Louw, A. I. (1983). The effect of cardiotoxin on (Ca2+ + Mg2+)-ATPase of the erythrocyte and sarcoplasmic reticulum. *Biochemistry international*, 6(5), 581-91, Epub 1983/05/01.

[31] Jagodzinski, N. A., Weerasinghe, C., & Porter, K. (2010). Crush injuries and crush syndrome- a review. Part 2: the local injury. *Trauma*, 12(3), 133-48.

[32] Bassaglia, Y., & Gautron, J. (1995). Fast and slow rat muscles degenerate and regenerate differently after whole crush injury. *Journal of muscle research and cell motility*, 16(4), 420-9.

[33] Järvinen, T. A. H., Järvinen, T. L. N., Kääriäinen, M., Kalimo, H., & Järvinen, M. (2005). Muscle Injuries. *The American Journal of Sports Medicine*, 33(5), 745-64.

[34] Winkler, T., von Roth, P., Matziolis, G., Schumann, M. R., Hahn, S., Strube, P., et al. (2011). Time course of skeletal muscle regeneration after severe trauma. *Acta orthopaedica*, 82(1), 102-11, Epub 2010/12/15.

[35] Ghaly, A., & Marsh, D. R. (2010). Ischaemia-reperfusion modulates inflammation and fibrosis of skeletal muscle after contusion injury. *International journal of experimental pathology*, 91(3), 244-55, Epub 2010/04/01.

[36] Iwata, A., Fuchioka, S., Hiraoka, K., Masuhara, M., & Kami, K. (2010). Characteristics of locomotion, muscle strength, and muscle tissue in regenerating rat skeletal muscles. *Muscle & nerve*, 41(5), 694-701, Epub 2010/04/21.

[37] Zhang, L. Y., Ding, J. T., Wang, Y., Zhang, W. G., Deng, X. J., & Chen, J. H. (2011). MRI quantitative study and pathologic analysis of crush injury in rabbit hind limb muscles. *The Journal of surgical researche*, 357-63, Epub 2010/11/03.

[38] Pereira, T., Gärtner, A., Amorim, I., Ribeiro, J., França, M., Armada-da-Silva, P., et al. Development of a skeletal muscle injury model in the rat and in vivo evaluation of the use of Human Mesenchymal Stem Cells (HMSCs) from the umbilical cord matrix in myectomy injury treatment. *17th annual Congress of the European College of Sport Science*, Bruges- Belgium.

[39] Merritt, E. K., Hammers, D. W., Tierney, M., Suggs, L. J., Walters, T. J., & Farrar, R. P. (2010). Functional assessment of skeletal muscle regeneration utilizing homologous extracellular matrix as scaffolding. *Tissue Engineering Part A*, 16(4), 1395-405.

[40] Page, R. L., Malcuit, C., Vilner, L., Vojtic, I., Shaw, S., Hedblom, E., et al. (2011). Restoration of Skeletal Muscle Defects with Adult Human Cells Delivered on Fibrin Microthreads. *Tissue Engineering Part A*, 17(21-22), 2629-2640.

[41] Coppi, P. D., Bellini, S., Conconi, M. T., Sabatti, M., Simonato, E., Gamba, P. G., et al. (2006). Myoblast-acellular skeletal muscle matrix constructs guarantee a long-term repair of experimental full-thickness abdominal wall defects. *Tissue engineering*, 12(7), 1929-36.

[42] Sato, K., Li, Y., Foster, W., Fukushima, K., Badlani, N., Adachi, N., et al. (2003). Improvement of muscle healing through enhancement of muscle regeneration and prevention of fibrosis. *Muscle & nerve*, 28(3), 365-72, Epub 2003/08/21.

[43] Menetrey, J., Kasemkijwattana, C., Day, C., Bosch, P., Vogt, M., Fu, F., et al. (2000). Growth factors improve muscle healing in vivo. *Journal of Bone and Joint Surgery-British*, 82(1), 131-7.

[44] Li, Y., & Huard, J. (2002). Differentiation of Muscle-Derived Cells into Myofibroblasts in Injured Skeletal Muscle. *The American journal of pathology*, 161(3), 895-907.

[45] Valero, M. C., Huntsman, H. D., Liu, J., Zou, K., & Boppart, MD. (2012). Eccentric exercise facilitates mesenchymal stem cell appearance in skeletal muscle. *PloS one*, 7(1), e29760, Epub 2012/01/19.

[46] Boppart, MD, Volker, S. E., Alexander, N., Burkin, D. J., & Kaufman, S. J. (2008). Exercise promotes $\alpha 7$ integrin gene transcription and protection of skeletal muscle. *American Journal of Physiology-Regulatory, Integrative and Comparative Physiology*, 295(5), R1623.

[47] Dedkov, E. I., Kostrominova, T. Y., Borisov, A. B., & Carlson, B. M. (2001). Reparative myogenesis in long-term denervated skeletal muscles of adult rats results in a reduction of the satellite cell population. *Anat Rec*, 263(2), 139-54, Epub 2001/05/22.

[48] Bruusgaard, J. C., & Gundersen, K. (2008). In vivo time-lapse microscopy reveals no loss of murine myonuclei during weeks of muscle atrophy. *The Journal of clinical investigation*, 118(4), 1450-7, Epub 2008/03/05.

[49] IJ-P, J., Meek, M. F., & Gramsbergen, A. (2005). Long-term reinnervation effects after sciatic nerve lesions in adult rats. *Annals of anatomy = Anatomischer Anzeiger : official organ of the Anatomische Gesellschaft*, 187(2), 113-20, Epub 2005/05/20.

[50] Edgerton, V. R., & Roy, R. R. (2002). How selective is the reinnervation of skeletal muscle fibers? *Muscle & nerve*, 25(6), 765-7, 2002/07/13.

[51] Ijkema-Paassen, J., Meek, M. F., & Gramsbergen, A. (2001). Muscle differentiation after sciatic nerve transection and reinnervation in adult rats. *Annals of anatomy = Anatomischer Anzeiger : official organ of the Anatomische Gesellschaft*, 183(4), 369-77, 2001/08/18.

[52] Cebasek, V., Radochova, B., Ribaric, S., Kubinova, L., & Erzen, I. (2006). Nerve injury affects the capillary supply in rat slow and fast muscles differently. *Cell Tissue Res*, 323(2), 305-12, 2005/09/15.

[53] Mc Geachie, J., & Grounds, M. (1989). The onset of myogenesis in denervated mouse skeletal muscle regenerating after injury. *Neuroscience*, 28(2), 509-14.

[54] Gramsbergen, A., van Eykern, L. A., & Meek, M. F. (2001). Sciatic nerve transection in adult and young rats: abnormal EMG patterns during locomotion. *Equine veterinary journal*, 33(S33), 36-40.

[55] Siu, P. M., & Always, S. E. (2005). Mitochondria-associated apoptotic signalling in denervated rat skeletal muscle. *The Journal of physiology*, 565(Pt 1), 309-23, 2005/03/19.

[56] Bulken-Hoover, JD, Jackson, W. M., Ji, Y., Volger, J. A., Tuan, R. S., & Nesti, L. J. (2011). Inducible Expression of Neurotrophic Factors by Mesenchymal Progenitor Cells Derived from Traumatically Injured Human Muscle. *Molecular biotechnology*, 2011/09/10.

[57] Gärtner, A., Pereira, T., Simões, M. J., Armada-da-Silva, P., França, M. L., Sousa, R., et al. (2012). Use of hybrid chitosan membranes and MSC cells for promoting nerve regeneration in an axonotmesis rat model. *Neural Regeneration Research*, (in press).

[58] Gärtner, A., Pereira, T., Amorim, I., Ribeiro, J., & França, M. L. P. A-d-S. (2012). Use of poly(DL-lactide-ε-caprolactone) PLC membranes and MSC cells for promoting nerve regeneration in an axonotmesis rat model: in vitro and in vivo analysis. *Differentiation*, (submitted).

[59] Maurício, A. C., Gärtner, A., Armada-da-Silva, P., Amado, S., Pereira, T., Veloso, A. P., et al. (2011). Cellular Systems and Biomaterials for Nerve Regeneration in Neurotmesis Injuries. *Pignatello R, editor. Biomaterials Applications for Nanomedicine*, 978-9-53307-661-4, Available from: InTech.

[60] Luis, A. L., Rodrigues, J. M., Geuna, S., Amado, S., Shirosaki, Y., Lee, J. M., et al. (2008). Use of PLGA 90:10 scaffolds enriched with in vitro-differentiated neural cells for repairing rat sciatic nerve defects. *Tissue Eng Part A*, 14(6), 979-93, 2008/05/02.

[61] Luis, A. L., Rodrigues, J. M., Geuna, S., Amado, S., Simoes, M. J., Fregnan, F., et al. (2008). Neural cell transplantation effects on sciatic nerve regeneration after a standardized crush injury in the rat. *Microsurgery*, 28(6), 458-70, 2008/07/16.

[62] Amado, S., Rodrigues, J. M., Luis, A. L., Armada-da-Silva, P. A., Vieira, M., Gartner, A., et al. (2010). Effects of collagen membranes enriched with in vitro-differentiated N1E-115 cells on rat sciatic nerve regeneration after end-to-end repair. *J Neuroeng Rehabil*, 7, 7, 2010/02/13.

[63] Simoes, M. J., Amado, S., Gartner, A., Armada-Da-Silva, P. A., Raimondo, S., Vieira, M., et al. (2010). Use of chitosan scaffolds for repairing rat sciatic nerve defects. *Ital J Anat Embryol*, 115(3), 190-210, 2011/02/04.

[64] Maurício, A. C. Cellular Systems and Biomaterials for Nerve Regeneration in Neurotmesis Injuries.

[65] Brandl, F., Sommer, F., & Goepferich, A. (2007). Rational design of hydrogels for tissue engineering: impact of physical factors on cell behavior. *Biomaterials*, 28(2), 134-46, 2006/10/03.

[66] Fedorovich, N. E., Alblas, J., de Wijn, J. R., Hennink, W. E., Verbout, A. J., & Dhert, W. J. A. (2007). Hydrogels as Extracellular Matrices for Skeletal Tissue Engineering: State-of-the-Art and Novel Application in Organ Printing. *Tissue engineering*, 13(8), 1905-25.

[67] Coelho, J., Ferreira, P., Alves, P., Cordeiro, R., Fonseca, A., Góis, J., et al. (2010). Drug delivery systems: Advanced technologies potentially applicable in personalized treatments. *The EPMA Journal*, 1(1), 164-209.

[68] Rehfeldt, F., Engler, A. J., Eckhardt, A., Ahmed, F., & Discher, D. E. (2007). Cell responses to the mechanochemical microenvironment--implications for regenerative medicine and drug delivery. Adv Drug Deliv Rev '2007/09/29 , 59(13), 1329-39.

[69] Lee, M., Wu, B. M., & Dunn, J. C. (2008). Effect of scaffold architecture and pore size on smooth muscle cell growth. *Journal of biomedical materials research Part A*, 87(4), 1010-6, 2008/02/08.

[70] Seliktar, D. (2012). Designing cell-compatible hydrogels for biomedical applications. *Science*, 336(6085), 1124-8, 2012/06/02.

[71] Garg, T., Singh, O., Arora, S., & Murthy, R. (2012). Scaffold: a novel carrier for cell and drug delivery. *Critical reviews in therapeutic drug carrier systems*, 29(1), 1-63, 2012/02/24.

[72] Rihova, B. (2000). Immunocompatibility and biocompatibility of cell delivery systems. *Adv Drug Deliv Rev*, 42(1-2), 65-80, 2000/08/16.

[73] Lyons, F., Partap, S., & O'Brien, F. J. (2008). Part 1: scaffolds and surfaces. *Technology and health care : official journal of the European Society for Engineering and Medicine*, 16(4), 305-17, 2008/09/09.

[74] Martino, M. M., Mochizuki, M., Rothenfluh, D. A., Rempel, S. A., Hubbell, J. A., & Barker, T. H. (2009). Controlling integrin specificity and stem cell differentiation in 2D and 3D environments through regulation of fibronectin domain stability. *Biomaterials*, 30(6), 1089-97, 08/11/26.

[75] Schmidt, J. J., Rowley, J., & Kong, H. J. (2008). Hydrogels used for cell-based drug delivery. *Journal of Biomedical Materials Research Part A*, 87(4), 1113-22.

[76] Freyman, T. M., Yannas, I. V., & Gibson, L. J. (2001). Cellular materials as porous scaffolds for tissue engineering. *Progress in Materials Science*, 46(3-4), 273-82.

[77] Tejas, Shyam. K., & Mauli, A. (2008). Functions and Requirements of Synthetic Scaffolds in Tissue Engineering. *Nanotechnology and Tissue Engineering: CRC Press*, 53-86.

[78] Hoffman, A. S. (2002). Hydrogels for biomedical applications. *Advanced drug delivery reviews*, 54(1), 3-12.

[79] Tan, H., & Marra, K. G. (2010). Injectable, Biodegradable Hydrogels for Tissue Engineering Applications. *Materials*, 3(3), 1746-67.

[80] Yu, L., & Ding, J. (2008). Injectable hydrogels as unique biomedical materials. *Chemical Society reviews*, 37(8), 1473-81.

[81] Nguyen, M. K., & Lee, D. S. (2010). Injectable Biodegradable Hydrogels. *Macromolecular Bioscience*, 10(6), 563-79.

[82] Goessl, A., Tirelli, N., & Hubbell, J. A. (2004). A hydrogel system for stimulus-responsive, oxygen-sensitive in situ gelation. *Journal of biomaterials science*, Polymer edition, 15(7), 895-904, 2004/08/21.

[83] Nguyen, K. T., & West, J. L. (2002). Photopolymerizable hydrogels for tissue engineering applications. *Biomaterials*, 23(22), 4307-14.

[84] Shim, W. S., Yoo, J. S., Bae, Y. H., & Lee, D. S. (2005). Novel injectable pH and temperature sensitive block copolymer hydrogel. *Biomacromolecules*, 6(6), 2930-4, 2005/11/15.

[85] Borschel, G. H., Dennis, R. G., & Kuzon, W. M. Jr. (2004). Contractile skeletal muscle tissue-engineered on an acellular scaffold. *Plastic and reconstructive surgery*, 113(2), 595-602, discussion 3-4, 2004/02/06.

[86] Ahmed, T. A. E., Dare, E. V., & Hincke, M. (2008). Fibrin: a versatile scaffold for tissue engineering applications. *Tissue Engineering Part B: Reviews*, 14(2), 199-215.

[87] Huang, Y. C., Dennis, R. G., Larkin, L., & Baar, K. (2005). Rapid formation of functional muscle in vitro using fibrin gels. *J Appl Physiol*, 98(2), 706-13, 2004/10/12.

[88] Koning, M., Harmsen, M. C., van Luyn, M. J. A., & Werker, P. M. N. (2009). Current opportunities and challenges in skeletal muscle tissue engineering. *Journal of tissue engineering and regenerative medicine*, 3(6), 407-15.

[89] Lee, K. Y., & Mooney, D. J. (2012). Alginate: properties and biomedical applications. *Progress in polymer science*, 37(1), 106-26, 2011/11/30.

[90] Pawar, S. N., & Edgar, K. J. (2012). Alginate derivatization: a review of chemistry, properties and applications. *Biomaterials*, 33(11), 3279-305, 2012/01/28.

[91] Zorlutuna, P., Jeong, J. H., Kong, H., & Bashir, R. (2011). Stereolithography-Based Hydrogel Microenvironments to Examine Cellular Interactions. *Adv Funct Mater*, 21(19), 3642-51.

[92] Lei, Y., Gojgini, S., Lam, J., & Segura, T. (2011). The spreading, migration and proliferation of mouse mesenchymal stem cells cultured inside hyaluronic acid hydrogels. *Biomaterials*, 32(1), 39-47, 2010/10/12.

[93] Prestwich, G. D. (2011). Hyaluronic acid-based clinical biomaterials derived for cell and molecule delivery in regenerative medicine. *Journal of controlled release : official journal of the Controlled Release Society*, 155(2), 193-9, 2011/04/26.

[94] Gauvin, R., Parenteau-Bareil, R., Dokmeci, M. R., Merryman, W. D., & Khademhosseini, A. (2012). Hydrogels and microtechnologies for engineering the cellular microenvironment. *Wiley interdisciplinary reviews Nanomedicine and nanobiotechnology*, 4(3), 235-46, 2011/12/07.

[95] Fossett, E., & Khan, W. S. (2012). Optimising human mesenchymal stem cell numbers for clinical application: a literature review. *Stem cells international*, 2012, 465259, 2012/03/27.

[96] Le Blanc, K., & Ringden, O. (2005). Immunobiology of human mesenchymal stem cells and future use in hematopoietic stem cell transplantation. *Biol Blood Marrow Transplant*, 11(5), 321-34, 2005/04/23.

[97] Wakitani, S., Imoto, K., Yamamoto, T., Saito, M., Murata, N., & Yoneda, M. (2002). Human autologous culture expanded bone marrow mesenchymal cell transplanta-

tion for repair of cartilage defects in osteoarthritic knees. *Osteoarthritis and cartilage OARS, Osteoarthritis Research Society*, 10(3), 199-206, 2002/03/01.

[98] Wang, L., Ott, L., Seshareddy, K., Weiss, M. L., & Detamore, MS. (2011). Musculoskeletal tissue engineering with human umbilical cord mesenchymal stromal cells. *Regen Med*, 6(1), 95-109, 2010/12/24.

[99] Friedenstein, A. J., Petrakova, K. V., Kurolesova, A. I., & Frolova, G. P. (1968). Heterotopic of bone marrow. Analysis of precursor cells for osteogenic and hematopoietic tissues. *Transplantation*, 6(2), 230-47, 1968/03/01.

[100] Togel, F., Weiss, K., Yang, Y., Hu, Z., Zhang, P., & Westenfelder, C. (2007). Vasculotropic, paracrine actions of infused mesenchymal stem cells are important to the recovery from acute kidney injury. *American journal of physiology, Renal physiology*, 292(5), F1626-35, 2007/01/11.

[101] Zhang, M., Mal, N., Kiedrowski, M., Chacko, M., Askari, A. T., Popovic, Z. B., et al. (2007). SDF-1 expression by mesenchymal stem cells results in trophic support of cardiac myocytes after myocardial infarction. *FASEB journal : official publication of the Federation of American Societies for Experimental Biology*, 21(12), 3197-207, 2007/05/15.

[102] Mezey, E., & Chandross, K. J. (2000). Bone marrow: a possible alternative source of cells in the adult nervous system. *European journal of pharmacology*, 405(1-3), 297-302, 2000/10/18.

[103] Brazelton, T. R., Rossi, F. M., Keshet, G. I., & Blau, H. M. (2000). From marrow to brain: expression of neuronal phenotypes in adult mice. *Science*, 290(5497), 1775-9, 2000/12/02.

[104] Bongso, A., Fong-Y, C., & Gauthaman, K. (2008). Taking stem cells to the clinic: Major challenges. *Journal of Cellular Biochemistry*, 105(6), 1352-60.

[105] Conconi, M. T., Burra, P., Di Liddo, R., Calore, C., Turetta, M., Bellini, S., et al. (2006). CD105(+) cells from Wharton's jelly show in vitro and in vivo myogenic differentiative potential. *Int J Mol Med*, 18(6), 1089-96, 2006/11/08.

[106] Kadivar, M., Khatami, S., Mortazavi, Y., Shokrgozar, MA, Taghikhani, M., & Soleimani, M. (2006). In vitro cardiomyogenic potential of human umbilical vein-derived mesenchymal stem cells. *Biochemical and Biophysical Research Communications*, 340(2), 639-47.

[107] Rantanen, J., Hurme, T., Lukka, R., Heino, J., & Kalimo, H. (1995). Satellite cell proliferation and the expression of myogenin and desmin in regenerating skeletal muscle: evidence for two different populations of satellite cells. *Laboratory investigation; a journal of technical methods and pathology*, 72(3), 341-7, 1995/03/01.

Chapter 14

Alignment of Cells and Extracellular Matrix Within Tissue-Engineered Substitutes

Jean-Michel Bourget, Maxime Guillemette,
Teodor Veres, François A. Auger and Lucie Germain

Additional information is available at the end of the chapter

http://dx.doi.org/10.5772/54142

1. Introduction

Most of the cells in our body are in direct contact with extracellular matrix (ECM) components which constitute a complex network of nano-scale proteins and glycosaminoglycans. Those cells constantly remodel the ECM by different processes. They build it by secreting different proteins such as collagen, proteoglycans, laminins or degrade it by producing factors such as matrix metalloproteinase (MMP). Cells interact with the ECM via specific receptors, the integrins [1]. They also organize this matrix, guided by different stimuli, to generate patterns, essential for tissue and organ functions. Reciprocally, cells are guided by the ECM, they modify their morphology and phenotype depending on the protein types and organization via bidirectional integrin signaling [2-4]. In the growing field of tissue engineering [5], control of these aspects are of the utmost importance to create constructs that closely mimic native tissues. To do so, we must take into account the composition of the scaffold (synthetic, natural, biodegradable or not), its organization and the dimension of the structure.

The particular alignment patterns of ECM and cells observed in tissues and organs such as the corneal stroma, vascular smooth muscle cells (SMCs), tendons, bones and skeletal muscles are crucial for organ function. SMCs express contraction proteins such as alpha-smooth-muscle (SM)-actin, desmin and myosin [6] that are essential for cell contraction [6]. To result in vessel contraction, the cells and ECM need to be organized in such a way that most cells are elongated in the same axis. For tubular vascular constructs, it is suitable that SMCs align in the circumferential direction, as they do in vivo [7, 8]. Another striking example of alignment is skeletal muscle cells that form long polynuclear cells, all elongated in the same axis. Each cell generates a weak and short contraction pulse but collectively, it results in a strong, long and sustained contraction of the muscle and, in term, a displacement of the member. In the corneal stroma, the particular arrangement of the corneal fibroblasts (keratocytes) and ECM is essential to keep the transparency of this tissue [9-13]. Tendons also present a peculiar matrix alignment relative to the muscle axis. It gives a substantial resistance and exceptional mechanical properties to the tissue in that axis [14, 15]. Intervertebral discs [16], cartilage [17], dental enamel [18], and basement membrane of epithelium are other examples of tissues/organs that present peculiar cell and matrix organization. By reproducing and controlling those alignment patterns within tissue-engineered substitutes, a more physiological representation of human tissues could be achieved.

Taking into account the importance of cell microenvironment on the functionality of tissue-engineered organ substitutes, one can assume the importance of being able to customise the 3D structure of the biomaterial or scaffold supporting cell growth. To do so, some methods have been developed and most of them rely on topographic or contact guidance. This is the phenomenon by which cells elongate and migrate in the same axis as the ECM. Topographic guidance was so termed by Curtis and Clark [19] to include cell shape, orientation and movement in the concept of contact guidance described by Harrison [20] and implemented by Weiss [21, 22]. Therefore, if one can achieve ECM alignment, cells will follow the same pattern. Inversely, if cells are aligned on a patterned culture plate, the end result would be aligned ECM deposition [23].

The specific property of tissues or materials that present a variation in their mechanical and structural properties in different axis is called anisotropy. This property can be evaluated either by birefringence measurements [24, 25], mechanical testing in different axis [26], immunological staining of collagen or actin filaments [23] or direct visualisation of collagen fibrils using their self-fluorescence around 488 nm [27, 28].

Several techniques have been recently developed to mimic the specific alignment of cells within tissues to produce more physiologically relevant constructs. In this chapter, we will describe five different techniques, collagen gel compaction, electromagnetic field, electrospinning of nanofibers, mechanical stimulation and microstructured culture plates.

2. Methods to align cells and ECM in tissue-engineered constructs

2.1. Collagen gel

Collagen is the main constituent of the ECM [1], it is therefore logical to use it as a scaffold for tissue engineering [29, 30]. Collagen is produced by cells as a protocollagen strand that, upon enzymatic modification, will be able to assemble into procollagen triple helix and then into tropocollagen triple helix. Self-assembly of tropocollagen into collagen microfibrils is followed by lysyl oxidase crosslinking to finally form a collagen fiber [31]. For tissue engineering applications, collagen can be extracted from dermis of different species including bovine, porcine, avian and human [32, 33] and tissues such as human placenta [34]. Collagen monomer can be bought commercially from different providers. However, these solutions do not always contain the full-length collagen molecule, which can affect its properties.

Preparation of a collagen gel from these solutions is quite simple, the collagen solution, provided at pH 2, self-assemble at 37°C and at a neutral pH. Cells can be added to the solution after neutralisation and before casting of the gel, allowing for a uniform distribution of cells in the construct. As cells migrate and elongate into the gel, they will try to anchor themselves to the collagen fibers, but since the collagen gels are soft, cells will deform the fibril network causing the gel to compact [35]. If this contraction is controlled by applying a static constraint in one direction, the gel will contract differentially in the constrained and unconstrained axis. This will result in an anisotropic construct because collagen fibers and cells will become aligned in the constrained axis [34, 36, 37].

Thomopoulos et al. [28] used fibroblast populated collagen gels that were constrained either uniaxially or biaxially to evaluate the anisotropy generated in the gel. Uniaxial static strain resulted in gel contraction in the unconstrained axis and lead to a structural and mechanical anisotropy. They found no difference between tendon fibroblasts and cardiac fibroblasts in anisotropy generated in the construct. They also demonstrated that active remodeling of the gel by cells is not necessary for the development of anisotropy in collagen gel. Indeed, uniaxially constrained collagen gel without cells also become anisotropic. This surprising result could be explained by the force generated by collagen polymerisation [38]. They also developed a mathematical model to predict anisotropy in fibroblast-populated collagen gels [39]. They found that mechanical anisotropy could not be explained solely by collagen fibers alignment, but also take into account the redistribution of collagen fibers upon remodeling, nonaffine fiber kinematics [40, 41] and fiber lengths. Costa et al. [42] also investigated the mechanism of cell and matrix alignment in constrained collagen gels. They constrained fibroblast-containing collagen gels with different shapes (square, triangle and circle) and liberated one or more of the edges to create anisotropy into the gel. Contrasting with a previous report of Klebe et al. [43], they showed that fully constrained gels present random cell and matrix orientation. Nevertheless, on the basis of the results obtained in their study, Costa et al. proposed that rather than aligning along the local direction of greatest tension, cells orient parallel to the local free boundary. An interesting result was obtained with the round shape constrain. As expected, no alignment were present when gel was uniformly constrained, but when they cut a central hole in the construct, the gel contracted away from the central hole and cells aligned in the circumferential orientation. This result is consistent with previous results obtained in blood vessel reconstruction in which a collagen gel is contracting around a central mandrel causing circumferential cell alignment [34]. Grinnell and Lamke [44] cultured fibroblasts on hydrated collagen lattices. They found that cells reorganized the network and aligned the collagen fibrils in the plane of cell spreading, becoming more densely packed. They noted that the lattice has thinned to one-tenth of its original thickness.

Weinberg and bell [45] used collagen gel seeded with bovine cells to produce the first tissue-engineered blood vessel. This weak construct was supported by a Dacron mesh in order to sustain physiological pressure. This method was improved later by other groups to enhance mechanical properties of the collagen gel to get rid of the synthetic material, but those constructs were still too weak to be implanted. L'heureux et al. [34] produced such a tissue-engineered blood vessel using collagen gel. The gel was made of a mixture of human type I and III collagen at a final concentration of 3 mg/ml, it was seeded with human SMCs and finally casted around a cylindrical mandrel, forming the media layer. The anisotropic strain generated by the mandrel constraining gel compaction, combined to a manual detachment of the gel adhesion to the mandrel, caused a progressive circumferential alignment of the SMC. The same process was repeated to produce the adventitia, using human dermal fibroblast embedded in collagen gel and casted around the media layer. Contrasting with the result obtained with the SMC media layer, fibroblasts of the adventitial layer did not get self-oriented. This type of construct was still too weak to be implanted. This paper also showed that gel compaction speed is influenced by initial cell seeding concentration in the construct.

Collagen gel has also been used to engineered intervertebral disc. The highly organized annulus fibrosus which present a cell and matrix alignment, is contrasting with the nucleus pulposus showing random organization. Bowles et al. [16] isolated cells from both parts of the intervertebral disc and seeded them into collagen gels. Interestingly, cells kept their capacities to become organized or not when cultured in vitro and seeded into collagen gel constructs. Robert Tranquillo published several papers using collagen gel, mostly on engineered vascular constructs. Some of his work was done using magnetic alignment of fibers, it will be discussed in the following section. In 1992, they [46] described mathematical theories to understand the complex coupling of cell and matrix deformation in collagen gel populated with cells.

Thus, alignment of cells in collagen gel was one of the first cell alignment method applied to tissue engineering. It is easy to perform, relatively inexpensive and gives interesting results for specific applications. On the other hand, collagen gel do not show mechanical properties sufficient for load bearing application such as bone, cartilage, ligament and blood vessels, at least if it is used alone.

2.2. Electromagnetic field

Electromagnetic field (EMF) can affect multiple aspects of cell behavior and ECM remodeling. We are always in contact with EMF, either coming from the earth, high voltage lines or mobile phones [47, 48]. It is still unclear whether EMF are linked or not to health problems such as cancer, but this hypothesis seems unlikely in most cases [49]. Even if mobile phone usage is probably not linked with brain cancer, the capability of EMF to influence cells and extracellular matrix are clearly demonstrated [50-55]. It is therefore crucial to investigate this relation to first prevent potential deleterious exposure that might lead to health problem and second, to understand mechanisms and eventually develop novel medical therapies [56-58] as well as tissue engineering applications [59-63]. When exposed to a strong EMF, collagen, as well as other biomolecules such as fibrin, will align perpendicularly to the field. This phenomenon is caused by the negative diamagnetic anisotropy of the collagen molecules [64]. This property causes the polymerisation process to take place in a particular orientation. Tissue engineers have used this property to produce alignment of ECM scaffolds and cells in different reconstructed tissues. EMF was also shown to induce orientation of cells, including epithelial cells [65-67], fibroblasts [65, 68], erythrocytes [69] and osteoblasts [61, 70].

http://dx.doi.org/10.5772/54142

The tendency of biomolecules to align in an EMF has been demonstrated more than 30 years ago, but tissue engineering applications have arisen more recently. Strong EMF were proposed as a method to align fibrin polymer by Torbet et al. [71]. They reported that polymerisation of fibrin gels under a strong EMF resulted in oriented fibrin polymerisation. They also speculate right when proposing that this technique could be extended to other polymers and to living cells as it was done afterward. Twenty-six years later, the same researcher [72, 73] used EMF to align collagen fibers, in order to replicate the physiologic structure of the corneal stroma. This structure possesses a particular arrangement of aligned collagen fibers that cross each other orthogonally. This pattern was recreated by cycle of gelation-rotation-gelation of type I collagen under a 7 Tesla (T) EMF. When corneal fibroblasts, or keratocytes, are seeded in the construct, they align themselves by contact guidance in the local orientation of the scaffold. This technique was used to produce a hemi-cornea, composed of the stroma and the epithelial layer and showed promising results when grafted on an animal model [62]. Even if this represents a significant advance in corneal tissue engineering, the resulting corneal stroma substitute presents a slightly different geometry than native cornea. In a non-pathological cornea, the arrangement of orthogonal lamellas within the stroma consist of a mesh of orthogonal collagen lamellas that are woven together in multilayer. This arrangement provides the cornea with a strong mechanical resistance while remaining a transparent structure [74-77]. Reproducing this geometry over the normal thickness of the stroma seems quite difficult using this layer-by-layer technique. Kotani et al [70] studied the effect of EMF on bone formation and orientation. They have shown that, when exposed to a strong EMF of 8 T, osteoblasts oriented parallel to the field. In contrast, when osteoblasts and collagen are mix together, the alignment of both constituents is perpendicular to the field, as for collagen alone. This supposed that contact guidance is a stronger inducer of cell alignment than EMF. Two years later, [61] they showed that exposure of mouse osteoblasts to a strong EMF improved differentiation and matrix synthesis in vitro. They also demonstrated that ectopic bone formation in vivo is stimulated by EMF. When pellets of collagen (1.2 mg/pellet, bovine) containing bone morphogenic protein 2 (5 μg/pellet, human recombinant) are implanted subcutaneously and exposed to an 8 T EMF, bone formation was exacerbated and aligned parallel to the EMF.

Robert Tranquillo used EMF as a method to align cells and ECM in tissue-engineered constructs [59, 63, 78, 79], with a particular focus on media substitutes. In 1993, Guido et al. [24] developed a quantitative method to study cells and fibrils orientation when submitted to an EMF. This method used time-lapse image analysis and live automated birefringence measurements to quantify this phenomenon. Tranquillo et al. [59] showed that a 4.7 T EMF could orient collagen fibrillogenesis resulting in a circonferential orientation, in order to create a media equivalent. Barocas et al. [80] compared four different fabrication conditions for a media equivalent composed of SMC embedded in a collagen gel that were either submitted to EMF and/or mandrel compaction. Compaction of the gel around a central mandrel by SMCs induced a circumferential alignment of cells and collagen fibers, as demonstrated previously [34]. Magnetic circumferential alignment was performed prior gel compaction to produce prealigned gels. When those gels were allowed to contract freely, the circumferential alignment was lost, but when a mandrel was present, the alignment was better than with EMF alone. This method can also be used to guide neurite outgrowth of neural cells. Dubey et al. used collagen [81] and fibrin [82] aligned gel to control outgrowth of neurite. When fibers were aligned, neurite outgrowths were stimulated and could therefore grow longer than random aligned controls.

Retrospectively, this method is effective for biological scaffold such as collagen and fibrin, but to our knowledge, alignment of synthetic materials such as poly(glycolic acid) (PGA) or poly(D,L-lactide-co-glycolide) (PLGA) has not been performed yet. This technique requires a special apparatus capable of generating a strong EMF. Cell viability does not seem to be affected by EMF, allowing for a uniform cell distribution in the construct.

2.3. Electrospun nanofiber

Electrospinning of nanofibers is an interesting approach to produce scaffold for tissue engineering [83-89]. This technique can be used to produce aligned scaffold that will dictate cell elongation by contact guidance [90]. The process of producing polymer microfiber using electrostatic forces was patented in 1934 by Formhals [91] but tissue engineering applications such as musculoskeletal [92] and vascular [93] has been developed recently. Electrospinning can be performed with simple setup consisting of a syringe pump, a high voltage source, and a rotating collector [85]. Precise description of the different possible setups and techniques have been reviewed in details previously [94]. Briefly, a polymer solution is hanging at the tip of a syringe needle by surface tension. When an electric current is applied, EMF results in charge repulsion within the polymer solution, causing the initiation of a jet. Solvent evaporate while jet is traveling, resulting in polymerisation into fibers, which are captured by a collector [94]. Depending on settings and polymers used, those fibers can range from 3 nm to greater than 5 μm in diameter [95]. This technique has been used to engineer many types of scaffolds for tissue engineering [90] including synthetic polymer such as poly(D,L-lactide-co-glycolide) (PLGA) [96], poly-(ε-caprolactone) (PCL) [97], 50:50 poly(L-lactic acid-co-ε-caprolactone) (PLCL) fibers [98], or natural polymer such as collagen [99] or fibrin [100] for various tissue applications. It is also possible to create composite scaffolds by spinning different polymer solution either together or consecutively on the same target. Due to the great plasticity of the technique, it is simple to engineer different patterns to guide cell fate in the desired direction. In order to do so, a rotating mandrel can be used to collect the fiber, resulting in aligned nanofibers [101].

Jose et al. [102] developed an aligned nanofibrous scaffold for bone tissue engineering. This scaffold is a nanocomposite copolymer of PLGA and nano-hydroxyapatite (HA). Fiber diameters, glass transition temperature, storage modulus and degradation rate were characterized for different concentration of nano-HA from 0 to 20%. Briefly, average fiber diameters were augmented from 300 nm for PLGA to 700 nm for 20% nano-HA but formed aggregate at those high concentration. Fiber alignment capability was not influenced by nano-HA concentration. Mechanical properties of the composite material were modulated by nano-HA concentration, it acted as a reinforcement agent at lower concentration (1% and 5 %) but induced defects in the structure at higher concentrations (10 % and 20 %). Influences were also seen in degradation rate and storage modulus. In order to show cell compati-

http://dx.doi.org/10.5772/54142

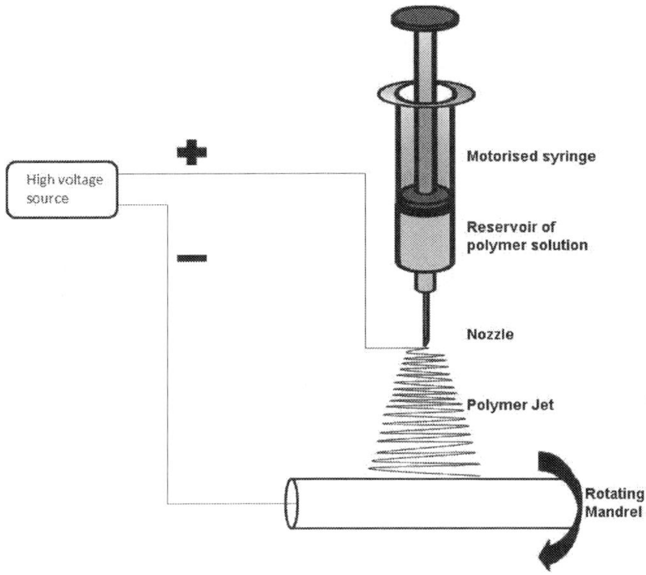

Figure 1. Schematization of an electrospinning setup. The polymer solution is positively charged while the rotating mandrel is negative. The solution is pushed through the nozzle at constant speed. The solvent evaporates while the jet is travelling, resulting in fibers formation.

bility, a collagen component was added to the PLGA (20:80) [103], and human mesenchymal stem cells (MSCs) were used for the demonstration. Thomas et al. [104] investigated the effect of rotation speed on scaffold alignment, mechanical properties and morphologies. They used a nanofibrous mesh of PCL collected at zero, 3,000 and 6,000 rpm for bone tissue engineering. When the collector rotation speed is increased, more aligned fibers were produced. This resulted in a modification of the morphology and mechanical properties of individual fibers and of the resulting scaffold. Interestingly, the hardness and Young's modulus of individual fibers were diminished while increasing rotation speed, but opposite results were obtained when the whole resulting scaffold was analysed. Ultimate tensile strength of the scaffold in the axis of alignment rises from 2.2 to 9.6 MPa when rotation speed was increased from 0 to 6,000 rpm. This result could be explained by increasing fiber alignment and packing as well as a decrease in inter-fiber pore size when rotation speed is increased. Li et al. [92] performed a similar study on rotation speed. They used electrospun aligned nanofibrous scaffold to control anisotropy into tissue-engineered musculoskeletal constructs. Their scaffolds were also made of a biodegradable PCL polymer, casted on a rotating target to align fibers. Increasing the rotation speed of the shaft from 0 to 9.3 m/s (0 to 7,000 rpm) produced more aligned fibers, to a maximum of 94 % of fibers aligned within a 20° angle range, at maximum speed. Alignment of the ECM leads to an increase in the isotropic tensile modulus ranging from 2.1 MPa for unaligned controls to 11.6 MPa for the aligned substrate at 9.3 m/s. Human MSCs and meniscus fibrochondrocytes, seeded on these aligned scaffolds, attached and elongated in the fibers direction. It is also possible with this technique to produce successive layers at different angles by changing the rotation vector.

Teh et al. [105] produced a silk fibroin (SF) hybrid scaffold for ligament regeneration. Silk is an interesting material for tissue engineering, but the hyper-allergenic sericin component must be removed. It has an interesting degradation rate and show remarkable mechanical properties. Silk was previously used to culture fibroblasts and keratinocytes [106]. The new design proposed by Teh et al. is composed of a bilayered SF, contained a knitted SF fibrous mesh layer and an aligned SF electrospun fibers cast on rotating rods. This construct was seeded with MSCs before rolling it into a cylindrical ligament analog. They found that MSCs differentiation into ligament fibroblast was enhanced by the alignment of the fibers. This resulted in an improvement of tensile properties from 125 to 158 N after 14 days and in an increase of ligament-related protein levels such as collagen I and III as well as Tenascin-c. Xu et al. [93] produced an aligned polymer nanofibrous scaffold for blood vessel reconstruction. They used a biodegradable copolymer of poly(L-lactid-co-ε-caprolactone) [P(LLA-CL)] (75:25), aligned using a rotating disc. They demonstrated that human coronary SMCs elongated and migrated in the direction of the fibers and express a spindle-like contractile phenotype with better adherence and proliferation than on control polymer. Yang et al. [107] used a fibrous scaffold to guide neural cells. They used mouse neural stem cells seeded in a poly(L-Lactic acid) (PLLA) nano/micro fibrous scaffolds made by electrospinning. By using two different concentrations of the PLLA solution, 2% and 5%, they were able to cast nanofibers of 150 to 500 nm and microfibers of 800 to 3000 nm, respectively. They have shown that cell elongation and neurite outgrowth is parallel to fiber's direction. They found that the differentiation of neural precursor cells was higher on nanofibers than on microfibers but independent of cell alignment.

Collagen scaffold produced by electrospinning was done by Matthews et al. [108]. In this paper, they set the basis for electrospinning of collagen fibers for tissue engineering application by varying different parameters (collagen source and concentration, solvent, input voltage). With optimal conditions found in this study, they obtained a matrix containing collagen fibers of 100 nm of diameter that exhibited a 67 nm banding pattern, characteristic of native collagen fibers. They cultured aortic SMCs into this collagen scaffold and obtained uniform distribution of cells into the construct. Zhong et al. [109] went further with collagen nanofibrous scaffolds for fibroblast culture. They used calf skin type I collagen, at 80 mg/ml, electrospunned over a wheel rotating at 15 m/s and formed fibrils of 180 nm, which is smaller than the unaligned ones that have an average diameter of 250 nm. They treated their collagen construct with glutaraldehyde vapor (30%) to enhance the biostability of the scaffold. After seeding the construct with rabbit conjunctiva fibroblasts, they measured cell adhesion, proliferation, morphology and interaction with the scaffold. In addition to cell alignment, they noted a lower cell adhesion but higher cell proliferation on aligned constructs.

Given that electrospinning is a versatile technique to produce aligned ECM for tissue engineering, by modifying the casting parameters (rotation speed, input voltage, distance from target, dimension of the tip) or the composition of the solution (type of polymer, solvent, concentration) it is possible to produce different structures in terms of fiber diameters and composition. It is also possible to cast successive layers with different orientations to obtain a complex scaffold. This technique will certainly be perfected in the future using computerized and robotized set-up to produce reproducible and complex scaffolds for various tissue engineering applications.

2.4. Microstructured culture plates

As explained above, topographic guidance is the process by which cells respond to the particular arrangement of their environment by modifying their shape and migration vectors [110]. In vivo, cell environment is composed mostly of native collagen molecules and proteoglycans, in vitro, researchers tried to mimic the cues given by proteins to influence cell fate. Those cues can come from different structures resembling or not collagen molecules and have been discuss in previous pages. In the following section, it is the very structure of the culture plate that dictates the organization of cells. To do so, gratings of various dimensions and shapes are created by various methods into a cell-compatible plastic and cells are seeded over it.

Among the different methods to create and control the type and the shape of the guiding structures, there are very interesting and versatile approaches using microfabrication processes analog to the ones developed by the microelectronic industry [111, 112].

Structures with features ranging from nano- to tens of microns scale matching both the ECM proteins or cells dimension can now be created. These techniques can be adapted to different polymers for various applications and use either polymer casting, micromachining or thermoforming. This section will focus on the most common process that uses photolithography and hot embossing as an example. Briefly, a pattern is printed in chromium on a quartz plate to form the mask for photolithography. A photoresist is poured over a silicon plate and exposed to UV light under the mask. The resist is then developed in organic solvent to reveal the pattern and obtain the first wafer (master). This master wafer is then replicated in a polydimethylsiloxan (PDMS) mold that is cured into the silicon wafer. PDMS can be used for cell culture or be replicated on another substrate. In order to replicate it, hot embossing with an intermediate epoxy wafer is used to finally obtain the desired pattern in the chosen polymer.

Numerous types of materials could be used for the final cell culture step. Polystyrene is an interesting material due to its great biocompatibility, being used regularly for cell culture flask [113]. Teixeira et al. [114] have shown that corneal epithelial cells take an elongated shape and aligned themselves along with grooves and ridges as small as 70 nm. They tested various lengths of grooves and ridges to show that cell alignment is similar from 70 nm to 850 nm ridges but is reduced with 1900 nm ridges. Those lengths were chosen to match the approximate dimension of the basement membrane features. In another paper [115], they showed similar results using corneal stromal fibroblasts (keratocytes). They obtain 70% of aligned fibroblasts with pitch (sum of groove and ridge) larger than 800 nm where as few as 35% of epithelial cells were aligned in the features direction. This is a logical result considering how fibroblastic cells elongate and migrate more than epithelial cells, they are therefore more prone or easy to align.

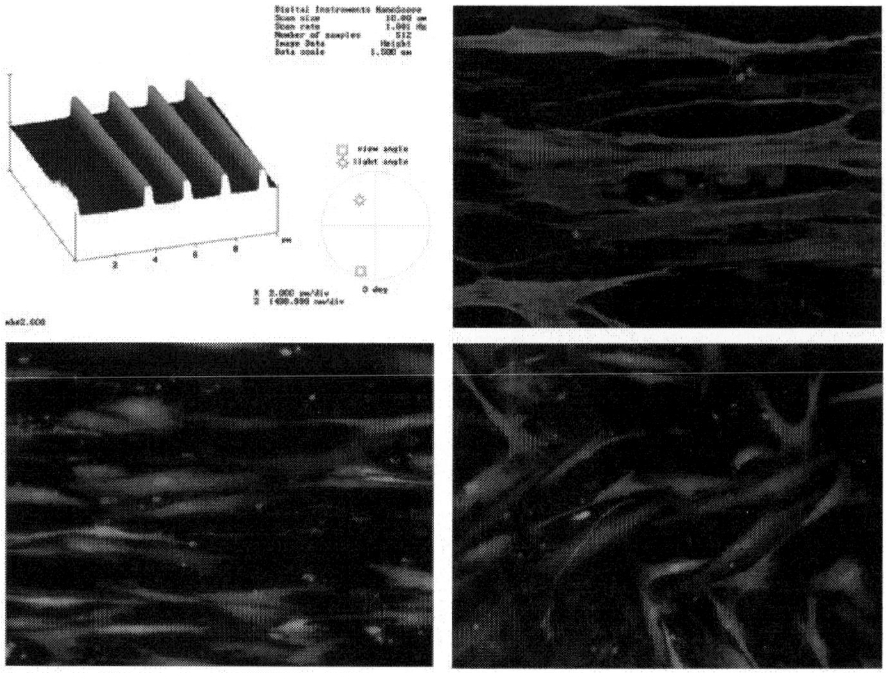

Figure 2. A) Atomic force microscopy of a sample used for replica molding. The master wafer represents the inverse (the negative) of the final results in polystyrene. B) Smooth muscle cells on a microstructured substrate of grooves (1 um) and ridges (4 um) after 1 day of culture. Cells are elongated in the direction of the microstructured pattern. C-D) Human dermal fibroblasts (GFP+) cultured on the same microstructured substrate (C) or on flat plastic (D) after 1 day.

Isenberg et al. [116] cultured SMCs on 50 μm wide and 5 μm deep gratings. They used a combination of photolithography and hot embossing [117] to reproduce patterns of ridges and grooves on polystyrene substrate coated with a thermoresponsive polymer, poly(N-isopropylacrylamide) (PIPAAm) compatible with cell culture [118]. They seeded human aortic SMC on these microtextured culture substrates and allow them to form cell sheets. At room temperature, the cell sheet spontaneously detach from the culture substrate and can be manipulated. In this paper, they have shown that cells elongated and migrated in the direction of the grooves but did not evaluate mechanical or structural anisotropy. In 2012, [119] they tested those parameters and showed mechanical anisotropy in the resulting media layer that mimic the organization of the native vessel.

http://dx.doi.org/10.5772/54142

Guillemette et al. [23] used photolithography and hot embossing to create a pattern of ridges and grooves (4 μm wide, 500 nm deep) in a thermoplastic elastomer, the styrene-(ethylene/butylene)-styrene block co-polymer (SEBS) [120]. Aligned sheets of SMCs, dermal fibroblasts and corneal fibroblasts were produced using the self-assembly or cell sheet engineering approach [121, 122]. All three cell types aligned in the direction of the long axis of the ridges and grooves, for the cells in direct contact with the substrate. Cells grown with ascorbic acid produced ECM and formed tissues comprising many cell layers [119]. Interestingly, suprabasal cell layer that is in contact with the first cell layer aligned themselves with a shift angle from the previous layer characteristic of the cell type. Guillemette et al. [123] used a combination of micromolding and laser microablation to culture cardiac muscle cells. A pattern of grooves and ridges was designed so that wide grooves could be ablated using a laser to create a porous mesh for cardiac cells to grow. Two different pore shapes were studied, square pores and rectangle pores to increase anisotropy. It was shown that square holes are less effective than rectangle holes in aligning cells within the long axis, and that the combination of rectangle holes and grooves gives better alignment than rectangle holes alone. This demonstration pertains to the classic action of cell alignment on an ECM, and improvement of cell functionality as an end result.

2.5. Mechanical strain

Many cells in the human body are subjected to mechanical stress that dictates their phenotype and orientation. This adaptive process starts within the embryo, but is driven by stress and controlled by morphogen gradients or other cues still to be identified. Mechanical stress is probably an inducer for some cell alignment, such as in vasculogenesis [124].

Cell alignment, induced following compaction of collagen gel around a mandrel [34], is driven by mechanical strain anisotropy. For blood vessel reconstruction, a circumferential constraint causes cell and ECM to align circumferentially, as long as the scaffolding gel could move freely in the longitudinal direction. Nevertheless, there is a particular interest in applying mechanical stimulation to different tissue-engineered constructs including vessel, bone, cartilage and ligament, for the induction of alignment and maturation.

Mechanical strain, as a method to align cells and matrix in order to improve functionality, was extensively used in vascular tissue engineering. Mostly via transluminal pressurisation or via a circulating medium in a specifically designed bioreactor. It has been shown by Kanda et al.[125] that a dynamic deformation of 5% at 60 Hertz of a ring of collagen gel containing SMCs will cause cells and ECM to align in the circumferential direction. Niklason et al. [126] has shown that vascular substitutes, made using a PGA scaffold, submitted to pulsed flow for weeks improved mechanical properties, resulting in a burst pressure of more than 2000 mmHg. In this study, cell and matrix alignment was not investigated. Nerem's team has shown that collagen gels present improved mechanical properties and circumferential alignment of the ECM when submitted to a cyclic deformation of 10% at 1 Hz frequency [127] and this process was driven by the ECM remodeling by matrix metalloproteinase 2 [128-130]

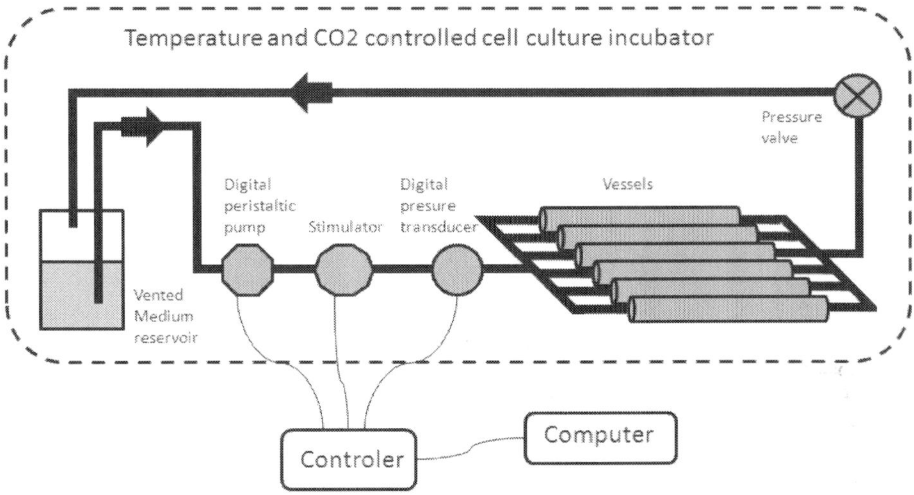

Figure 3. Design of a bioreactor for mechanical strain stimulation of tissue-engineered cylindrical constructs. The culture medium is pumped from the medium reservoir, which is vented through a 0.22 μm sterile air filter, with a computer controlled peristaltic pump. The stimulator is composed of a computer-controlled piston that can generate a cyclic strain in the vessels. The pressure transducer monitors the variation in pressure. The pressure valve ensures an adequate pressure to be built in the vessels. Such devices are commercially available.

Grenier et al. [131, 132] demonstrated that SMCs sheets produced by self-assembly [26, 121, 133] can undergo collagen reorganization following an uniaxial static stretch. More recently, Gauvin et al. [134] extended this experiment using human dermal fibroblast sheets with dynamic stimulation. Constructs were subjected for 3 days to either a static strain of 10%, or a dynamic mechanical stimulation of 10% at 1 Hz. As expected, cells and ECM aligned in the strain axis when submitted to either static or dynamic mechanical stimulation. Ultimate tensile strength and tensile modulus were increased in dynamic mechanical stimulation as compared to unstrained controls and to static strain. Therefore, it demonstrated that mechanical stimulation contributed to establishment of anisotropy in the structure and improves mechanical properties of cell sheet based constructs.

Isenberg et al. [135] evaluated the effects of mechanical strain on media equivalent with an experimental design that isolate its effect from other confounding factors such as gel compaction, creep fatigue and fiber alignment. Therefore, they waited 2 weeks for gel compaction to stabilize, used ribose cross links to fix fiber orientation and to prevent creep deformation. Finally, they placed the construct on distensible latex tubing in a custom designed chamber to produce a pure circumferential strain. They found that modulus and ultimate tensile strength were augmented when the constructs were stimulated for 5 weeks (no change at 2 weeks), at 5% strain, 0,5 Hz and 12.5% duty cycle. Therefore, they demonstrated that independently of the effect on cell and ECM alignment, mechanical strain increased the mechanical properties of collagen based media substitutes.

2.6. Other techniques

There are other techniques that have been developed to align cells and ECM that were not discussed yet. Some of these techniques were too specific or not enough common to give them a complete section, however a quick overview of some of these methods will be given below.

The first method that will be discussed in this section is shear stress. This force is of great importance for blood vessels where it dictates the longitudinal alignment of endothelial cells. This longitudinal configuration is really important for cell function and disturbance in blood flow cause the alignment to be lost. In vessel branching for example, it is associated with endothelial dysfunction, atherosclerosis development and thrombus formation [136]. This capability of endothelial cells to sense the shear stress and align themselves in response to the stimuli rely on specialized cell component called primary cilia [137-139]. This capability can be reproduced in vitro using either a laminar flow chamber [140, 141] or a bioreactor for transluminal flow, either on reconstructed [121] or native [142] endothelium. To induce an alignment response of endothelial cells in vitro, it is important to take into account the parameters that influence the resulting shear stress on endothelium. Those parameters are: vessel section surface, fluid flow and fluid viscosity. Alignment of endothelial cells is influenced by wall shear stress intensity, exposure time and turbidity [110].

A second method, also dependant on shear, was recently developed by McClendon et al. [143] and takes advantage of the biocompatibility of the peptide amphiphile (PA) [144, 145] to produce circumferentially aligned tubes acting as scaffold for arterial tissue engineering. The PA is contained in a liquid crystalline solution that form aligned domains that can be trapped in a gel showing macroscopic alignment by applying low shear rates and ionic crosslink. The interesting point about this technique is the possibility of incorporating cells while forming the scaffold, allowing for full cell penetration into the culture substrate. Therefore, they encapsulated SMCs into their construct and showed that cells proliferated and aligned themselves in the direction of the scaffold, without depending on external stimuli or gel compaction.

Another method for cell alignment is by printing a protein pattern on a culture surface. Cells seeded on these surfaces will adhere and migrate preferentially on the printed pattern resulting in an aligned culture. Thakar et al. [146] used this technique to align vascular SMCs on collagen strip of various width, in order to investigate the relationship between morphology and function of these cells. This interesting technique is difficult to use in a tissue engineering perspective because cells will gradually invade the unprinted region of the surface and therefore alignment of cells will be gradually lost over time.

3. Discussion

The field of tissue engineering has moved forward at great speed in the last two decades, as shown by the rapid augmentation of the number of publication with the terms "tissue engineering" on Pubmed. Researchers have built models for in vitro testing and substitutes that work well within different animal models. On the other hand, there are only a few tissue engineering products that have been tested in humans and even fewer that have moved on as FDA approved product for the market. The words "tissue engineering" allowed retrieval of 46 records on clinicaltrials.gov at the end of august 2012. Beside complications regarding regulatory affairs to get a product accepted, falling in a category between medical device and pharmaceutical drugs [147, 148], some substitutes failed to show adequate mechanical properties to do the job right. Creating more physiological substitutes by recreating the geometry of native ECM and cells is a quite interesting way to improve resistance without introducing a new material or create a thicker construct.

In this chapter, some of the existing techniques that have been published to produce tissue-engineered constructs showing a customised geometry were reviewed. Most of these techniques have been developed for other applications, and adapted later for tissue engineering. Alignment of collagen fibers in collagen gel constrained uniaxially is probably the oldest one, it is a quite simple technique where cell align themselves in the axis of the constraint. This model has been combined with EMF alignment of biomolecules, a more complex technique that direct ECM assembly in a desired orientation. Electrospun nanofibers are becoming more popular in the field and the simple modification of adding a rotating target make it an interesting technique for ECM alignment. The recent advances in microfabrication have made it easy to produce the custom culture substrates that present nanoscale structures at the ECM level. While contact or topographic guidance has been studied quite a while ago, this capability of cells to align themselves in the direction of grooves and ridges of certain dimension is interesting for tissue engineering applications, especially with cell sheet engineering. Finally, mechanical strain is a strong inducer of cell alignment that dictates the geometry of cells in our body. This technique is of great interest for load bearing applications such as cartilage, bone or vascular tissue engineering. Each of the technique mentioned above have advantages and drawbacks, and some of them are dependent on the type of tissue desired.

Constrained collagen gel compaction is a simple technique that is compatible with cell seeding prior gelation, allowing for a uniform cell distribution throughout the construct. On the other hand, collagen gel presents poor mechanical properties, making them unsuitable for load bearing applications such as vascular tissue engineering. The development of hydrogels has helped to partially overcome this problem [29]. Alignment of ECM and cells in EMF is an effective technique but it necessitate state-of-the-art apparatus, not commonly available in a lab, in order to generate an EMF between 4 T and 8 T. As for collagen gels, cells can be added into the solution to have a uniform cell distribution. This technique seems to be restricted at this time to biomolecules, therefore limiting its application field. It can be combined with controlled collagen gel compaction to produce a more potent alignment. Electrospinning of polymer fibers is a very interesting technique that can be applied to a lot of different kinds of polymers by simply modifying the parameters of casting. This technique is particularly effective for tubular constructs since using a cylindrical mandrel as the target will directly create the desired construct. However, it is not possible to seed cells

http://dx.doi.org/10.5772/54142

while casting the construct. Therefore, cells will need to penetrate the construct on their own, causing cell distribution to be potentially non homogenous if the construct is too thick or not enough porous to allow for cell migration. Microstructured culture plate is an effective technique, especially for cell sheet engineering or self-assembly. This approach relies on the capacity of cells to secrete and organize ECM. When grown on structured substrate, cells will align the extracellular matrix in the direction of the grooves. With this technique, it is possible to create complex design of cell alignment by modifying the surface topography. This is the only method described here that does not rely on a pre-existing scaffolding material. On the other hand, producing those cell culture substrates is long and costly, needing access to a microfabrication clean room containing expensive apparatus. Mechanical strain can be applied by a number of different setups depending on the construct type. This technique is effective to align the whole construct in a preferred direction, but complex patterns cannot be created. The cost of this technique depends on the setting that could be computer-controlled or not. Mechanical stress must be precisely controlled since a strain that would be too strong could lead to creeping of the construct or affect cell viability or ECM integrity.

4. Conclusion

The future of tissue engineering relies on the production of more complex structures, composed of many cell types that will interact together. In order to do so, cells must be assembled together in a structure that mimics their native microenvironment. Techniques to align and organize scaffolds will continue to go forward and new technologies will arise, pushed by the constant need for tissue-engineered constructs for organ transplantation.

Author details

Jean-Michel Bourget[1,2,3], Maxime Guillemette[4], Teodor Veres[5], François A. Auger[1,2] and Lucie Germain[1,2]

1 Laval University LOEX center, Tissue Engineering And Regenerative Medecine : LOEX – FRQS Research Center of the "Centre Hospitalier Affilié Universitaire de Québec", Canada

2 Department of Surgery, Faculty of Medicine, Laval University, Quebec, Canada

3 National Research Council Canada, Boucherville, PQ, Canada

4 Physics Department, Faculty of Science and Engineering, Laval University, and Medical Physics Unit, "Centre Hospitalier Universitaire de Québec", Québec, QC, Canada

5 Life Sciences Division, National Research Council Canada, Biomedical Engineering, McGill University, National Research Council Canada, Boucherville, PQ, Canada

References

[1] Alberts B. Molecular biology of the cell. 5th ed. New York: Garland Science; 2008.

[2] Delon I, Brown NH. Integrins and the actin cytoskeleton. Curr Opin Cell Biol. 2007;19:43-50.

[3] Arnaout MA, Mahalingam B, Xiong JP. Integrin structure, allostery, and bidirectional signaling. Annu Rev Cell Dev Biol. 2005;21:381-410.

[4] Wiesner S, Legate KR, Fassler R. Integrin-actin interactions. Cell Mol Life Sci. 2005;62:1081-99.

[5] Langer R, Vacanti JP. Tissue engineering. Science. 1993;260:920-6.

[6] Koubassova NA, Tsaturyan AK. Molecular mechanism of actin-myosin motor in muscle. Biochemistry (Mosc). 2011;76:1484-506.

[7] Holzapfel GA, Sommer G, Gasser CT, Regitnig P. Determination of layer-specific mechanical properties of human coronary arteries with nonatherosclerotic intimal thickening and related constitutive modeling. Am J Physiol Heart Circ Physiol. 2005;289:H2048-58.

[8] Wolinsky H, Glagov S. Structural Basis for the Static Mechanical Properties of the Aortic Media. Circ Res. 1964;14:400-13.

[9] Robert L, Legeais JM, Robert AM, Renard G. Corneal collagens. Pathol Biol (Paris). 2001;49:353-63.

[10] Pinsky PM, van der Heide D, Chernyak D. Computational modeling of mechanical anisotropy in the cornea and sclera. J Cataract Refract Surg. 2005;31:136-45.

[11] Kamma-Lorger CS, Hayes S, Boote C, Burghammer M, Boulton ME, Meek KM. Effects on collagen orientation in the cornea after trephine injury. Mol Vis. 2009;15:378-85.

[12] Boote C, Elsheikh A, Kassem W, Kamma-Lorger CS, Hocking PM, White N, et al. The influence of lamellar orientation on corneal material behavior: biomechanical and structural changes in an avian corneal disorder. Invest Ophthalmol Vis Sci. 2011;52:1243-51.

[13] Boote C, Kamma-Lorger CS, Hayes S, Harris J, Burghammer M, Hiller J, et al. Quantification of collagen organization in the peripheral human cornea at micron-scale resolution. Biophys J. 2011;101:33-42.

[14] Lewis G, Shaw KM. Modeling the tensile behavior of human Achilles tendon. Biomed Mater Eng. 1997;7:231-44.

[15] Lynch HA, Johannessen W, Wu JP, Jawa A, Elliott DM. Effect of fiber orientation and strain rate on the nonlinear uniaxial tensile material properties of tendon. J Biomech Eng. 2003;125:726-31.

[16] Bowles RD, Williams RM, Zipfel WR, Bonassar LJ. Self-assembly of aligned tissue-engineered annulus fibrosus and intervertebral disc composite via collagen gel contraction. Tissue Eng Part A. 2010;16:1339-48.

[17] Jeffery AK, Blunn GW, Archer CW, Bentley G. Three-dimensional collagen architecture in bovine articular cartilage. J Bone Joint Surg Br. 1991;73:795-801.

[18] Paine ML, Snead ML. Protein interactions during assembly of the enamel organic extracellular matrix. J Bone Miner Res. 1997;12:221-7.

[19] Curtis ASG, Clark P. The Effects of Topographic and Mechanical-Properties of Materials on Cell Behavior. Critical Reviews in Biocompatibility. 1990;5:343-62.

[20] Harrison RG. The reaction of embryonic cells to solid structures. Journal of Experimental Zoology. 1914;17:521-44.

[21] Weiss P. In vitro experiments on the factors determining the course of the outgrowing nerve fiber. Journal of Experimental Zoology. 1934;68:393-448.

[22] Weiss P. Experiments on Cell and Axon Orientation Invitro - the Role of Colloidal Exudates in Tissue Organization. Journal of Experimental Zoology. 1945;100:353-86.

[23] Guillemette MD, Cui B, Roy E, Gauvin R, Giasson CJ, Esch MB, et al. Surface topography induces 3D self-orientation of cells and extracellular matrix resulting in improved tissue function. Integr Biol (Camb). 2009;1:196-204.

[24] Guido S, Tranquillo RT. A methodology for the systematic and quantitative study of cell contact guidance in oriented collagen gels. Correlation of fibroblast orientation and gel birefringence. J Cell Sci. 1993;105 (Pt 2):317-31.

[25] Vidal BC. Form birefringence as applied to biopolymer and inorganic material supra-organization. Biotech Histochem. 2010;85:365-78.

[26] Gauvin R, Guillemette M, Galbraith T, Bourget JM, Larouche D, Marcoux H, et al. Mechanical properties of tissue-engineered vascular constructs produced using arterial or venous cells. Tissue Eng Part A. 2011;17:2049-59.

[27] Brightman AO, Rajwa BP, Sturgis JE, McCallister ME, Robinson JP, Voytik-Harbin SL. Time-lapse confocal reflection microscopy of collagen fibrillogenesis and extracellular matrix assembly in vitro. Biopolymers. 2000;54:222-34.

[28] Thomopoulos S, Fomovsky GM, Holmes JW. The development of structural and mechanical anisotropy in fibroblast populated collagen gels. J Biomech Eng. 2005;127:742-50.

[29] Parenteau-Bareil R, Gauvin R, Berthod F. Collagen-Based Biomaterials for Tissue Engineering Applications. Materials. 2010;3:1863-87.

[30] Yarlagadda PK, Chandrasekharan M, Shyan JY. Recent advances and current developments in tissue scaffolding. Biomed Mater Eng. 2005;15:159-77.

[31] Shoulders MD, Raines RT. Collagen structure and stability. Annu Rev Biochem. 2009;78:929-58.

[32] Cliche S, Amiot J, Avezard C, Gariepy C. Extraction and characterization of collagen with or without telopeptides from chicken skin. Poultry Science. 2003;82:503-9.

[33] Parenteau-Bareil R, Gauvin R, Cliche S, Gariepy C, Germain L, Berthod F. Comparative study of bovine, porcine and avian collagens for the production of a tissue engineered dermis. Acta Biomater. 2011;7:3757-65.

[34] L'Heureux N, Germain L, Labbe R, Auger FA. In vitro construction of a human blood vessel from cultured vascular cells: a morphologic study. J Vasc Surg. 1993;17:499-509.

[35] Tranquillo RT, Durrani MA, Moon AG. Tissue engineering science: consequences of cell traction force. Cytotechnology. 1992;10:225-50.

[36] Lopez Valle CA, Auger FA, Rompre P, Bouvard V, Germain L. Peripheral anchorage of dermal equivalents. Br J Dermatol. 1992;127:365-71.

[37] Barocas VH, Tranquillo RT. An anisotropic biphasic theory of tissue-equivalent mechanics: the interplay among cell traction, fibrillar network deformation, fibril alignment, and cell contact guidance. J Biomech Eng. 1997;119:137-45.

[38] Eastwood M, Porter R, Khan U, McGrouther G, Brown R. Quantitative analysis of collagen gel contractile forces generated by dermal fibroblasts and the relationship to cell morphology. J Cell Physiol. 1996;166:33-42.

[39] Thomopoulos S, Fomovsky GM, Chandran PL, Holmes JW. Collagen fiber alignment does not explain mechanical anisotropy in fibroblast populated collagen gels. J Biomech Eng. 2007;129:642-50.

[40] Chandran PL, Barocas VH. Affine versus non-affine fibril kinematics in collagen networks: theoretical studies of network behavior. J Biomech Eng. 2006;128:259-70.

[41] Chandran PL, Barocas VH. Deterministic material-based averaging theory model of collagen gel micromechanics. J Biomech Eng. 2007;129:137-47.

[42] Costa KD, Lee EJ, Holmes JW. Creating alignment and anisotropy in engineered heart tissue: role of boundary conditions in a model three-dimensional culture system. Tissue Eng. 2003;9:567-77.

[43] Klebe RJ, Caldwell H, Milam S. Cells transmit spatial information by orienting collagen fibers. Matrix. 1989;9:451-8.

[44] Grinnell F, Lamke CR. Reorganization of hydrated collagen lattices by human skin fibroblasts. J Cell Sci. 1984;66:51-63.

[45] Weinberg CB, Bell E. A blood vessel model constructed from collagen and cultured vascular cells. Science. 1986;231:397-400.

[46] Tranquillo RT, Murray JD. Continuum model of fibroblast-driven wound contraction: inflammation-mediation. J Theor Biol. 1992;158:135-72.

[47] Takebayashi T, Varsier N, Kikuchi Y, Wake K, Taki M, Watanabe S, et al. Mobile phone use, exposure to radiofrequency electromagnetic field, and brain tumour: a case-control study. Br J Cancer. 2008;98:652-9.

[48] Swerdlow AJ, Feychting M, Green AC, Leeka Kheifets LK, Savitz DA. Mobile phones, brain tumors, and the interphone study: where are we now? Environ Health Perspect. 2011;119:1534-8.

[49] Miyakoshi J. Effects of static magnetic fields at the cellular level. Prog Biophys Mol Biol. 2005;87:213-23.

[50] Zhong C, Zhao TF, Xu ZJ, He RX. Effects of electromagnetic fields on bone regeneration in experimental and clinical studies: a review of the literature. Chin Med J (Engl). 2012;125:367-72.

[51] Costin GE, Birlea SA, Norris DA. Trends in wound repair: cellular and molecular basis of regenerative therapy using electromagnetic fields. Curr Mol Med. 2012;12:14-26.

[52] Liu YX, Tai JL, Li GQ, Zhang ZW, Xue JH, Liu HS, et al. Exposure to 1950-MHz TD-SCDMA Electromagnetic Fields Affects the Apoptosis of Astrocytes via Caspase-3-Dependent Pathway. PLoS One. 2012;7:e42332.

[53] Ongaro A, Varani K, Masieri FF, Pellati A, Massari L, Cadossi R, et al. Electromagnetic fields (EMFs) and adenosine receptors modulate prostaglandin E(2) and cytokine release in human osteoarthritic synovial fibroblasts. J Cell Physiol. 2012;227:2461-9.

[54] Vincenzi F, Targa M, Corciulo C, Gessi S, Merighi S, Setti S, et al. The anti-tumor effect of a(3) adenosine receptors is potentiated by pulsed electromagnetic fields in cultured neural cancer cells. PLoS One. 2012;7:e39317.

[55] Lu YS, Huang BT, Huang YX. Reactive Oxygen Species Formation and Apoptosis in Human Peripheral Blood Mononuclear Cell Induced by 900 MHz Mobile Phone Radiation. Oxid Med Cell Longev. 2012;2012:740280.

[56] Dube J, Rochette-Drouin O, Levesque P, Gauvin R, Roberge CJ, Auger FA, et al. Restoration of the transepithelial potential within tissue-engineered human skin in vitro and during the wound healing process in vivo. Tissue Eng Part A. 2010;16:3055-63.

[57] Dube J, Rochette-Drouin O, Levesque P, Gauvin R, Roberge CJ, Auger FA, et al. Human keratinocytes respond to direct current stimulation by increasing intracellular

calcium: preferential response of poorly differentiated cells. J Cell Physiol. 2012;227:2660-7.

[58] Sun LY, Hsieh DK, Yu TC, Chiu HT, Lu SF, Luo GH, et al. Effect of pulsed electromagnetic field on the proliferation and differentiation potential of human bone marrow mesenchymal stem cells. Bioelectromagnetics. 2009;30:251-60.

[59] Tranquillo RT, Girton TS, Bromberek BA, Triebes TG, Mooradian DL. Magnetically orientated tissue-equivalent tubes: application to a circumferentially orientated media-equivalent. Biomaterials. 1996;17:349-57.

[60] Tsai MT, Chang WH, Chang K, Hou RJ, Wu TW. Pulsed electromagnetic fields affect osteoblast proliferation and differentiation in bone tissue engineering. Bioelectromagnetics. 2007;28:519-28.

[61] Kotani H, Kawaguchi H, Shimoaka T, Iwasaka M, Ueno S, Ozawa H, et al. Strong static magnetic field stimulates bone formation to a definite orientation in vitro and in vivo. J Bone Miner Res. 2002;17:1814-21.

[62] Builles N, Janin-Manificat H, Malbouyres M, Justin V, Rovere MR, Pellegrini G, et al. Use of magnetically oriented orthogonal collagen scaffolds for hemi-corneal reconstruction and regeneration. Biomaterials. 2010;31:8313-22.

[63] Morin KT, Tranquillo RT. Guided sprouting from endothelial spheroids in fibrin gels aligned by magnetic fields and cell-induced gel compaction. Biomaterials. 2011;32:6111-8.

[64] Worcester DL. Structural origins of diamagnetic anisotropy in proteins. Proc Natl Acad Sci U S A. 1978;75:5475-7.

[65] Méthot S, Moulin V, Rancourt D, Bourdages MG, D., Plante M, Auger FA, et al. Morphological changes of human skin cells exposed to a DC electric field in vitro using a new exposure system. Can J Chem Eng. 2001;79:668-77.

[66] Messerli MA, Graham DM. Extracellular electrical fields direct wound healing and regeneration. Biol Bull. 2011;221:79-92.

[67] Zhao M, McCaig CD, Agius-Fernandez A, Forrester JV, Araki-Sasaki K. Human corneal epithelial cells reorient and migrate cathodally in a small applied electric field. Curr Eye Res. 1997;16:973-84.

[68] Sunkari VG, Aranovitch B, Portwood N, Nikoshkov A. Effects of a low-intensity electromagnetic field on fibroblast migration and proliferation. Electromagn Biol Med. 2011;30:80-5.

[69] Higashi T, Yamagishi A, Takeuchi T, Kawaguchi N, Sagawa S, Onishi S, et al. Orientation of erythrocytes in a strong static magnetic field. Blood. 1993;82:1328-34.

[70] Kotani H, Iwasaka M, Ueno S, Curtis A. Magnetic orientation of collagen and bone mixture. Journal of Applied Physics. 2000;87:6191-3.

[71] Torbet J, Freyssinet JM, Hudry-Clergeon G. Oriented fibrin gels formed by polymerization in strong magnetic fields. Nature. 1981;289:91-3.

[72] Torbet J, Malbouyres M, Builles N, Justin V, Roulet M, Damour O, et al. Orthogonal scaffold of magnetically aligned collagen lamellae for corneal stroma reconstruction. Biomaterials. 2007;28:4268-76.

[73] Torbet J, Dickens MJ. Orientation of skeletal muscle actin in strong magnetic fields. FEBS Lett. 1984;173:403-6.

[74] Meek KM, Fullwood NJ. Corneal and scleral collagens--a microscopist's perspective. Micron. 2001;32:261-72.

[75] Meek KM, Boote C. The organization of collagen in the corneal stroma. Exp Eye Res. 2004;78:503-12.

[76] Hayes S, Boote C, Lewis J, Sheppard J, Abahussin M, Quantock AJ, et al. Comparative study of fibrillar collagen arrangement in the corneas of primates and other mammals. Anat Rec (Hoboken). 2007;290:1542-50.

[77] Meek KM, Boote C. The use of X-ray scattering techniques to quantify the orientation and distribution of collagen in the corneal stroma. Prog Retin Eye Res. 2009;28:369-92.

[78] Dickinson RB, Guido S, Tranquillo RT. Biased cell migration of fibroblasts exhibiting contact guidance in oriented collagen gels. Ann Biomed Eng. 1994;22:342-56.

[79] Girton TS, Dubey N, Tranquillo RT. Magnetic-induced alignment of collagen fibrils in tissue equivalents. Methods Mol Med. 1999;18:67-73.

[80] Barocas VH, Girton TS, Tranquillo RT. Engineered alignment in media equivalents: magnetic prealignment and mandrel compaction. J Biomech Eng. 1998;120:660-6.

[81] Dubey N, Letourneau PC, Tranquillo RT. Guided neurite elongation and schwann cell invasion into magnetically aligned collagen in simulated peripheral nerve regeneration. Exp Neurol. 1999;158:338-50.

[82] Dubey N, Letourneau PC, Tranquillo RT. Neuronal contact guidance in magnetically aligned fibrin gels: effect of variation in gel mechano-structural properties. Biomaterials. 2001;22:1065-75.

[83] Kumar PR, Khan N, Vivekanandhan S, Satyanarayana N, Mohanty AK, Misra M. Nanofibers: effective generation by electrospinning and their applications. J Nanosci Nanotechnol. 2012;12:1-25.

[84] Jin L, Wang T, Zhu ML, Leach MK, Naim YI, Corey JM, et al. Electrospun fibers and tissue engineering. J Biomed Nanotechnol. 2012;8:1-9.

[85] Castano O, Eltohamy M, Kim HW. Electrospinning technology in tissue regeneration. Methods Mol Biol. 2012;811:127-40.

[86] Nisbet DR, Forsythe JS, Shen W, Finkelstein DI, Horne MK. Review paper: a review of the cellular response on electrospun nanofibers for tissue engineering. J Biomater Appl. 2009;24:7-29.

[87] Sill TJ, von Recum HA. Electrospinning: applications in drug delivery and tissue engineering. Biomaterials. 2008;29:1989-2006.

[88] Murugan R, Ramakrishna S. Design strategies of tissue engineering scaffolds with controlled fiber orientation. Tissue Eng. 2007;13:1845-66.

[89] Teo WE, Ramakrishna S. A review on electrospinning design and nanofibre assemblies. Nanotechnology. 2006;17:R89-R106.

[90] Liao S, Li B, Ma Z, Wei H, Chan C, Ramakrishna S. Biomimetic electrospun nanofibers for tissue regeneration. Biomed Mater. 2006;1:R45-53.

[91] Formhals A. Process and apparatus for preparing artificial threads. In: Patent U, editor. USA1934.

[92] Li WJ, Mauck RL, Cooper JA, Yuan X, Tuan RS. Engineering controllable anisotropy in electrospun biodegradable nanofibrous scaffolds for musculoskeletal tissue engineering. J Biomech. 2007;40:1686-93.

[93] Xu CY, Inai R, Kotaki M, Ramakrishna S. Aligned biodegradable nanofibrous structure: a potential scaffold for blood vessel engineering. Biomaterials. 2004;25:877-86.

[94] Pham QP, Sharma U, Mikos AG. Electrospinning of polymeric nanofibers for tissue engineering applications: a review. Tissue Eng. 2006;12:1197-211.

[95] Subbiah T, Bhat GS, Tock RW, Pararneswaran S, Ramkumar SS. Electrospinning of nanofibers. Journal of Applied Polymer Science. 2005;96:557-69.

[96] Li WJ, Laurencin CT, Caterson EJ, Tuan RS, Ko FK. Electrospun nanofibrous structure: a novel scaffold for tissue engineering. J Biomed Mater Res. 2002;60:613-21.

[97] Li WJ, Danielson KG, Alexander PG, Tuan RS. Biological response of chondrocytes cultured in three-dimensional nanofibrous poly(epsilon-caprolactone) scaffolds. J Biomed Mater Res A. 2003;67:1105-14.

[98] Geng X, Kwon OH, Jang J. Electrospinning of chitosan dissolved in concentrated acetic acid solution. Biomaterials. 2005;26:5427-32.

[99] Venugopal J, Ramakrishna S. Biocompatible nanofiber matrices for the engineering of a dermal substitute for skin regeneration. Tissue Eng. 2005;11:847-54.

[100] Peretti GM, Randolph MA, Zaporojan V, Bonassar LJ, Xu JW, Fellers JC, et al. A biomechanical analysis of an engineered cell-scaffold implant for cartilage repair. Annals of Plastic Surgery. 2001;46:533-7.

[101] Boland ED, Bowlin GL, Simpson DG, Wnek GE. Electrospinning of tissue engineering scaffolds. Abstracts of Papers of the American Chemical Society. 2001;222:U344-U.

[102] Jose MV, Thomas V, Johnson KT, Dean DR, Nyairo E. Aligned PLGA/HA nanofibrous nanocomposite scaffolds for bone tissue engineering. Acta Biomater. 2009;5:305-15.

[103] Jose MV, Thomas V, Xu Y, Bellis S, Nyairo E, Dean D. Aligned bioactive multi-component nanofibrous nanocomposite scaffolds for bone tissue engineering. Macromol Biosci. 2010;10:433-44.

[104] Thomas V, Jose MV, Chowdhury S, Sullivan JF, Dean DR, Vohra YK. Mechano-morphological studies of aligned nanofibrous scaffolds of polycaprolactone fabricated by electrospinning. J Biomater Sci Polym Ed. 2006;17:969-84.

[105] Teh TK, Toh SL, Goh JC. Aligned hybrid silk scaffold for enhanced differentiation of mesenchymal stem cells into ligament fibroblasts. Tissue Eng Part C Methods. 2011;17:687-703.

[106] Min BM, Lee G, Kim SH, Nam YS, Lee TS, Park WH. Electrospinning of silk fibroin nanofibers and its effect on the adhesion and spreading of normal human keratinocytes and fibroblasts in vitro. Biomaterials. 2004;25:1289-97.

[107] Yang F, Murugan R, Wang S, Ramakrishna S. Electrospinning of nano/micro scale poly(L-lactic acid) aligned fibers and their potential in neural tissue engineering. Biomaterials. 2005;26:2603-10.

[108] Matthews JA, Wnek GE, Simpson DG, Bowlin GL. Electrospinning of collagen nanofibers. Biomacromolecules. 2002;3:232-8.

[109] Zhong S, Teo WE, Zhu X, Beuerman RW, Ramakrishna S, Yung LY. An aligned nanofibrous collagen scaffold by electrospinning and its effects on in vitro fibroblast culture. J Biomed Mater Res A. 2006;79:456-63.

[110] Nelson CM, Tien J. Microstructured extracellular matrices in tissue engineering and development. Curr Opin Biotechnol. 2006;17:518-23.

[111] Clark P, Connolly P, Curtis AS, Dow JA, Wilkinson CD. Topographical control of cell behaviour. I. Simple step cues. Development. 1987;99:439-48.

[112] Kumar A, Whitesides GM. Features of Gold Having Micrometer to Centimeter Dimensions Can Be Formed through a Combination of Stamping with an Elastomeric Stamp and an Alkanethiol Ink Followed by Chemical Etching. Applied Physics Letters. 1993;63:2002-4.

[113] Walboomers XF, Ginsel LA, Jansen JA. Early spreading events of fibroblasts on microgrooved substrates. J Biomed Mater Res. 2000;51:529-34.

[114] Teixeira AI, Abrams GA, Bertics PJ, Murphy CJ, Nealey PF. Epithelial contact guidance on well-defined micro- and nanostructured substrates. J Cell Sci. 2003;116:1881-92.

[115] Teixeira AI, Nealey PF, Murphy CJ. Responses of human keratocytes to micro- and nanostructured substrates. J Biomed Mater Res A. 2004;71:369-76.

[116] Isenberg BC, Tsuda Y, Williams C, Shimizu T, Yamato M, Okano T, et al. A thermoresponsive, microtextured substrate for cell sheet engineering with defined structural organization. Biomaterials. 2008;29:2565-72.

[117] Sarkar S, Dadhania M, Rourke P, Desai TA, Wong JY. Vascular tissue engineering: microtextured scaffold templates to control organization of vascular smooth muscle cells and extracellular matrix. Acta Biomater. 2005;1:93-100.

[118] Okano T, Yamada N, Okuhara M, Sakai H, Sakurai Y. Mechanism of cell detachment from temperature-modulated, hydrophilic-hydrophobic polymer surfaces. Biomaterials. 1995;16:297-303.

[119] Isenberg BC, Backman DE, Kinahan ME, Jesudason R, Suki B, Stone PJ, et al. Micropatterned cell sheets with defined cell and extracellular matrix orientation exhibit anisotropic mechanical properties. J Biomech. 2012;45:756-61.

[120] Guillemette MD, Roy E, Auger FA, Veres T. Rapid isothermal substrate microfabrication of a biocompatible thermoplastic elastomer for cellular contact guidance. Acta Biomater. 2011;7:2492-8.

[121] L'Heureux N, Paquet S, Labbe R, Germain L, Auger FA. A completely biological tissue-engineered human blood vessel. Faseb J. 1998;12:47-56.

[122] Auger FA, Rémy-Zolghadri M, Grenier G, Germain L. The Self-Assembly Approach for Organ Reconstruction by Tissue Engineering. e-biomed: The Journal of Regenerative Medicine. 2000;1:75-86.

[123] Guillemette MD, Park H, Hsiao JC, Jain SR, Larson BL, Langer R, et al. Combined technologies for microfabricating elastomeric cardiac tissue engineering scaffolds. Macromol Biosci. 2010;10:1330-7.

[124] Risau W, Flamme I. Vasculogenesis. Annu Rev Cell Dev Biol. 1995;11:73-91.

[125] Kanda K, Matsuda T. Mechanical stress-induced orientation and ultrastructural change of smooth muscle cells cultured in three-dimensional collagen lattices. Cell Transplant. 1994;3:481-92.

[126] Niklason LE, Gao J, Abbott WM, Hirschi KK, Houser S, Marini R, et al. Functional arteries grown in vitro. Science. 1999;284:489-93.

[127] Seliktar D, Black RA, Vito RP, Nerem RM. Dynamic mechanical conditioning of collagen-gel blood vessel constructs induces remodeling in vitro. Ann Biomed Eng. 2000;28:351-62.

[128] Seliktar D, Nerem RM, Galis ZS. The role of matrix metalloproteinase-2 in the remodeling of cell-seeded vascular constructs subjected to cyclic strain. Ann Biomed Eng. 2001;29:923-34.

[129] Seliktar D, Nerem RM, Galis ZS. Mechanical strain-stimulated remodeling of tissue-engineered blood vessel constructs. Tissue Eng. 2003;9:657-66.

[130] Seliktar D, Zisch AH, Lutolf MP, Wrana JL, Hubbell JA. MMP-2 sensitive, VEGF-bearing bioactive hydrogels for promotion of vascular healing. J Biomed Mater Res A. 2004;68:704-16.

[131] Grenier G, Remy-Zolghadri M, Bergeron F, Guignard R, Baker K, Labbe R, et al. Mechanical loading modulates the differentiation state of vascular smooth muscle cells. Tissue Eng. 2006;12:3159-70.

[132] Grenier G, Remy-Zolghadri M, Larouche D, Gauvin R, Baker K, Bergeron F, et al. Tissue reorganization in response to mechanical load increases functionality. Tissue Eng. 2005;11:90-100.

[133] Pricci M, Bourget JM, Robitaille H, Porro C, Soleti R, Mostefai HA, et al. Applications of human tissue-engineered blood vessel models to study the effects of shed membrane microparticles from T-lymphocytes on vascular function. Tissue Eng Part A. 2009;15:137-45.

[134] Gauvin R, Parenteau-Bareil R, Larouche D, Marcoux H, Bisson F, Bonnet A, et al. Dynamic mechanical stimulations induce anisotropy and improve the tensile properties of engineered tissues produced without exogenous scaffolding. Acta Biomater. 2011;7:3294-301.

[135] Isenberg BC, Tranquillo RT. Long-term cyclic distention enhances the mechanical properties of collagen-based media-equivalents. Ann Biomed Eng. 2003;31:937-49.

[136] Resnick N, Yahav H, Shay-Salit A, Shushy M, Schubert S, Zilberman LC, et al. Fluid shear stress and the vascular endothelium: for better and for worse. Prog Biophys Mol Biol. 2003;81:177-99.

[137] Van der Heiden K, Egorova AD, Poelmann RE, Wentzel JJ, Hierck BP. Role for primary cilia as flow detectors in the cardiovascular system. Int Rev Cell Mol Biol. 2011;290:87-119.

[138] Lu D, Kassab GS. Role of shear stress and stretch in vascular mechanobiology. J R Soc Interface. 2011;8:1379-85.

[139] Egorova AD, van der Heiden K, Poelmann RE, Hierck BP. Primary cilia as biomechanical sensors in regulating endothelial function. Differentiation. 2012;83:S56-61.

[140] Tremblay PL, Huot J, Auger FA. Mechanisms by which E-selectin regulates diapedesis of colon cancer cells under flow conditions. Cancer Res. 2008;68:5167-76.

[141] Ishida T, Peterson TE, Kovach NL, Berk BC. MAP kinase activation by flow in endothelial cells. Role of beta 1 integrins and tyrosine kinases. Circ Res. 1996;79:310-6.

[142] Muller JM, Chilian WM, Davis MJ. Integrin signaling transduces shear stress--dependent vasodilation of coronary arterioles. Circ Res. 1997;80:320-6.

[143] McClendon MT, Stupp SI. Tubular hydrogels of circumferentially aligned nanofibers to encapsulate and orient vascular cells. Biomaterials. 2012;33:5713-22.

[144] Hartgerink JD, Beniash E, Stupp SI. Self-assembly and mineralization of peptide-amphiphile nanofibers. Science. 2001;294:1684-8.

[145] Webber MJ, Kessler JA, Stupp SI. Emerging peptide nanomedicine to regenerate tissues and organs. J Intern Med. 2010;267:71-88.

[146] Thakar RG, Ho F, Huang NF, Liepmann D, Li S. Regulation of vascular smooth muscle cells by micropatterning. Biochem Biophys Res Commun. 2003;307:883-90.

[147] Hellman KB, Johnson PC, Bertram TA, Tawil B. Challenges in tissue engineering and regenerative medicine product commercialization: building an industry. Tissue Eng Part A. 2011;17:1-3.

[148] Johnson PC, Bertram TA, Tawil B, Hellman KB. Hurdles in tissue engineering/regenerative medicine product commercialization: a survey of North American academia and industry. Tissue Eng Part A. 2011;17:5-15.

Chapter 15

Autograft of Dentin Materials for Bone Regeneration

Masaru Murata, Toshiyuki Akazawa,
Masaharu Mitsugi, Md Arafat Kabir, In-Woong Um,
Yasuhito Minamida, Kyung-Wook Kim,
Young-Kyun Kim, Yao Sun and Chunlin Qin

Additional information is available at the end of the chapter

http://dx.doi.org/10.5772/53665

1. Introduction

Regenerative medicine is based on advanced and applied biomaterials science. Biomaterials have a major impact on the patient cure for improving the quality of life. We have been challenging to develop bioabsorbable dentin materials (Murata et al, 2011; Murata et al, 2012), harmonized with bone remodelling, by using the supersonic and acid-etching technology (Akazawa et al, 2012).

While human bone autograft was done in 19th century, human dentin autograft for bone augmentation was reported in IADR 2003. The first clinical case was a sinus lifting using auto-dentin for bone augmentation (Murata, 2003). Dentin is acellular matrix, while bone include osteocytes. Very interestingly, biochemical components in dentin and bone are almost simillar. They consist of body fluid (10%), collagen (18%), non-collagenous proteins (NCPs: 2%) and hydroxyapatite (HAp: 70%) in weight volume (Fig. 1). Demineralized dentin matrix (DDM) and demineralized bone matrix (DBM) are mainly type I collagen with growth factors such as bone morphogenetic proteins (BMPs) (Urist, 1965) and fibroblast growth factors (FGFs) (Fig. 2) (Butler et al.,1977; Murata et al, 2010a,b).

Korea Tooth Bank (KTB) was established in Seoul 2009 for an unique service of tooth-derived graft materials. The medical service system is the preparation and delivery of the tooth-derived materials on demand (Kim et al,2010; Kim et al, 2012). The tooth-derived materials were named as auto-tooth graft materials, which divided into the block-type and powder-type (Park et al., 2010). The block-type material, which is hydrated in 0.9% NaCl solution for 15-30 min before use, can be cut by operators with surgical knife or scissors. Recently, the enamel-dentin grafting has been becoming a realistic alternative to the bone grafting in Korea. We have thought the non-functional teeth as native resources of various graft materials and have achieved the medical recycle of patient-own teeth as novel materials for bone regeneration in Japan and Korea. This matrix-based bone therapy is *Dental Innovation* early in 21st century. Our innovative technique will expand from East Asia to the world.

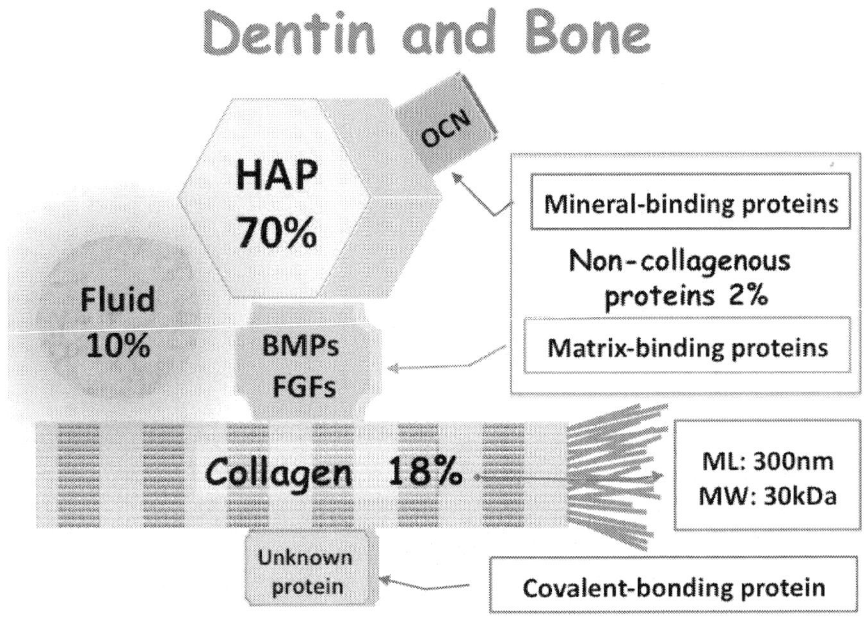

BMPs, FGFs: matrix-binding proteins in NCPs. OCN: mineral-binding proteins in NCPs; Collagen: mainly type I collagen

Figure 1. Chemical components (w/v%) of human dentin and bone;

2. Biochemistry of human dentin

Dentin and bone are mineralized tissues and almost similar in chemical components. They consist of body fluid, collagen, non-collagenous proteins (NCPs) and hydroxyapatite (HAp) in weight volume (Fig. 1). The NCPs in dentin and bone are secreted into the ECM in the process of biomineralization. The category is termed the SIBLING (Small Integrin-Binding Ligand, N-linked Glycoprotein) family that includes dentin sialophosphoprotein (DSPP), dentin matrix protein 1 (DMP1), bone sialoprotein (BSP) and osteopontin (OPN) (Fisher et al, 2001; Qin et al, 2007; Sun et al, 2010; Qin et al, 2011).

Both DDM and DBM are composed of predominantly type I collagen (95%) and matrix-binding proteins such as BMPs (Murata et al., 2000; Akazawa et al., 2006; Murata et al., 2007). BMPs, transforming growth factor-beta (TGF-β), insulin growth factor-I (IGF-I) and IGF-II were detected in human dentin (Finkelman et al., 1990). In the rabbit study, completely demineralized dentin matrix induced bone in the muscle at 4 weeks, while calcified dentin induced bone at 8-12 weeks after implantation (Yeoman & Urist, 1967; Bang & Urist, 1967). Many researchers made effort to discover dentin-derived BMPs. (Butler et al., 1977; Urist et al., 1982; Kawai & Urist., 1989; Bessho et al., 1990). In our study, human DDM and human DBM induced bone and cartilage independently in the subcutaneous tissues at 4 weeks (Murata et al, 2010b). These results indicated that highly calcified tissues such as cortical bone and calcified dentin are not earlier in osteoinduction and osteoconduction than spongy bone, DBM, and DDM. The delayed inductive properties of the calcified dentin and bone may be related to the inhibition of BMPs-release by HAp crystals (Huggins et al., 1970).

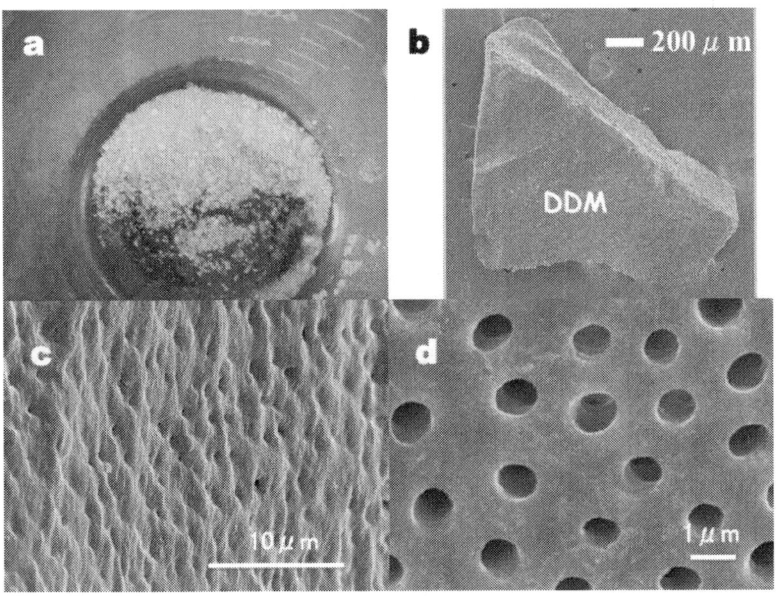

a: wet granules, b,c,d: SEM of DDM granule. Note: dentinal tubes

Figure 2. Crushed tooth granules and SEM photos of demineralized dentin matrix (DDM)

DDM is defined as an acid-insoluble dentin collagen that is absorbable, but hard to digest in human body (Fig. 2). DDM is acellular biomatrix with the micro-tube structure. DDM and DBM possess the ability to coagulate blood plasmas (Huggins & Reddi., 1973). The coagulation action of blood plasma by DDM should become advantageous for surgical operations.

Dentin formation is a dynamic and complicated process, involving interplays among a number of molecules including type I collagen, NCPs and prtoteoglycans, which work collectively to precisely control the site and rate of apatite formation. Type I collagen secreted by odontoblasts forms the scaffold, upon which HAp crystals are deposited. In addition to type I collagen, the extracelluar matrix contains a number of NCPs which play critical roles in the initiation and regulation of HAp crystals (Qin et al., 2011).

3. Clinical study of human dentin

3.1. Case 1: Bone augmentation, 17 year-old female

Patient: A 17-year-old female presented with missing teeth (#11). Clinical and radiological examinations revealed atrophied bone and fractured root residue in the region (Fig. 3a,b). Her medical history was unremarkable.

Surgical procedure 1: Four wisdom teeth were extracted for the preparation of tooth-derived materials (block-type, powder-type).

Preparations of dentin materials: The extracted molar was divided into the crown portion and the root portion. The crown portion was crushed under the cooling. The crushed granules were decalcified in 0.6N HCl solution, rinsed in cold distilled water and freeze-dried. On the other hand, the root portion was perforated by using a round bar to create a porous structure. The root with many holes was decalcified in 0.6N HCl solution, rinsed and freeze-dried. These biomaterials are named as auto-tooth bone (ATB) by KTB.

Surgical procedure 2: This patient-own blood sample was centrifuged and the middle layer was collected as fibrin glue (so called concentrated growth factors: CGF) (Fig. 4a,b). The different ATB materials were immersed in 0.9% NaCl solution before use (Fig. 4c). Additionally, ATB granules were mixed with the fibrin glue (CGF) prepared from autologous blood (Fig. 4d,e). The root-dentin material was divied into 2 parts by using a knife. A titanium fixture (Nobel Replace Tapered NP: 16mm) was implanted into the atrophied bone under local anesthesia (Fig. 3c,d). The root-dentin wall was grafted into the bone defect (fixture-exposed region) as veneer graft (Fig. 4f). The composite of ATB and fibrin contributed to the attachment between the grafted root-dentin and the muco-periosteal flap (Fig. 5a,b,c).

Results and discussion: This patient was successfully restored with the dental implant and the autograft of 2 types of ATB (root-on, powders) with autologous fibrin glue (Fig. 5d). Properly hydrated ATB should facilitate its adaption to the bone defect due to its elasticity and flexibility. The results demonstrated that autogenous tooth could be recycled as the innovative biomaterials.

http://dx.doi.org/10.5772/53665

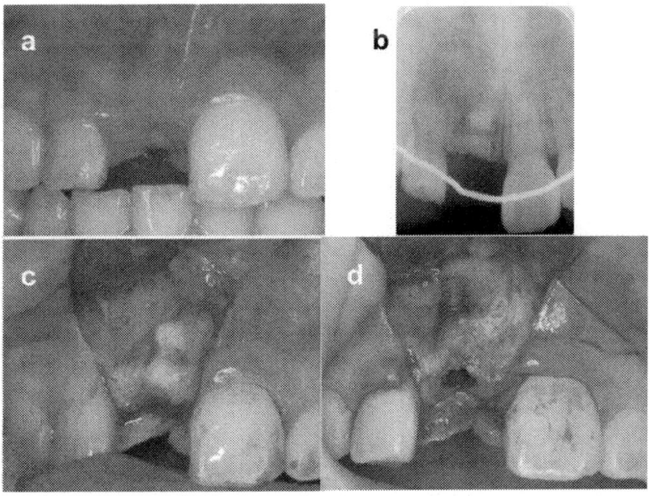

a: intraoral initial view (before operation), Note: a missing tooth (#11) b: X-ray photo, Note: radio-opacity of residual root c: exposed bone, Note: concave shape d: view just after Ti. fixture implantation, Note: labial bone defect

Figure 3. Case 1: Auto-tooth bone graft for implant placement, 17 year-old girl;

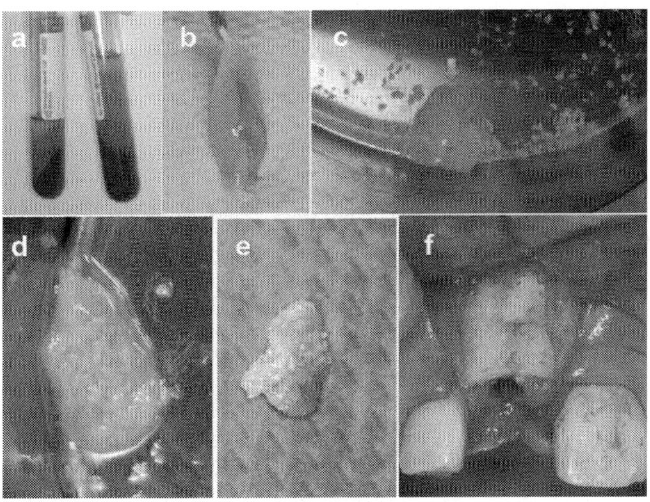

a: blood after centrifugation, Note: 3 layers b: fibrin glue; middle layer in 4a c: wettable ATB materials (block-type ⇩, powder-type) d,e: composite of powder and fibrin glue f: covering with block-type of dentin

Figure 4. Case 1: Auto-tooth bone (ATB) graft for implant placement, 17 year-old girl

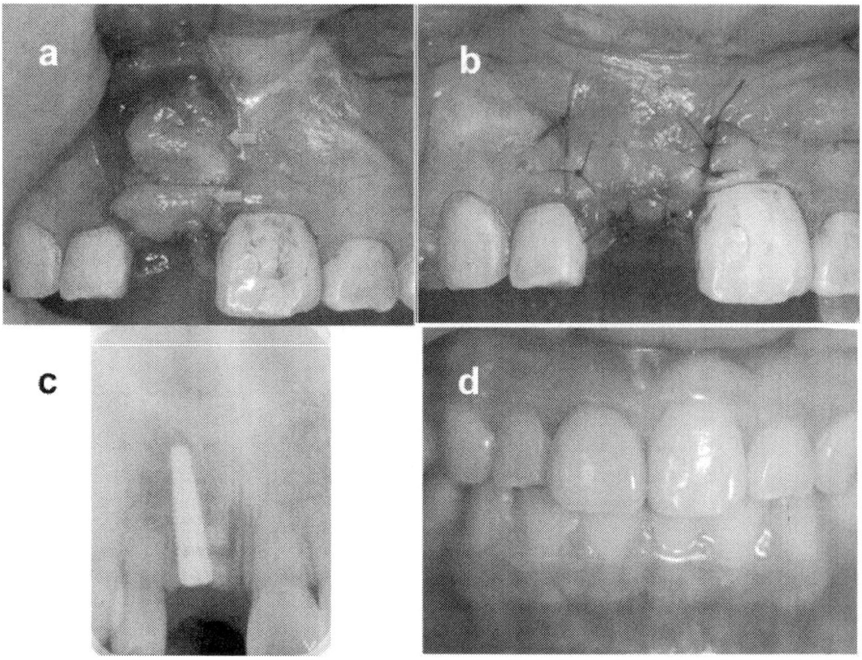

a: fibrin glue including ATB powders (⇐) b: repositioned flap. Note: suture with nylon c: X-ray photo just after operation d: final view after prosthetic restoration

Figure 5. Case 1: Auto-tooth bone graft for implant placement, 17 year-old girl

3.2. Case 2: DDM onlay graft and tooth autograft, 25 year-old female

Patient: A 25-year-old female presented with missing teeth (#46). She lost the first molar about 12 years ago. A clinical examination revealed an atrophied bone in the region. Her medical history was unremarkable.

Surgical procedure 1: A non-functional vital tooth (#28) was extracted and immediately crushed with saline ice by our newly developed tooth- mill (Osteo-Mill®, Tokyo Iken Co., Ltd) at 12000rpm for 30 sec (Fig. 6) (Patent: 4953276). Briefly, vessel and blade were made in ZrO_2. The crushed tooth-granules were decalcified in 2% HNO_3 solution for 20 min (Murata et al., 2009). The DDM granules including cementum were rinsed in cold distilled water. Cortical perforations were performed in the atrophied bone, and DDM were immediately autotransplanted on the perforated bone under local anesthesia.

Surgical procedure 2: At 4 months after the first operation, a non-functional vital tooth (#18) was extracted and received the immediate root canal filling (RCF), using a new fixation device (Fig. 7). The device was developed for tooth transplantation and replantation (Patent: 4866994).

http://dx.doi.org/10.5772/53665

After the bone biopsy for the tissue observation and the preparation of transplated cavity, tooth autograft was carried out into the host bone (Fig. 8a,b,d).

Results and discussion: The biopsy tissue showed that DDM granules were received to host, and partially replaced by new bone (Fig. 8e). This case was onlay graft of DDM on perforated cortical bone (Murata et al, 1999; Murata et al, 2000). Though RCF is generally carried out at more than 4 weeks after tooth transplantation, we did immediate RCF, using the medical device. This patient was successfully restored with her own 2 teeth. This case was the immediate tooth autotransplantition with the immediate root canal filling at 4 months after DDM autograft in 2009.

a: mill, b: tooth with ice blocks, c: ZrO_2 vessel, d: crushed tooth, e: DDM granules before clinical use.

Figure 6. Preparation of DDM using automatic tooth mill (Osteo-Mill®, Tokyo Iken)

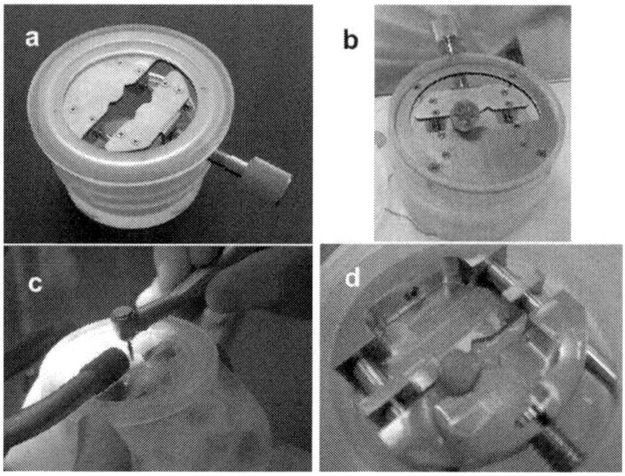

a: whole view, Note: the device developed for tooth transplantation and replantation b: fixed tooth, Note: correspondence to all teeth c: crown treatment, Note: periodontal ligament tissue protected from infected fine particles d: root view, Note: keeping blood even after cutting and root canal filling

Figure 7. New device for protecting periodontal ligament cells (Mr.FIX®, Tokyo Iken)

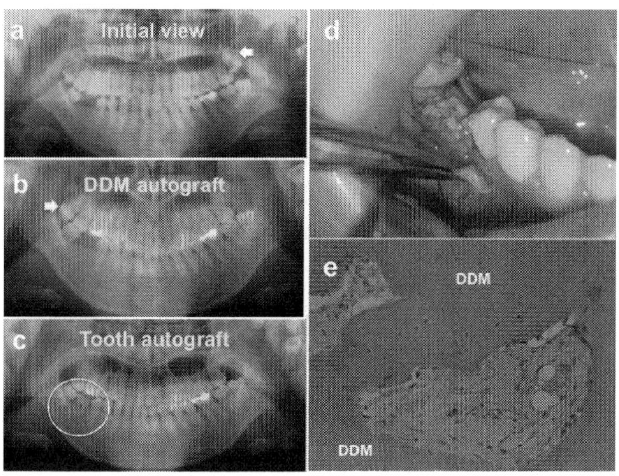

a: initial X-ray photo: missing tooth (#46) and atrophied bone. Non-functional tooth (⇦) for DDM b: just after DDM graft. Non-functional tooth (⇨) for next tooth autograft c: tooth auto-transplantation at 4 months after DDM graft d: DDM autograft on perforated cortical bone before suture e: biopsy: mature bone connected with DDM residue (HE section)

Figure 8. Case 2: 24 year-old woman

4. Supersonic and acid-etching method for Dentin geometry

Compact structure inhibits the body fluid permeation and the cell invasion into the inside of the materials. Generally, this situation is called a material wall. Dentin and cortical bone have compact structure. We have been challenging to develop new dentin materials, using a supersonic and acid-etching technology (Akazawa et al., 2009; Akazawa et al., 2010; Akazawa et al., 2012). The surface structure design of dentin by the supersonic treatment might easily produce new functional scaffolds, which control the bio-absorption rate and the adsorption ability for protein and cells. Figure 9 shows the dissolution efficiencies of human dentin granules, which were demineralized for 5-45 min in 2.0%-HNO_3 solutions by the supersonic treatment at 600W. A photograph inside Fig.9 is Digital microscopic view of DDM, dissolution for 30 min in 2.0%- HNO_3 by the supersonic treatment at 600W and 28 kHz.

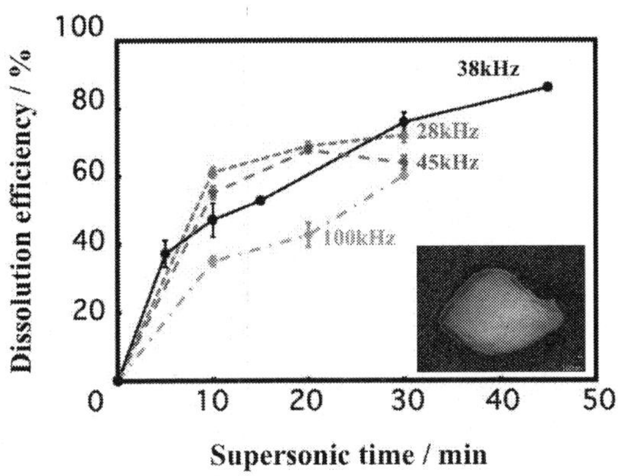

Figure 9. Dissolution efficiencies of human dentin granules, demineralized for 5-45 min in 2.0%-HNO_3 by supersonic treatment at 600W. Inside photo: Digital microscopic DDM view, dissolution for 30 min in 2.0%- HNO_3 by supersonic treatment at 600W and 28 kHz.

The innovative technology can create the adequate geometry and the surface structure of commercially available materials (Akazawa et al., 2012). Geometrical factors will improve the performance of biomaterials for bone regeneration (Reddi, 1974: Kuboki et al, 1995; Murata et al, 1998). Biomaterials science should support and develop the advanced regenerative therapy using tooth-derived materials for patients in the near future.

Acknowledgment

This project was greatly supported by the grant (consortium: 2004-5) of Japan Ministry of Economy, Trade and Industry, and Korea Tooth Bank Co. Ltd. The authors would like to thank WISM Mutoh Co. Ltd., and Tokyo Iken Co. Ltd., for developing the devices (Patents: 4866994, 4953276).

Author details

Masaru Murata[1], Toshiyuki Akazawa[2], Masaharu Mitsugi[3], Md Arafat Kabir[1], In-Woong Um[4], Yasuhito Minamida[1], Kyung-Wook Kim[5], Young-Kyun Kim[6], Yao Sun[7,8] and Chunlin Qin[8]

1 Health Sciences University of Hokkaido, Japan

2 Hokkaido Organization, Japan

3 Takamatsu Oral and Maxillofacial Surgery, Japan

4 Korea Tooth Bank Co. Ltd, Korea

5 Dankok University, Korea

6 Seoul National University, Korea

7 Harbin Medical University, China

8 Texas A&M Health Science Center Baylor College of Dentistry, USA

References

[1] Akazawa, T., Murata, M., Sasaki, T., Tazaki, J., Kobayashi, M., Kanno, T., Matsushima, K., & Arisue, M. (2006). Biodegradation and bioabsorption innovation of the functionally graded cattle-bone-originated apatite with blood compatibility. J Biomed Mater Res, 76A., 1., 44-51.

[2] Akazawa, T., Murata, M., Hino, J., Nakamura, K., Tazaki, J., Kikuchi, M., & Arisue, M. (2007). Materials design and application of demineralized dentin/apatite composite granules derived from human teeth. Archives of Bioceramics Research, 7., 25-28.

[3] Akazawa, T., Murata, M., Tazaki, J., Nakamura, K., Hino, J., Ito, I., Yamamoto, M., Tabata, Y., Takahata, M., & Ito, M.(2009). Biomimetic microstructure and biocompatibility of functionally graded hydroxyapatite derived from animal bone by a supersonic dissolution - Precipitation method- . Bioceramics 22, 22., 155-158.

[4] Akazawa, T., Murata, M., Takahata, M., Xianjun, D., Abe, Y., Nakamura, K., Hino, J., Tazaki, J., Ito, K., Ito, M., Iwasaki, N., Minami, A., Nakajima, T., & Sakamoto, M. (2010). Characterization of microstructure and bio-absorption of the hydroxyapatite

ceramics modified by a partial dissolution-precipitation technique using supersonic treatment. Journal of the Ceramic Society of Japan, 118., 6., 535-540.

[5] Akazawa, T., Murata, M., Tabata, Y., & Ito, M.(2012). Bone regeneration (ISBN) Chapter: biomimetic microstructure and biocompatibility of hydroxyapatite porous ceramics designed by a partial dissolution-precipitation technique with supersonic treatment. INTECK Publisher, Croatia, pp275-292

[6] Akazawa, T., Murata, M., Hino, J., Nagano, F., Shigyo, T, Nomura, T, Inano, H., Itabashi, K., Yamagishi, T, Nakamura, K., Takahashi, T., Iida, S., Kashiwazaki, H. (2012). Surface structure and biocompatibility of demineralized dentin matrix granules soaked in a simulated body fluid. Applied Surface Science, in press.

[7] Bang, G. & Urist, MR. (1967). Bone induction in excavation chambers in matrix of decalcified dentin. Arch Surg, 94., 6., 781-789.

[8] Bessho, K., Tagawa, T., & Murata, M. (1990). Purification of rabbit bone morphogenetic protein derived from bone, dentin, and wound tissue after tooth extraction. J Oral Maxillofac Surg, 48., 162-169.

[9] Butler, WT., Mikulski, A., Urist, MR., Bridges, G., & Uyeno, S. (1977). Noncollagenous proteins of a rat dentin matrix possessing bone morphogenetic activity. J Dent Res, 56., 228-232.

[10] Finkelman, RD., Mohan, S., Jennings, JC., Taylor, AK., Jepsen, S., & Baylink, DJ. (1990). Quantitation of growth factors IGF-I, SGF/IGF-II, and TGF-beta in human dentin. J Bone Miner Res., 5., 7., 717-23.

[11] Fisher L.W., Torchia D.A., Fohr B., Young M.F., & Fedarko N.S. (2001). Flexible structures of SIBLING proteins, bone sialoprotein, and osteopontin. Biochem Biophys Res Commun., 280: 460-465.

[12] Huggins, CB., Wiseman, S., & Reddi, AH. (1970). Transformation of fibroblasts by allogeneic and xenogeneic transplants of demineralized tooth and bone. J Exp Med, 132., 1250-1258.

[13] Huggins, CB., & Reddi, AH. (1973). Coagulation of blood plasma of guinea pig by the bone matrix. Proc Natl Acad Sci U S A., 70., 3., 929-33.

[14] Inoue, T., Deporter, DA., & Melcher, AH. (1986). Induction of chondrogenesis in muscle, skin, bone marrow, and periodontal ligament by demineralized dentin and bone matrix in vivo and in vitro. J Dent Res, 65., 12-22.

[15] Ito, K., Arakawa, T., Murata, M., Tazaki, J., Takuma, T., & Arisue, M. (2008). Analysis of bone morphogenetic protein in human dental pulp tissues. Archives of Bioceramics Research, 8., 166-169.

[16] Kawai, T., & Urist, MD. (1989). Bovine tooth-derived bone morphogenetic protein. J Dent Res, 68., 1069-1074.

[17] Kawakami, T., Kuboki, Y., Tanaka, J., Hijikata, S., Akazawa, T., Murata, M., Fujisawa, R., Takita, H., & Arisue, M. (2007). Regenerative Medicine of Bone and Teeth. Journal of Hard Tissue Biology, 16(3),95-113.

[18] Kim, YK., Kim, SG., Byeon, JH., Lee, HJ., Um, IU., Lim, SC., & Kim, SY. (2010). Development of a novel bone grafting material using autogenous teeth. Oral Surg Oral Med Oral Pathol Oral Radiol Endod., 109., 4., 496-503.

[19] Kim, YK.(2012). Bone graft material using teeth. Journal of the Korean Association of Oral and Maxillofacial Surgens, 38., 3, 134-138.

[20] Kuboki, Y., Saito, T., Murata, M., Takita, H., Mizuno, M., Inoue, M., Nagai, N. & Poole, R. (1995). Two distinctive BMP-carriers induce zonal chondrogenesis and membranous ossification, respectively; geometrical factors of matrices for cell-differentiation. Connective Tissue Research, 31., 1-8.

[21] Murata, M., Inoue, M., Arisue, M., Kuboki, Y., & Nagai, N. (1998). Carrier-dependency of cellular differentiation induced by bone morphogenetic protein (BMP) in ectopic sites. Int J Oral Maxillofac Surg, 27., 391-396.

[22] Murata, M., Huang, BZ., Shibata, T., Imai, S., Nagai, N., & Arisue, M. (1999). Bone augmentation by recombinant human BMP-2 and collagen on adult rat parietal bone. Int J Oral Maxillofac Surg, 28., 232-237.

[23] Murata, M., Maki, F., Sato, D., Shibata, T., & Arisue, M. (2000). Bone augmentation by onlay implant using recombinant human BMP-2 and collagen on adult rat skull without periosteum. Clin Oral Impl Res, 11., 289-295.

[24] Murata, M. (2003). Autogenous demineralized dentin matrix for maxillary sinus augmentation in human. The first clinical report. 81th International Association for Dental Research, Geteburg, Sweden, 2003, June.

[25] Murata, M., Akazawa, T., Tazaki, J., Ito, K., Sasaki, T., Yamamoto, M., Tabata, Y., & Arisue, M. (2007). Blood permeability of a novel ceramic scaffold for bone morphogenetic protein-2. J Biomed Mater Res, 81B., 2., 469-475.

[26] Murata, M., Akazawa, T., Tazaki, J., Ito, K., Hino, J., Kamiura, Y., Kumazawa, R., & Arisue, M. (2009). Human Dentin autograft for bone regeneration - Automatic pulverizing machine and biopsy –. Bioceramics 22, 22., 745-748.

[27] Murata, M., Kawai, T., Kawakami, T., Akazawa, T., Tazaki, J., Ito, K., Kusano, K., & Arisue, M. (2010a). Human acid-insoluble dentin with BMP-2 accelerates bone induction in subcutaneous and intramuscular tissues. Journal of the Ceramic Society of Japan, 118., 6., 438-441.

[28] Murata, M., Akazawa, T., Takahata, M., Ito, M., Tazaki, J., Hino, J., Nakamura, K., Iwasaki, N., Shibata,T., & Arisue, M. (2010b). Bone induction of human tooth and bone crushed by newly developed automatic mill. Journal of the Ceramic Society of Japan, 118., 6., 434-437.

[29] Murata, M., Akazawa, T., Mitsugi, M., Um, IW., Kim, KW., & Kim, YK. (2011). Human Dentin as Novel Biomaterial for Bone Regeneration, Biomaterials - Physics and Chemistry, Rosario Pignatello (Ed.), ISBN: 978-953-307-418-4, INTECK publisher, Croatia, pp127-140.

[30] Murata, M., Sato, D., Hino, J., Akazawa, T., Tazaki, J., Ito, K., & Arisue, M. (2012). Acid-insoluble human dentin as carrier material for recombinant human BMP-2. J Biomed Mater Res, 100A., 571-577.

[31] Park, SM., Um, IW., Kim, YK., & Kim, KW. (2012). Clinical application of auto-tooth bone garaft material. Journal of the Korean Association of Oral and Maxillofacial Surgens, 38.,1., 2-8.

[32] Qin C., D'Souza R., & Feng J.Q. (2007). Dentin matrix protein 1 (DMP1): new and important roles for biomineralization and phosphate homeostasis. J Dent Res 86:1134-1141

[33] Qin C., Brunn J.C., Jones J., George A., Ramachandran A., Gorski J.P., & Butler W.T. (2011). A comparative study of sialic acid-rich proteins in rat bone and dentin. Eur J Oral Sci 109: 133-141.

[34] Reddi, AH. (1974). Bone matrix in the solid state:geometric influence on differentiation of fibroblasts. Adv Biol Med Phys, 15., 1-18.

[35] Sun, Y., Lu, Y., Chen, S., Prasad, M., Wang, X., Zhu, Q., Zhang, J., Ball, H., Feng, J., Butler, WT., & Qin, C. (2010). Key proteolytic cleavage site and full-length form of DSPP. J Dent Res 89: 498-503.

[36] Togari, K., Mitazawa, K., Yagihashi, K., Tabuchi, M., Maeda, H., Kawai, Y., & Goto, S. (2011). Bone regeneration by demineralized dentin matrix in skull defects of rats. Journal of Hard Tissue Biology, 21(1),25-33.

[37] Urist, MR. (1965). Bone: Formation by autoinduction. Science, 150., 893-899.

[38] Urist, MR., Mizutani, H., Conover, MA., Lietze, A., & Finerman, GA. (1982) Dentin, bone, and osteosarcoma tissue bone morphogenetic proteins. Prog Clin Biol Res, 101., 61-81.

[39] Yeomans, JD. & Urist, MR. (1967). Bone induction by decalcified dentine implanted into oral, osseous and muscle tissues. Arch Oral Biol, 12., 999-1008.

Chapter 16

Healing Mechanism and Clinical Application of Autogenous Tooth Bone Graft Material

Young-Kyun Kim, Jeong Keun Lee,
Kyung-Wook Kim, In-Woong Um and
Masaru Murata

Additional information is available at the end of the chapter

http://dx.doi.org/10.5772/53200

1. Introduction

Autogenous bone, allogenic bone, xenogenic bone, and alloplastic materials are bone graft materials that are presently used in dental clinics. According to bone healing mechanism, they can be categorized into materials that induce osteogenesis, osteoinduction, and osteoconduction. Among the many different types of bone graft materials, autogenous bone is the most ideal since it is capable of osteogenesis, osteoinduction, and osteoconduction. Its advantage is the rapid healing time without immune rejection. As its biggest shortcomings, however, the harvest amount is limited, bone resorption after graft is unavoidable, and second defect is generated in the donor area. Therefore, to overcome such shortcomings, allogenic bone and synthetic bone were developed and used in clinics, and efforts have been made to develop more ideal bone substitution materials [1]. Lately, researchers and clinicians have become interested in the use of human dentin from extracted teeth in the context of autogenous bone grafts [2,3]. Dentin has inorganic and organic components that are very similar to those of human bone. In dentin, the inorganic content is 70 ~ 75%, whereas the organic content is about 20%. In alveolar bone, the inorganic content is 65%, and the organic content is 25%. At least 90% of organic content of dentin is type I collagen, which plays an important role in bone formation and mineralization. Dentin also contains bone morphogenetic proteins (BMP), which promote the differentiation of mesenchymal stem cells into chondrocytes and consequently enhance bone formation. In addition, both alveolar bone and teeth are derived from neural crest cells [4-6]. Thus, studies have been done to use fresh tooth in the form of demineralized dentin matrix (DDM) as a biocompatible autogenous bone graft material in alveolar bone repair. Butler, et al [7] and Conover and Urist, et al [8] successfully extracted bone BMP from rabbit DDM, and Bessho, et al [9] secured new bone formation *in situ* by BMP from human DDM. Furthermore, Ike and Urist [10] used dentin root matrix as a carrier of recombinant human bone morphogenetic protein (rhBMP). Starting in 1993, we developed bone graft materials using human teeth with which we conducted experimental studies [11-22]. In 2008, we developed an autogenous tooth bone graft material (AutoBT; Korea Tooth Bank Co., Seoul, Korea) from extracted teeth prepared as powder and grafted it to the donor patient himself. The mineral components of autogenous tooth bone graft materials have 4 stages (types) of calcium phosphate (HA, TCP, OCP, and ACP). Under scanning electron microscopic examination, HA crystalline structures and collagen fibers around the dentinal tubules were detected. Short-term clinical studies reported that, even when wounds became dehiscent, the bone graft materials were not infected, and good secondary healing was achieved [3,23].

2. Osteoinduction of AutoBT

Many researchers have examined tooth dentin as a potential carrier for human proteins and as grafting material because its biological composition is very similar to that of alveolar bone [9, 24-28]. Both tooth and alveolar bone are derived from neural crest cells and are made up of the same Type I collagen. Furthermore, dentin contains BMPs, which induce bone formation and noncollagenous proteins such as osteocalcin, osteonectin, and dentin phosphoprotein [29, 30]. Since its investigation by Urist in 1965, BMP has been widely studied and used in clinical applications [31]. As a result, Yeoman and Urist, et al (1967) and Bang and Urist, et al (1967) showed the osteoinductivity of rabbit DDM by BMP [32, 33]. Bessho, et al extracted BMP from bone matrix, dentin matrix, and wound tissue after extracting teeth from rabbits. Each BMP was confirmed to have induced the formation of new bone when xenogenic implantation was performed [9]. Bessho, et al extracted human dentin matrix containing 4mol/L guanidine HC1 and refined it into liquid chromatography and found out based on SDS-PAGE and IEF that purified BMP is homogenous, inducing the formation of new bone within 3 weeks of implantation in muscle pouches in Wistar rats. Dentin matrix-derived BMP is not exactly same as bone matrix-derived BMP, but they are very similar. In other words, two types of BMP exhibit the same action in the body [34]. The organic component accounts for about 20% of dentin weight and mostly consists of type I collagen. Moreover, it was proven to have BMP promoting cartilage and bone formation, and differentiating undifferentiated mesenchymal stem cells into chondrocytes and osteogenic cells [30, 35-37]. Noncollagenous proteins of dentin such as osteocalcin, osteonectin, phosphoprotein, and sialoprotein are known to be involved in bone calcification [38,39].

Patterns of matrix protein in teeth must have osteoinductive potential even though it does not perfectly match the protein in alveolar bone. Moreover, the apatite in teeth has long been known to play the role of protecting proteins [40]. According to Boden, et al, LIM mineralization protein 1 (LMP-1) is an essential positive regulator of osteoblast differentiation and maturation and bone formation [41]. Wang, et al found that LIM-1 was expressed primarily in predentin, odontoblasts, and endothelial cells of the blood vessels of teeth [42].

Many researchers have observed that alveolar bone formation occurs around bone graft materials as a result of experiments on animals [43-47]. Chung registered the patent for the technology of extracting proteins from teeth in 2002 and 2004; this carries an important, serving as evidence that teeth contain bone morphogenic protein [48,49]. Ike and Urist suggested that root dentin prepared from extracted teeth may be recycled for use as carrier of rhBMP-2 because it induces new bone formation in the periodontium [10]. Murata, et al reported that demineralized dentin matrix (DDM) does not inhibit BMP-2 activity but shows better release profile of BMP-2. Human recycled DDM is an unique, absorbable matrix with osteoinductivity, and DDM should be an effective graft material as a carrier of BMP-2 and a scaffold for bone-forming cells for bone engineering [2].

Lee [50] performed quantitative analysis of proliferation and differentiation of the MG-63 cell line on the bone grafting material using human tooth. This study demonstrated that the cellular adhesion and proliferation activity of the MG-63 cell on partially demineralized dentin matrix (PDDM) were comparable to control with enhanced osteogenic differentiation (Figure 1). Kim & Choi [51] reported a case on tooth autotransplantation with autogenous tooth bone graft. The extracted right mandibular third molar of a 37-year-old man was transplanted into the first molar area, and a bone graft procedure using autogenous tooth-bone graft material was performed for the space between the root and the alveolar socket. Reattachment was achieved (Figure 2). Therefore, the autogenous tooth bone graft material is considered reasonable for bone inducement and healing in the autotransplantation of teeth.

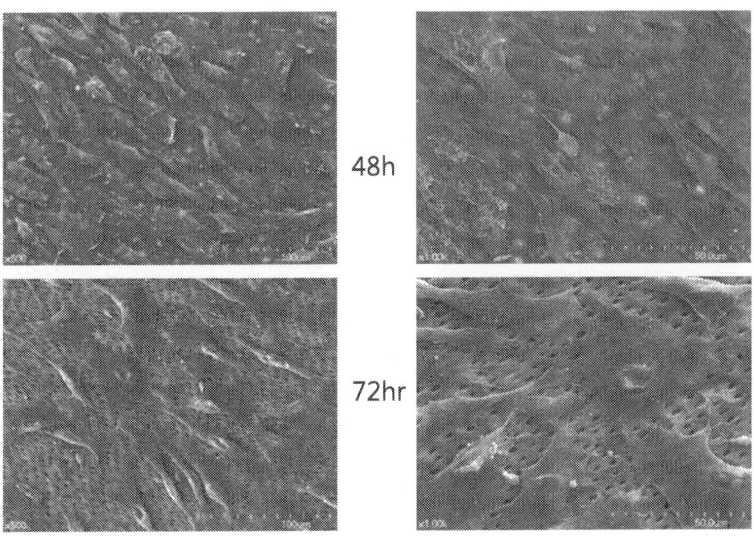

Figure 1. MG-63 cells adhered to PDDM, and they were spread out. This means excellent biocompatibility between cells and PDDM. (Lee H.J. Quantitative Analysis of Proliferation and Differentiation of MG-63 Cell Line on the Bone Grafting Material Using Human Tooth. PhD Thesis. School of Dentistry, Seoul National University, 2011.)

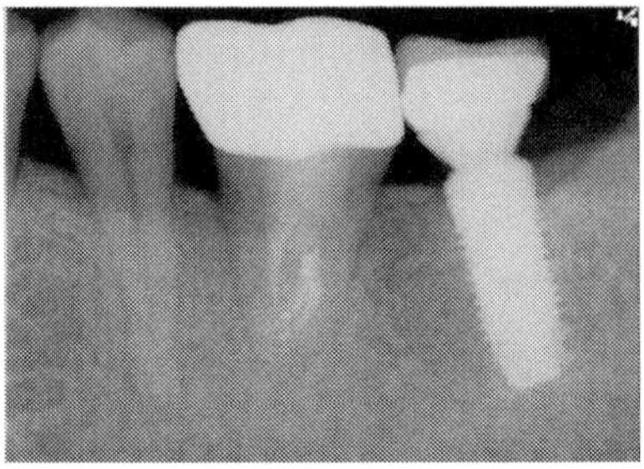

Figure 2. Periapical radiograph 2 years after autotransplantation.

Recently, we conducted a study to demonstrate the osteoinductivity of AutoBT when fabricated from bio-recycled dysfunctional teeth after patented processing. A total of 46 extracted dysfunctional teeth samples were collected from actual patients. *In vivo* study was done on 15 athymic mice by inserting AutoBT in dorsal subcutaneous muscular tissues. Samples were then biopsied in 2, 5, and 8 weeks. For additional analyses, Bradford assay, SDS-PAGE, and western blotting were performed *in vitro*. Histologic analyses *in vivo* showed new active bone formation as early as 2 weeks later (Figure 3,4,5). The Bradford assay indicated the existence of noncollagenous proteins in AutoBT. Nonetheless, rhBMP-2 was not extractable from AutoBT according to electrophoresis and immunoblotting analyses (Figure 6). In conclusion, this study provided an evidence of osteoinductivity of AutoBT th rough noncollagenous proteins.

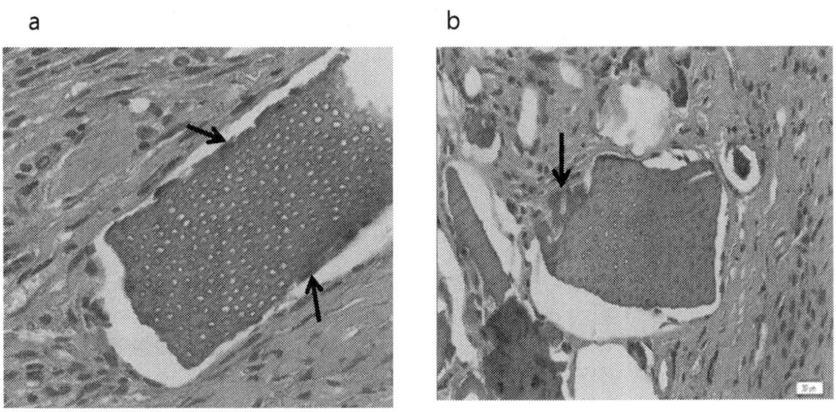

Figure 3. Histologic analyses of 2-week biopsy sample. a) The new cell lining and attachment to AutoBT powder and b) Newly deposited osteoid formations were observed. (H&E staining, X 200).

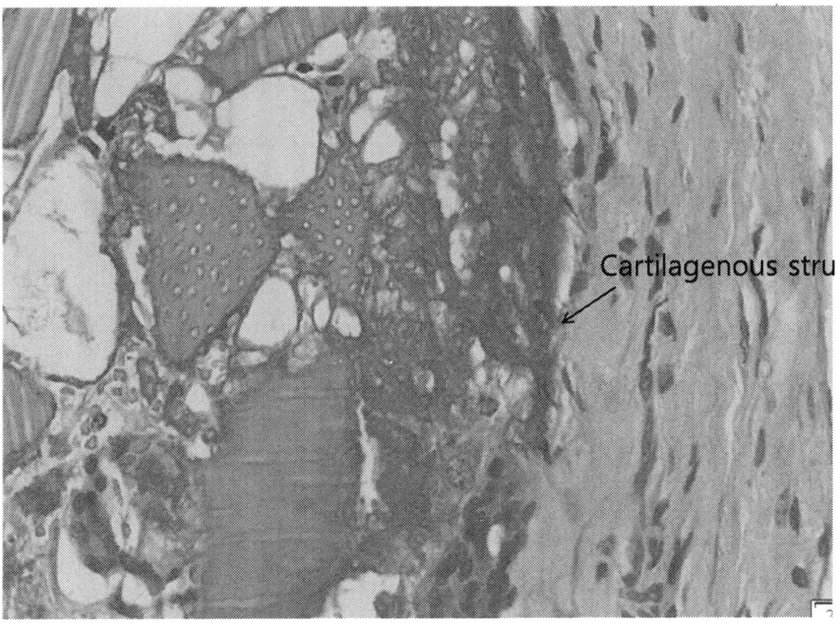

Figure 4. Cartilages were formed at the periphery of AutoBT in 5-week biopsy sample (H&E staining X 200).

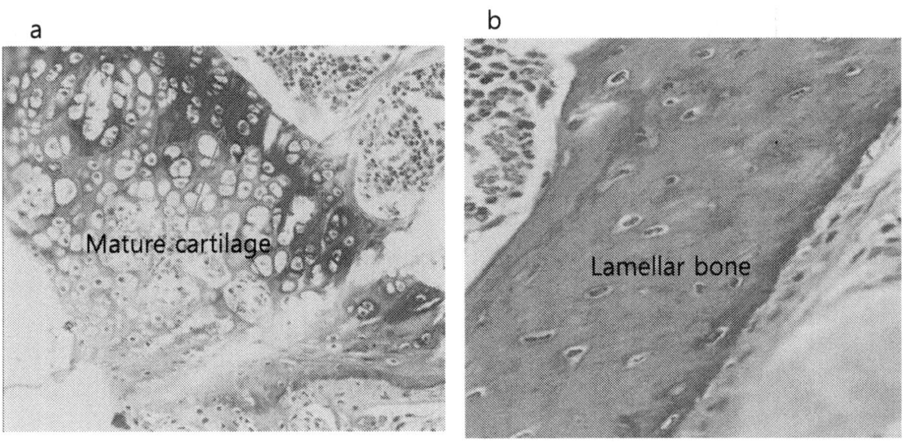

Figure 5. a) Endochondral ossification and b) lamellar bone formation were identified 8 weeks after the insertion of AutoBT powder in the intramuscular pouch of athymic mice (H&E, staining, X 200).

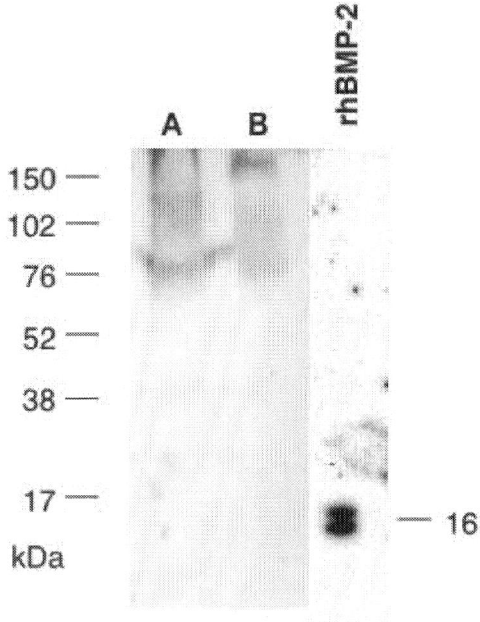

Figure 6. SDS-PAGE of purified fractions from AutoBT powder fabricated from a dried tooth in 25 °C (A) and from wisdom tooth in fresh state (B).

3. Osteoconduction of AutoBT

The analytic results showed that AutoBT consisted of low-crystalline hydroxyapatite (HA) and possibly other calcium phosphate minerals (ß-tricalcium phosphate (ß-TCP), ACP, and OCP), similar to the minerals of human bone tissues. Note, however, that the level of HA crystallization and the amount of HA differed greatly depending on the area of the tooth. The XRD pattern was much stronger in the crown portion with enamel than in the root portion (Figure 7). Likewise, the dental crown portion consisted of high-crystalline calcium phosphate minerals (mainly HA) with higher Ca/P ratio, whereas the root portion was mainly made up of low-crystalline calcium phosphates with relatively low Ca/P ratio [3, 23]. Kim, et al [52] performed the study to evaluate the surface structures and physicochemical characteristics of a novel autogenous tooth bone graft material currently in clinical use. The material's surface structure was compared with a variety of other bone graft materials via scanning electron microscope (SEM). The crystalline structure of the autogenous tooth bone graft material from the crown (AutoBT crown) and root (AutoBT root), xenograft (BioOss), alloplastic material (MBCP), allograft (ICB), and autogenous mandibular cortical bone were compared using x-ray diffraction (XRD) analysis. The solubility of each material was measured with the Ca/P dissolution test. The result of the SEM analysis showed that the pattern associated with AutoBT was similar to that from autogenous cortical bone (Figure 8). In the XRD analysis, AutoBT root and allograft showed a low crystalline structure similar to that of autogenous cortical bone (Figure 9). In the CaP dissolution test, the amount of calcium and phosphorus dissolution in AutoBT was significant from the beginning, displaying a pattern similar to that of autogenous cortical bone (Tables 1, 2). In conclusion, autogenous tooth bone graft materials can be considered to have physicochemical characteristics similar to those of autogenous bone.

Day	MBCP	ICB	BioOss	AutoBT Crown	AutoBT Root	Auto Bone
3 d	54.2	97.7	35.5	230.7	280.0	246.8
7 d	48.6	71.7	33.6	162.7	255.2	189.2
14 d	62.7	97.6	35.1	144.5	180.6	180.6

m/z: mass-to-charge ratio

Table 1. Ca (m/z; 42.959) ion dissolution (Kim Y.K., et al. Autogenous teeth used for bone grafting: a comparison to traditional grafting materials. Oral Surg. Oral Med. Oral Pathol. Oral Radiol., 2013, in press)

Day	MBCP	ICB	BioOss	AutoBT Crown	AutoBT Root	Auto Bone
3 d	301.7	217.8	174.0	269.8	269.4	260.5
7 d	311.4	191.2	151.7	282.8	230.2	282.8
14 d	302.5	165.4	148.7	253.8	229.0	245.3

m/z: mass-to-charge ratio

Table 2. P (m/z; 30.994) ion dissolution (Kim Y.K., et al. Autogenous teeth used for bone grafting: a comparison to traditional grafting materials. Oral Surg. Oral Med. Oral Pathol. Oral Radiol., 2013, in press)

In an in vitro dissolution test, AutoBT showed excellent biodegradability, whereas apatite re-precipitation was actively visible immediately after transplantation. We conjecture that this material plays an effective role in inducing bone regrowth [52]. Priya, et al [53] reported that the extensive dissolution of calcium phosphate composites, which release calcium and phosphorus ions, induces the re-precipitation of the apatite onto the surfaces. According to them, the combination of dissolution and re-precipitation was the mechanism behind apatite formation. Apatite layer formation was expected to encourage the osseointegration of bioceramic composites.

Both the organic and inorganic compositions differ between the crown and root of autogenous tooth bone graft materials. Thus, when the material is grafted, crown and root show different healing mechanisms. Apatites present in bone tissues form a ceramic/high-molecular weight

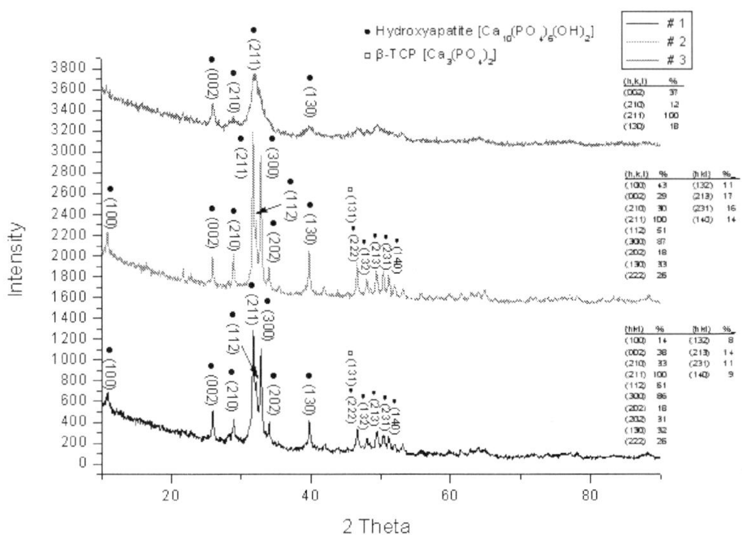

Figure 7. X-ray diffraction patterns of human tooth. (#1: root portion, #2: crown portion, 3: whole tooth) (Kim Y.K., et al. Development of a novel bone grafting material using autogenous teeth. Oral Surg. Oral Med. Oral Pathol. Oral Radiol. Endod., 2010.)

nanocomplex pattern [54]. In particular, apatites present in human bone tissues have low crystallinity and crystal size that are several tens of nanometers. On the other hand, hydroxyapatites prepared via the sintering process at high temperatures have high crystallinity. Grain growth occurs during the sintering process, resulting in sizes that are at least ten times larger than those apatites present in bone tissues [55]. The biodegradation of large particles with high crystallinity is almost impossible. Their osteoconduction capacity is very low, and osteoclasts cannot degrade them. Low-crystalline carbonic apatites show the best osteoconduction effects [56,57].

Nampo, et al introduced alveolar bone repair using extracted teeth for the graft material. DSP is a dentin-specific noncollagenous protein involved in the calcification of dentin. Based on immunohistochemical staining with anti-DSP antibody, the positive reaction was localized to the dentin of the extracted tooth fragments; thus suggesting that dentin has high affinity for and marked osteoconductive effect on the jaw bone [58].

Kim, et al reported bone healing capacity of demineralized dentin matrix materials in a minipig cranium defect [59]. A defect was induced in the cranium of mini-pigs, and those without defect were used as control. In the experimental group, teeth extracted from the mini-pig were manufactured into autogenous tooth bone graft material and grafted to the defect. The minipigs were sacrificed at 4, 8, and 12 weeks to evaluate histologically the bone healing ability and observe the osteonectin gene expression pattern with RT-PCR. At 4 weeks, the inside of the bur hole showed fibrosis, and there was no sign of bone formation in the control group. On the other

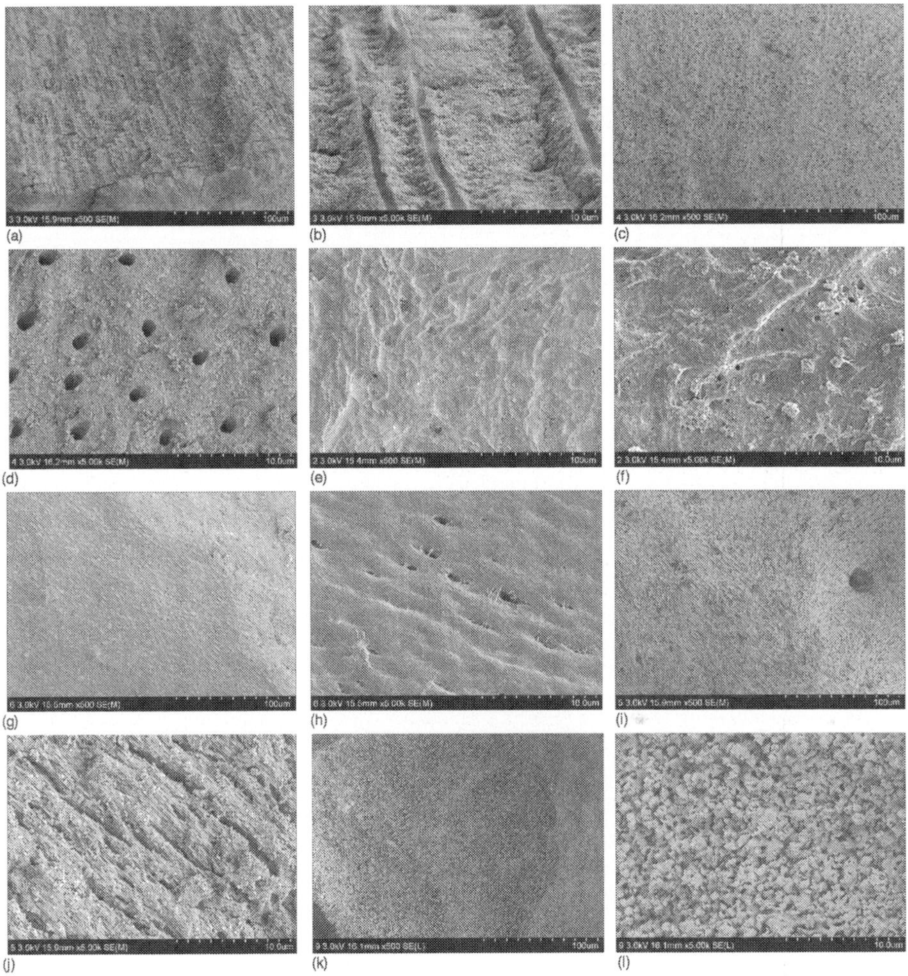

Figure 8. SEM views of the different types of bone graft materials. (Kim Y.K., et al. Autogenous teeth used for bone grafting: a comparison to traditional grafting materials. Oral Surg. Oral Med. Oral Pathol. Oral Radiol., 2013, in press) a): AutoBT crown (x500), b): AutoBT crown (x5,000), c): AutoBT root (x500), d): AutoBT root (x5,000), e): Autogenous cortical bone (x500), f): Autogenous cortical bone (x5,000), g): ICB (x500), h): ICB (x5,000), i): BioOss (x500), j): BioOss (x5,000), k): MBCP (x500), l): MBCP (x5,000)

hand, bone formation surrounding the tooth powder granule was observed at 4 weeks in the experimental group wherein the bur hole was filled with tooth powder. There was practically no osteonectin expression in the control group, whereas active osteonectin expression was observed from 4 to 12 weeks in the experimental group. In this study, excellent osteoconductive healing of autogenous tooth bone graft material was confirmed (Figure 10, 11).

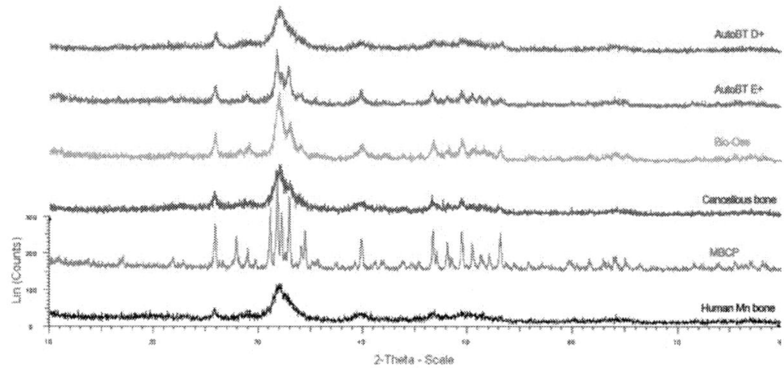

Figure 9. Results of the X-ray diffraction analysis. AutoBT D+: AutoBT root, AutoBT E+: AutoBT crown, Cancellous bone: ICB. (Kim Y.K., et al. Autogenous teeth used for bone grafting: a comparison to traditional grafting materials. Oral Surg. Oral Med. Oral Pathol. Oral Radiol., 2013, in press)

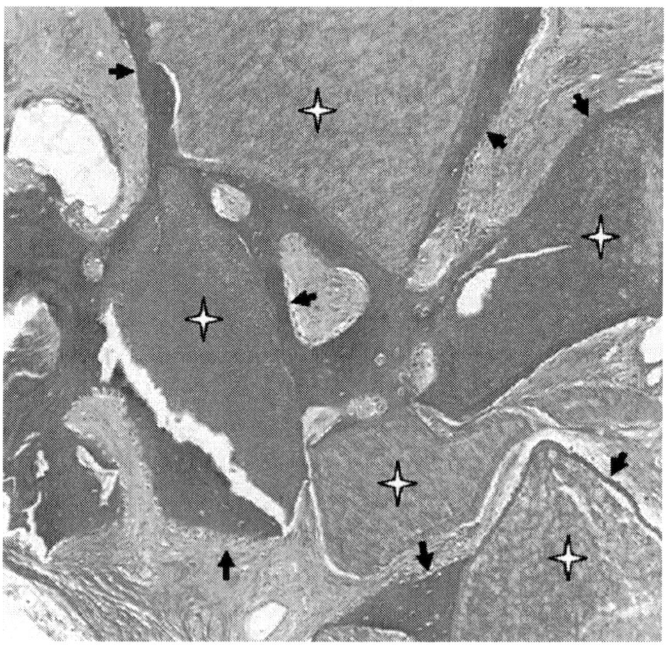

Figure 10. Experimental group of 8 weeks. New bone is actively formed around the tooth powder granules. Asterisks and arrows indicate graft tooth granule materials and new bone formation around the tooth granules, respectively. Hematoloxylin and eosin staining (×100). (Kim J.Y., et al. Bone healing capacity of demineralized dentin matrix materials in a mini-pig cranium defect. J. Korean Dent. Sci., 2012.)

http://dx.doi.org/10.5772/53200

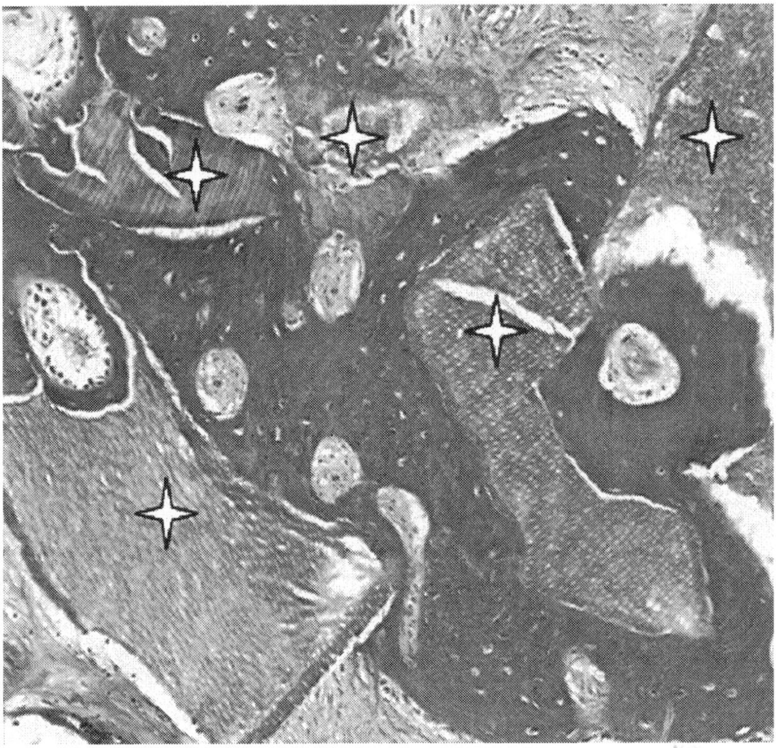

Figure 11. Experimental group of 12 weeks. Extensive new bone formation was noted around the bone powder granules in the bur hole. Asterisks indicate tooth powder materials. Hematoloxylin and eosin staining (×100). (Kim J.Y., et al. Bone healing capacity of demineralized dentin matrix materials in a mini-pig cranium defect. J. Korean Dent. Sci., 2012.)

4. Clinical application of AutoBT

Kim, et al developed a novel bone grafting material using autogenous teeth (AutoBT) in 2008 and provided the basis for its clinical application. Having organic and inorganic mineral components, AutoBT is prepared from autogenous grafting material; thus eliminating the risk of immune reaction that may lead to rejection. AutoBT was used at the time of implant placement -- simultaneously with guided bone regeneration -- and excellent bone healing by osteoinduction and osteoconduction was confirmed [3]. In a total of 6 patients, guided bone regeneration was performed simultaneously at the time of implant placement, and tissue samples were then harvested at the time of the second surgery with the patient's consent. In the histomorphometric analysis of the samples collected from 6 patients during the 3 ~ 6 months' healing period, new bone formation was detected in 46 ~ 87% of the area of interest, and excellent bone remodeling was achieved (Table 3) (Figure 12). Clinically available AutoBT consists of powder, chips, and block (Figure 13).

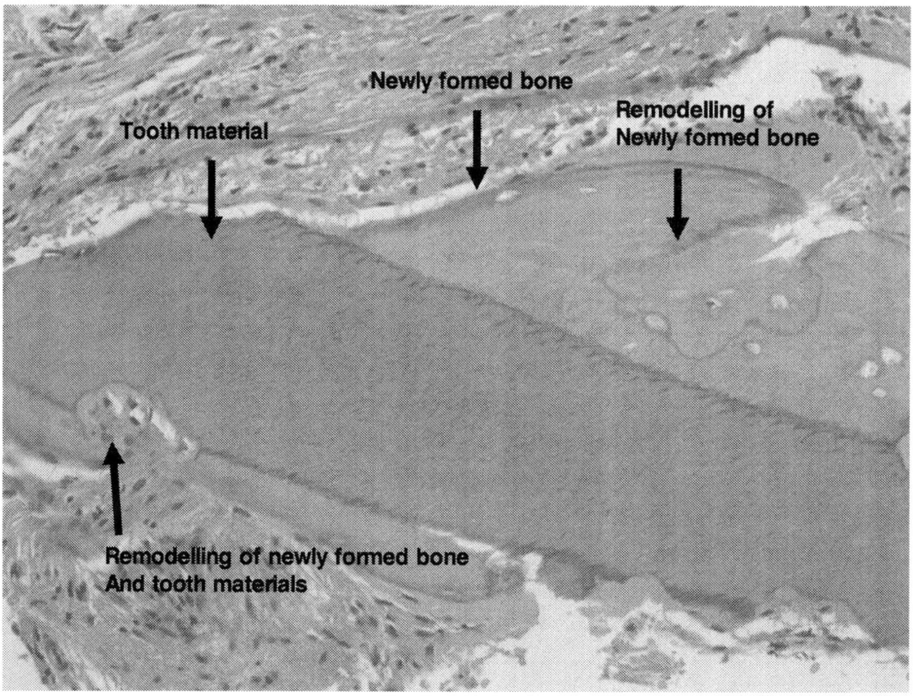

Figure 12. Newly formed bone and tooth materials showing remodeling were identified around the implant chip and at the periphery of the implant chip, respectively (H&E staining, X 100). (Kim Y.K., et al. Development of a novel bone grafting material using autogenous teeth. Oral Surg. Oral Med. Oral Pathol. Oral Radiol. Endod., 2010.)

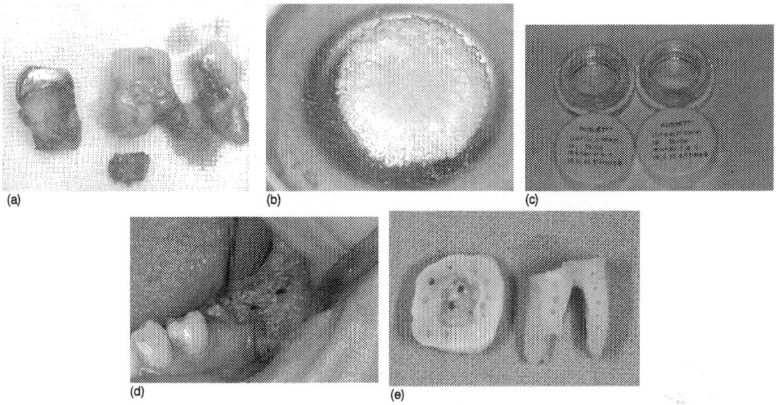

Figure 13. Three types of AutoBT can be fabricated from extracted teeth. a): Extracted teeth. Foreign body such as prosthetic crown, cements, calculus, and soft tissue are removed. AutoBT is then fabricated through pulverization, defatting, demineralization, and lyophilization. b): AutoBT one-powder. Crown and root portion are mixed. c): AutoBT crown and root powder. d): AutoBT chips. e): AutoBT block.

Case	Age/Sex	Site	Healing period	WB:LB:IM ratio	New bone-forming area (%)
1	**40/M**	#24	3	43:11:46	74
2	**28/F**	#17	4	85:14:1	87
3	**47/F**	#17	6	56:39:5	46
4	**50/M**	#24	5	84:12:4	73
5	**43/F**	#36	3	51:1:48	52
6	**61/M**	#25-27	6	65:0:35	68

WB: woven bone; LB: lamellar bone; IM: residual implant material

Table 3. Histomorphometric finding (Kim Y.K., et al. Development of a novel bone grafting material using autogenous teeth. Oral Surg. Oral Med. Oral Pathol. Oral Radiol. Endod., 2010.)

Lee and Kim [60]performed a retrospective study to evaluate the clinical efficacy of AutoBT. This study included 37 patients (54 implants) into which AutoBT was grafted between Oct. 2008 and Dec. 2009. The mean follow-up period was 31 months. Postoperative complications and marginal bone status around the implants were evaluated using medical records and dental radiography. Wound dehiscence and hematoma developed in 7 patients (8 implants). Osseointegration failure in 2 patients (4 implants) was recorded. These complications were well managed through conservative treatment and re-implantation. Mean peri-implant marginal bone loss 1 year after implant placement was 0.33±0.63mm. Autogenous tooth bone graft was confirmed to be a safe procedure, showing excellent bone healing through a 2-year retrospective study (Tables 4, 5, 6).

Type	Number of implants
GBR	29 (53.7%)
Sinus graft (lateral approach)	14 (25.9%)
Sinus lifting (crestal approach)	7 (13.0%)
Ridge augmentation	4 (7.4%)
Total	54 (100%)

Table 4. Types of surgery

Type	Number of patients
Powder	32 (86.5%)
Block	2 (5.4%)
Powder + Block	3 (8.1%)
Total	37 (100%)

Table 5. Types of AutoBT

Type	Number of implants
Wound dehiscence	7
Hematoma	1
Osseointegration failure	4
Total	12

Table 6. Types of complications

5. Sinus bone graft

If there is any material whose resorption speed is not too high and whose bone healing process approximates that of autogenous bone graft, it may be useful in maxillary sinus bone grafting. Likewise, more excellent clinical achievement may be expected when these materials are used in mixture with other bone substitutes with slow resorption properties [61,62,63]. With evidence presented in the foregoing paragraphs, AutoBT® developed by the author, et al was proven to exhibit bone healing ability through osteoinduction and osteoconduction, demonstrating a histological healing process similar to that of free bone grafting being resorbed over 3~6 months [3]. Accordingly, AutoBT® is regarded as a possible substitute when autogenous bone is needed for sinus bone graft, and it may wield a useful effect on increasing the volume of bone graft materials and minimizing repneumatization (Figure 14).

A retrospective study on sinus bone graft was performed. One hundred implants in 51 patients were selected, with the patients receiving maxillary sinus augmentation and implant placement using autogenous tooth graft materials at Chosun University Dental Hospital and Seoul National University Bundang Hospital (SNUBH) between July 2009 and November 2010. In cases of using autogenous tooth bone graft alone or together with other graft material, the implant survival rate was 96.15%. Based on the histomorphologic examination, autogenous tooth bone graft materials showed gradual resorption and new bone formation through osteoconduction and osteoinduction. The results suggest that autogenous tooth bone graft materials are appropriate for use in maxillary sinus augmentation [64].

Lee, et al [65] conducted a study to evaluate histomorphometrically and compare the efficiency of various bone graft materials and autogenous tooth bone graft material used in the sinus bone graft procedure. The subjects were 24 patients who had been treated with sinus bone graft using the lateral approach from October 2007 to September 2009 at SNUBH. The average age was 52.51±11.86 years. All cases were taken after 4 months of procedure and divided into 3 groups according to bone graft material: Group 1 for autogenous tooth bone graft material (AutoBT), Group 2 for OrthoblastII (Integra Lifescience Corp., Irvine, US)+Biocera (Osscotec, Cheonan, Korea), and Group 3 for DBX (Synthes, West Chester, PA, USA), BioOss (Geistlich Pharm AG, Wolhusen, Switzerland). A total of 37 implant placement areas was included and evaluated (7 in group 1, 10 in group 2, 20 in group 3). The evaluation of new bone formation, ratio of woven bone to lamellar bone, and ratio of new bone to graft material was performed on each tissue section. The Kruskal-Wallis test was used for statistical analysis (SPSS Ver. 12.0, USA). New bone formation was 52.5±10.7 % in group 1, 52.0±23.4% in group 2, and 51.0±18.3% in group 3 (Table 7) (Figure 15-18). There were no statistically significant differences between groups, however. The ratio of woven bone to lamella bone was 82.8±15.3% in group 1, 36.7 ±59.3% in group 2, and 31.0±51.2% in group 3. The ratio of new bone to graft material was 81.3±10.4% in group 1, 72.5±28.8% in group 2, and 80.3±24.0% in group 3. After a 4-month healing period, all groups showed favorable new bone formation and around the graft material and implant. Within the limitation of our study, autogenous tooth bone graft material may be used as a novel bone graft material for sinus bone graft. Kim, et al and Lee, et al performed sinus bone graft and guided bone regeneration using autogenous tooth bone from humans and took the tissue specimen 2 months and 4 months later for histomorphometric analysis. They found favorable new bone formation as a result and suggested that autogenous tooth bone graft materials could be used in various bone grafts [65,66].

Group	New bone formation
I	52.5±10.7%
II	52.0±23.4%
III	51.0±18.3%

*Kruskal-Wallis test: P-value>0.05

Table 7. Histomorphometric data on new bone formation (Mean±SD)

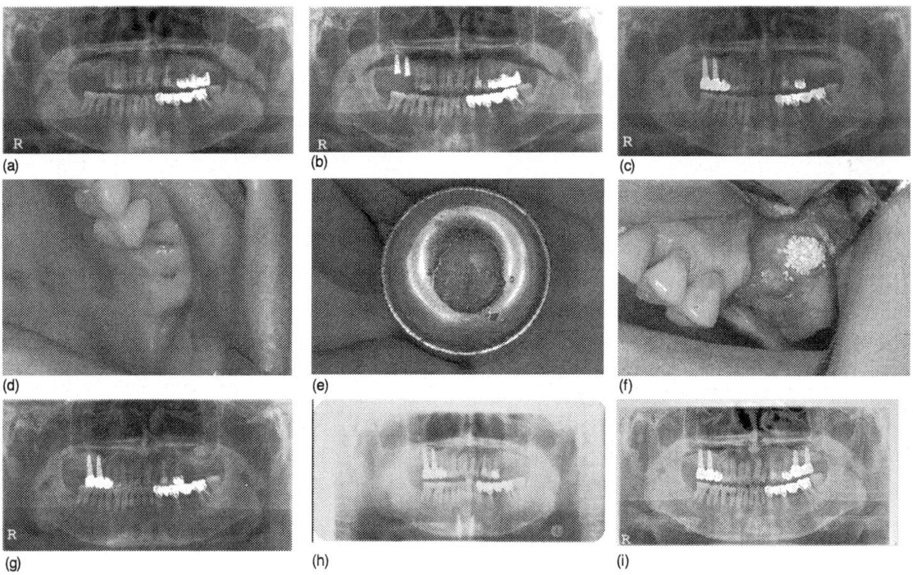

Figure 14. A case of sinus bone graft performed by the mixture of AutoBT, autogenous maxillary tuberosity bone and synthetic bone. a) Panoramic radiography of a 64-year-old man at the first examination. b) Radiography after placing implants simultaneously with the sinus bone graft on the right side. c) Panoramic radiography after 2 weeks of maxillary left 1st molar extraction. The prosthodontic therapy for the upper right maxillary bone was completed, and the extracted tooth was replaced with bone graft materials. d) Intraoral photography before operation. e) View of mixture of AutoBT and maxillary tuberosity bone. f) Grafted in the mixture with a synthetic bone, OSTEON (GENOSS, Suwon, Korea). g) Panoramic radiography after sinus bone graft. h) Panoramic radiography taken in a private dental clinic after 3 months of bone grafting. Performing implant placement in a private dental clinic was decided due to the medical costs. i) Panoramic radiography one year after final prosthetic delivery. The bone materials grafted on the maxillary sinus are maintained stably.

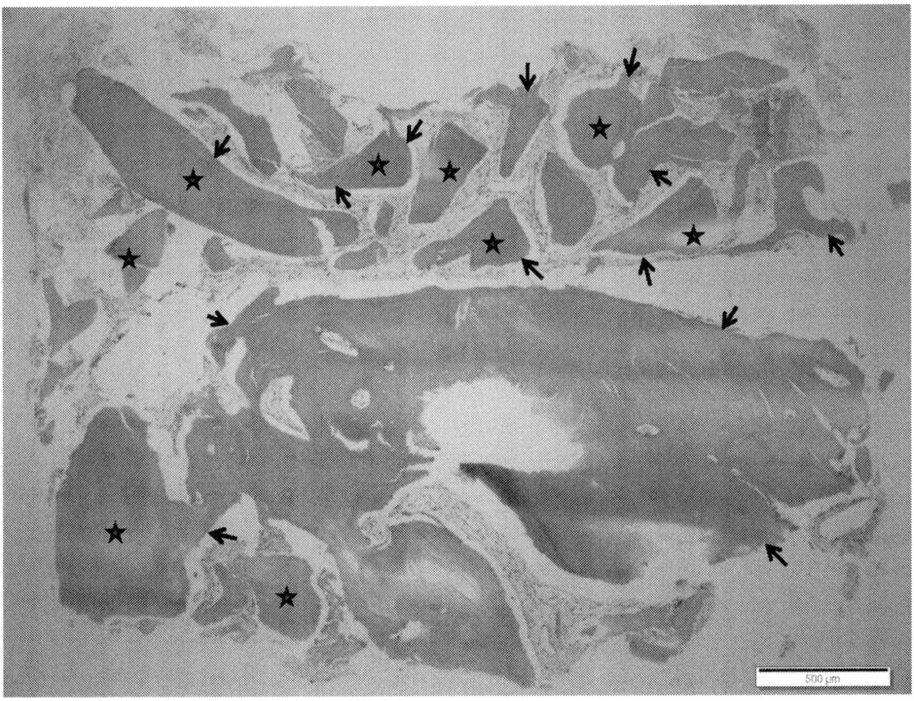

Figure 15. Overview of biopsy of Group I (Auto BT®). New bone formation (arrows) was identified around the graft material (asterisks). (Hematoxylin & Eosin stain, x40. scale bar measures 500um)

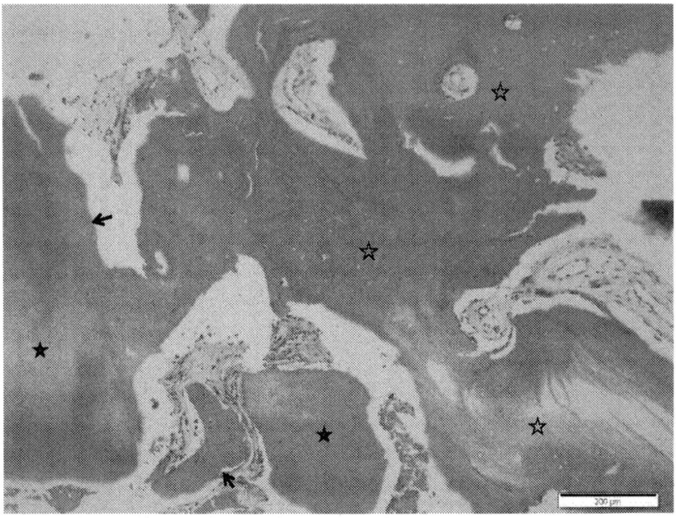

Figure 16. Histomorphometric image of Group I (Auto BT®). New bone formation (arrows) was identified around the graft material (asterisks). Confluent new bone formation was observed (open asterisk) (Hematoxylin & Eosin stain, x200. scale bar measures 200um)

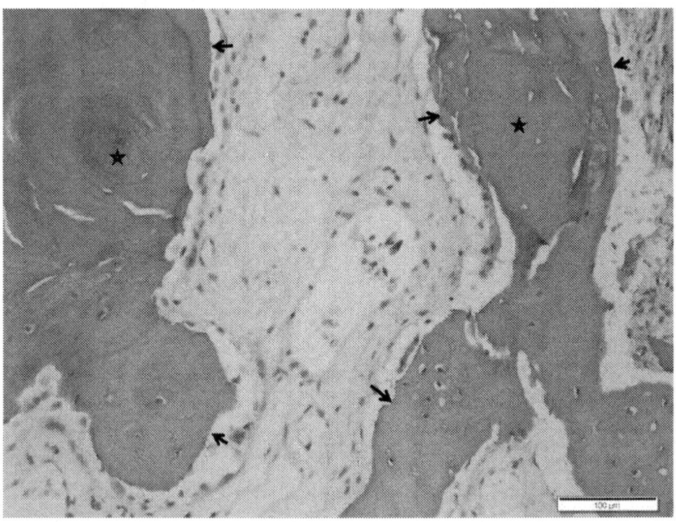

Figure 17. Microphotograph 4 months after Orthoblast/Biocera transplantation (Group II). Higher magnification demonstrated new bone formation (arrows) around the implant chips (asterisks). (Hematoxylin & Eosin stain, x200. scale bar measures 100um)

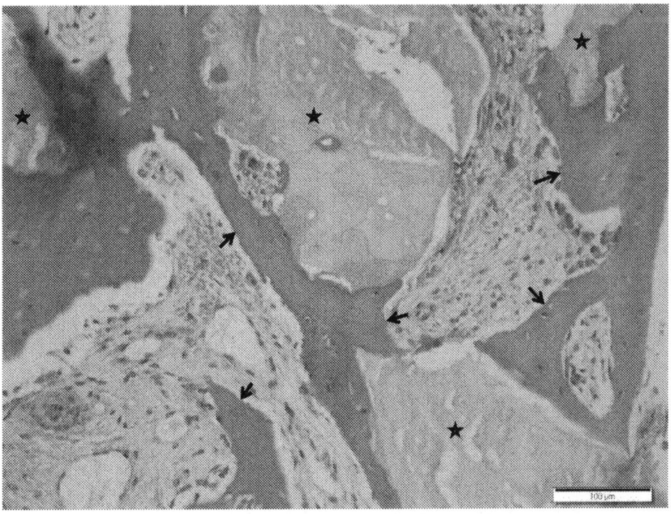

Figure 18. Microphotograph 4 months after DBX/BioOss transplantation (Group III). Higher magnification demonstrated new bone formation (arrows) around the implant chips (asterisks). (Hematoxylin & Eosin stain, x200. scale bar measures 100um)

6. Guided bone regeneration

Bone dehiscence or bone fenestration often develops after dental implant placement, and guided bone regeneration using bone graft materials has become a popular method. The most ideal material for guided bone regeneration is autogenous bone, but autogenous bone graft has limited sources and high risk of complications at the donor site and causes high resorption after bone graft. Therefore, alternative bone materials have been developed and used clinically, such as allogenic bone, xenogenic bone, and synthetic bone. Note, however, that they are often mixed with autogenous bone to maximize their advantages.

Autogenous teeth bone graft materials have very good osteoinductive and osteoconductive properties due to the organic and inorganic contents of the teeth, such as collagen, bone growth factors, and various forms of calcium phosphate. In our study, we achieved 46~74% new bone formation in 3~6 months compared with the results of Babbush [3,67]. Considering the histological healing of the sites where autogenous teeth bone graft materials were applied, bone graft materials were replaced with new bone following resorption, and new bone directly fused with the remaining autogenous teeth bone graft materials. A healing process associated with excellent osteoinduction and osteoconduction was observed in every sample, including abundant lamella bone; thus indicating that rapid bone reconduction was occurring [50,51,59,65,66]. Kim, et al [68] installed implants combined with guided bone regeneration using autogenous tooth bone graft material in 6 patients. In the 6 months' histological examination after operation, excellent osteoconductive bone healing was noted. A clinically favorable outcome was obtained (Figure 19~21).

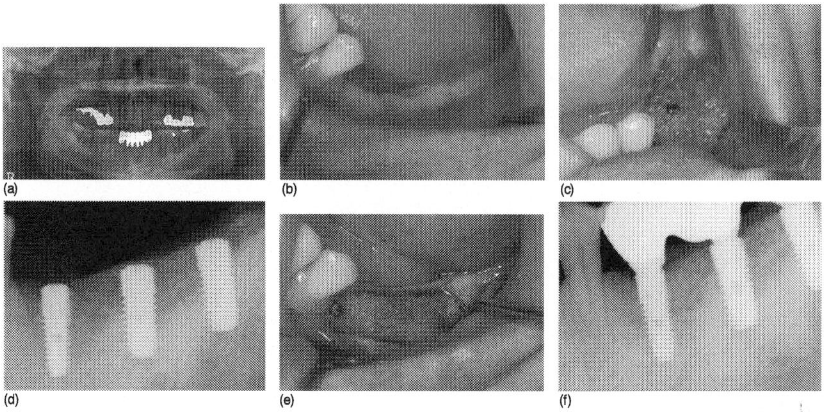

Figure 19. Guided bone regeneration using AutoBT powder (Kim Y.K., et al. Guided bone regeneration using autogenous teeth: case reports. J. Korean Assoc. Oral Maxillofac. Surg., 2011.) a): Initial panoramic radiography of a 44-year-old male patient. b): Preoperative intraoral view. Teeth were extracted 2 months ago. c): Implants were placed, and dehiscence defects were covered with autogenous tooth bone graft material. d): Periapical radiography 6 months after implant placement. e): Secondary surgery was performed, and flap was elevated. Excellent bone healing was observed. f): Periapical radiography 6 months after the final prosthetic delivery.

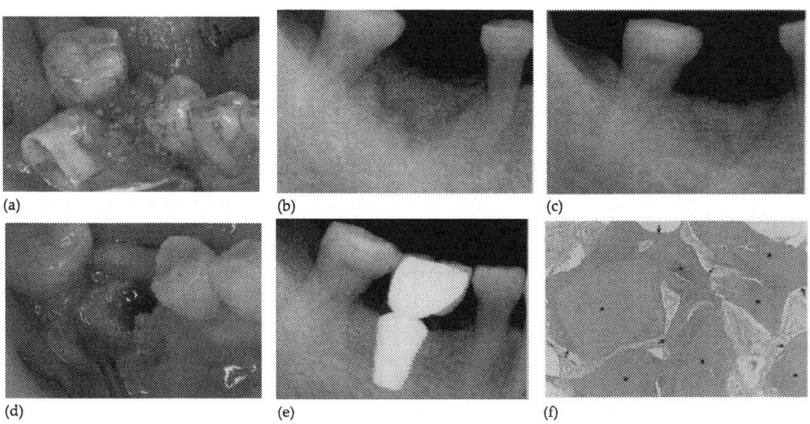

Figure 20. GBR was performed on the right mandibular 1st molar area of a 49-year-old female patient. (Kim Y.K., et al. Guided bone regeneration using autogenous teeth: case reports. J. Korean Assoc. Oral Maxillofac. Surg., 2011.) a): Autogenous tooth bone graft material and collagen membrane (BioGuide) were used. b): Periapical radiography 3 weeks after bone graft. c): Periapical radiography 6 months after bone graft. The alveolar crestal level was stable. d): Implant was installed 6 months after bone graft. Bone quality was type I. e): Periapical radiography after the final prosthetic delivery. f): Microphotograph 6 months after AutoBT transplantation. Higher magnification demonstrated new bone formation (arrows) around the implant chips (asterisks). Hematoxylin & Eosin stain, x100.

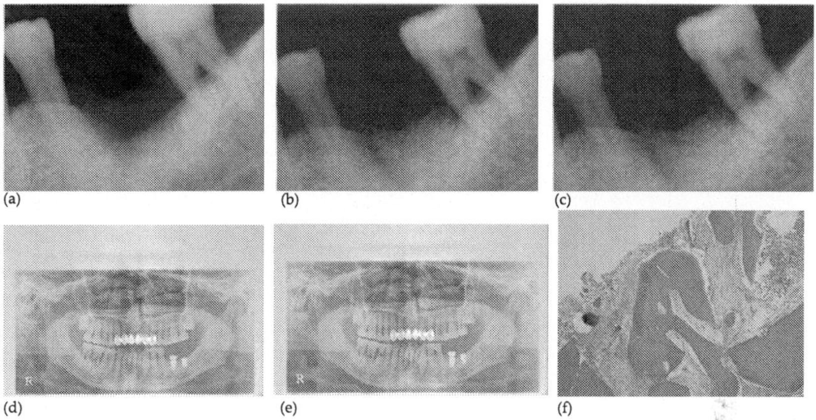

Figure 21. GBR was performed on the mandibular left 1st molar. (Kim Y.K., et al. Guided bone regeneration using autogenous teeth: case reports. J. Korean Assoc. Oral Maxillofac. Surg., 2011.) a): Periapical radiography of a 50-year-old male patient 2 months after the extraction of the mandibular left 1st molar. b): Periapical radiography 2 weeks after autogenous tooth bone graft. c): Periapical radiography 5 months after autogenous tooth bone graft. Alveolar crestal bone level was stable. d): Implant was placed 6 months after bone graft. The adjacent 2nd molar was extracted. e): Second surgery was performed at the #36 area. Additional implant was placed at the #37 area. f): Periapical radiography after the final prosthetic delivery. g): Microphotograph 6 months after AutoBT transplantation. Higher magnification demonstrated new bone formation around the implant chips. Hematoxylin & Eosin stain, x200

7. Ridge augmentation (Figure 22)

Autogenous bone grafting produces the best results in case a large volume of bone increase is required, as in the reconstruction of a site with lots of bone defects or ridge augmentation. The autograft may be taken from the endochondral bone such as ilium, rib, tibia, etc., and from the intramembranous bone such as calvaria, facial bone, etc. Alveolar ridge augmentation is a method of augmenting the height or width of the alveolar ridge by implementing bone grafting on the upper part or lateral part of the ridge in particulate or block type in case bone volume is insufficient vertically or horizontally; vertical and horizontal augmentation may be done simultaneously, but it may also be carried out individually. Since it is a kind of onlay graft, bone resorption occurs considerably after grafting, and dehiscence on the upper soft tissue easily arises [69]. Meanwhile, as for the autogenous bone graft, there may be some complications on the donor site, and doing the grafting takes time. Likewise, there are several problems such as limit to the volume of collection. Consequently, patients and clinical doctors are inclined to avoid it in many cases. As substitutes for autograft, bone graft materials such as allograft, xenograft, synthetic bone, etc., were developed, but the single use of each is not recommended in the method of augmenting bone tissue vertically or horizontally [69,70]. For the vertical or horizontal ridge augmentation, AutoBT may be a substitute method for autogenous bone graft and may be very useful in clinical practices when used in mixture with other graft materials in case of insufficient volume. Kim, et al. [71,72] reported the successful case of alveolar ridge augmentation using various autogenous tooth bone graft materials.

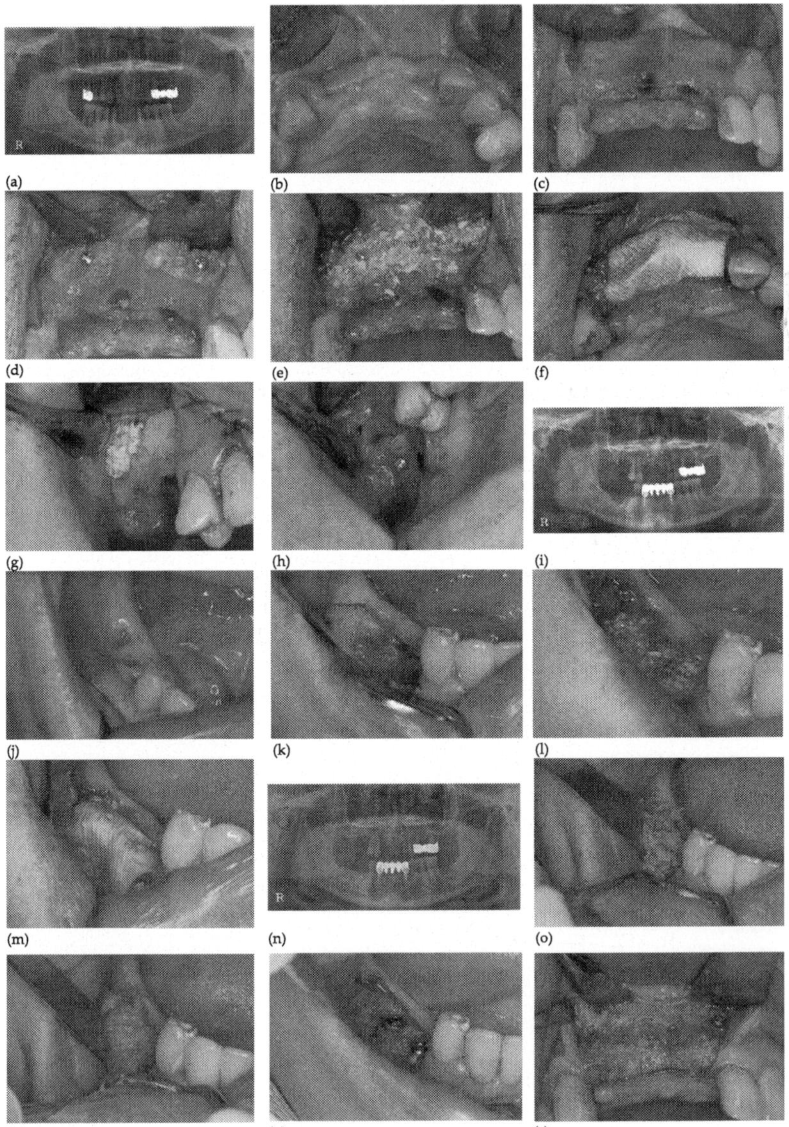

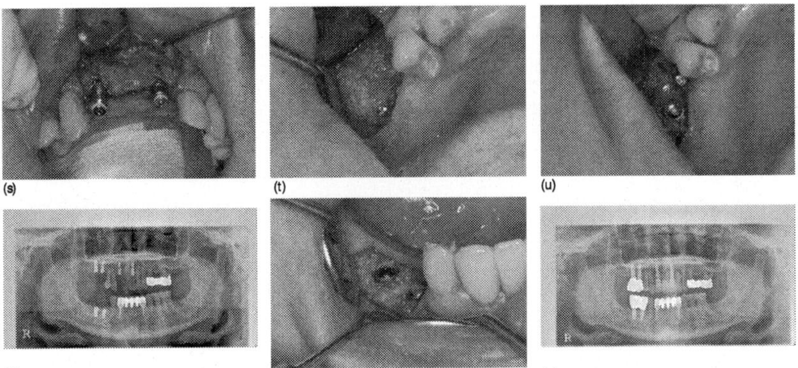

Figure 22. Placement of implants after the ridge augmentation of the maxillary anterior and maxillary/mandibular posterior area. a): Panoramic radiography at the first examination. Alveolar resorption was in considerable progress on the whole. b): Intraoral photograph taken just before the ridge augmentation of the maxillary anterior area. One month passed after extraction. c): View of the elevated mucoperiosteal flap. The labial side concavity is observed. d): View of fixation with titanium screws after applying the AutoBT block on the labial side. e): View of grafting the AutoBT powder additionally. f): Sutured after covering the resorbable collagen membrane (Ossix plus). g): View of the sinus bone graft on the right side using the AutoBT powder. h): View of fixation with titanium screws after vertical ridge augmentation with the AutoBT block. i): Panoramic radiography after grafting the bone on the maxillary anterior area and the right posterior area. j): Intraoral photograph prior to the right mandibular posterior bone grafting. One month passed after extraction. k): View of the elevated mucoperiosteal flap. The vertical bone defects on the ridge is observed. l): After applying the AutoBT block on the #45 area, the AutoBT powder was grafted on the surrounding sites. The AutoBT block was hydrated in saline solution for 15 ~ 30 minutes and operated. m): After covering the Ossix plus, the wound was closed. n): Panoramic radiography after bone graft. o): View of the elevated mucoperiosteal flap on the #45 and 46 sites after 2 months of ridge augmentation. Some Ossix plus that were not resorbed is observed. p): After removing Ossix plus, very excellent bone healing was observed. q): View of #45 and 46 implant placement. r):View of the elevated mucoperiosteal flap on the maxillary anterior area 4 months after bone grafting. Good bone healing is observed. There was not much bone resorption when the state of titanium screws was examined. s): After removing the titanium screws, the implants were placed. t): Exposed #15 and16 areas. The titanium screws fixing the block is observed, and bone healing was very good. u): View of implants placed on the site. v): Panoramic radiography after the #12, 21, 15, and 16 implants were placed. w): View of the #45 and 46 implants exposed while doing the secondary surgery after 2 months. x): Panoramic radiography 6 months after the final prosthetic delivery.

8. Extraction socket preservation or reconstruction (Figure 23)

The resorption of the residual alveolar bone in the vicinity of extraction sockets reportedly occurs primarily during the initial period after tooth extraction; in cases wherein teeth are infected with periodontal diseases, it shows more severe resorption [73]. Severe resorption of the alveolar bone may cause aesthetic problems in the anterior teeth. In addition, normal, natural healing may be difficult since the soft tissues may fall down into the defective area if there is progressive periodontal disease or periapical inflammatory lesion, or in case of serious defects of the surrounding bone wall after tooth extraction. Therefore, the preservation or reconstruction of the extraction sockets should be considered positively in case of serious defects after tooth extraction [74]. Ridge preservation methods using various bone graft materials were introduced and reported to be effective in preventing vertical and horizontal ridge resorption [75-77]. Kim, et al [78] reported an actual case of extraction socket preservation and reconstruction using autogenous tooth bone powder and block. They reported good healing of extraction socket after 3~3.5 months, and they could successfully perform the placement of implants.

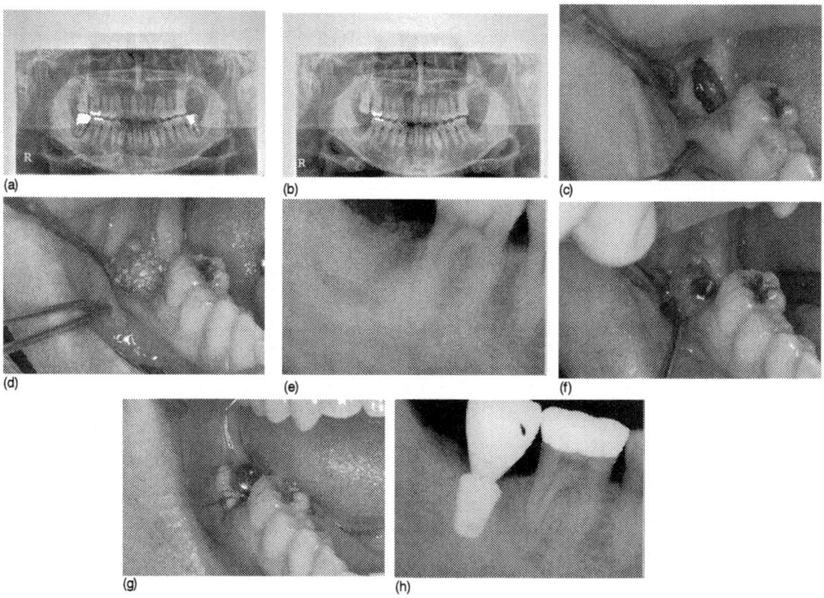

Figure 23. Extraction socket graft and delayed implant placement were performed on a 48-year-old male patient. (Kim Y.K., et al. Extraction socket preservation and reconstruction using autogenous tooth bone graft. J. Korean Assoc. Maxillofac. Plast. Reconstr. Surg., 2011.). a): Initial panoramic radiographic view, Periapical radiolucent lesion was observed at #37, 47 area. Radiolucent lesion was extended to the vicinity of the inferior alveolar canal. b):. Panoramic radiograph 3 months after extraction. c): The mucoperiosteal flap was elevated for implant placement 3 months after #47 extraction. The healing of extraction socket was poor. It was impossible to install the implants because of inadequate stability. d): Autogenous tooth bone graft powder was grafted into the socket. e): Postoperative periapical radiograph. f): Implant was installed 3 months after socket graft. Primary implant stability was excellent. g): Second surgery was performed 2.5 months after implant placement. h): Periapical radiograph 14 months after the final prosthetic delivery.

9. Conclusion

It is obvious that autogenous tooth bone graft materials(AutoBT) are safer than allogeneic and xenogeneic bon egraft materials; the fact that they are compared with the healing performance of free autogenous bone graft in histological view is clear evidence. AutoBT can be used safely

http://dx.doi.org/10.5772/53200

in a variety of bone reconstructive procedures such as sinus bone graft, GBR, ridge augmentation and extraction socket graft.

Author details

Young-Kyun Kim[1], Jeong Keun Lee[2], Kyung-Wook Kim[3], In-Woong Um[4] and Masaru Murata[5]

1 Department of Oral and Maxillofacial Surgery, Section of Dentistry, Seoul National University Bundang Hospital, Seongnam, Korea

2 Department of Dentistry Oral & Maxillofacial Surgery, Ajou University School of Medicine, Suwon, Korea

3 Department of Oral and Maxillofacial Surgery, College of Dentistry, Dankook University, Cheonan, Korea

4 Director, Korea Tooth Bank, R&D Institute, Seoul, Korea

5 Division of Oral and Maxillofacial Surgery, University of Hokkaido, Hokkaido, Japan

References

[1] Misch, C. E, & Dietsh, F. Bone-grafting materials in implant dentistry. Implant Dent (1993)., 2(3), 158-167.

[2] Murata, M, Akazawa, T, Mitsugi, M, Um, I. W, Kim, K. W, & Kim, Y. K. Human Dentin as Novel Biomaterial for Bone Regeneration. In: Rosario Pignatello R. (ed.) Biomaterials- Physics and Chemistry, InTech; (2011)., 127-140.

[3] Kim, Y. K, Kim, S. G, Byeon, J. H, Lee, H. J, Um, I. U, Lim, S. C, & Kim, S. Y. Development of a novel bone grafting material using autogenous teeth. Oral Surg Oral Med Oral Pathol Oral Radiol Endod. (2010)., 109(4), 496-503.

[4] Nanci, A. Ten Cate's Oral Histology, 7thedi. Elsevier Inc. (2008)., 202-211.

[5] Min, B. M. Oral Biochemistry. Daehan Narae Pub Co. Seoul.(2007)., 22-26.

[6] Bhaskar, S. N. Orban's Oral histology and embryology. 9th edition. Mosby Co. USA. (1980).

[7] Butler, W. T, Mikulski, A, & Urist, M. R. Noncollagenous proteins of a rat dentin matrix possessing bone morphogenetic activity. J Dent Res (1977)., 56(3), 228-232.

[8] Bang, G. Induction of heterotopic bone formation by demineralized dentin in guinea pigs: antigenicity of the dentin matrix. J Oral Pathol (1972). , 1(4), 172-185.

[9] Bessho, K, Tagawa, T, & Murata, M. Purification of rabbit bone morphogenetic protein derived from bone, dentin, and wound tissue after tooth extraction. J Oral Maxillofac Surg (1990). , 48(2), 162-169.

[10] Ike, M, & Urist, M. R. Recycled dentin root matrix for a carrier of recombinant human bone morphogenetic protein. J Oral Implantol (1998). , 24(3), 124-132.

[11] Kim, Y. K, Yeo, H. H, Ryu, C. H, Lee, H. B, Byun, U. R, & Cho, J. E. An experimental study on the tissue reaction of toothash implanted in mandible body of the mature dog. J Korean Assoc Maxillofac Plast Reconstr Surg (1993). , 15, 129-136.

[12] Kim, Y. K, Yeo, H. H, Yang, I. S, Seo, J. H, & Cho, J. O. Implantation of toothash combined with plaster of paris: Experimental study. J Korean Assoc Maxillofac Plast Reconstr Surg (1994). , 16, 122-129.

[13] Kim, Y. K. The experimental study of the implantation of toothash and plaster of Paris and guided tissue regeneration using Lyodura. J Korean Assoc Oral Maxillofac Surg (1996). , 22, 297-306.

[14] Kim, Y. K, & Yeo, H. H. Transmitted electronic microscopic study about the tissue reaction after the implantation of toothash. J Korean Assoc Oral Maxillofac Surg (1997). , 23, 283-289.

[15] Kim, Y. K, Kim, S. G, Lee, J. G, Lee, M. H, & Cho, J. O. An experimental study on the healing process after the implantation of various bone substitutes in the rats. J Korean Assoc Oral Maxillofac Surg (2001). , 27, 15-24.

[16] Kim, Y. K, Kim, S. G, & Lee, J. H. Cytotoxicity and hypersensitivity test of toothash. J Korean Maxillofac Plast Reconstr Surg (2001). , 23, 391-395.

[17] Kim, S. G, Yeo, H. H, & Kim, Y. K. Grafting of large defects of the jaws with a particulate dentin-plaster of Paris combination. Oral Surg Oral Med Oral Pathol Oral Radiol Endod (1999). , 88(1), 22-25.

[18] Kim, S. G, Kim, H. K, & Lim, S. C. Combined implantation of particulate dentine, plaster of Paris, and a bone xenograft (Bio-Oss) for bone regeneration in rats. J Craniomaxillofac Surg (2001). , 29(5), 282-288.

[19] Kim, S. G, Chung, C. H, Kim, Y. K, Park, J. C, & Lim, S. C. Use of particulate dentin-plaster of Paris combination with/without platelet-rich plasma in the treatment of bone defects around implants. Int J Oral Maxillofac Implants (2002). , 17(1), 86-94.

[20] Kim, S. Y, Kim, S. G, Lim, S. C, & Bae, C. S. Effects on bone formation in ovariectomized rats after implantation of toothash and plaster of Paris mixture. J Oral Maxillofac Surg (2004). , 62(7), 852-857.

[21] Park, S. S, Kim, S. G, Lim, S. C, & Ong, J. L. Osteogenic activity of the mixture of chitosan and particulate dentin. J Biomed Mater Res A (2008). , 87(3), 618-623.

[22] Kim, W. B, Kim, S. G, Lim, S. C, Kim, Y. K, & Park, S. N. Effect of Tisseel on bone healing with particulate dentin and plaster of Paris mixture. Oral Surg Oral Med Oral Pathol Oral Radiol Endod (2010). ee40., 34.

[23] Kim, Y. K, Kim, S. G, Oh, J. S, Jin, S. C, Son, J. S, Kim, S. Y, & Lim, S. Y. Analysis of the inorganic component of autogenous tooth bone graft material. J Nanosci Nanotechnol. (2011). , 11(8), 7442-7445.

[24] Bang, G, & Urist, M. R. Bone induction in excavation chambers in matrix of decalcified dentin. Arch Surg (1967). , 94(6), 781-789.

[25] Butler, W. T, Mikulski, A, & Urist, M. R. Noncollagenous proteins of a rat dentin matrix processing bone morphogenetic activity. J Dent Res (1977). , 56(3), 228-232.

[26] Conover, M. A, & Urist, M. R. Transmembrane bone morphogenesis by implanted of dentin matrix. J Dent Res (1979).

[27] Conover, M. A, & Urist, M. R. Dentin matrix morphogenetic protein. The Chemistry and Biology of Mineralized Connective Tissues. Northwestern University. New York;Elsevier-North Holland Inc.; (1981). , 597-606.

[28] Ike, M, & Urist, M. R. Recycled dentin toot matrix for a carrier of recombinant human bone morphogenetic protein. J Oral Implantol (1998). , 24(3), 124-132.

[29] Morotome, Y, Goseki-sone, M, Ishikawa, I, & Oida, S. Gene expression of growth and differentiation factors-5,-6, and-7 in developing bovine tooth at the root forming stage. Biochem Biophys Res Commun (1988). , 244(1), 85-90.

[30] Urist, M. R, & Strates, B. S. Bone morphogenetic protein. J Dent Res (1971). , 50, 1393-406.

[31] Urist, M. R. Bone: formation by autoinduction. Science (1965). , 150(3698), 893-899.

[32] Yeomans, D. J, & Urist, M. R. Bone induction by decalcified dentin implanted into oral osseous and muscle tissues. Arch Oral Biol (1967). , 12(8), 999-1008.

[33] Bang, G, & Urist, M. R. Bone induction in excavation chambers in matrix of decalcified dentin. Arch Surg (1967). , 94(6), 781-789.

[34] Bessho, K, Tanaka, N, Matsumoto, J, Tagawa, T, & Murata, M. Human dentin-matrix-derived bone morphogenetic protein. J Dent Res (1991). , 70, 171-175.

[35] Turnbull, R. S, & Freeman, E. Use of wounds in the parietal bone of the rat for evaluating bone marrow for grafting into periodontal defects. J Periodontal Res (1974). , 9(1), 39-43.

[36] Inoue, T, Deporter, D. A, & Melcher, A. H. Induction of cartilage and bone by dentin demineralized in citric acid. J Periodontal Res (1986). , 21(3), 243-255.

[37] Kawai, T, & Urist, M. R. A bovine tooth derived bone morphogenetic protein. J Dent Res (1989). , 68(6), 1069-1074.

[38] Feng, J. Q, Luan, X, Wallace, J, Jing, D, Ohshima, T, Kulkarni, A. B, Souza, D, & Kozak, R. N. CA, MacDougall Ml. Genomic organization, chromosomal mapping, and promoter analysis of the mouse dentin sialophosphoprotein (Dspp) gene, which codes for both dentin sialoprotein and dentin phosphoprotein. J Biol Chem (1998). , 273(16), 9457-9464.

[39] Ritchie, H. H, Ritchie, D. G, & Wang, L. H. Six decades of dentinogenesis research. Historical and prospective views on phosphophoryn and dentin sialprotein. Eur J Oral Sci (1998). Suppl , 1, 211-220.

[40] Schmidt-schultz, T. H, & Schultz, M. Intact growth factors are conserved in the extracellular matrix of ancient human bone and teeth: a storehouse for the study of human evolution in health and disease. Biol Chem (2005). , 386(8), 767-776.

[41] Boden, S. D, Liu, Y, Hair, G. A, Helms, J. A, Hu, D, Racine, M, Nanes, M. S, & Titus, L. LMP-1, aLIM-domain protein, mediates BMP-6 effects on bone formation. Endocrinology (1998). , 139(12), 125-134.

[42] Wang, X, Zhang, Q, Chen, Z, & Zhang, L. Immunohistochemical localization of LIM mineralization protein 1 in pulp-dentin complex of human teeth with normal and pathologic conditions. J Endod (2008). , 34(2), 143-147.

[43] Al-talabani, N. G, & Smith, C. J. Continued development of 5-day old tooth-germs transplanted to syngeneic hamster (Mesocricetus Auratus) cheek pouch. Arch Oral Biol (1978). , 23(12), 1069-1076.

[44] Steidler, N. E, & Reade, P. C. An histological study of the effects of extra-corporeal time on murine dental isografts. Arch Oral Biol (1979). , 24(2), 165-169.

[45] Barrett, A. P, & Reade, P. C. Changes in periodontal fibre organization in mature bone/tooth isografts in mice. J Oral Pathol (1981). , 10, 276-283.

[46] Barrett, A. P, & Reade, P. C. The relationship between degree of development of tooth isografts and the subsequent formation of bone and periodontal ligament. J Periodontal Res (1981). , 16, 456-465.

[47] Barrett, A. P, & Reade, P. C. A histological investigation of isografts of immature mouse molars to an intrabony and extrabony site. Arch Oral Biol (1982). , 27, 59-63.

[48] Chung, P. H. Method for extracting tooth protein from extracted tooth. Korea Intellectual Property Rights Information Service. Patent. Application (2002). (10-2002)

[49] Chung, P. H. Tooth protein extracted from extracted tooth and method for using the same. Korea Intellectual Property Rights Information Service. Patent. Application (2004). (10-2004)

[50] Lee, H. J. Quantitative analysis of proliferation and differentiation of MG-63 cell line on the bone grafting material using human tooth. PhD thesis. School of Dentistry, Seoul National University; (2011).

[51] Kim, Y. K, & Choi, Y. H. Tooth autotransplantation with autogenous tooth-bone graft: A case report. J Korean Dent Sci (2011). , 4(2), 79-84.

[52] Kim, Y. K, Kim, S. G, Yun, P. Y, et al. Autogenous teeth used for bone grafting: A comparison to traditional grafting materials. Oral Surg Oral Med Oral Pathol Oral Radiol. (2013). in press.

[53] Priya, A, Nath, S, Biswas, K, & Bikramjit, B. In vitro dissolution of calcium phosphate-mullite composite in simulated body fluid. J Mater Sci Mater Med (2010). , 21, 1817-1828.

[54] Glimcher, M. J. Molecular biology of mineralized tissues with particular reference to bone. Rev Mod Phys (1959). , 31, 359-393.

[55] Lee, S. H. Low Crystalline hydroxyl carbonate apatite. J Korean Dental Assoc (2006). , 44, 524-533.

[56] Lu, J, Descamps, M, Dejou, J, Koubi, G, Hardouin, P, Lemaitre, J, & Proust, J. P. The biodegradation mechanism of calcium phosphate biomaterials in bone. J Biomed Mat Res Appl Biomater (2002). , 63(4), 408-412.

[57] Fulmer, M. T, & Ison, I. C. Hanker mayer CR, Constantz BR, Ross J. Measurements of the solubilities and dissolution rates of several hydroxyapatites. Biomaterials (2002). , 23(3), 751-755.

[58] Nampo, T, Watahiki, J, Enomoto, A, Taguchi, T, Ono, M, Nakano, H, Tamamoto, G, Irie, T, Tachikawa, T, & Maki, K. A new method for alveolar bone repair using extracted teeth for the graft material. J Periodontol (2010). , 81(9), 1264-1272.

[59] Kim, J. Y, Kim, K. W, Um, I. W, Kim, Y. K, & Lee, J. K. Bone healing capacity of demineralized dentin matrix materials in a mini-pig cranium defect. J Korean Dent Sci (2012). , 5(1), 21-8.

[60] Lee, J. Y, & Kim, Y. K. Retrospective cohort study of autogenous tooth bone graft. Oral Biol Res (2012). , 36(1), 39-43.

[61] Hatano, N, Shimizu, Y, & Ooya, K. A clinical long-term radiographic evaluation of graft height changes after maxillary sinus floor augmentation with a 2:1 autogenous bone/xenograft mixture and simultaneous placement of dental implants. Clin Oral Impl Res. (2004). , 15(3), 339-345.

[62] Velich, N, Nemeth, Z, Toth, C, & Szabo, G. Long-term results with different bone substitutes used for sinus floor elevation. J Craniofacial Surg (2004). , 15(1), 38-41.

[63] Garg, A. K. Current concepts in augmentation grafting of the maxillary sinus for placement of dental implants. Dent Implantol Update. (2001). , 12(3), 17-22.

[64] Jeong, K. I, Kim, S. G, Kim, Y. K, Oh, J. S, Jeong, M. A, & Park, J. J. Clinical Study of Graft Materials Using Autogenous Teeth in Maxillary Sinus ugmentation. Implant Dent (2011)., 20(6), 471-475.

[65] Lee, J. Y, Kim, Y. K, Kim, S. G, & Lim, S. C. Histomorphometric study of sinus bone graft using various graft material. J Dental Rehabilitation and Applied Science (2011)., 27, 141-147.

[66] Kim, S. G, Kim, Y. K, Lim, S. C, Kim, K. W, & Um, I. W. Histomorphometric analysis of bone graft using autogenous tooth bone graft. Implantology (2011)., 15, 134-141.

[67] Babbush, C. Histologic evaluation of human biopsies after dental augmentation with a demineralized bone matrix putty. Implant Dent (2003)., 12, 325-332.

[68] Kim, Y. K, Lee, H. J, & Kim, K. W. Kim SGm Um IW. Guided bone regeneration using autogenous teeth: case reports. J Korean Assoc Oral Maxillofac Surg (2011)., 37, 142-147.

[69] Kim, Y. K, & Kim, S. G. Lee BG: Bone Graft and Implant. Clinical application of a variety of bone graft, Seoul: Narae Pub Co; (2007)., 2-2, 86-134.

[70] Kim, M. J, Kim, Y. K, & Kim, S. G. A variety of grafting biomaterial used in dental surgery, Seoul: Narae Pub Co.; (2004)., 19-24.

[71] Kim, Y. K, & Yi, Y. J. Horizontal Ridge Augmentation using Ridge expansion and Autogenous Tooth Bone Graft: A Case Report. J Dent Rehabil Appl Sci, (2011)., 2011(1), 109-115.

[72] Kim, Y. K, & Um, I. W. Vertical and horizontal ridge augmentation using autogenous tooth bone graft materials: A case report. J Korean Assoc Maxillofac Plast Reconstr Surg (2011)., 33(2), 166-170.

[73] Becker, W, Urist, M, Becker, B. E, et al. Clinical and histological observation of sites implanted with intraoral autologous bone grafts or allografts. 15 human case reports. J Periodontol (1996)., 67, 1025-1033.

[74] Kim, Y. K, & Um, I. W. Autogenous tooth bone graft. Seoul: Charmyun: (2011)., 192-211.

[75] Kim, Y. K, Yun, P. Y, Lee, H. J, Ahn, J. Y, & Kim, S. G. Ridge preservation of the molar extraction socket using collagen sponge and xenogeneic bone grafts. Implant Dent (2011)., 20(4), 267-272.

[76] Sclar, A. The Bio-Col technique. In: Bowyers LC, ed. Soft Tissue and Esthetic Considerations in Implant Therapy. Chicago, IL: Quintessence; (2003)., 2003, 163-187.

[77] Iasella, J. M, Greenwell, H, Miller, R. L, Hill, M, Drisko, C, Bohra, A. A, & Scheetz, J. P. Ridge preservation with freeze-dried bone allograft and a collagen membrane compared to extraction alone for implant site development: A clinical and histologic study in humans. J Periodontol. (2003)., 74(7), 990-999.

[78] Kim, Y. K, Kim, S. G, Kim, K. W, & Um, I. W. Extraction socket preservation and reconstruction using autogenous tooth bone graft. J Korean Assoc Maxillofac Plast Reconstr Surg (2011). , 33, 264-269.

Chapter 17

The Integrations of Biomaterials and Rapid Prototyping Techniques for Intelligent Manufacturing of Complex Organs

Xiaohong Wang, Jukka Tuomi, Antti A. Mäkitie, Kaija-Stiina Paloheimo, Jouni Partanen and Marjo Yliperttula

Additional information is available at the end of the chapter

http://dx.doi.org/10.5772/53114

1. Introduction

In the human body, an organ is a composite of different tissues in an ordered structural unit to serve a common function [1]. Ordinarily, cells self-assemble into tissues before forming an organ. There are at least three different tissues in a complex organ, such as the liver, heart, and kidney. Currently, complex organ failures are the first cause of mortality in developed countries despite advances in pharmacological, interventional, and surgical therapies [2]. Orthotopic organ transplantation is severely limited by the problems of donor shortage and immune rejections [3]. Extracorporeal support systems perform some specific functions within a limited time period [4]. Cell encapsulation techniques face the problems of capsule loss, low stability, and poor efficiency [5]. Cell sheet technique cannot rescue tissues with increased thicknesses above 80 μm [6]. Decellularized matrices are hard to be repopulated by multiple cell types [7]. On the other hand, stem cell research has emerged as one of the most high-profile and promising areas of 21st century science [8-10]. Typically, autologous adipose-derived stem cells (ADSCs) represent one of the most abundant, easily cultured, rapidly expanded, and multipotent cell source [11]. It has been a long-term goal in this field to manufacture complex organs from biocompatible materials (including non-immune patient derived cells) and computer-aided design (CAD) models in a fast, easy, cheap and automatic manner.

To manufacture a complex organ, cells act like building blocks and have special functions. A comprehensive multidisciplinary effort from biology, implantable biomaterials, and rapid prototyping (RP) technology is extraordinarily needed. A biomaterial is defined as any matter, surface, or construct that interacts with biological systems [12]. It may be an autograft, allograft or xenograft transplant material, or a nature derived or laboratory synthesized chemical component. Biomaterials are often used and/or adapted for a medical application, and thus comprise whole or part of a living structure or biomedical device which performs, augments, or replaces a natural function [13]. RP, also referred to as additive manufacturing (AM) or solid freeform fabrication (SFF), is a set of manufacturing processes which can deposit materials layer-by-layer until a CAD model with freeform geometry has been built. RP technology, which has been widely used in the automatic fabrication of complex geometric structure areas, carries the promise to become the most convenient and reliable technique for manufacturing of complex organs in the coming years [14-18].

Over the last two decades, tissue-engineering researchers have devoted themselves to seeding cells onto a porous biodegradable scaffold material to direct cell differentiation and functional assembly into three-dimensional (3D) tissues [19]. This strategy has achieved a great success in simple tissue/organ regeneration [20]. However, it is extremely difficult for this strategy to be used in creating a branched vascular system or a complex organ regenerative template mimicking the native ones with similar mechanical and biological properties. Similar to building a nuclear power plant for complex organ manufacturing, there is a significant gap between simple tissue/organ engineering and complex organ manufacturing approaches both in fabrication technique employed and ultimate goal achieved (Table 1) [14-18].

Complex organ manufacturing	**Nuclear power plant building**
Cells	Bricks, nuclear reactors
Synthetic, natural polymers	Steel
Crosslinking agents	Cements
Vascular systems	Water and light pipes
Nerve system	Electric control system
Multi-nozzle rapid prototyping machines	Cranes
CAD models	Blueprints
Construction	Architecture

Table 1. Analogues between complex organ manufacturing and nuclear power plant building.

The ultimate goal of complex organ manufacturing is to fabricate hybrid biomaterial (including living cells, even gene/protein) structures over a range of size scales (i.e. from a few micrometers to a few millimeters). We herein provide insights into some special integrations of biomaterials and RP techniques towards the purpose of intelligent freeform manufacturing of complex organs. The most successful and promising integrations have been highlighted; meanwhile the future development directions have been highlighted.

http://dx.doi.org/10.5772/53114

2. Biomaterials and RP techniques in thousands of postures

As stated above, biomaterials, usually acting as synthetic frameworks (referred as scaffolds, matrices, or constructs), can be categorized into different groups according to their supply sources, existence states, chemical properties as well as biomedical applications. Typically, patient specific blood, cells (especially stem cells), acellular matrices, tissues and organs are a kind of biomaterials with no immune reactions. More than 100 implantable biomaterials have been reported in different forms, such as bulks, blocks, membranes, sheets, beads, hydrogels, fibers, sutures, plates, nets, meshs, tubes, non-woven fabrics, porous scaffolds (or sponges), heart valves, intraocular lenses, dental implants, pacemakers, biosensors, etc [21-23]. However, very few of them are suitable for complex organ manufacturing purposes. For biomedical applications, biocompatibility, biodegradability and processing ability are among the most crucial issues one should consider. In most cases the implantable biomaterial has to be nontoxic, biocompatible, and biodegradable. Therefore, stringent criteria must be met before proceeding to clinical applications.

Especially, hydrogels are a family of natural or synthetic polymers with high water contents. During the last twenty years, hydrogels have been an important class of soft tissue repair materials or cell delivering vehicles that can be fabricated in the form of 3D micro-periodic structures by colloidal templating [24], interference lithography [25], direct-writing [26], ink-jet printing [27], and two-photon polymerization (2PP) [28].

In the last four decades, significant advances have been made in the progress of scaffold fabrication techniques for biomedical applications. For example, synthetic and natural biodegradable polymers, such as polylactic acid (PLA) [29], poly(lactic/glycolic) acid (PLGA) [30], collagen [31], hyaluronic acid [32] and chitosan [33], are often used as pure implantable biomaterials or tissue engineering scaffolds.

In parallel with the development of biomaterials, the number of commercial RP techniques has expanded rapidly during the last decade. More than 30 different RP techniques have been applied in the most diverse industries. Several companies are now using RP technologies for plastic, wood and metal product manufacturing. For example, Siemens, Phonak Widex, and other hearing aid manufactures use selective laser sintering (SLS) techniques to produce hearing aid shells. Align technology uses SLS techniques to fabricate molds for producing clear braces ("aligners"). And Boeing and its suppliers use SLS techniques to produce ducts and similar parts for F-18 fighter jets [34]. Around 20 of the RP techniques have been adapted in the field of regenerative medicine [35]. Basically, these adaptations can be classified into three major groups hinged on the RP working principles (Figure 1): (i) nozzle-based extruding/assembling/deposition systems, e.g. fused deposition modeling (FDM) (Figure 1A) [36], pressure assisted manufacturing (PAM), low-temperature deposition manufacturing (LDM), and bio-plotters (3DB) (Figure 1B) [37,38], which deposit materials either thermally or chemically through pens/syringes/nozzles; (ii) laser/photolithography-based writing systems, e.g. laser-guided direct writing, which arrange meterials/cells by laser beams [39,40] or photopolymerize a liquid (resin, powder, or wax) in stereolithography (SLA or STL) (Figure 1C) [36,41], or sinter powdered material in SLS systems (Figure 1D)

[42]; (iii) printing-based inkjeting systems, e.g. 3D printing (3DP) systems and wax-based systems, which print a chemical binder onto a powder bed and print two types of wax materials in sequence (Figure 1 E) [36].

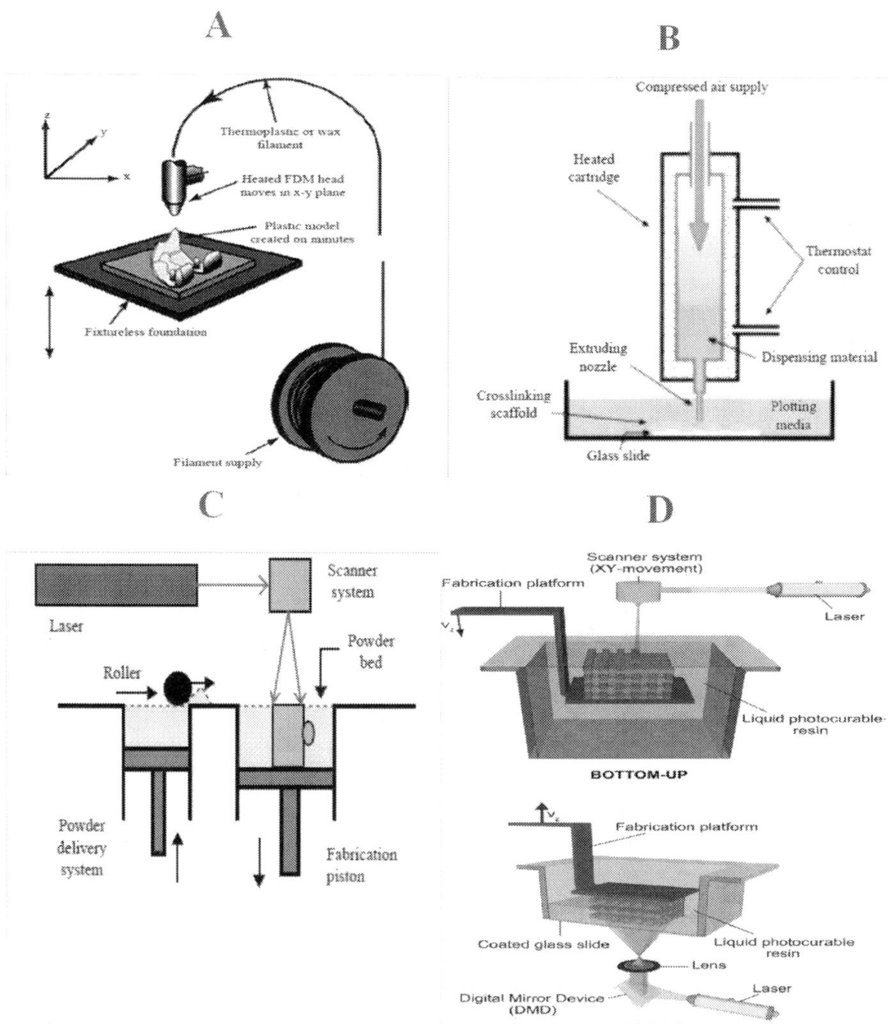

http://dx.doi.org/10.5772/53114

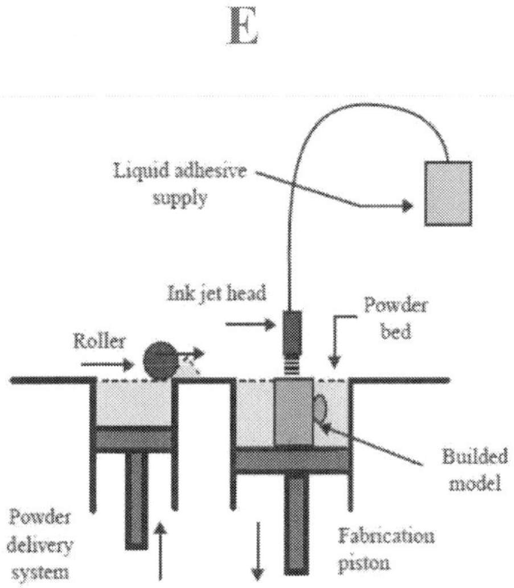

Figure 1. Working principles of various rapid prototyping systems: A) Schematic illustration of the nozzle-based FDM process [36]. B) Scheme of a nozzle-based 3D-Bioplotter heated cartridge setup [37,38]. C) Schematic of the laser-based SLS techniques [36]. D) Schemes of two laser-based of stereolithography (SLA or STL) setups [39]. Upper: a bottom-up system whereby the laser scans the surface for the curing of the photosensitive materials. Bottom: a top-down setup with dynamic digital light projection to cure a complete 2D layer at once. E) Schematic of the 3DP systems [36].

Although most of the adapted techniques can be used in building complex geometrical shapes with CAD modelling, every technique group is subjected to a limited biomaterial incorporation ability and has its own drawbacks in creating 3D living organs. For example, Chu and coworkers have developed design-for-manufacturing rules for their lattice mesostructure fabrication technique with a STL system. Lattice structures tend to have geometry variations in three dimensions [43]. However, this system is not fully capable of creating a branched vascular system, which is vitally important in the context of organ manufacturing to direct spatially heterogeneous tissue development. On the other hand, Arcaute and coworkers have encapsulated human dermal fibroblasts in a synthetic poly(ethylene glycol)-dimethacrylate hydrogel by a SLA technique. Without porous structures and biodegradable properties of the synthetic polymers, it is hard for the cells to form tissues inside the hydrogel [44]. The integrations of biomaterials and RP techniques can form a huge "family tree" with many different combinations. Figure 2 summarizes the integrations of biomaterials with RP techniques and their potential usages in complex organ manufacturing.

Currently, as the concepts of "factory in a box" and "desktop manufacturing" are expanding, new applications of RP techniques in architectural design and 3D construct building increase speedily. Among the most popular RP techniques, the Fab@Home equipments with an average price of about 3000 US dollars are among the most convenient and cost effective RP instruments used in biomaterial fabrication field [45].

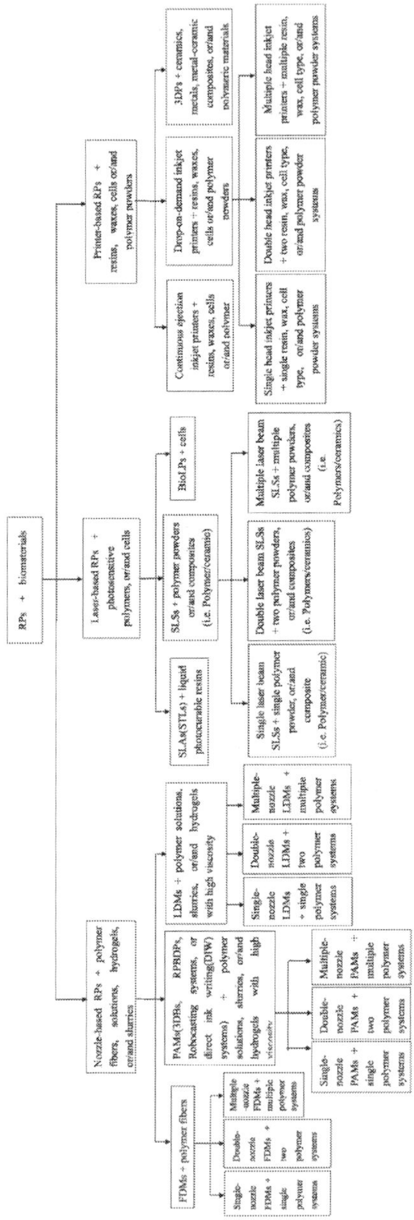

Figure 2. A "family tree" indicates various integrations of biomaterials and rapid prototyping techniques.

At present, the concepts of "scaffolds", "tissues", and "organs" are rather confused both in scientific and industrial areas. Most researchers and manufacturers in the area of tissue engineering like to label their RP products as "scaffolds", "tissues", or "organs". It is reasonable to describe an accellular porous 3D structure with a micro-scale internal architecture but without cells as a "scaffold". However, those with living cells incorporated should be defined as "constructs". Especially, those with cells have already connected to each other to perform special functions should be called "tissues". As described in the beginning of the introduction section, those with more than three different tissue types inside a construct should be called "organs". Simple organs, such as the bladder and blood vessels, should have less than or equal to three tissue types, while complex organs, such as the liver, heart, and kidney, should posses more than three tissue types. With these definitions, it is easier to distinguish which RP technique will be useful in complex organ manufacturing.

3. The integrations of biomaterials with RP techniques for complex organ manufacturing

The ability to put material only into a specific location where it is desired could have a profound impact on how parts are designed and manufactured [46]. Similarly, the ability to put different biomaterials (including different cell types) to exact sites where they are desired could have a profound impact on how complex organs are designed and manufactured. For example, in a complex organ, such as the liver, at least three different cell types (hepatocytes, stellate cells, and Kupffer cells) are required that function in a construct along with the three common cell types of a vascular system. The fundamental unit of the liver, the acinus, has a typical radius of 500 μm. Within this structure at least six cell types interact with one another to coordinate the diversity of liver functions [46]. The spatially heterogeneous arrangements of multi-tissues make all the traditional, or existing techniques incapable of completing this ambitious task.

Over the last ten years, the integration of biomaterials with RP techniques in creating special 3D constructs for various biomedical applications has emerged. The ability to use data from clinical imaging techniques like magnetic resonance imaging (MRI), computerized tomography (CT) or patient-specific data makes RP techniques particularly useful for biomedical applications. Several research groups have adapted different RP techniques to assemble (or print) cell-laden constructs directly from computer-programmed design models with high resolution (Table 2) [47-64]. Six unique intelligent RP devices as well as their primary products are shown in Figure 3[47-52, 65-67]. These processes have demonstrated some possibilities in the area of complex organ manufacturing. The pros and cons of these techniques in complex organ manufacturing are outlined in Table 2. Those with only porous 3D scaffolds are not included here because these integrations have been reported extensively in the former reviews [36,37,44,68-79].

Technique	Pros	Cons	Refs
3D inkjet bioprinting (3DP) in and Pittsburgh Clemson University, USA	Several thermosensitive hydrogels can be used as biopaper; low viscosity cell suspensions or aggregates can be used as bioink; Cell viability greater than 85%.	Complex 3D constructs are difficult to realize; limited feature height (< 5 μm); lack of structural support for cell layer or cell aggregates; tissue formation into lines depends on cell or cell aggregate fusion or assembling; poor mechanical properties.	[47, 48]
3D direct-write bioprinting in University of Cornell and Arizona, USA.	Low and high viscosity hydrogels, including type 1 collagen and alginate can be used; high cell viability (up to 98%); flexible geometric shapes.	Cell viability depends largely on the inner diameter of the gauge tip, collagen concentration and extraction environments; difficult to control the collage gel state.	[49, 50]
3D fiber deposition (3DF) in University Medical Center Utrecht, The Netherlands.	High viscosity hydrogels, such as Pluronic F127, Matrigel, alginate and agarose, can be used; multiple cell types can be incorporated; homogeneous and heterogeneous structures can be created.	Limited materials can be used; limited height of 3D construct (< 10 μm); difficult cell-cell interactions; poor mechanical properties.	[51, 52]
3D single/double syringe cell assembling (or pressure assisted manufacturing (PAM)) in Tsinghua University, China.	Gelatin-based hydrogels can be used; a wide range of biological components can be incorporated; variable and hybrid geometric shapes; high cell viability (more than 98%); easy for long-term storage and transportation.	Limited materials can be used; weaker mechanical properties.	[53 - 60]
Double-nozzle low-temperature deposition manufacturing (DLDM) in Tsinghua University, China.	A wide range of biomaterials including both synthetic and natural polymers can be used; a wide range of biological components can be incorporated; arbitrarily hybrid geometric shapes; high mechanical properties; easy for long-term storage and transportation.	Material viscosity and temperature dependent.	[61 - 64]

Table 2. Comparison of different cell-laden rapid prototyping techniques in complex organ manufacturing

Due to the heterogeneous properties of complex organs both in geometrical structures and material components, emphases should be given to those RP techniques with further development possibilities in the further integrations of biomaterials and equipments. In the following part of this section, some special two or multiple syringe/nozzle techniques are highlighted. In Harvard Medical School, Lee and coworkers have printed a collagen hydrogel precursor, fibroblasts and keratinocytes into a quasi 3D structure for skin repair using a robotic platform (Figure 3D) [65]. The procedure involves printing a layer of liquid collagen to act as a hydrogel precursor. The liquid collagen is crosslinked with a nebulized aqueous crosslinking agent (sodium bicarbonate) to form a hydrogel that provides structural integrity for the subsequent cell suspensions. In fact, this technique is an extension of the above mentioned 3DP or 3DB robotic system with additional syringes as "cartridges" to load two cell suspensions and hydrogel precursors. Highly viable proliferation of each cell layer (85% for keratinocytes and 95% for fibroblasts) was observed on both planar and non-planar surfaces. For thin tissue/organ (such as skin/bladder) manufacturing, this technique is a right choice. However, for complex organ manufacturing, some intrinsic shortcomings, such as limited printing height, and difficult to control the collagen gelation process, made this technique almost incapable.

In University Medical Center Utrecht, The Netherlands, Prof. J Alblas's group, a special bioscaffolder pneumatic dispensing system (SYS + ENG) was adapted for printing cell-laden bone tissue repair hydrogels. High viscosity alginate (10% w/v) and BD Matrigel™ (10% w/v) hydrogels were employed. A limited ten-layer rectangular 3D construct of 10 ×10 mm with spacing between fibers of 0.8-2.5 mm and a thickness of 100 μm was fabricated and subsequently crosslinked in a $CaCl_2$ solution [51]. In spite of the limited height, the interconnected channels are still necessary for oxygen and nutrient delivery, as well as for tissue formation and vascular ingrowth. There are two critical drawbacks of this technique in complex organ manufacturing. The first is the poor mechanical properties of the cell-laden alginate or matrigel hydrogel for use as vascular systems. The second is that low viscosity hydrogels (including alginate and matrigel) are hard to be assembled into 3D constructs.

In university of Missouri, Norotte and coworkers used agarose rods as a molding template to print multicellular spheroids with their special bioprinter to form a tubular cell-laden structure (Figure 3E) [66]. After the fabrication stage, they manually pulled the agarose rods out of the tube, and concluded that it is a time consuming and labor-/spheroid- intensive procedure.

4. Some outstanding achievements made in Tsinghua University

In parallel with the above mentioned RP approaches, a series of RP technologies have been explored extensively by Professor XH Wang's group at the Center of Organ Manufacturing, Department of Mechanical Engineering, Tsinghua University, China. State-of-the-art of the layer-by-layer modeling, material incorporation, and manufacturing principles of these techniques can be found in some of the pertaining references. The advantages and disadvantages of these approaches to be used in complex organ manufacturing have also been listed in Table 2. Previous studies have demonstrated their abilities to engineer complex 3D tissues using various single/double nozzle/syringe RP systems. In the following section some technical specifications are highlighted.

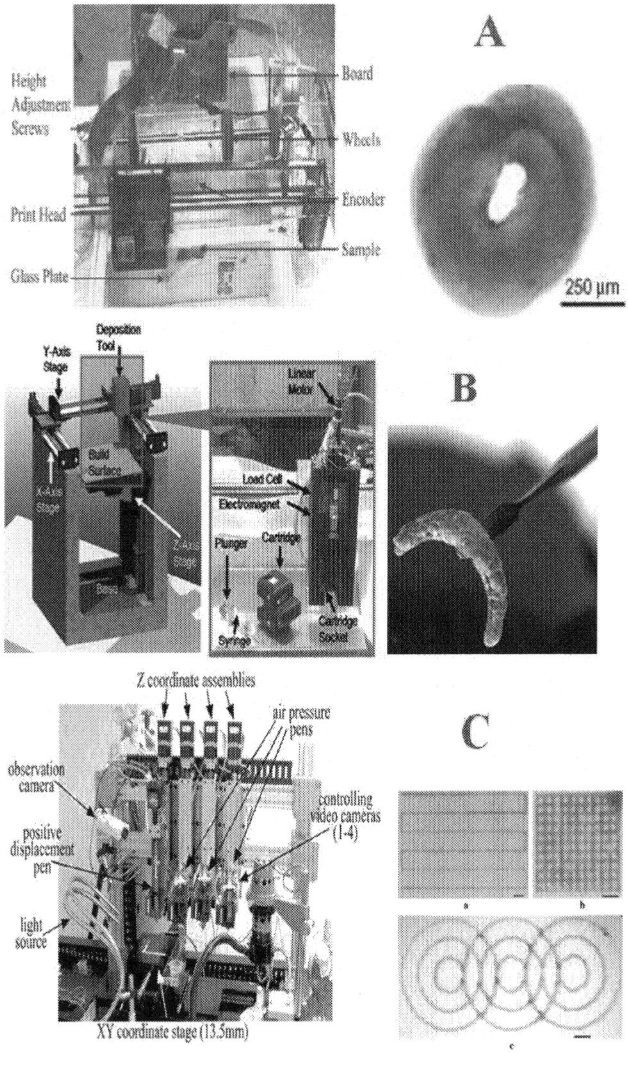
Height Adjustment Screws
Board
Wheels
Encoder
Print Head
Sample
Glass Plate
A
250 µm
Deposition Tool
Y-Axis Stage
Build Surface
X-Axis Stage
Z-Axis Stage
Base
Linear Motor
Load Cell
Electromagnet
Plunger
Cartridge
Syringe
Cartridge Socket
B
Z coordinate assemblies
air pressure pens
observation camera
controlling video cameras (1-4)
positive displacement pen
light source
XY coordinate stage (13.5mm)
C
a
b
c

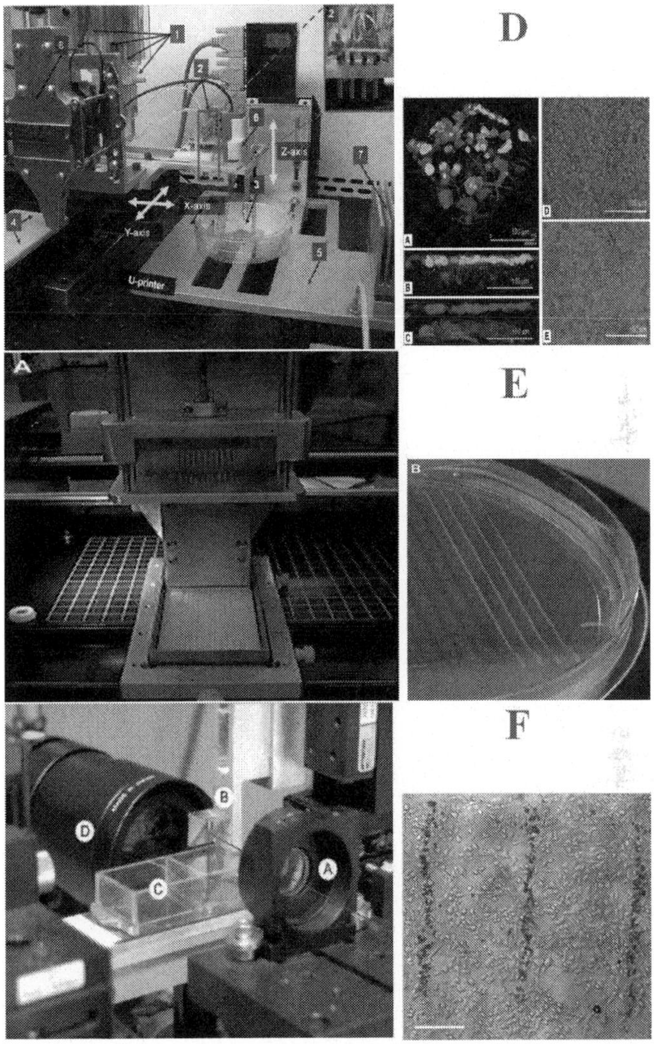

Figure 3. Several unique intelligent rapid prototyping devices and their functional cell-laden products: A) The inkjet cell printer and its bagel-like quasi-3D structure developed in Clemson University, prof. T Boland's group [47]. B) The robotic printing platform and its crescent construct made in Cornell University, prof. LJ Bonassar's group [49]. C) The direct-write system and its preliminary 3D figures developed in University of Arizona, prof. SK Williams' group [50]. D) A modular tissue printing platform with 4 'cartridges' to load cell suspensions and hydrogel precursors developed in Brigham and Women's Hospital, Harvard Medical School, Prof. S.-S. Yoo's group [65]. E) A bioprinting tubular structure with cellular cylinders developed in University of Missouri, Columbia, USA, Prof. G Forgacs' group [66]. F) A laser-guided direct writing (LGDW) system and its patterned factor-linked beads on a stem cell monolayer with micrometer accuracy (Bar = 200 μm) developed in University of Minnesota, prof. D.J. Odde's group [67].

4.1. The single syringe cell assembling technique

Figure 4 shows some of the cell assembling results using our first generation cell assembling system. A gelatin based hydrogel system, such as gelatin, gelatin/chitosan, gelatin/hyaluronan, gelatin/alginate, gelatin/fibrinogen or gelatin/alginate/fibrinogen, was integrated with a single syringe cell-assembling machine to obtain the necessary space and stabilizing factors for cell survival and tissue formation [53-58]. A single cell type was deposited at an ambient temperature (1~10℃) layer by layer in a chamber as the sol state material was transferred into a hydrogel. Grid hepatic tissues, endodermis, and adipose tissues have been regenerated by using this single syringe cell-assembly machine at about 8℃. The gelatin based hydrogel network provided stabilization support for the 3D constructs during the fabrication and post culture stages. This mild deposition temperature is favorable for biological property preservation as increased Joule heating can result in loss of cell viabilities and bioactivities. During the culture period, the gelatin based hydrogel served as both a mass transportation template for tissue development and an extracellular matrix accommodation mimicking the microenvironment in native organs. The use of the natural gelatin based hydrogels was clearly highlighted the distinct advantage of this cell assembly technique for fabricating living tissue analogs. A shortage of the single nozzle/syringe systems was that, these systems lack the ability to easily create parts with spatial heterogeneous materials. Consequently, two double nozzle/syringe RP systems have been explored to deposit different materials at different temperatures.

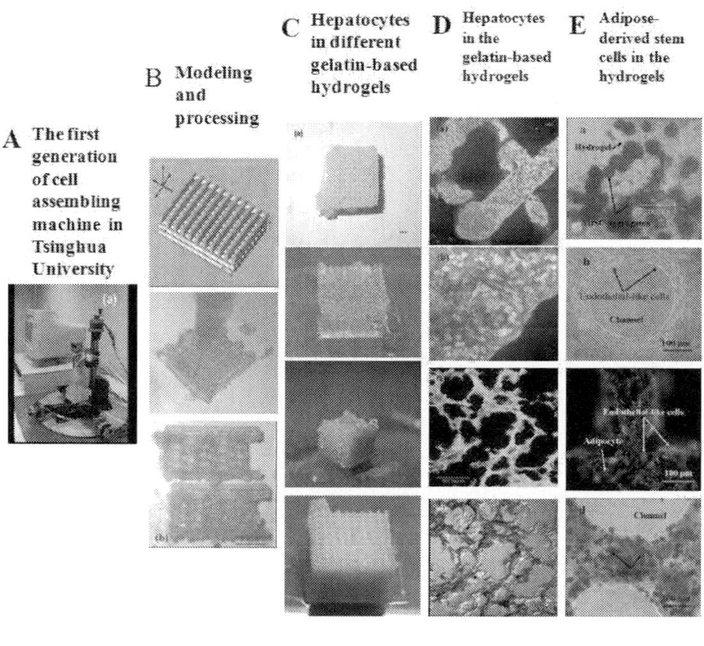

Figure 4. Hepatocyte and adipose-derived stem cell (ADSC) assembling based on the first generation of cell assembling technique developed in Tsinghua University, prof. XH Wang' group [53-58]

http://dx.doi.org/10.5772/53114

4.2. The double syringe cell assembling technique

Different from the above single syringe cell assembling technique, a double syringe cell assembling technique was developed in Tsinghua Unversity with a updated software and hardware. Gradient and cylindrical architectures consist of two different cell-laden hydrogels have been fabricated at a temperature range of 8 – 10℃ [59,60]. Two cell lines encapsulated in the similar gelatin-based hydrogels were put into different regions or compartment in a construct (Figure 5). The embedded branched networks enable culture medium to flow through the entire construct with unparalleled geometric complexity. However, there is a fatal shortcoming of this system to be used in complex organ manufacturing. The mechanical weak properties of the gelatin-based hydrogel made it impossible to connect the branched construct to an *in vivo* vascular system to endure anti-suture anastomosis and blood pressure even after a long-term *in vitro* culture period.

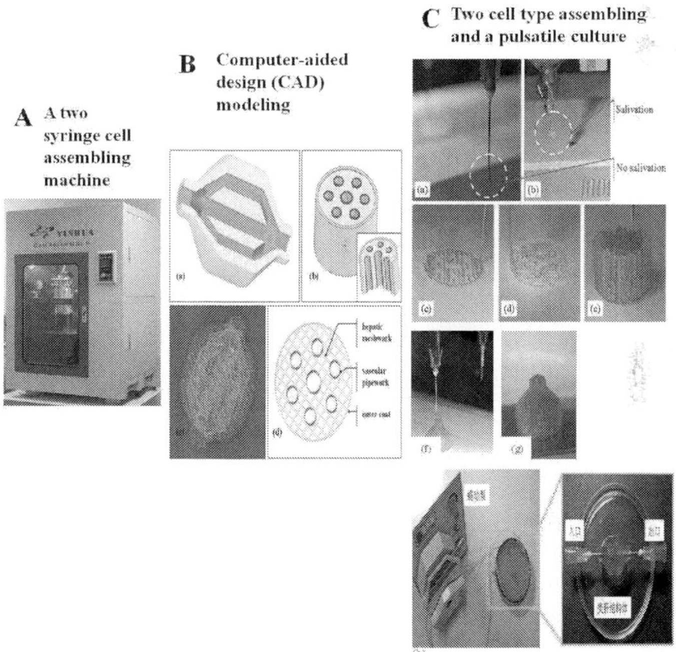

Figure 5. Cell assembling based on a two syringe RP technique developed in Tsinghua University, prof. XH Wang' group. Two different cell types in the gelatin-based hydrogels can be assembled simultaneously into a construct [59-60].

4.3. The combination of cell assembly and cryopreservation techniques

With the advantages of the gelatin-based hydrogel, cryoprotectants (e.g. dimethyl sulfoxide (DMSO), glycerol, and dextran-40) can be incorporated into the cell/hydrogel system and the constructs can be stored at low temperature (below -80℃) directly after the fabrication stage (Figure 6). This incorporation technique represents a significant advancement towards the cell-laden product storage and transport, potentially resulting in labor and resource saving, clinical availability and medical convenience [80-82]. With the gelatin-based hydrogel various bio-factors including macromolecular cell growth factors, small chemical regulators, and even genes/drugs can be easily incorporated to the deposition or assembling systems. This approach is suitable for some special natural thermosetting polymers' (e.g. gelatin and agarose) deposition and opens a new avenue for complex organ manufacturing.

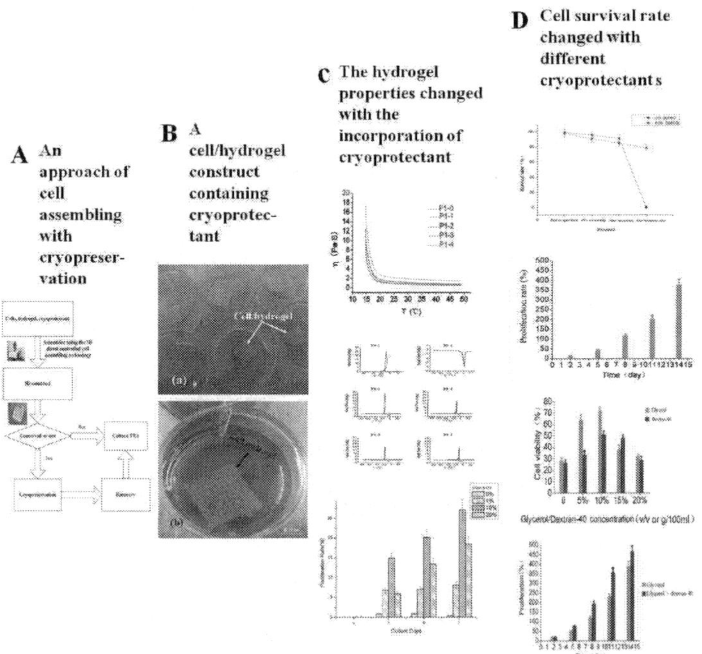

Figure 6. The combination of cell assembly and cryopreservation techniques, developed in prof. XH Wang' group [80-82].

4.4. The double-nozzle low-temperature deposition manufacturing (DLDM) system

The creation of a geometrically complex branched vascular system is a subject of broad fundamental and technological interest in complex organ manufacturing. With the DLDM system it is easy to deposit two different material systems, especially both synthetic and natural polymer systems simultaneously in a construct (Figure 7). Grid, tubular and elliptic structures with both synthetic and natural polymers, such as PU/gelatin and PU/collagen, have been produced at a low-temperature range of -20 - -30 ℃ [61-64]. As shown in Figure 7C, PU and collagen were successfully assembled into a tubular double layer construct. In Figure 7D, an elliptic hybrid hierarchical PU-cell/hydrogel construct with branched and grid internal channels was realized. Cells can survival the heterogeneous fabrication, polymerization/crosslinking, and even storage stages with a high recovery proliferation ability. Figure 7D demonstrates that the external out coat was made of a PU/tetraglycol solution to provide mechanical support for the whole construct. The internal branched and grid channels were made of a cell/dimethyl sulfoxide (DMSO) containing gelatin/alginate/fibrinogen hydrogel to encapsulate ADSCs. Both the out coat PU and compartment cell/hydrogel layers possess microporous, which permit water, oxygen and other small molecules to pass. During the fabrication stage, a low temperature in the range of -20 - -30 ℃ around the nozzles is an important factor to control the sol-gel transformation of the material systems. If the temperature is set too high, the deposited fiber (strand) cannot solidify to form a stable 3D structure. On the other hand, if the temperature is set too low, the fiber is frozen too quickly to fuse with the previous deposited layer. An optimum deposition temperature has played a central role in putting the heterogeneous material systems at the desired locations in the construct.

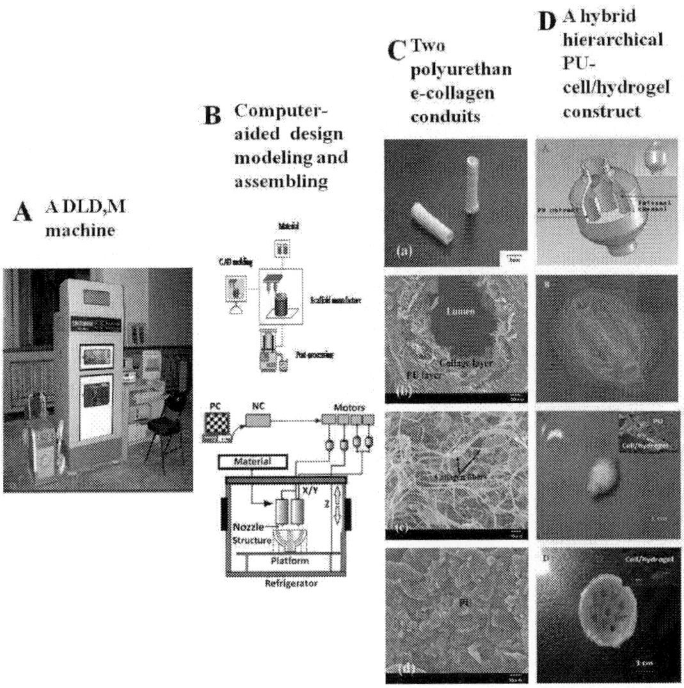

Figure 7. A DLDM technique developed in Tsinghua University, prof. XH Wang' group. An elliptical hybrid hierarchical polyurethane and cell/hydrogel construct was fabricated using the DLDM system [61-64].

This DLDM technique has demonstrated some outstanding merits in complex organ manufacturing with two different material systems that are technologically and biologically incapable to produce using the other existing or traditional RP techniques. The potential applications of the assembled elliptic hybrid hierarchical constructs are diverse, such as cell-cell interaction analyses, stem cell differentiation tracing (pursuing), chemical drug screening, and pathogenic mechanism studies. The synthetic PU system can provide elaborate compartments for cell/hydrogel accommodation. In these compartments, the composition of the cell/hydrogel mixture becomes the key factor in ensuring spatially uniform cell distribution, survival, proliferation and differentiation. By encapsulating the cell/hydrogel mixture in the PU compartments, the composition and proportion of hydrogel components can be easily adjusted to meet the necessary requirements for mimicking the natural cellular arrangements. A maximal cell density (hydrogel-poor and cell-rich) can be easily achieved in the compartment. The use of gelatin-based hydrogel can even be avoided completely in this system, irrespective of stabilization of the construct. Compared with the pure cell/gelatin/alginate/fibrin construct made by the single/double RP systems, the hybrid hierarchical network can provide much higher mechanical stability and pressure resistance abilities when it is applied to *in vitro* pulsatile cultures and *in vivo* blood vessel anastomoses. Some experiments have proved that the 3D constructs with intrinsic interconnected branched and grid channels were easily adapted to an *in vitro* pulsatile culture and *in vivo* implantation system [83-86].

5. A four-nozzle low-temperature deposition manufacturing (FLDM) system

At present, a FLDM system is under development in professor XH Wang's group [18]. Figure 8 demonstrates the outlook of the machine and a primary try on a liver lobe like structure construction. Compared with the DLDM RP system, two more nozzles have been equipped. Thus, two more cell types can be incorporated simultaneously into a construct. This amplified integration possesses some outstanding advantages towards complex organ manufacturing: (i) hierarchically organization of multiple population of cells and growth factors in a more intricate physiologically mimicking geometry; (ii) simultaneously deposition of one scaffold material, a vascular system with two main cell types, and one parenchymal cell type in a more elegant native tissue-specific phenotype; (iii) computer definition of the fluid paths and macro/microstructures in a more patient specific manner; and (iv) spatial distribution of multi-tissue boundaries and fluorescent biomarkers in a more controllable pattern. This FLDM RP system makes it possible to partially control over the design, modeling and fabrication of a highly hierarchical liver lobe like construct in a rapid, convenient, and cost effective manner.

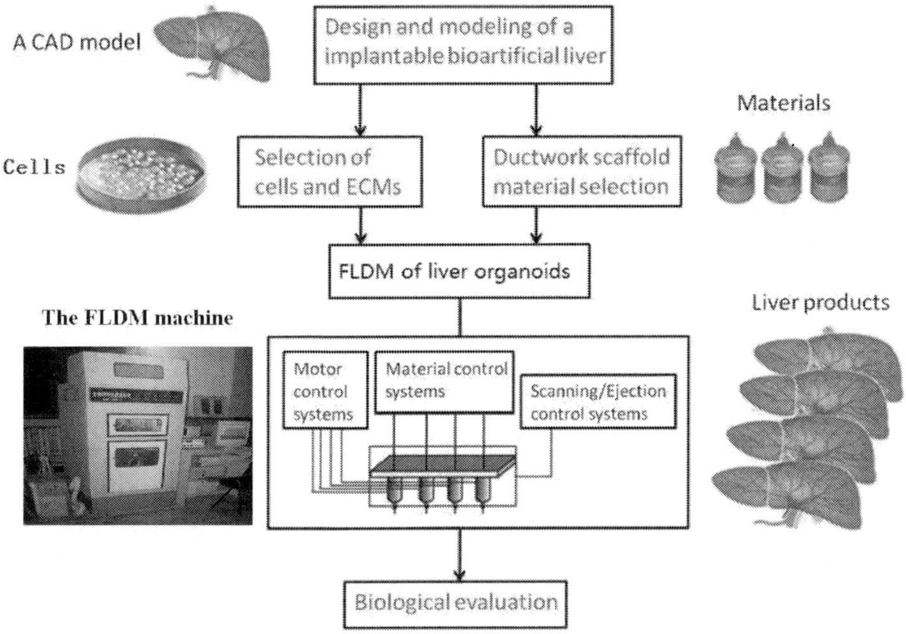

Figure 8. A schematic description of the modeling and manufacturing processes of four liver lobe-like constructs with a FLDM RP system developed in Tsinghua University, prof. XH Wang' group [18].

6. Emphases to some technical specifications

Theoretically, RP technology is able to produce any required complex shape. The standard modeling and deposition technologies enable the hybrid hierarchically ordered patterns to be generated in an automatic, convenient, and inexpensive manner. Again, we use the liver as an example. In a liver lobe at least 6 different cell types are structured as repeated units. These units can achieve high oxygen exchange and nutrient supply for a mass of cells where the cell sizes are in the range of ~20 μm. This geometry enables a high degree of processing optimization, which provides the opportunity for RP designers and manufacturers to control readily the distribution of different cells in a construct. Stimulated by this motivation, many groups have tried different RP systems with only thin or quasi-3D cell containing structures so far. Someone even claims to use scaffold free cell aggregates to print organs. This has been proven to be a time-consuming process and cells can not find their respective places in a complex organ without the support of scaffold materials.

Different from all the existing techniques, the gelatine-based hydrogel has been explored extensively as an internal scaffold material with the single/double syringe/nozzle RP techniques in the author's own group Tsinghua University. Aqueous gelatin solution is an amorphous natural hydrogel in which cells can be encapsulated, extruded and deposited at desired positions [87-90]. This solution is flexible with a gelation temperature of 20℃ and allow the diffusion of hydrophilic substrates. The sol-gel transform property makes it possible to deposit the gelatin-based cell-laden hydrogel at a large range of temperature (from 20℃ to -30℃). However, this hydrogel is not stable at 37℃. The mechanical properties of the gelatin-based hydrogels are notoriously inadequate and prohibit the use in stress-loaded implants. To improve the stability of the gelatin-based hydrogels, alginate and fibrinogen were incorporated. Sodium alginate (composed of mannuronic and guluronic (G) dimmers) is a biocompatible and biodegradable polymer, and has been widely used in cell encapsulation technology, although the biocompatibility of the alginates in relation to their composition is a matter of debate. Sodium alginate can be reversibly cross-linked by divalent cations, such as Ca^{+2} and Mg^{+2}, to form a relatively stable hydrogel. Fibrin, derived from blood fibrinogen, is another natural biocompatible and biodegradable polymer, which has been widely used as sealant and adhesive during surgery. With the catalyzing of thrombin smaller fibrinogen molecules can polymerize to form a stable fibrin hydrogel.

In addition to being able to build complex structures with precision and accuracy, it is equally important that the mechanical properties of the supporting materials are suitable for the intended applications. A novel linear elastomeric polyurethane from soft polycaprolactone (PCL) and polyethylene glycol (PEG) segments, and hexamethylene diisocyanate (HDI) chain extender has been synthesized in the authors' own group and used as an external scaffold material. This PU possesses tunable biodegradability, excellent biocompatibility and compatible mechanical properties with animal veins [91,92]. Long-term *in vivo* biocompatibility and biodegradability of the PU have been proven with a rabbit model. It has successfully repaired nerve and vein defects without any detected side effects, such as thrombosis, inflammation, intimal hyperplasia, and calcification. The excellent mechanical properties, biocompatibilities, adjust abilities and processing abilities have made this kind of polymer to be outstanding from the other existing synthetic scaffold biomaterials, such as polyhydroxybutyrate (PHB) [93], poly(D,L-lactic-co-glycolic acid) (PLGA) [94], and poly(tetrafluoroethylene) (PTFE) [95].

To date, the most widely recognized advantage of the RP technology (i.e. layered manufacturing methodology) is the relative ease of automatically manufacture of complex geometric shapes with heterogeneous structures composed of multi-material regions. Complex organ manufacturing aims to automatically produce complex organs directly from CAD models with high sophisticated RP techniques. Since the earlier concept of complex organ manufacturing using both synthetic and natural scaffold biomaterials and multi-nozzle RP techniques was first introduced in 2007, the present technique was developed gradually [13-18]. As displayed in Figure 7D, the virtual elliptic construct with branched fluidic network has been designed and fabricated according to a pre-defined CAD software. The integration of the DLDM RP technique with the cell cryopreservation technique together with the mechan-

http://dx.doi.org/10.5772/53114

ically strong enough synthetic PU scaffold enables us to efficiently produce spatially heterogeneous cell-laden tissue/organ substitutes that would otherwise be challenging to achieve [61-64]. This integrated technique therefore has the potential to lead a big revolution in the fields of tissue engineering and regenerative medicine.

It is expected that in the following several years these integrated RP technologies will see their major break-through development stage and play a key role in complex organ manufacturing area. With the proper integrations of biomaterials and enabling RP techniques, it is possible for us to address all the challenges involved in complex organ manufacturing and to make the realization of complex organ manufacturing both feasible and practical. These proper integrations also benefit some of other related areas, such as high throughput drug screening, stem cell differentiation induction, fluorescent dye discovering, energy metabolite model establishing and cancer/stem cell behavior controlling.

7. Concluding remarks

The goal of complex organ manufacturing is to directly fabricate multiple cell types into an organ substitute using a multiple nozzle RP system. Analogous to the process of building a nuclear power plant, complex organ manufacturing requires the ability to control the positions of many cell types, internal/external scaffold materials, and even cell growth factors on the nano-, micro-, and macro-scales with respect to each others. The integrations of biomaterials and RP techniques have significantly improved the ability to manufacture cell-laden constructions with predefined geometries under the instructions of CAD models or medical data (for example, patient-specific images). Especially, recent advances in DLDM and FLDM RP techniques in Tsinghua University have leveraged these progresses. Although still at its infant stages and associated with numerous problems, ever-increasing evidence supports the intriguing hypothesis that the integrations of multiple biomaterials (including multiple cell types) and multiple nozzle RP techniques will eventually change the traditional practices and make the dreams of complex organ manufacturing come true. It is expected that in the future, most of the reconstructive disciplines of complex organ manufacturing will be fully revised by the development of new multiple nozzle RP systems with optimal safety, easy manipulation ability and maximum reliability. Multiple nozzle RP techniques will undoubtedly play an important role in the future complex organ manufacturing area. Cells in the engineered construct will potentially behave as comfortably as in their natural *in vivo* environment. Further studies are therefore needed to elucidate and determine the fundamental structure-function relationships of diverse tissues in a complex organ, the nutrition supply systems and the heterogeneous structural cues to promote full functional realization in a complex organ. Ever increasing evidences have indicated that with the right integrations of biomaterials and RP techniques, a brand-new era of complex organ manufacturing like the rising sun, is on the horizon.

Acknowledgment

The Project was Supported by the State Key Laboratory of Materials Processing and Die & Mould Technology, Huazhong University of Science and Technology (No. 2012 - P03), the National Natural Science Foundation of China / the Research Grants Council of Hong Kong (NSFC/RGC, No. 50731160625), the National Natural Science Foundation of China (No. 81271665), the National Natural Science Foundation of China (No. 30970748), the Cross-Strait Tsinghua Cooperation Basic Research (No.2012THZ02-3), the National High Tech 863 Grant (No. 2009AA043801), the Finland Distinguished Professor program (FiDiPro) of Tekes (No. 40041/10).

Author details

Xiaohong Wang[1,2,3*], Jukka Tuomi[1], Antti A. Mäkitie[1,4], Kaija-Stiina Paloheimo[1], Jouni Partanen[1] and Marjo Yliperttula[5]

*Address all correspondence to: wangxiaohong@tsinghua.edu.cn

1 Business Innovation Technology (BIT) Research Centre, School of Science and Technology, Aalto University, Aalto, Finland

2 Key Laboratory for Advanced Materials Processing Technology, Ministry of Education & Center of Organ Manufacturing, Department of Mechanical Engineering, Tsinghua University, Beijing, P.R. China

3 State Key Laboratory of Materials Processing and Die & Mould Technology, Huazhong University of Science and Technology, Wuhan, P.R. China

4 Department of Otolaryngology - Head & Neck Surgery, Helsinki University Hospital and University of Helsinki, Helsinki, Finland

5 Division of Biopharmaceutics and Pharmacokinetics, Faculty of Pharmacy, University of Helsinki, Helsinki, Finland

References

[1] Widmaier EP, Raff H, Strang KT. Vander's Human Physiology. 2003 11th Ed. McGraw-Hill. ISBN 9870073049625.

[2] Walther G, Gekas J, Bertrand DF. Amniotic stem cells for cellular cardiomyoplasty: promises and premises. Catheterization and Cardiovascular Inter Ventions 2009;73:917-924.

[3] Aikawa E, Nahrendorf M, Sosnovik D, Lok VM, Jaffer FA, Aikawa M, Weissleder R. Multimodality molecular imaging identifies proteolytic and osteogenic activities in early aortic valve disease. Circulation 2007;115L:377-386.

[4] Demetriou AA, Brown RS Jr, Busuttil RW, Fair J, McGuire BM, Rosenthal P, Am Esch JS II, Lerut J, Nyberg SL, Salizzoni M, Fagan EA, de Hemptinne B, Broelsch CE, Muraca M, Salmeron JM, Rabbin JM, Metselaar HJ, Pratt D, De La Mata M, McChesney LP, Everson GT, Lavin PT, Sterens AC, Pitkin Z, Solomon BA. Prospective, randomized multicenter, controlled trial of a bioartificial liver in treating acute liver failure. Ann Surg 2004;239(5):660-670.

[5] Orive G, Hernández RM, Gascón AR, Calafore R, Chang TMS, de Vos P. Hortelano G, Hunkeler D, Lacík I, Pedraz JL. History, challenges and perspectives of cell microencapsulation. Trends Biotechnol 2004;22:87-92.

[6] Sekine H, Shimizu T, Yang J, Kobayashi E, Okano T. Pulsatile myocardial tubes fabricated with cell sheet engineering. Circulation 2006;114[suppl I]:I-87- I-93.

[7] Moniaux N, Faivre J. A reengineered liver for transplantation. Journal of Hepatology 2011;54:386-387.

[8] Planat-Benard V, Silvestre J-S, Cousin B, André M, Nibbelink M, Tamarat R, Clergue M, Manneville C, Saillan-Barreau C, Duriez M, Tedgui A, Levy B, Pénicaud L, Casteilla L. Plasticity of human adipose lineage cells toward endothelial cellls. Circulation 2004;109:656-663.

[9] Simper D, Stalboerger PG, Panetta CJ, Wang S, Caplice NM. Smooth muscle progenitor cells in human blood. Circulation 2002;106:1199-1204.

[10] Rowley JA, Sun Z, Goldman D, Mooney DJ. Biomaterials to spatially regulate cell fate. Adv Mater 2002;14(12):886-889.

[11] Caplan AI. Mesenchymal stem cell. J Orthop Res 1991; 9(5): 641-650.

[12] en.wikipedia.org/wiki

[13] Wang XH, Ma JB, Wang YN, He BL. Bone repair in radii and tibias of rabbits with phosphorylated chitosan reinforced calcium phosphate cements. Biomaterials. 2002;23(21):4167-4176

[14] Wang XH, Yan YN, Zhang RJ. Gelatin-based hydrogels for controlled cell assembly. In: Ottenbrite RM, ed. Biomedical Applications of Hydrogels Handbook. New York: Springer, 2010;269-284.

[15] Wang XH, Yan YN, Zhang RJ Rapid prototyping as a tool for manufacturing bioartificial livers. Trends Biotechnol 2007;25:505-513.

[16] Wang XH, Yan YN, Zhang RJ. Recent trends and challenges in complex organ manufacturing. Tissue Eng Part B 2010;16:189-197.

[17] Wang XH, Zhang QQ. Overview on "Chinese–Finnish workshop on biomanufacturing and evaluation techniques. Artificial Organs 2011;35(10):E191- E193.

[18] Wang XH. Intelligent freeform manufacturing of complex organs. Artificial Organs. 2012; doi:10.1111/j.1525-1594.2012.01499.x

[19] Langer R, Vacanti JP. Tissue Engineering. Science 1993;260:920-926.

[20] Oberpenning F, Meng J, Yoo J J, Atala A. De novo reconstitution of a functional mammalian urinary bladder by tissue engineering. Nature Biotechnol 1999;17:149-155.

[21] Ramakrishna S, Mayer J, Wintermantel E, Leong KW. Biomedical applications of polymer-composite materials: a review. Comp Sci Technol 2001;61(9):1189-1224.

[22] Vert M. Aliphatic polyesters: great degradable polymers that cannot do everything. Biomacromolecules 2005;6(2):538-546.

[23] Piskin E. Biodegrdable polymers as biomaterials. J Biomater Sci Polym Edi 1994;6:775-795.

[24] Lee Y-J, Braun PV. Tunable inverse opal hydrogel PH sensors. Adv Mater 2003; 5(7-8): 563-566.

[25] Kang JH, Moon JH, Lee SK, Park SG, Jang SG, Yang S. Thermoresponsive hydrogel photonic crystals by three-dimensional holographic lithography. Adv Mater 2008;20(16):3061-3065.

[26] Barry 111 RA, Shepherd RF, Hanson JN, Nuzzo RG, Wiltzius P, Lewis JA. Direct-write assembly of 3D hydrogel scaffolds for guided cell growth. Adv. Mater. 2009; 21(23): 2407-2410.

[27] Calvert P. Inkjet printing for materials and devices. Chem Mater 2001;13(10): 3299-3305.

[28] Liska R, Schuster M, Infuhr R, Turecek C, Fritscher C, Seidl B, Schmidt V, Kuna L, Haase A, Varga F. Photopolymers for rapid prototyping. 2007;4(4):505-510.

[29] Mikos AG, Thorsen AJ, Czerwonka LA, Bao Y, Langer R, Winslow DN, Vacanti JP. Preparation and characterization of poly(l-lactic acid) foams. Polymer 1994;35(5): 1068-1077.

[30] Karp JM, Shoichet MS, Davies JE. Bone formation on two-dimensional poly(DL-lactide-co-glycolide) (PLGA) films and three-dimensional PLGA tissue engineering scaffolds in vitro. J Biomed Mater Res A 2003;64A(2):388-396.

[31] Sai P, Babu M. Collagen based dressings - a review. Burn 2000;26(1):54-62.

[32] Khademhosseini A, Eng G, Yeh J, Fukuda J, Blumling 111 J, Langer R, Burdick JA. J Biomed Mater Res A 2006; 79(3):522-532.

[33] Wang XH, Ma JB, Wang YN, He BL. Structural characterization of phosphorylated chitosan and their applications as effective additives of calcium phosphate cements. Biomaterials 2001;22(16): 2247-2255.

[34] Chu C, Graf G, Posen DW. Design for additive manufacturing of cellular structures. Compter-aided design and applications 2008;5(5):680-696.

[35] Chua LK, Leong KF, Lim CS. Rapid prototyping: principles and applications. Singapore: World Scientific Publishing 2004.

[36] Azari A, Nikzad S. The evolution of rapid prototyping in dentistry: a review. Rapid Prototyping J 2009;15(3): 216-225.

[37] Pfister A, Landers R. Laib A, Hübner U, Schmelzeisen R. Biofunctional rapid prototyping for tissue engineering applications: 3D bioplotting versus 3D printing. J Polymer Science Part A Polymer Chem 2004;42(3):624-638.

[38] Maher PS, Keatch RP, Donnelly K, Mackay RE, Paxton JZ. Construction of 3D biological matrices using rapid prototyping technology. Rapid Prototyping J 2009;15(3): 204-210.

[39] Yeong WY, Chua CK, Leong KF, Chandrasekaran M, Lee MW. Indirect fabrication of collagen scaffold based on inkjet printing technique. Rapid Prototyping J 2006;12(4): 229-237.

[40] Odde DJ, Renn MJ. Laser-guided direct writing of living cells. Biotechnol Bioeng 2000;67:312-318.

[41] Tan KH, Chua CK, Leong KF, Cheah CM, Gui WS, Tan WS, Wiria FE. Selective laser sintering of biocompatible polymers for applications in tissue engineering. Bio-Medical Materials and Engineering 2005;15(1-2):113-124.

[42] Billiet T, Vandenhaute M, Schelfhout J, Vlierberghe SV, Dubruel P. A review of trends and limitations in hydrogel-rapid prototyping for tissue engineering. Biomaerials 2012;33(26): 6020-6041.

[43] Chu J, Engelbrecht S, Graf G, Rosen DW. A comparison of synthesis methods for cellular structures with application to additive manufacturing. Rapid Prototyping J 2010;16:275-283.

[44] Arcaute K, Mann BK, Wicker RB. Stereolithography of three-dimensional bioactive poly(ethylene glycol) constructs with encapsulated cells. Ann Biomed Eng 2006;34(9): 1429-1441.

[45] Malone E, Lipson H. Fab@Home: the personal desktop fabricator kit. Rapid Prototyping J 2007;13(4):245-255.

[46] Bhatia SN, Chen CS. Tissue engineering at the micro-scale. Biomedical Microdevices 1999;2:131-144.

[47] Boland T, Xu T, Damon B, Cui X. Application of inkjet printing to tissue engineering. Biotechnol J 2006;1(9):910-917.

[48] Cooper GM, Miller ED, DeCesare GE, Usas A, Lensie EL, Bykowski MR, Huard J, Weiss LE, Losee JE, Campbell PG. Inkjet-based biopatterning of bone orphogenetic protein-2 to spatially control calvarial bone formation. Tissue Eng Part A 2010;16:1749-1759.

[49] Cohen DL, Malone E, Lipson H, Bonassar LJ. Direct freeform fabrication of seeded hydrogels in arbitrary geometries. Tissue Eng 2006;12:1325-1335.

[50] Smith CM, Stone AL, Parkhill RL, Stewart RL, Simpkins MW, Kachurin AM, Warren WL, Williams SK. Three-dimensional bioassembly tool for generating viable tissue-engineered constructs. Tissue Eng 2004;10:1566-1576.

[51] Fedorovich NE, Schuurman W, Wijnberg HM, Prins H-J, van Weeren PR, Malda J, Alblas J, Dhert WJA. Biofabrication of osteochondral tissue equivalents by printing topologically defined, cell-laden hydrogel scaffolds. Tissue Eng Part C 2012;18:33-44.

[52] Fedorovich NE, Alblas J, Hennink WE, Öner FC, Dhert WJA. Organ printing: the future of bone regeneration? Trends Biotechnol 2011;29:601-606.

[53] Yan YN, Wang XH, Xiong Z, Liu HX, Liu F, Lin F, Wu RD, Zhang RJ, Lu QP. Direct construction of a three-dimensional structure with cells and hydrogel. J Bioact Compat Polym 2005;20:259-69.

[54] Yan YN, Wang XH, Pan YQ, Liu HX, Cheng J, Xiong Z, Lin F, Wu RD, Zhang RJ, Lu QP. Fabrication of viable tissue-engineered constructs with 3D cell-assembly technique. Biomaterials 2005;26:5864-5871.

[55] Wang XH, Yan YN, Pan YQ, Wang, Xiong Z, Liu HX, Cheng J, Liu F, Lin F, Wu RD, Zhang RJ, Lu QP. Generation of three-dimensional hepatocyte/gelatin structures with rapid prototyping system. Tissue Eng 2006;12:83-90.

[56] Xu W, Wang XH, Yan YN, Zhang W, Xiong Z, Lin F, Wu RD, Zhang RJ. Rapid prototyping three-dimensional cell/gelatin/fibrinogen constructs for medical regeneration. J Bioact Compat Polym 2007;22 (4):363-377.

[57] Zhang T, Yan YN, Wang XH, Xiong Z, Lin F, Wu RD, Zhang R.J. Three-dimensional gelatin and gelatin/hyaluronan hydrogel structures for traumatic brain injury. J Bioact Compat Polym 2007;22(1):19-29.

[58] Xu W, Wang XH, Yan YN, Zhang RJ. Rapid Prototyping of Polyurethane for the Creation of Vascular Systems. J Bioact Compat Polym 2008;23:103-114. (Featured by Nature China on May 3, 2008).

[59] Li SJ, Yan YN, Xiong Z, Weng CY, Zhang RJ, Wang XH. Gradient hydrogel construct based on an improved cell assembling system. J Bioact Compat Polym 2009;24 (S1): 84-99.

[60] Li SJ, Xiong Z, Wang XH, Yan YN, Liu HX, Zhang RJ. Direct fabrication of a hybrid cell/hydrogel construct via a double-nozzle assembling technology. J Bioact Compat Polym 2009;24:249-264.

[61] Cui TK, Yan YN, Zhang RJ, Liu L, Xu W, Wang XH. Rapid prototyping of a double layer polyurethane-collagen conduit for peripheral nerve regeneration. Tissue Eng Part C Methods 2009;15:1-9.

[62] Cui T K, Wang XH, Yan YN, Zhang RJ. Rapid prototyping a new polyurethane-collagen conduit and its Schwann cell compatibility. J Bioact Compat Polym 2009; 24(S1): 5-7.

[63] Wang XH, Cui T K, Yan YN, Zhang RJ. J Bioact Compat Polym 2009;24(2):109-127.

[64] He K, Wang XH. Rapid prototyping of tubular polyurethane and cell/hydrogel construct. J Bioact Compat Polym 2011;26(4):363-374.

[65] Lee W, Debasitis JC, Lee VK, Lee J-H, Fischer K, Edminster K, Park J-K, Yoo S-S. Multi-layered culture of human skin fibroblasts and kerainocytes through three-dimensional freeform fabrication. Biomaterials 2009;30(8):1587-1595.

[66] Norotte C, Marga FS, Niklason LE, Forgacs G. Scaffold-free vascular tissue engineering using bioprinting. Biomaterials 2009; 30(23):5910-5917.

[67] Nahmias Y, Schwartz RE, Verfaillie CM, Odde DJ. Laser-guided direct writing for three-dimensional tissue engineering. Biotech Bioeng 2005; 92(2):129-136.

[68] Leong KF, Cheah CM, Chua CK. Solid freeform fabrication of three-dimensional scaffolds for engineering replacement tissues and organs. Biomaterials 2003;24(13): 2363-2378.

[69] Yeong W-Y, Chua C-K, Leong K-F, Chandrasekaran M. Rapid prototyping in tissue engineering: challenges and potential. Trends Biotechnol 2004;22(12)643-652.

[70] Hutmacher DW, Sittinger M, Risbud MV. Scaffold-based tissue engineering: rationale for cimputer-aided design and solid free-form fabrication systems. Trends Biotechnol 2004;22(7):354-362.

[71] Hollister SJ. Porous scaffold design for tissue engineering. Nature Mater 2005;4:518524.

[72] Kou XY, Tan ST. Heterogeneous object modeling: a review. Computer-Aided Design 2007a; 39:284-301.

[73] Peltola SM, Melchels FPW, Grijpma DW, Kellomaki M. A review of rapid prototyping techniques for tissue engineering purposes. Ann Med 2008; 40:268-280.

[74] Uetla BR, Storti D, Anderson RC, Ganter M. A review of process development steps for new material-systems in three dimensional printing (3DP). J Manufacturig Process 2008;10:96-104.

[75] Grayson WL, Chao PH, Marolt D, Kaplan DL, Vunjak-Novakovic G. Engineering custom-designed osteochondral tissue grafts. Trends Biotechnol 2008;26:181-189.

[76] Bibb R, Eggbeer D, Evans P. Rapid prototyping technologies in soft tissue facial prosthetics: current state of the art. Rapid Prototyping J 2010;16(2):130-137.

[77] Melchels FPW, Feijen J, Grijpma DW. A review on stereolithography and its applications in biomedical engineering. Biomaterials 2010; 31(24):6121-6130.

[78] Neugebauer J, Stachulla G, Ritter L, Dreiseidler T, Mischkowski RA, Keeve E, Zöller JE. Computer-aided manufacturing technologies for guided implant placement. Expert. Rev. Med. Devices 2010;7(1):113-129.

[79] Dhandayuthapani B, Yoshida Y, Maekawa T, Kumar DS. Polymeric scaffolds in tissue engineering application: a review. In J Polym Science 2011, doi: 10.1155/2011/290602.

[80] Sui SC, Wang XH, Liu PY, Yan YN, Zhang RJ. Cryopreservation of cells in 3D constructs based on controlled cell assembly processes. J Bioact Compat Polym 2009;24(5):473-487.

[81] Wang XH, and Xu HR. Incorporation of DMSO and dextran-40 into a gelatin/alginate hydrogel for controlled assembled cell cryopreservation, Cryobiology 2010;61:345-351.

[82] Wang XH, Paloheimo K-S, Xu HR, Liu C. Cryopreservation of Cell/Hydrogel Constructs Based on a New Cell-assembling Technique. J BioactCompat Polym 2010;25(6): 634-653.

[83] Wang XH, Mäkitie AA, Paloheimo K-S, Tuomi J, Paloheimo M, Sui SC, Zhang QQ. Characterization of a PLGA sandwiched cell/fibrin construct and induction of the adipose stem cells (ADSCs) into smooth muscle cells. Materials Science and Engineering C 2011;31:801-808.

[84] Wang XH, Sui SC, Liu C. Optimizing the step-by-step forming processes for fabricating a poly(DL-lactic-co-glycolic acid)-sandwiched cell/hydrogel construct. J Appl Polym Sci 2011;120:1199-1207.

[85] Wang XH, Mäkitie AA, Paloheimo K-S, Tuomi, J., Paloheimo, M. A tubular PLGA-sandwiched cell/hydrogel fabrication technique based on a step-by-step mold/extraction process. Advances in Polymer Technology. 2011;30:163-173.

[86] Wang XH, Sui SC. Pulsatile culture of a PLGA sandwiched cell/hydrogel construct fabricated by a step-by step mold/extraction method. Artificial Organs 2011;35(6): 645-655.

[87] Yao R, Zhang RJ, Wang XH. Design and evaluation of a cell microencapsulating device for cell assembly technoloty. J Bioact Compat Polym 2009;24(1):48-62.

[88] Yao R, Zhang RJ, Yan YN, Wang XH. In vitro angiogenesis of 3D tissue engineered adipose tissue. J Bioact Compat Polym 2009;24(1):5-24.

[89] Xu ME, Yan YN, Liu HX, Yao R, Wang XH. Control adipose-derived stromal cells differentiation into adipose and endothelial cells in a 3-D structure established by cell-assembly technique. J Bioact Compat Polym 2009;24(S1):31-47.

[90] Xu ME, Wang XH, Yan YN, Yao R, Ge YK. An cell-assembly derived physiological 3D model of the metabolic syndrome, based on adipose-derived stromal cells and a gelatin/alginate/fibrinogen matrix. Biomaterials 2010;31(14):3868-3877.

[91] Yin DZ, Wang XH, Yan YN, Zhang RJ. Preliminary Studies on Peripheral Nerve Regeneration Using a New Polyurethane Conduit. J Bioact Compat Polym. 2007;22:143-159.

[92] Yan YN, Wang XH, Yin DZ, Zhang RJ. A New Polyurethane/Heparin Vascular Graft for Small-caliber Vein Repair. J Bioact Compat Polym. 2007;22:323-341.

[93] Pouton CW, Akhtar S. Biosynthetic polyhydroxyalkanoates and their potential in drug delivery. Advanced Drug Delivery Reviews. 1996;18(2):133-162.

[94] Kim SS, Utsunomiya H, Koski JA, Wu BM, Cima MJ, Sohn J, Mukai K, Griffith LG, Vacanti JP. Survival and function of hepatocytes on a novel three-dimensional synthetic biodegradable polymer scaffold with an intrinsic network of channels. Annals of Surgery 1998;228:8-13.

[95] Krupnick AS, Kreisel D, Engels FH, Szeto WY, Plappert T, Popma SH, Flake AW, Rosengard BR. A novel small animal of left ventricular tissue engineering. J Heart Lung Transpl 2002;21(2):233-243.

Chapter 18

Mesenchymal Stem Cells from Extra-Embryonic Tissues for Tissue Engineering – Regeneration of the Peripheral Nerve

Andrea Gärtner, Tiago Pereira, Raquel Gomes,
Ana Lúcia Luís, Miguel Lacueva França,
Stefano Geuna, Paulo Armada-da-Silva and
Ana Colette Maurício

Additional information is available at the end of the chapter

http://dx.doi.org/10.5772/53336

1. Introduction

Recent advances in Regenerative Biology and Regenerative Medicine are impressive and in the last years the scientific community has witnessed the emergence of many new concepts and discoveries. Until a few years ago, biological tissues were regarded as unable of extensive regeneration, but nowadays organs and tissues like the brain, spinal cord or cardiac muscles appear as capable to be reconstructed, based on "stem cells" [1].

Stem cell research has sparked an international effort due to the variety of possible uses in clinical procedures to treat diseases and improve health and life expectancy. Stem cell research has crossed a century journey and has evolved greatly even in its own definition. In 1967, Lajtha defined that adult stem cells could only be found in regenerative organs, such as blood, intestine, cartilage, bone and skin. Nowadays, these cells are considered to exist even in tissues with no commitment to regeneration such as the central nervous system [1, 2].

2. Stem cells

Stem cells are undifferentiated cells, with endless self-renewal sustained proliferation *in vitro* and multilineage differentiation capacity [3]. This *in vitro* multilineage differentiation capacity has targeted these cells with extreme importance for use in tissue and cell-based therapies.

The first stem cell appearance is in the early zygotic cells, which are totipotent and give rise to the blastocyst. They are capable to differentiate into all cell and tissue types. With differentiation, cells become less capable of self-renewal and differentiation in other cell type becomes more limited [1]. Stem cells can be loosely classified into 3 broad categories based on their growth behavior and isolation time during ontogenesis: embryonic, fetal and adult.

Embryonic stem cells (ESCs) were first observed in a pre-implantation embryo by Bongso and colleagues in 1994 [4]. Since then, many cell lines and a multiplicity of tissues have been successfully derived from ESCs and tested in several animal disease models [5-7]. Nevertheless, post-transplantation immune-rejection has been a major problem. Many studies are being conducted to avoid this major issue. This could be resolved by personalizing tissues through somatic nuclear transfer (NT) or induced pluripotent stem cells (iPSC) techniques [8], but the teratoma development in animals is still a concern and a serious problem [9]. In order to overcome the limitations placed by ESCs and iPSCs, a variety of adult stem cell populations have been recently isolated and characterized for their potential clinical use. While still multipotent, adult stem cells have long been considered restricted, giving rise only to progeny of their resident tissues [9]. *In vivo,* adult stem cells exist in a quiescent state, located in almost all tissues, until mediators activate them to restore and repair injured tissues. These cells are surrounded by mature cells that have reached the end line in terms of differentiation and proliferation [10]. Stem cell research focuses on the development of cell and tissue differentiation, so as characterization techniques, for tissue and cell identification with marker patterns. Such protocols are essential for regenerative therapies [11].

2.1. Mesenchymal stem cells

The development of cell-based therapies for cartilage [12] and skin [13] reconstruction marks the beginning of a new age in tissue regeneration. Mesenchymal stem cells (MSCs) have become one of the most interesting targets for tissue regeneration due to their high plasticity, proliferative and differentiation capacity together with their attractive immunosuppressive properties. MSCs present low immunogenicity and high immunosuppressive properties due to a decreased or even absence of Human Leucocyte Antigen (HLA) class II expression [14]. Research in this field has brought exciting promises in many disorders and therefore in tissue regeneration. Currently the differentiation potential of MSCs in multilineage end-stage cells is already proven, and their potential for treatment of cardiovascular [15], neurological [16], musculoskeletal [17, 18], and cutaneous [19] diseases is now well established. Fibroblast colony-forming units or marrow stromal cells, currently named MSCs, were first isolated in 1968 from rat bone marrow [20]. These cells were clonogenic, formed colonies when cultured, and were able to differentiate *in vitro* into bone, cartilage, adipose tissue, tendon, muscle and fibrous tissue. Since then many other tissues have been used to isolate these cells. MSCs can be obtained from many different tissues, including bone marrow, adipose tissue, skeletal muscle, umbilical cord matrix and blood, placental tissue, amniotic fluid, synovial membranes, dental pulp, fetal blood, liver, and lung [21]. The concept of MSCs is based on their ability to differentiate into a variety of mesodermal tissues and was first proposed by Caplan in 1991 [22] and further validated by additional research in 1999 [23]. Due to the many different methods and approaches used for MSCs culture, the Mesenchymal and Tissue Stem Cell Committee, of the International Society for Cellular Therapy (ISCT), recommended several standards do define MSCs [24]. Therefore, MSCs are defined as presenting: i) plastic adherent ability; ii) absence of definitive hematopoietic lineage markers, such as CD45, CD34, CD14, CD11b, CD79α, CD19 and class-II Major Histocompatibility Complex (MHC) molecules, specially HLA-DR; and expression of nonspecific markers CD105, CD90 and CD73 iii) ability to differentiate into mesodermal lineage cells, osteocytes, chondrocytes and adipocytes. Along with mesodermal differentiation, it has been demonstrated the capacity of MSCs to differentiate into ectodermal cell lines, as neurons [25, 26], keratocytes [27] and keratinocytes [28], so as endodermal cell line, like hepatocytes [29, 30] and pancreatic β-cells [31]. Moreover, they also possess anti-inflammatory and immunomodulation properties and trophic effects [32, 33]. Increasing evidence now demonstrates that the therapeutic effects of MSCs do not lay only on the ability to repair damage tissue, but also on the capacity of modulating surrounding environment, by secretion of multiple factors and activation of endogenous progenitor cells [34, 35]. Compared with ESCs and other tissue specific stem cells, MSCs are more advantageous. Moreover some studies have demonstrated that MSCs have a higher chromosomal stability and lower tendency to form tumors and teratomas, compared to other stem cells [36, 37].

Although they present similar biological characteristics, it cannot be ignored the existing of some disparities, as differences in, expansion potential under same culture conditions and age-related functional properties [38]. Compared to ESCs, MSCs isolated from the umbilical cord matrix (Wharton's jelly) have many advantages, such as shorter population doubling time, easy culture in plastic flasks, good tolerance towards the immune system, so that transplantation into non-immunesuppressed animals does not induce acute rejection, anticancer properties, [9] and most important absence of tumorigenic activity. As well as ESCs, these cells are originated from the inner cell mass of the blastocyst but with a major difference: they do not raise ethical controversies, since they are collected from tissues usually discarded at birth [39].

2.1.1. MSCs sources and validation of transport and processing protocols

Bone marrow, adipose tissue, umbilical cord blood and umbilical cord matrix have been considered the main sources of MSCs for tissue engineering purposes. Among these sources, bone marrow represents the main source of MSCs for cell therapy. However, the proliferative capacity [40-43], differentiation potential and clonal expandability [44] of MSCs derived from bone marrow decrease significantly with age, gender and seeding density, and the number of cells per marrow aspirate is usually quite low [3, 45]. It is still a mystery if MSCs ageing is due to factors intrinsic or extrinsic to the cells. Many possible reasons have been described in an attempt to explain MSCs ageing. Possible extrinsic factors include: reduced synthesis of proteoglycans and glycosaminoglycans reducing proliferation and viability [46], and production of glycosylated end products, inducing apoptosis and reactive oxygen species [47]. Intrinsic factors causing MSCs ageing might include: cell senescence-associated β-galactosidase and higher expression of p53 and pathway genes p21 and BAX, resulting in blunted proliferation potential [43]. Regarding seeding density, many authors suggest that lower seeding densities induce faster proliferation rates [48, 49]. This has been explained by contact inhibition in higher seeding densities [49], and higher nutrient availability per cell in lower seeding densities [49]. Use of bone marrow MSCs has disadvantages; donors are submitted to invasive harvest of bone marrow. This raises the need to find alternative sources of MSCs for autologous and allogenic use. Candidate tissue sources should provide MSCs displaying high proliferative and differentiation potency [50].

Extra-embryonic tissues are a good alternative to adult donor. This tissues, such as, amnion, microvillus, Wharton's jelly and umbilical cord perivascular cells, are routinely discarded at child-birth, so little ethical and religious controversy attends the harvesting of the resident stem cell populations. The comparatively large volume of extra-embryonic tissues increases the chance of isolating suitable amounts of stem cells, despite the complex and expensive procedures needed for their isolation. Some protocols use enzymatic digestion while others use enzyme-free tissue explant methods that require longer culture time [51]. There are also MSCs in cord blood (CB), but many studies report low frequency of these cells and unsuccessful isolation. However, Zhang and colleagues were able to isolate MSCs from CB with a 90% successful rate when CB volume was ≥ 90ml and a transport time until storage was ≤ 2 hours [51].

In recent years, MSCs derived from umbilical cord matrix Wharton's jelly, have attracted much interest. Wharton's jelly is a mature mucous tissue and the main component of the umbilical cord, connecting the umbilical vessels to the amniotic epithelium. Umbilical cord derives from extra-embryonic or embryonic mesoderm; at birth it weighs about 40g and measures approximately 30-65cm in length and 1.5cm in width [52]. Anyway, individual differences are observed within newborn babies. Fong and colleagues characterized Wharton's Jelly stem cells and found the presence of both embryonic and MSCs, targeting this source as unique and of valuable use for clinical applications. MSCs from the Wharton's jelly can be cultured with little or even no major loss trough at least 50 passages [53].

CB and more recently, umbilical cord tissue (UCT) have been stored cryopreserved in private and public cord blood and tissue banks worldwide in order to obtain hematopoietic and MSCs and, although guidelines exist (Netcord – Foundation for the Accreditation of Cellular Therapy), standardized procedures for CB and UCT transport from the hospital / clinic to the laboratory, storage, processing, cryopreservation and thawing are still awaited.

These may be critical in order to obtain higher viable stem cells number after thawing and limit microbiological contamination.

Our research group focused in determining whether UCT storage and transport from the hospital / clinical to the laboratory at room temperature (RT) or refrigerated (4-6°C) and immersed in several sterile saline solutions affects the UCT integrity in order to be cryopreserved. The umbilical cord contains two arteries and one vein, which are surrounded by mucoid connective tissue, and this is called the Wharton's jelly. The cord is covered by an epithelium derived from the enveloping amnion. The interlaced collagen fibers and small, woven bundles are arranged to form a continuous soft skeleton that encases the umbilical vessels. In the Wharton's jelly, the most abundant glycosaminoglycan is hyaluronic acid, which forms a hydrated gel around the fibroblasts and collagen fibrils and maintains the tissue architecture of the umbilical cord by protecting it from pressure [54].

One centimeter-long fragments of umbilical cords (N = 12) were collected from healthy donors after written informed consent and following validated procedures according to the clinical and technical guidelines of the Private Bank Biosckin, Molecular and Cell Therapies, SA (authorized for processing and cryopreserving CB and UCT units by the Portuguese Minister of Health, ASST – Autoridade para os Serviços de Sangue e de Transplantação). The 1 cm fragments were immersed for 168 hours in 4 different sterile saline solutions at RT (22-24°C) and refrigerated (4-6°C): NaCl 0.9% (Labesfal, Portugal), AOSEPT®-PLUS (Ciba Vision, Portugal), Dulbecco's Phosphate-Buffered Saline without calcium, magnesium and phenol red (DPBS, Gibco, Invitrogen, Portugal) and Hank's Balanced Salt Solution (HBSS, Gibco, Invitrogen, Portugal). The preservative-free, aqueous AOSEPT® PLUS solution contains hydrogen peroxide 3%, phosphonic acid (stabiliser), sodium chloride, phosphate (buffer system), and poloxamer (surfactant), and is usually used to transport and wash contact lenses. After 168 hours, the fragments were collected in 4% of paraformaldehyde and processed for light microscopy. The samples were fixed in 4% paraformaldehyde for 4 hours and then washed and conserved in phosphate buffer saline (PBS) until embedding. The specimens were dehydrated and embedded in paraffin and cut at 10 μm perpendicular to the main umbilical cord axis. For light microscope analysis, sections were stained with haematoxylin and eosin (HE) and observed with a Leica DM400 microscope equipped with a Leica DFC320 digital camera. The UCT integrity was evaluated through the following parameters:

i. detachment of vessels and retraction of vascular structures;

ii. loss of detail and integrity of the endothelium;

iii. connective tissue degradation;

iv. autolysis of fat (impossible to assess, due to histological technique); and

v. loss of detail and integrity of the mesothelium.

It was concluded that the best transport solutions were HBSS or DPBS at a temperature of 4-6°C since those maintained the histological structure of UC evaluated through those 5 parameters previously referred (Figure 1 and Figure 2).

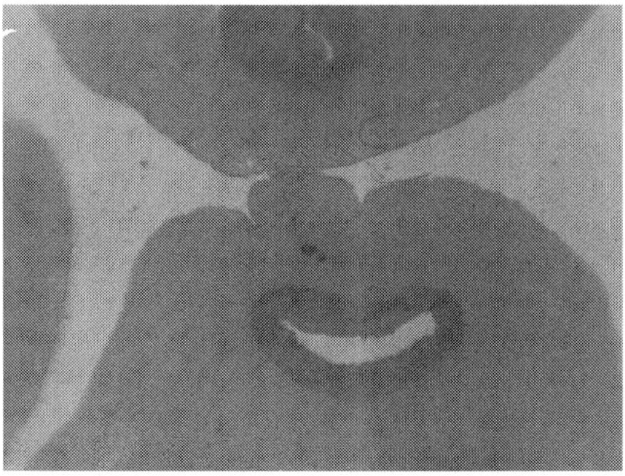

Figure 1. Cross section of an umbilical cord transported immersed in DPBS at the refrigerated temperature of 4-6°C. Samples were stained with haematoxylin and eosin (HE). Magnification: 10X.

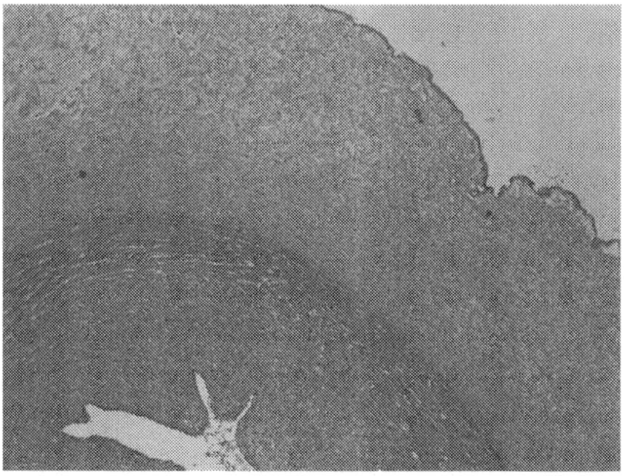

Figure 2. Cross section of an umbilical cord transported immersed in DPBS at the refrigerated temperature of 4-6°C. Samples were stained with haematoxylin and eosin (HE). The UCT integrity was quality evaluated through the following parameters: i) detachment of vessels and retraction of vascular structures; ii) loss of detail and integrity of the endothelium; iii) connective tissue degradation; iv) autolysis of fat (impossible to assess, due to histological technique); and v) loss of detail and integrity of the mesothelium. Magnification: 40X.

As a matter of fact, the UC immersed for 168 hours in DPBS and HBSS at refrigerated temperature presented integrity of the histological structure comparable to a UC collected and processed for histological analysis immediately after birth (Figure 3). With DPBS, a slight retraction of the vessels was noted, which is advantageous since the vessels are stripped and discarded before cryopreservation of the UCT. It was concluded that the transport of the UC from the hospital / clinic to the cryopreservation laboratory should be performed with the UC immersed in DPBS or HBSS at refrigerated temperatures.

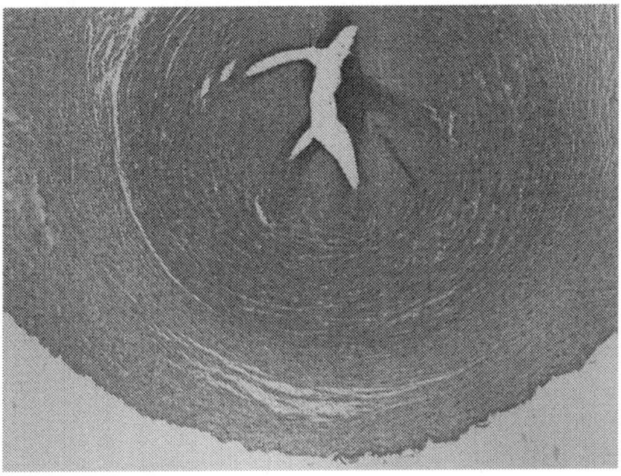

Figure 3. Cross section of an umbilical cord collected and processed for histological analysis immediately after birth under optimal conditions according to Netcord guidelines. Stained with haematoxylin and eosin (HE). Magnification: 40X.

The isolation and culture of MSCs from the Wharton's jelly was performed by our research group in order to obtain undifferentiated MSCs and *in vitro* differentiated into neural-like cells to be tested in axonotmesis and neurotmesis lesions of the rat sciatic nerve. The isolation has been performed by enzyme-free tissue explant and enzymatic isolation. Despite our standard approaches, we are aware that there are still significant variations that exist between laboratory protocols, which must be taken into account when comparing results using other methodologies. There is a wide range of individual differences among donor tissues also and our protocols usually use 15 - 20 cm of UC. While most UC samples will provide a reasonable number of MSCs using the provided protocols, some samples may result in sub-optimal cell isolation and expansion. The reasons behind this phenomenon still remain to be clarified, but as we have previously mentioned, the temperature and the time of transport from the hospital / clinic to the cryopreservation laboratory is crucial.

Irrespective of the specific protocol, the washing procedure of the umbilical cord fragments is crucial in order to avoid microbiological contamination of the cultures. After obtaining the written informed consent from the parents, fresh human umbilical cords are obtained after birth and collected in HBSS or DPBS at 4-6°C, as it was previously described. After washing the umbilical cord unit 4 times in rising DPBS, disinfection is performed in 75% ethanol for 30 seconds. Finally, and before the dissection step, umbilical cord unit is washed in DPBS. The vessels are usually stripped with UC unit still immersed in DPBS. Once washing step in MSCs isolation and culture is essential to achieve good UCT units for cryopreservation and future clinical use, washing protocol was validated. DPBS from the first washing step (used immediately after collection for transportation of the unit to the laboratory – *washing step 1 solution*) and

DPBS used in washing step after disinfection in 75% ethanol (*washing step 6 solution*) from 14 umbilical cord units (N = 14) collected from healthy donors and transported from the hospital/ clinic at 4-6°C in less than 96 hours were tested for microbiological contamination using BacT/ ALERT® (bioMérieux). Each unit was tested for aerobic and anaerobic microorganisms and fungi using 10 ml of the *washing step 1 solution* and *washing step 6 solution* which were aseptically introduced into the BacT/ALERT® testing flasks. All procedures were performed in a laminar flow tissue culture hood under sterile conditions. All the units that presented microbial contamination in DPBS obtained from the first washing step (*washing step 1 solution*) presented no contamination in the analysis performed to DPBS from the last washing step immediately performed before MSCs isolation or UCT cryopreservation (*washing step 6 solution*). The following microorganisms were identified in the DPBS solution from the first washing step: *Staphylococcus lugdunensis* (N = 2); *Staphylococcus epidermidis* (N = 1); *Staphylococcus coagulase* (N = 2); *Escherichia coli* (N = 4); *Enterococcus faecalis* (N = 1); and *Streptococcus sanguinis* (N = 1). The DPBS solution from the first washing step (*washing step 1 solution*) from 3 units was negative for microbial contamination (N = 3). These results permitted us to conclude that the washing protocol was 100% efficient in what concerns microbiological elimination (including aerobic and anaerobic bacteria, yeast and fungi).

Once the transport and washing protocols were validated, it was important to isolate and expand *in vitro* the MSCs from the UCT units for pre-clinical trials.

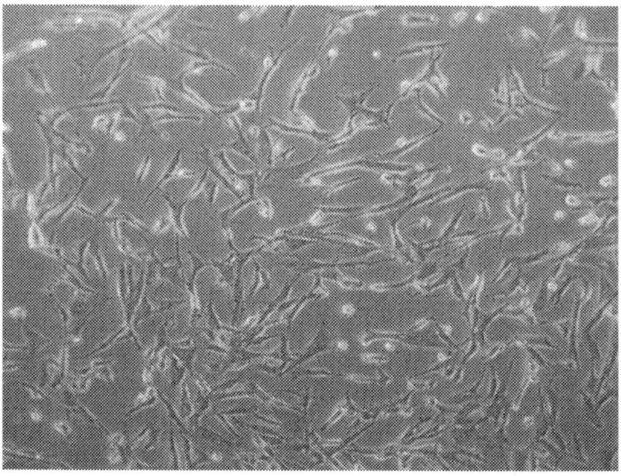

Figure 4. MSCs isolated from Wharton's jelly using the "enzymatic protocol" exhibiting a mesenchymal-like shape with a flat polygonal morphology. Magnification: 100x.

In the "enzymatic procedure" we use collagenase type I (Sigma-Aldrich). With the written informed consent from the parents, fresh human umbilical cords were obtained after birth and stored in HBSS (Gibco, Invitrogen, Portugal) for 1–48 hours before tissue processing to obtain MSCs. After removal of blood vessels, the mesenchymal tissue is scraped off from the Wharton's jelly with a scalpel and centrifuged at 250 g for 5 minutes at room temperature and the pellet is washed with serum-free Dulbecco's modified Eagle's medium (DMEM,

Gibco, Invitrogen, Portugal). Next, the cells are centrifuged at 250 g for 5 minutes at room temperature and then treated with collagenase (2 mg/ml) for 16 hours at 37°C,washed, and treated with 2.5% trypsin-EDTA solution (Sigma-Aldrich) for 30 minutes at 37°C with agitation. Finally, the cells are washed and cultured in DMEM (Gibco, Invitrogen, Portugal) supplemented with 10% fetal bovine serum (FBS), glucose (4.5 g/l), 1% (w/v) penicillin and streptomycin (Sigma), and 2.5 mg/ml amphotericin B (Sigma) in 5% CO_2 in a 37°C incubator (Nuaire). Around 2×10^5 cells are plated into each T75 flask in 10 ml culture medium. Cells are allowed to attach and grow for 3 days. To remove the non-adherent cells or fragments, the flasks are gently washed using pre-warmed DPBS after which 10 ml of pre-warmed culture medium is added. The culture medium is changed every third day (or twice per week). Confluence (80-90%) is normally reached at day 12–16, and the cells are removed with pre-warmed trypsin-EDTA solution (4 ml per flask), for 10 min at 37°C. The cells are plated onto poly-l-lysine coated glass coverslips (in 6- or 24-well tissue culture plates) or on biomaterials used in the nerve reconstruction. Normally, 5000 cells/cm^2 are plated on the coverslips or on the membranes (Figure 4).

In our "enzyme-free tissue explant protocol" for isolation of MSCs, enzymatic digestion is not employed. The mesenchymal tissue (Wharton's jelly) is diced into cubes of about 0.5 cm^3 and the remaining vessels are removed by dissection. Using a sterile scalp, the cubes are diced in 1-2 mm fragments and transferred to a Petri dish pre-coated with poly-l-lysine (Sigma) with Mesenchymal Stem Cell Medium (PromoCell, C-28010) supplemented with 1% (w/v) penicillin and streptomycin (Sigma), and 2.5 mg/ml amphotericin B (Sigma) and cultured in 5% CO_2 in a 37°C incubator (Nuaire). Some tissue fragments will allow cell migration from the explants in 3-4 days incubation. Confluence is normally obtained 15-21 days after.

The laboratory's processing and cryopreservation protocols of the UCT units following the technical procedures of Biosckin, Molecular and Cell Therapies S.A. (BSK.LCV.PT.7) were validated for the ability of isolating and expanding *in vitro* MSCs after cryopreserved UCT thawing. The protocols of processing and cryopreservation of the UCT are protected by a Confidentiality Agreement between Biosckin, Molecular and Cell Therapies S.A. and all the involved researchers. Briefly, the UCT collected from healthy donors (N = 60), and according to Netcord guidelines and following the Portuguese law 12/2009 (*Diário da República, lei 12/2009 de 26 de Março de 2009*) is diced into cubes of about 0.5 cm^3 and the remaining vessels are removed by dissection. In order to ensure the viability of the UCT after parturition and limit the microbiological contamination of the samples, the umbilical cords were transported from the hospital / clinic to the laboratory at refrigerated temperatures monitored by a datalloger in less than 72 hours. The UCT units from 15-20 centimeters-long umbilical cords and after the blood vessels dissection are treated and processed for cryopreservation using a cryoprotective solution (freezing medium). The UCT units are transferred to a computer-controlled slow rate freezer (Sylab, Consensus, Portugal) and a nine-step freezing program is used to set up the time, temperature, and rates specifically optimized for the human umbilical cord-MSCs cooling. To thaw frozen cells, the cryovials are transferred directly to a 37°C water bath. Upon thawing in less than a minute, the cell suspension is centrifuged at 150 × g for 10 min, and the supernatant is gently removed and the cell pellet is resuspended in culture medium. It was possible to obtain MSCs in culture from 52 out of 60 thawed UCT units.

In some UCT cryopreserved units (N = 8) it was not possible to isolate MSCs due to increase number of erythrocytes' lysis or microbiological contamination during cell culture. The MSCs morphology was observed in an inverted microscope (Zeiss, Germany) at different points of expansion. The MSCs exhibited a mesenchymal-like shape with a flat and polygonal morphology. The MSCs obtained were characterized by flow cytometry (FACSCalibur®, BD Biosciences) analysis for a comprehensive panel of markers, such as PECAM (CD31), HCAM (CD44), CD45, and Endoglin (CD105). In the presence of neurogenic medium, the MSCs were able to, became exceedingly long and there was a formation of typical neuroglial-like cells with multi-branches and secondary branches. These results permitted to conclude that the processing and cooling protocols used for UCT units' cryopreservation were adequate to preserve the UCT viability since it was possible to isolate and expand MSCs after appropriate thaw and in presence of adequate cell culture conditions.

An established and ready-to-use Human MSC cell line was also employed for promoting axonotmesis and neurotmesis lesions regeneration. Human MSCs from Wharton's jelly umbilical cord were purchased from PromoCell GmbH (C-12971, lot-number: 8082606.7). Cryopreservated cells are cultured and maintained in a humidified atmosphere with 5% CO_2 at 37°C. Mesenchymal Stem Cell Medium (PromoCell, C-28010) is replaced every 48 hours. At 80-90% confluence, cells are harvested with 0.25% trypsin with EDTA (Gibco) and passed into a new flask for further expansion. MSCs at a concentration of 2500 cells/ml are cultured on poli-D-lysine coverslips (Sigma) or on biomaterials membranes and after 24 hours cells exhibit 30-40% confluence. Differentiation into neuroglial-like cells is induced with MSC neurogenic medium (Promocell, C-28015). Medium is normally replaced every 24 hours during 3 days. The formation of neuroglial-like cells can be observed after 24 hours in an inverted microscope (Zeiss, Germany) (Figure 5 and Figure 6).

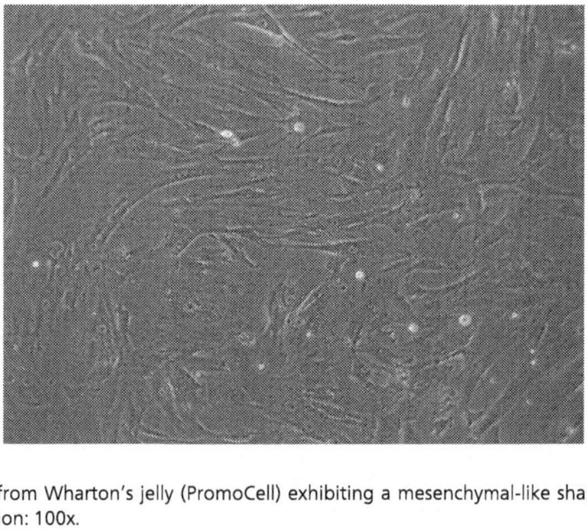

Figure 5. MSC cell line from Wharton's jelly (PromoCell) exhibiting a mesenchymal-like shape with a flat polygonal morphology. Magnification: 100x.

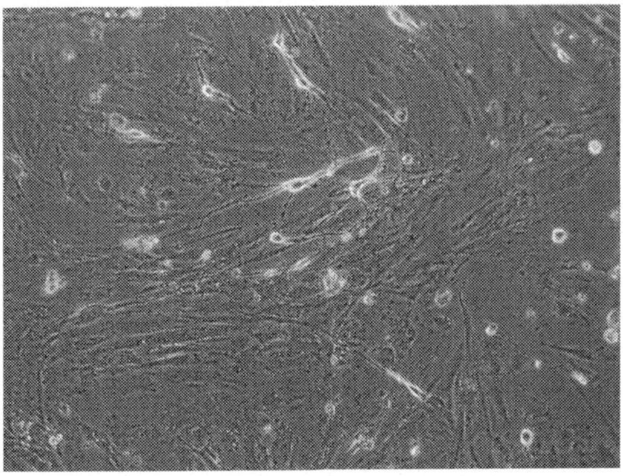

Figure 6. MSC cell line from Wharton's jelly (PromoCell) after 72h of incubation in neurogenic medium. The cells became exceedingly long and there is a formation of typical neuroglial-like cells with multibranches. Magnification: 100x.

This established human MSC cell line is preferred for *in vivo* testing in rats, since the number of MSCs obtained is higher in a shorter culture time, it is not dependent on donors availability and ethic committee authorization, and the protocol is much less time consuming which is advantageous for pre-clinical trials with a large number of experimental animals. As a matter of fact, there is no need of administrating immunosuppressive treatment to the experimental animals during the entire healing period after the surgical procedure. The phenotype of MSCs was assessed by PromoCell. Rigid quality control tests are performed for each lot of PromoCell MSCs isolated from Wharton's jelly of umbilical cord. MSCs are tested for cell morphology, adherence rate and viability. Furthermore, each cell lot is characterized by flow cytometry analysis for a comprehensive panel of markers.

The MSCs isolated with the two protocols described (from fresh and the cryopreserved UCT units) and from the established Promocell cell line exhibited a mesenchymal-like shape with a flat and polygonal morphology. During expansion the cells became long spindle-shaped and colonized the whole culturing surface. After 96 hours of culture in neurogenic medium, cells changed in morphology. The cells became exceedingly long and there was a formation of typical neuroglial-like cells with multi-branches and secondary branches. Giemsa-stained cells of differentiated MSC cell line at passage 5 were analyzed for cytogenetic characterization. However, no metaphases were found, therefore the karyotype could not be established. The karyotype of undifferentiated HMSCs was determined previously and no structural alterations were found demonstrating absence of neoplastic characteristics in these cells, as well as chromosomal stability to the cell culture procedures [55, 56]. The differentiated MSCs karyotype could not be established, since no dividing cells were obtained at passage 5, which can be in agreement with the degree of differentiation. The karyotype analysis of undifferentiated MSCs previously determined, excluded the presence of neoplastic cells, thus supporting the suitability of our cell culture and differentiation procedures. This concern also resulted from our previous experience with N1E-115 neoplastic cell line and the negative results we obtained in the treatment of axonotmesis and neurotmesis injuries [57-59]. Nevertheless, undifferentiated MSCs from the Wharton's jelly culture (obtained from either protocol or from the Promocell cell line) showed normal morphology when inspected with an inverted microscope (Figure 7).

The differentiation was tested based on the expression of typical neuronal markers such as GFAP, GAP-43 and NeuN by neural-like cells attained from MSCs. Undifferentiated MSCs were negatively labeled to GFAP, GAP-43 and NeuN. After 96 hours of differentiation the attained cells were positively stained for glial protein GFAP and for the growth-associated protein GAP-43. All nucleus of neural-like cells were also labeled with the neuron specific nuclear protein called NeuN showing that differentiation of MSCs in neural-like cells was successfully achieved for MSCs obtained from UCT (fresh and cryopreserved) and for the Promocell MSC cell line (Figure 8) [55].

2.1.2. *Differentiation into neuroglial-like cells*

MSCs express nestin, a maker for neural and other stem cells [60, 61] and can be differentiated in adipose tissue, bone, cartilage, skeletal muscle cells, cardiomyocyte-like cells, and neuroglial-like cells [54, 55, 60, 62], presenting great potential to biomedical engineering applications. These cells fit into the category of primitive stromal cells and because they are abundant and inexpensive, they might be very useful for regenerative medicine and biotechnology applications.

By employing neuron-conditioned media, sonic hedgehog and fibroblast growth factor 8, MSCs isolated from the Wharton's jelly can be induced toward dopaminergic neurons. These cells have been transplanted into hemiparkinsonian rats where they prevented the progressive degeneration/behavioral deterioration seen in these rats [63]. Rat MSCs isolated from the Wharton's jelly when transplanted into brains of rats with global cerebral ischemia significantly reduced neuronal loss, apparently due to a rescue phenomenon [64]. Neuronal differentiation of human MSCs could also provide cells to replace neurons lost due to neurodegenerative diseases. Recent studies showed that transplanted MSCs-derived neurons become electrophysiologically integrated within the host neural tissue [65]. However, all these therapeutic applications need uniform and reproducible regulation.

A consequence of cell metabolism during *in vitro* expansion is that culture conditions are constantly changing. The comprehension and optimization of the expansion and differentiation process will contribute to maximization of cell yield, reduced need of cell culture, and a decrease in total processing costs [66, 67]. Elucidation of regulatory mechanisms of MSCs differentiation will allow optimization of *in vitro* culture and their clinical use in the treatment of neural-related diseases. Research is being performed to optimize expansion process parameters in order to grow MSCs in a controlled, reproducible, and cost-effective way [68]. Metabolism is certainly one of these parameters.

http://dx.doi.org/10.5772/53336

3. Regeneration and in vivo testing

With the world wide global increase in life expectancy, a variety of disabling diseases with large impact on human population are arising. This includes cardiovascular, neurological, musculoskeletal, and malignancies. Therefore, it is imperative that new and more effective treatment methods are developed to correct for these changes. Further research with experimental animal systems is required to translate to in vivo cell-based therapy that has been extensively investigated in vitro [1]. Stem cell biology is probably the golden key for cell therapies and regenerative medicine. Regeneration is the physical process where remaining tissues organize themselves to replace missing or injured tissues *in vivo* [39].

It has been speculated that once MSCs have the potential to differentiate into several tissues, they might be responsible for turnover and maintenance of adult tissues, just like hematopoietic stem cells have this role in blood cells [69]. First, it was believed that after injection of MSCs, these were able to migrate to the damaged site and to differentiate into ones with the appropriate function for repairing, so MSCs could mediate tissue repair through there multilineage capacity replacing damaged cells. Subsequent studies have suggested that the mechanism used by MSCs for tissue repairing is not really this way. This new idea was reinforced by the confirmation that this cells homed to damaged site, particularly to spots of hypoxia, inflammation and apoptosis [70, 71].

Recent studies demonstrated that transplanted MSCs modified the surrounding tissue microenvironment, promoting repair with functional improvement by secretion factors (known as paracrine effect), stimulation of preexisting stem cells in the original tissue and decreasing of inflammation and immune response [72]. Other studies have demonstrated that MSC-conditioned media by itself could have therapeutic effects. All this data suggest that MSC apply a reparative effect on injured side through its paracrine effects [73].

It is necessary to overcome some barriers before a cell-based therapy becomes routine in clinics, including the cell number and the administration way of treatment. MSCs are difficult to be maintained stable in culture for long time, but due to their short doubling time, if at the outset many cells are harvested they may be properly scaled up in primary culture, never forgetting the ideal seeding number [39].

MSCs are an attractive candidate for cell-based regenerative therapy; the evidence is that currently there are 139 trial registries for MSC therapy 27 of which are based on umbilical cord MSCs [74].

3.1. Nerve regeneration

After Central Nervous System (CNS) lesions, Peripheral Nervous System (PNS) injuries are the ones with minor successes in terms of functional recovery. These kinds of injuries are frequent in clinical practice. About two centuries ago it was assumed that these nerves would never regenerate. Indeed, scientific and clinical knowledge greatly increased in this area. Nevertheless, a full understanding of axonal recovery and treatment of nerve defects, especially complete functional achievement and organ reinnervation after nerve injury, still remains the principle challenge of regenerative biology and medicine [75, 76].

3.1.1. *Nerve repair*

Many peripheral nerve injuries can only be dealt through reconstructive surgical procedures. Despite continuous refinement of microsurgery techniques, peripheral nerve repair still stands as one of the most challenging tasks in neurosurgery, as functional recovery is rarely satisfactory in these patients [76]. Direct repair should be the procedure of choice whenever tension-free suturing is possible; however, patients with loss of nerve tissue, resulting in a nerve gap, are considered for a nerve graft procedure. In these cases, the donor nerves used for grafting are commonly expendable sensory nerves. This technique, however, has some disadvantages, with the most prominent being donor site morbidity, that may lead to a secondary sensory deficit and occasionally neuroma and pain. In addition, no donor and recipient nerve diameters often occurs which might be the basis for poor functional recovery. Alternatives to peripheral nerve grafts include cadaver nerve segments allografts, end-to-side neurorrhaphy, and entubulation by means of autologous non-nervous tissues, such as vein and muscles [76]. One advantage of these allografts compared with the autografts is the absence of donor site morbidity and theoretically the unlimited length of tissue available [77]. Experimental work from a number of laboratories has emphasized the importance of entubulation for peripheral nerve repair to manage nerve defects that cannot be bridged without tension (neurotmesis with loss of nerve tissue). Nerves will regenerate from the proximal nerve stump towards the distal one, whereas neuroma formation and ingrowth of fibrous tissue into the nerve gap are prevented [78]. The reliability of animal models is crucial for PN research, including therapeutic strategies using biomaterials and cellular systems. As a matter of fact, rodents, particularly the rat and the mouse, have become the most frequently used animal models for the study of peripheral nerve regeneration because of the widespread availability of these animals as well as the distribution of their nerve trunks which is similar to humans [79]. Because of its PN size, the rat sciatic nerve has been the most commonly experimental model used in studies concerning the PN regeneration and possible therapeutic approaches [80]. Functional recovery after PN injury is frequently incomplete, even with adequate microsurgery, so, many research and clinical studies have been performed including biomaterials for tube-guides. Since the 80's, Food and Drug Administration (FDA) has approved a variety of these biomaterials both natural and synthetic. The ideal biomaterial nerve graft should increase number, length and speed of axon regeneration [77]. It should be:

i. biocompatible, not toxic neither present undesired immunologic response;

ii. permeable enough to permit nutrient and oxygen diffusion and allows cell support systems;

iii. flexible and soft to avoid compression;

iv. biodegradable, the ideal rate is to remain intact during axon regeneration across nerve gap and after degrade softly and

v. technically reproducible, transparent, easy to manipulate, and sterilize [81].

Currently 3 types of materials are available for nerve reconstruction: non-resorbable, natural resorbable and synthetic resorbable. Polyvinil alcohol hydrogel (PVA) is an example of a non-absorbable biomaterial. It combines water in similar proportions to human tissue, with PVA providing a stable structure easy to sterilize, which is a main advantage of this materials, but has some limitations such as: nerve compression and suture tension after regeneration due to its non-resorbable nature [77]. Collagen type I from humans or animals, is an example of a natural resorbable device, which has some advantages such as:

i. easy to isolate and purify,

ii. good adhesiveness for supporting cell survival an proliferation,

iii. has been proven to be highly biocompatible and support nerve regeneration *in vivo*.

On the other hand, offers some immune response requiring the use of immunosuppressive drugs or pre-treatment of the material before clinical use [77]. Poly (DL-lactide-ε-caprolactone) (PLC) a synthetic resorbable material is the only transparent device approved by FDA, important characteristic for the surgeon that facilitates the insertion of the nerve stumps across the nerve gap, but, on the other hand it is not flexible [77]. Chitosan, PLC, collagen, poly(L-lactide) and poly(glycolide) copolymers (PLGA) and others, some of them, previously studied by our group [57, 58, 82] were associated to cellular systems, which are able to differentiate into neuroglial-like cells or capable of modulating the inflammatory process, improved nerve regeneration, in terms of motor and sensory recovery, and also shortening the healing period after axonotmesis and neurotmesis, avoiding regional muscular atrophy [57, 58, 82].

Researches with acellular nerve allografts, as alternative for repairing peripheral nerve defects have been reported. These nerve allografts remove the immunoreactive SCs and myelin however preserve the internal structure of original nerve, containing vital components such as collagen I, laminin and growth factors essential for repairmen of the lesions [83]. Acellular grafts remain insufficient, due to the increasing extent of nerve damages. Also, viable cells are necessary for debris removal and environmental regeneration reestablishment [83].

Cell transplantation, such as Schwann cells (SCs) transplantation has been proposed as a method of improving peripheral nerve regeneration [84]. SCs are peripheral glial cells that enwrap axons to form myelin with a central role in neuronal function. When there is damage in PNS, SCs are induced to mislay myelin, proliferate and segregate numerous factors, including cytokines responsible for reproducing a microenvironment suitable for supporting axon regeneration [85, 86]. They also have a vital participation in endogenous repair, reconstructing myelin, which are essential for functional recovery [85, 86]. SCs, MSCs, ESCs, marrow stromal cells are the most studied support cells candidates. SCs transplantation enhance axon outgrowth both *in vitro* [87] and *in vivo* [88]. Although to achieve an adequate amount of autologous SC, a donor nerve is necessary and a minimum of 4-8 weeks for *in vitro* expansion. Umbilical cord MSCs may be the perfect cell model as supplement for nerve grafts, once they are easily obtained, with no ethical controversy and can differentiate into neuglial-like cells [83]. Matuse and collaborators induced MSCs from the umbilical cord into SCs capable of supporting peripheral nerve regeneration and myelin reconstruction *in vivo*. They transplanted these SCs into injured sciatic nerve, and proved that these cells maintained their differentiated phenotype *in vivo*, and contributed for axonal regeneration and functional recovery [89].

In our studies we aimed to explore the therapeutic value of human umbilical cord matrix (Wharton's jelly) derived MSCs, undifferentiated and differentiated in neuroglial-like cells, both *in vitro* and *in vivo*, associated to a variety of biomaterials such as, Poly (DL-lactide-ε-caprolactone) PLC (Vivosorb®) membrane, and Chitosan type III on rat sciatic nerve axonotmesis and neurotmesis experimental model. For cell transplantation into injured nerves (with axonotmesis and neurotmesis injuries), there are two main techniques. The cellular system may be directly inoculated into the neural scaffold which has been interposed between the proximal and distal nerve stumps or around the crush injury (in neurotmesis and axonotmesis injuries, respectively); or the cells can be pre-added to the neural scaffold via inoculation or co-culture (in most of the cellular systems, it is allowed to form a monolayer) and then the biomaterial with the cellular system is implanted in the injured nerve [82].

Our PLC studies [55] demonstrated that this biomaterial does not interfere negatively with the nerve regeneration process, in fact, the information on the effectiveness of PLC membranes and tube-guides for allowing nerve regeneration was already provided experimentally and with patients [82]. PLC becomes hydrophilic by water uptake, which increases the permeability of the polymer. This is essential for the control of nutrient and other metabolite transportation to the surrounding healing tissue. A few weeks after implantation, the mechanical power gradually decreases and there is a loss of molecular weight as a result of the hydrolysis process. Nearly in 24 months, PLC degrades into lactic acid and hydroxycaproic acid which are both safely metabolized into water and carbon dioxide and/or excreted through the urinary tract. In contrast to other biodegradable polymers, PCL has the advantage of not creating an acidic and potentially disturbing micro-environment, which is favorable to the surrounding tissue [90]. Chitosan has attracted particular attention in medical areas due to its biocompatibility, biodegradability, and low toxicity, low cost, improvement of wound-healing and antibacterial properties. Moreover, the potential use of chitosan in nerve regeneration has been demonstrated both *in vitro* and *in vivo* [57, 91]. Chitosan is a partially deacetylated polymer of acetyl glucosamine obtained after the alkaline deacetylation of chitin [57, 82]. While chitosan matrices have low mechanical strength under physiological conditions and are unable to maintain a predefined shape after transplantation, their mechanical properties can be improved by modification with a silane agent, namely γ-glycidoxypropyltrimethoxysilane (GPTMS), one of the silane-coupling agents which has epoxy and methoxysilane groups. The epoxy group reacts with the amino groups of chitosan molecules, while the methoxysilane groups are hydrolyzed and form silanol groups. Finally, the silanol groups are subjected to the construction of a siloxane network due to the condensation. Thus, the mechanical strength of chitosan can be improved by the cross-linking between chitosan, GPTMS and siloxane network. By adding GPTMS and employing a freeze-drying technique, we have previously obtained chitosan type III membranes (hybrid chitosan membranes) with pores of about 110 μm diameter and about 90% of porosity, and which were successful in improving sciatic nerve regeneration after axonotmesis and neurotmesis [56, 57, 82].

The induction of a crush injury in rat sciatic nerve provides a very realistic and useful model of damage for the study of the role of numerous factors in regenerative processes [57]. Focal crush causes axonal interruption but preserves the connective sheaths (axonotmesis). After axonotmesis injury regeneration is usually successful, after a short (1-2 day) latency, axons regenerate at a steady rate towards the distal nerve stump, supported by the reactive SCs and the preserved endoneural tubules enhance axonal elongation and facilitate adequate reinnervation [92]. Our research group has been testing the efficacy of combining biomaterials and cellular systems in the treatment of sciatic nerve crush injury [57-59, 82, 90, 91, 93-95]. Following transection, axons show staggered regeneration and may take substantial time to actually cross the injury site and enter the distal nerve stump [60]. Although delayed axonal elongation might be caused by growth inhibition originating from the distal nerve itself, growth-stimulating influences may overcome axons stagger. More robust and fast nerve regeneration is expected to result in better reinnervation and functional recovery. As a potential source of growth promoting signals, MSCs transplantation is expected to have a positive outcome. Our results showed that the use of either undifferentiated or differentiated HMSCs enhanced the recovery of sensory and motor function in axonotmesis lesion of the rat sciatic nerve [56]. Neurotmesis must be surgically treated by direct end-to-end suture of the two nerve stumps or by a nerve graft harvested from elsewhere in the body in case of tissue loss. To avoid secondary damage due to harvesting of the nerve graft, a tube-guide can be used to bridge the nerve gap. Acutely after sciatic nerve transection there is a complete loss of both motor and thermal sensory function. Sensory and motor deficit then progressively decrease along the post-operative. From a morphological point of view, nerve regeneration occurs if Wallerian degeneration is efficient and is substituted by re-growing axons and the accompanying viable SCs [96, 97]. The axon regeneration pattern is improved by using appropriate biomaterials for the tube-guide design, like chitosan type III and PLC and cellular systems like MSCs from the Wharton jelly [57, 90, 91, 95]. The surgical technique and the time for the reconstructive surgery is also crucial for the nerve regeneration after neurotmesis [57, 90, 91, 95].

3.2. Assessment of nerve regeneration in the sciatic nerve rat model

Although both morphological and functional data have been used to assess neural regeneration after induced crush injuries, the correlation between these two types of assessment is usually poor [94, 98-100]. Classical and newly developed methods of assessing nerve recovery, including histomorphometry, retrograde transport of horseradish peroxidase and retrograde fluorescent labeling [79] do not necessarily predict the reestablishment of motor and sensory functions [100-103]. Although such techniques are useful in studying the nerve regeneration process, they generally fail in assessing functional recovery [100]. In this sense, research on peripheral nerve injury needs to combine both functional and morphological assessment. The use of biomechanical techniques and rat's gait kinematic evaluation is a progress in documenting functional recovery [104]. Indeed, the use of biomechanical parameters has given valuable insight into the effects of the sciatic denervation/reinnervation, and thus represents an integration of the neural control acting on the ankle and foot muscles, which is very useful and accurate to evaluate different therapeutic approaches [103-105].

3.2.1. Functional Assessment

After injury and treatment of animals, follow-up results are very important for analysis of functional recovery. Animals are tested preoperatively (week 0), and every week during 12 and 20 weeks, for axonotmesis and neurotmesis of the rat sciatic nerve, respectively. Motor performance and nociceptive function are evaluated by measuring extensor postural thrust (EPT) and withdrawal reflex latency (WRL), respectively [55, 58, 94]. For EPT test, the affected and normal limbs are tested 3 times, with an interval of 2 minutes between consecutive tests, and the 3 values are averaged to obtain a final result. The normal (unaffected limb) EPT (NEPT) and experimental EPT (EEPT) values are incorporated into an equation (Equation (1)) to derive the percentage of functional deficit, as described in the literature [106]:

$$\%\ \text{Motor deficit} = \left[(\text{NEPT} - \text{EEPT}) \ / \ \text{NEPT} \right] \times 100 \tag{1}$$

The nociceptive withdrawal reflex (WRL) was adapted from the hotplate test developed by Masters et al. [107]. Normal rats withdraw their paws from the hotplate within 4s or less. The cutoff time for heat stimulation is set at 12 seconds to avoid skin damage to the foot.

For Sciatic Functional Index (SFI), animals are tested in a confined walkway that they cross, measuring 42 cm long and 8.2 cm wide, with a dark shelter at the end. Several measurements are taken from the footprints:

i. distance from the heel to the third toe, the print length (PL);

ii. distance from the first to the fifth toe, the toe spread (TS); and

iii. distance from the second to the fourth toe, the intermediary toe spread (ITS).

In the static evaluation (SSI) only the parameters TS and ITS, are measured. For SFI and SSI, all measurements are taken from the experimental (E) and normal (N) sides. Prints for measurements are chosen at the time of walking based on precise, clear and completeness of footprints. The mean distances of three measurements are used to calculate the following factors (dynamic and static):

$$\text{Toe spread factor (TSF)} = (\text{ETS} - \text{NTS}) \ / \ \text{NTS} \tag{2}$$

$$\text{Intermediate toe spread factor (ITSF)} = (\text{EITS} - \text{NITS}) \ / \ \text{NITS} \tag{3}$$

$$\text{Print length factor}\ (\text{PLF}) = (\text{EPL} - \text{NPL}) \ / \ \text{NPL} \tag{4}$$

SFI is calculated as described by Bain et al. [108] according to the following equation:

$$\begin{aligned}\text{SFI} &= -38.3(\text{EPL} - \text{NPL}) \ / \ \text{NPL} + 109.5(\text{ETS} - \text{NTS}) \ / \ \text{NTS} + 13.3(\text{EIT} - \text{NIT}) \ / \ \text{NIT} - 8.8 \\ &= (-38.3 \times \text{PLF}) + (109.5 \times \text{TSF}) + (13.3 \times \text{ITSF}) - 8.8\end{aligned} \tag{5}$$

For SFI and SSI, an index score of 0 is considered normal and an index of -100 indicates total impairment. When no footprints are measurable, the index score of -100 is given [109]. In each walking track 3 footprints are analyzed by a single observer, and the average of the measurements is used in SFI calculations.

Ankle kinematics analysis is carried out prior nerve injury, at week-2 and every 4 weeks during the 12 or the 20-week follow-up time, for axonotmesis and neurotmesis lesions, respectively. The motion capture is performed with 2 digital high speed cameras (Oqus, Qualysis®) at a rate of 200 images per second, and Qualisys Track Manager software (QTM, Qualysis®). The cameras operate on a infra-red light frequency ensuring a high level of accuracy on the determination of reflective marker position and a position residual of less than 2.7 mm was obtained. Cameras are usually positioned to not recorder significant signal deflection during the test and four reflective markers were placed at the skin of the rat right hindlimb at the proximal edge of the tibia, the lateral malleolus and the fifth metatarsal head. Advanced analysis of the 2-D movement (sagittal plan) data is performed with Visual3D software (C-Motion®, Inc). The rats' ankle angle is determined using the scalar product between a vector representing the foot and a vector representing the lower leg. With this model, positive and negative values of position of the ankle joint (θ°) indicate dorsiflexion and plantarflexion, respectively. For each step cycle the following time points are identified: midswing, midstance, initial contact (IC) and toe-off (TO) [104, 109-113] and are time normalized for 100% of step cycle. The normalized temporal parameters are averaged over all recorded trials. Angular velocity of the ankle joint (Ω °/s) is also determined where negative values correspond to dorsiflexion. A total of 6 walking trials for each animal with stance phases lasting between 150 and 400 ms are considered for analysis, since this corresponds to the normal walking velocity of the rat (20–60 cm/s) [104]. Animals walk on a Perspex track with length, width and height of respectively 120, 12, and 15 cm. In order to ensure locomotion in a straight direction, the width of the apparatus is adjusted to the size of the rats during the experiments.

3.2.2. *Morphologic Assessment*

Nerve samples are processed for quantitative morphometry of myelinated nerve fibers [114]. Fixation is usually carried out using 2.5% purified glutaraldehyde and 0.5% saccarose in 0.1M Sorensen phosphate buffer for 6-8 hours and resin embedding is obtained following Glauerts' procedure (Scipio et al., 2008). Series of 2-μm thick semi-thin transverse sections are cut using a Leica Ultracut UCT ultramicrotome (Leica Microsystems, Wetzlar, Germany) and stained by Toluidine blue. Stereology is carried out on a DM4000B microscope equipped with a DFC320 digital camera and an IM50 image manager system (Leica Microsystems, Wetzlar, Germany). Systematic random sampling and D-disector is always adopted using a protocol previously described [115, 116]. Fiber density and total number of myelinated fibers is estimated together with fiber and axon diameter and myelin thickness.

3.3. Results

3.3.1. Differentiation and metabolism of MSCs from Wharton's jelly

In our experimental studies we expanded undifferentiated MSCs from human umbilical cord Wharton's jelly that exhibited a normal star-like shape with a flat morphology in culture (Figures 4 and 5). To prevent the possibility of eventual mutations due to expansion artifacts, a total of 20 Giemsa-stained metaphases of these cells, were analyzed for numerical aberrations. Sporadic, non-clonal aneuploidy was found in 3 cells (41-45 chromosomes). The other 17 metaphases had 46 chromosomes (Figure 7). The karyotype was determined in a completely analyzed G-banding metaphase. No structural alterations were found. The karyotype analysis to the MSCs cell line derived from Human Wharton jelly demonstrated that this cell line has not neoplastic characteristics and is stable during the cell culture procedures in terms of number and structure of the somatic and sexual chromosomes [55].

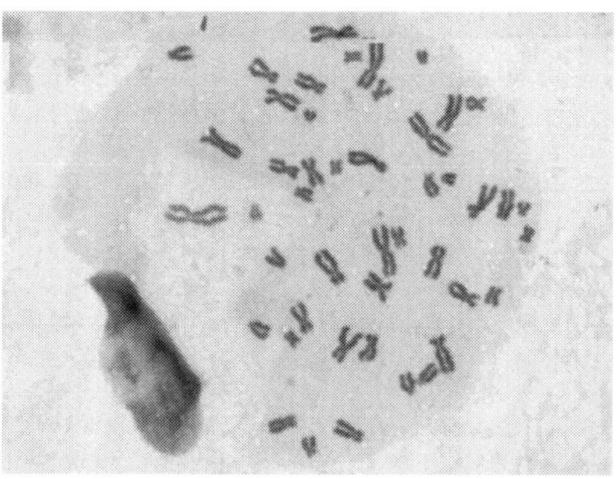

Figure 7. Selected metaphases from undifferentiated MSC cells isolated from Wharton's jelly, showing the normal number of chromosomes (46, XY). Magnification: 1000X.

We differentiated MSC from Wharton's Jelly into neuroglial-like cells. After 96 hours of incubation in neurogenic medium, we observed a morphological change. The cells became exceedingly long and there was a formation of typical neural-like cells with multi-branches and secondary branches (Figure 6). The differentiation was tested based on the expression of typical neuronal markers such as GFAP, GAP-43 and NeuN by neural-like cells attained from HwMSCs. Undifferentiated MSCs were negatively labeled to GFAP, GAP-43 and NeuN (Figure 8A,C,E). After 96 hours of differentiation the attained cells were positively stained for glial protein GFAP (Figure 8B) and for the growth-associated protein GAP-43 (Figure 8D). All nucleus of neural-like cells were also labeled with the neuron specific nuclear protein called NeuN (Figure 8F) showing that differentiation of MSCs in neural-like cells were successfully achieved [55].

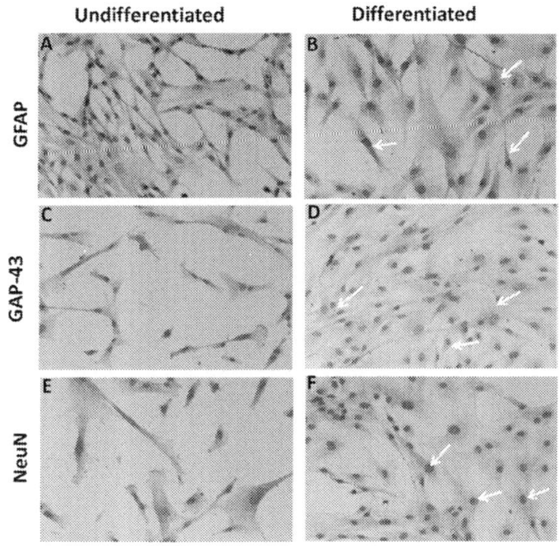

Figure 8. Undifferentiated MSC cells from the Wharton's jelly presenting a negative staining for: (A) GFAP which is a glial cell marker; (C) GAP-43 which is related with axonal outgrowth and (E) NeuN which is a marker for nucleus of neurons. Neuroglial-like cells obtained from HMSCs in vitro differentiated with neurogenic medium exhibiting a positive staining for: (B) GFAP; (D) GAP-43 and (F) NeuN. Magnification: 200x [55].

The *in vitro* expansion and differentiation of MSCs for clinical cell-based therapy is a very expensive and long process that needs standardization. Although pre-clinical and clinical data demonstrated the safety and effectiveness of MSCs therapy in some pathologies such as neurological, there are still questions surrounding the mechanism of action. In our research work we aimed to disclose the possible role of metabolism not only in the MSCs maintenance and expansion but also during the differentiation in neural-like cells [55]. MSCs maintenance and differentiation, to neural-like cells, depends on metabolic modulation. *In vitro,* glucose is the most widely used substrate for the generation ATP which is essential for cell growth and maintenance. It has been proposed that cells undergoing high proliferation rates depend on glycolysis to generate ATP, known as Warburg effect, although this pathway is less effective than the oxidative phosphorylation in terms of ATP production [117]. Our results showed that during expansion, the undifferentiated MSCs consume glucose and produce high concentration of lactate as a metabolic sub product which is consistent with the

Warburg effect and glycolysis stimulation. MSCs do not require oxidative phosphorylation to survive as alternative, hypoxia extends the lifespan, increases their proliferative ability and reduces differentiation [118]. The morphologic and biochemical characteristics of neural-like cells are already described but the mechanism by which stem cells differentiate into neural-like cells is still unknown. In our research work, MSCs that undergone differentiation into neural-like cells, consumed significantly less glucose and produced significantly less lactate than MSCs that undergone only expansion. These major differences allow us to conclude that during MSCs differentiation in neural-like cells the glycolytic process, which proved to be the crucial metabolic mechanism during MSCs expansion, is switched to oxidative metabolism [55].

Our results show clear evidences that MSCs expansion is dependent of glycolysis while their differentiation in neural-like cells requires the switch of the metabolic profile to oxidative metabolism. Also important may be the role of oxidative stress during this process. This work is a first step to identify key metabolic-related mechanisms responsible for human MSCs from the Wharton's jelly expansion and differentiation [55].

The lack of standardization of MSCs isolated from the Wharton's jelly culture conditions has limited some progress in scientific and clinical research. Understanding these MSCs metabolism during expansion, as well as determining molecular and biochemical mechanisms for differentiation is of great significance to develop new effective stem cell-based therapies.

4. Biomaterial and cellular system association – discussion and final remarks

Using the rat model, we recently tested *in vivo* the efficacy of biomaterials and cellular system association in treatment of sciatic nerve axonotmesis and neurotmesis injury. Following transection, axons show staggered regeneration and may take substantial time to cross the injured site and enter the distal nerve stump [119]. However delayed axonal elongation might be caused by growth inhibition originated from the distal nerve itself, growth-stimulating influences may overcome axons stagger. As a potential source of growth promoting signals, MSCs transplantation is expected to give a positive outcome. Our results showed that the use of either undifferentiated or differentiated MSCs in axonotmesis lesion boosted the recovery of sensory and motor function. In both cell-enriched experimental groups we observed that the myelin sheath was thicker, this suggests that MSCs might apply their positive effects on SCs, the key element in Wallerian degeneration and the following axonal regeneration [120]. Also results from *in vivo* testing previously performed by our research group showed that infiltration of MSCs from the Wharton's jelly, or the combination of chitosan type III membrane enwrapment and MSCs enrichment after nerve crush injury provide an advantage to post-traumatic nerve regeneration [56, 57]. Chitosan type III was developed as a hybrid of chitosan by adding GPTMS. A synergistic effect of an extra permeability and physicochemical properties of chitosan type III and the presence of silica ions may be responsible for the good results in post-traumatic nerve regeneration promotion observed in the sciatic nerve after axonotmesis and neurotmesis [57, 91]. The substantial improvement of axonal regeneration found in sciatic nerve crush enwrapped by chitosan type III membranes and for bridging nerve gaps after neurotmesis [57, 91], suggests that this biomaterial may not just work as a simple mechanical device but instead may induce nerve regeneration. The neuroregenerative properties of chitosan type III may be explained by the effect on SCs proliferation, axon elongation and myelinization [55, 91]. Our data also showed that PLC does not deleteriously interfere with the nerve regeneration process, as a matter of fact, the information on the effectiveness of PLC membranes and tube-guides for allowing nerve regeneration was already provided experimentally and with patients [82]. The MSCs from the Wharton's jelly may be a valuable source in the repair of the peripheral nervous system with capacity to differentiate into neuroglial-like cells. The transplanted MSCs are also able to promote local blood vessel formation and release the neurotrophic factors brain-derived neurotrophic factor (BDNF) and glial cell line-derived neurotrophic factor (GDNF) [55]. Previous results obtained by our research group using N1E-115 cells *in vitro* differentiated into neuroglial-like cells to promote regeneration of axonotmesis and neurotmesis lesions in the rat model showed that there was no significant effect in promoting axon regeneration and, when N1E-115 cells were cultured inside a PLGA scaffold used to bridge a nerve defect, they can even exert negative effects on nerve fiber regeneration. The presence of transplanted N1E-115 cells in nerve scaffolds competing for the local blood supply of nutrients and oxygen and by space-occupying effect could have hindered the positive effect of local neurotrophic factor release leading a negative outcome on nerve regeneration. Thus, N1E-115 cells did not prove to be a suitable candidate cellular system for treatment of nerve injury after axonotmesis and neurotmesis and their application is limited only to research purposes as a basic scientific step for the development of other cell delivery systems, due to its neoplastic origin [57-59, 91, 93]. The MSCs isolated from the Wharton's jelly through PLC and chitosan type III membranes might be a potentially valuable tool to improve clinical outcome especially after trauma to sensory nerves, such as digital nerves. The results from our experimental work [55, 56] showed that the use of either undifferentiated or neuroglial-like differentiated MSCs enhanced the recovery of sensory and motor function of the rat sciatic nerve. The observation that in both cell-enriched experimental groups myelin sheath was thicker, suggest that MSCs might exert their positive effects on SCs, the key element in Wallerian degeneration and the following axonal regeneration [120]. In addition, these cells represent a non-controversial source of primitive mesenchymal progenitor cells that can be harvested after birth, cryogenically stored, thawed, and expanded for therapeutic uses, including nerve injuries like axonotmesis and neurotmesis. The time and temperature of the transport (and the saline solution used) of the UC units from the hospital / clinic to the laboratory is crucial for a successful outcome considering MSCs isolation and proliferation from fresh and cryopreserved UCT. It is highly recommend that the transport from the clinic or hospital to the laboratory should be refrigerated, and the UC units should be immediately immersed in a sterile saline solution like HBSS or DPBS.

Acknowledgements

The authors would like to gratefully acknowledge the valuable support by Dr. José Manuel Correia Costa, from Laboratório de Parasitologia, Instituto Nacional de Saúde Dr. Ricardo Jorge (INSRJ), Porto, Portugal; and Biosckin, Molecular and Cell Therapies SA support for the umbilical cord units supply used in the experimental work and for the access of the authors to the GMP classified cell culture room and all the equipment used in cell culture and flow citometry analysis (Scientific Protocol between Porto University and Biosckin, Molecular and Cell Therapies SA). This work was supported by Fundação para a Ciência e Tecnologia (FCT), Ministério da Ciência e Ensino Superior (MCES), Portugal, through the financed research project PTDC/DES/104036/2008, and by QREN N° 1372 para Criação de um Núcleo I&DT para Desenvolvimento de Produtos nas Áreas de Medicina Regenerativa e de Terapias Celulares – Núcleo Biomat & Cell. A Gärtner has a Doctoral Grant from Fundação para a Ciência e Tecnologia (FCT), Ministério da Ciência e Ensino Superior (MCES), Portugal, SFRH/BD/70211/2010.

Author details

Andrea Gärtner[1,2], Tiago Pereira[1,2], Raquel Gomes[1,2], Ana Lúcia Luís[1,2], Miguel Lacueva França[1,2], Stefano Geuna[4], Paulo Armada-da-Silva[3*] and Ana Colette Maurício[1,2]

*Address all correspondence to: ana.colette@hotmail.com

1 Instituto de Ciências Biomédicas Abel Salazar (ICBAS), Universidade do Porto (UP), Portugal

2 Centro de Estudos de Ciência Animal (CECA), Instituto de Ciências e Tecnologias Agrárias e Agro-Alimentares (ICETA), Universidade do Porto (UP), Portugal

3 Faculdade de Motricidade Humana (FMH), Universidade Técnica de Lisboa (UTL), Portugal

4 Department of Clinical and Biological Sciences, University of Turin, Italy

References

[1] Triffitt, J. T. (2002). Stem cells and the philosopher's stone. *Journal of cellular biochemistry Supplement*, 38, 13-9, 2002/06/06.

[2] Lajtha, L. G. (1967). Stem cells and their properties. *Proceedings Canadian Cancer Conference*, 7, 31-9, 1967/01/01.

[3] Fossett, E., & Khan, W. S. (2012). Optimising human mesenchymal stem cell numbers for clinical application: a literature review. *Stem cells international*, 2012, 465259, 2012/03/27.

[4] Bongso, A., Fong, C. Y., Ng, S. C., & Ratnam, S. (1994). Isolation and culture of inner cell mass cells from human blastocysts. *Human reproduction (Oxford, England)*, 9(11), 2110-7, 1994/11/01.

[5] Shim, J. H., Kim, S. E., Woo, D. H., Kim, S. K., Oh, C. H., Mc Kay, R., et al. (2007). Directed differentiation of human embryonic stem cells towards a pancreatic cell fate. *Diabetologia*, 50(6), 1228-38, 2007/04/26.

[6] Yang, D., Zhang, Z. J., Oldenburg, M., Ayala, M., & Zhang, S. C. (2008). Human embryonic stem cell-derived dopaminergic neurons reverse functional deficit in parkinsonian rats. *Stem Cells*, 26(1), 55-63, 2007/10/24.

[7] Mohib, K., & Wang, L. (2012). Differentiation and characterization of dendritic cells from human embryonic stem cells. *Current protocols in immunology / edited by John E Coligan [et al]*, Chapter 22:Unit 22F 11, 12/08/03.

[8] Takahashi, K., & Yamanaka, S. (2006). Induction of pluripotent stem cells from mouse embryonic and adult fibroblast cultures by defined factors. *Cell*, 126(4), 663-76, 2006/08/15.

[9] Fong-Y, C., Chak-L, L., Biswas, A., Tan-H, J., Gauthaman, K., Chan-K, W., et al. (2010). Human Wharton's Jelly Stem Cells Have Unique Transcriptome Profiles Compared to Human Embryonic Stem Cells and Other Mesenchymal Stem Cells. *Stem Cell Reviews and Reports*, 7(1), 1-16.

[10] Jones, E. A., Kinsey, S. E., English, A., Jones, R. A., Straszynski, L., Meredith, D. M., et al. (2002). Isolation and characterization of bone marrow multipotential mesenchymal progenitor cells. *Arthritis and rheumatism*, 46(12), 3349-60, 2002/12/17.

[11] Wohlers, I., Stachelscheid, H., Borstlap, J., Zeilinger, K., & Gerlach, J. C. (2009). The Characterization Tool: A knowledge-based stem cell, differentiated cell, and tissue database with a web-based analysis front-end. *Stem Cell Res*, 3(2-3), 88-95, 2009/06/13.

[12] Brittberg, M., Lindahl, A., Nilsson, A., Ohlsson, C., Isaksson, O., & Peterson, L. (1994). Treatment of deep cartilage defects in the knee with autologous chondrocyte transplantation. *The New England journal of medicine*, 331(14), 889-95, 1994/10/06.

[13] Gentzkow, G. D., Iwasaki, S. D., Hershon, K. S., Mengel, M., Prendergast, J. J., Ricotta, J. J., et al. (1996). Use of dermagraft, a cultured human dermis, to treat diabetic foot ulcers. *Diabetes care*, 19(4), 350-4, 1996/04/01.

[14] Le Blanc, K., & Ringden, O. (2005). Immunobiology of human mesenchymal stem cells and future use in hematopoietic stem cell transplantation. *Biol Blood Marrow Transplant*, 11(5), 321-34, 2005/04/23.

[15] Leri, A., Kajstura, J., Anversa, P., & Frishman, W. H. (2008). Myocardial regeneration and stem cell repair. *Current problems in cardiology*, 33(3), 91-153, 2008/02/05.

[16] Sanchez-Ramos, J. R. (2002). Neural cells derived from adult bone marrow and umbilical cord blood. *J Neurosci Res*, 69(6), 880-93, 2002/09/03.

[17] Wakitani, S., Imoto, K., Yamamoto, T., Saito, M., Murata, N., & Yoneda, M. (2002). Human autologous culture expanded bone marrow mesenchymal cell transplantation for repair of cartilage defects in osteoarthritic knees. *Osteoarthritis and cartilage OARS, Osteoarthritis Research Society*, 10(3), 199-206, 2002/03/01.

[18] Wang, L., Ott, L., Seshareddy, K., Weiss, M. L., & Detamore, MS. (2011). Musculoskeletal tissue engineering with human umbilical cord mesenchymal stromal cells. *Regen Med*, 6(1), 95-109, 2010/12/24.

[19] Chen, J. S., Wong, V. W., & Gurtner, G. C. (2012). Therapeutic potential of bone marrow-derived mesenchymal stem cells for cutaneous wound healing. *Frontiers in immunology*, 3, 192, 2012/07/13.

[20] Friedenstein, A. J., Petrakova, K. V., Kurolesova, A. I., & Frolova, G. P. (1968). Heterotopic of bone marrow. Analysis of precursor cells for osteogenic and hematopoietic tissues. *Transplantation*, 6(2), 230-47, 1968/03/01.

[21] Phinney, D. G., & Prockop, D. J. (2007). Concise review: mesenchymal stem/multipotent stromal cells: the state of transdifferentiation and modes of tissue repair--current views. *Stem Cells*, 25(11), 2896-902, 2007/09/29.

[22] Caplan, A. I. (1991). Mesenchymal stem cells. *Journal of orthopaedic research : official publication of the Orthopaedic Research Society*, 9(5), 641-50, 1991/09/01.

[23] Pittenger, M. F. (1999). Multilineage Potential of Adult Human Mesenchymal Stem Cells. *Science*, 284(5411), 143-7.

[24] Dominici, M., Le Blanc, K., Mueller, I., Slaper-Cortenbach, I., Marini, F., Krause, D., et al. (2006). Minimal criteria for defining multipotent mesenchymal stromal cells. The International Society for Cellular Therapy position statement. *Cytotherapy*, 8(4), 315-7, 2006/08/23.

[25] Ahmadi, N., Razavi, S., Kazemi, M., & Oryan, S. (2012). Stability of neural differentiation in human adipose derived stem cells by two induction protocols. *Tissue & cell*, 44(2), 87-94, 2011/12/20.

[26] Chen, J., Liu, R., Yang, Y., Li, J., Zhang, X., Wang, Z., et al. (2011). The simulated microgravity enhances the differentiation of mesenchymal stem cells into neurons. *Neurosci Lett*, 505(2), 171-5, 2011/10/22.

[27] Du, Y., Roh, D. S., Funderburgh, M. L., Mann, M. M., Marra, K. G., Rubin, J. P., et al. (2010). Adipose-derived stem cells differentiate to keratocytes in vitro. *Molecular vision*, 16, 2680-9, 2010/12/24.

[28] Jin, G., Prabhakaran, M. P., & Ramakrishna, S. (2011). Stem cell differentiation to epidermal lineages on electrospun nanofibrous substrates for skin tissue engineering. *Acta Biomater*, 7(8), 3113-22, 2011/05/10.

[29] Al, Battah. F., De Kock, J., Vanhaecke, T., & Rogiers, V. (2011). Current status of human adipose-derived stem cells: differentiation into hepatocyte-like cells. *Scientific World Journal*, 11, 1568-81, 12/01/10.

[30] Ayatollahi, M., Soleimani, M., Tabei, S. Z., & Kabir, Salmani. M. (2011). Hepatogenic differentiation of mesenchymal stem cells induced by insulin like growth factor-I. *World journal of stem cells*, 3(12), 113-21, 2012/01/10.

[31] Bhandari, D. R., Seo, K. W., Sun, B., Seo, M. S., Kim, H. S., Seo, Y. J., et al. (2011). The simplest method for in vitro beta-cell production from human adult stem cells. *Differentiation; research in biological diversity*, 82(3), 144-52, 2011/07/26.

[32] Ankrum, J., & Karp, J. M. (2010). Mesenchymal stem cell therapy: Two steps forward, one step back. *Trends in molecular medicine*, 16(5), 203-9, 2010/03/26.

[33] Nauta, A. J., & Fibbe, W. E. (2007). Immunomodulatory properties of mesenchymal stromal cells. 2007/08/01. *Blood*, 110(10), 3499-506.

[34] Togel, F., Weiss, K., Yang, Y., Hu, Z., Zhang, P., & Westenfelder, C. (2007). Vasculotropic, paracrine actions of infused mesenchymal stem cells are important to the recovery from acute kidney injury. *American journal of physiology Renal physiology*, 292(5), F1626-35, 2007/01/11.

[35] Zhang, M., Mal, N., Kiedrowski, M., Chacko, M., Askari, A. T., Popovic, Z. B., et al. (2007). SDF-1 expression by mesenchymal stem cells results in trophic support of cardiac myocytes after myocardial infarction. *FASEB journal : official publication of the Federation of American Societies for Experimental Biology*, 21(12), 3197-207, 2007/05/15.

[36] Rao, M. S. (2006). Are there morally acceptable alternatives to blastocyst derived ESC? *J Cell Biochem*, 98(5), 1054-61, 2006/04/07.

[37] Vilalta, M., Degano, I. R., Bago, J., Gould, D., Santos, M., Garcia-Arranz, M., et al. (2008). Biodistribution, long-term survival, and safety of human adipose tissue-derived mesenchymal stem cells transplanted in nude mice by high sensitivity non-invasive bioluminescence imaging. *Stem Cells Dev*, 17(5), 993-1003, 2008/06/10.

[38] Si, Y. L., Zhao, Y. L., Hao, H. J., Fu, X. B., & Han, W. D. (2011). MSCs: Biological characteristics, clinical applications and their outstanding concerns. *Ageing research reviews*, 10(1), 93-103, 2010/08/24.

[39] Bongso, A., Fong-Y, C., & Gauthaman, K. Taking stem cells to the clinic: Major challenges. *Journal of Cellular Biochemistry*, 105(6), 1352-60.

[40] Baxter, MA, Wynn, R. F., Jowitt, S. N., Wraith, J. E., Fairbairn, L. J., & Bellantuono, I. (2004). Study of telomere length reveals rapid aging of human marrow stromal cells following in vitro expansion. *Stem Cells*, 22(5), 675-82, 2004/09/03.

[41] Mareschi, K., Ferrero, I., Rustichelli, D., Aschero, S., Gammaitoni, L., Aglietta, M., et al. (2006). Expansion of mesenchymal stem cells isolated from pediatric and adult donor bone marrow. *J Cell Biochem*, 97(4), 744-54, 2005/10/18.

[42] Stolzing, A., Jones, E., Mc Gonagle, D., & Scutt, A. (2008). Age-related changes in human bone marrow-derived mesenchymal stem cells: consequences for cell therapies. *Mechanisms of ageing and development*, 129(3), 163-73, 2008/02/05.

[43] Zhou, S., Greenberger, J. S., Epperly, M. W., Goff, J. P., Adler, C., Leboff, MS, et al. (2008). Age-related intrinsic changes in human bone-marrow-derived mesenchymal stem cells and their differentiation to osteoblasts. *Aging Cell*, 7(3), 335-43, 2008/02/06.

[44] Dexheimer, V., Mueller, S., Braatz, F., & Richter, W. (2011). Reduced reactivation from dormancy but maintained lineage choice of human mesenchymal stem cells with donor age. *PLoS One*, 6(8), e22980, 11/08/19.

[45] Gronthos, S., Zannettino, A. C., Hay, S. J., Shi, S., Graves, S. E., Kortesidis, A., et al. (2003). Molecular and cellular characterisation of highly purified stromal stem cells derived from human bone marrow. *Journal of cell science*, 116(Pt 9), 1827-35, 2003/04/01.

[46] Bi, Y., Stuelten, C. H., Kilts, T., Wadhwa, S., Iozzo, R. V., Robey, P. G., et al. (2005). Extracellular matrix proteoglycans control the fate of bone marrow stromal cells. *The Journal of biological chemistry*, 280(34), 30481-9, 2005/06/21.

[47] Kume, S., Kato, S., Yamagishi, S., Inagaki, Y., Ueda, S., Arima, N., et al. (2005). Advanced glycation end-products attenuate human mesenchymal stem cells and prevent cognate differentiation into adipose tissue, cartilage, and bone. *Journal of bone and mineral research : the official journal of the American Society for Bone and Mineral Research*, 20(9), 1647-58, 2005/08/02.

[48] Both, S. K., van der Muijsenberg, A. J., van Blitterswijk, C. A., de Boer, J., & de Bruijn, J. D. (2007). A rapid and efficient method for expansion of human mesenchymal stem cells. *Tissue Eng*, 13(1), 3-9, 2007/05/24.

[49] Colter, D. C. (2000). Rapid expansion of recycling stem cells in cultures of plastic-adherent cells from human bone marrow. *Proceedings of the National Academy of Sciences*, 97(7), 3213-8.

[50] Romanov, Y. A., Svintsitskaya, V. A., & Smirnov, V. N. (2003). Searching for alternative sources of postnatal human mesenchymal stem cells: candidate MSC-like cells from umbilical cord. *Stem Cells*, 21(1), 105-10, 2003/01/17.

[51] Zhang, X., Hirai, M., Cantero, S., Ciubotariu, R., Dobrila, L., Hirsh, A., et al. (2011). Isolation and characterization of mesenchymal stem cells from human umbilical cord blood: Reevaluation of critical factors for successful isolation and high ability to proliferate and differentiate to chondrocytes as compared to mesenchymal stem cells fro. *Journal of Cellular Biochemistry*, 112(4), 1206-18.

[52] Conconi, M. T., Di Liddo, R., Tommasini, M., Calore, C., & Parnigotto, P. P. (2011). Phenotype and Differentiation Potential of Stromal Populations Obtained from Various Zones of Human Umbilical Cord: An Overview. *Open Tissue Engineering & Regenerative Medicine Journal*, 4, 6-20.

[53] Fong, C. Y., Richards, M., Manasi, N., Biswas, A., & Bongso, A. (2007). Comparative growth behaviour and characterization of stem cells from human Wharton's jelly. *Reprod Biomed Online*, 15(6), 708-18, 2007/12/08.

[54] Wang, H. S., Hung, S. C., Peng, S., Huang, T. C. C., , C., Wei, H. M., Guo, Y. J., et al. (2004). Mesenchymal Stem Cells in the Wharton's Jelly of the Human Umbilical Cord. *Stem Cells*, 22(7), 1330-7.

[55] Gärtner, A., Pereira, T., Armada-da-Silva, P., Amorim, I., Gomes, R., Ribeiro, J., et al. Use of poly(DL-lactide-ε-caprolactone) membranes and Mesenchymal Stem Cells for promoting nerve regeneration in an axonotmesis rat model: in vitro and in vivo analysis. *Differentiation; research in biological diversity.Submitted.*

[56] Gärtner, A., Pereira, T., Simões, MJ, Armada da Silva, P., França, M. L., Sousa, R., et al. Use of hybrid chitosan membranes and human mesenchymal stem cells from the Wharton jelly of umbilical cord for promoting nerve regeneration in an axonotmesis rat model. *Neural Regeneration Research.In Press.*

[57] Amado, S., Simoes, MJ, Armada da Silva, P. A., Luis, A. L., Shirosaki, Y., Lopes, MA, et al. (2008). Use of hybrid chitosan membranes and N1E-115 cells for promoting nerve regeneration in an axonotmesis rat model. *Biomaterials*, 29(33), 4409-19, 2008/08/30.

[58] Luis, A. L., Rodrigues, J. M., Geuna, S., Amado, S., Shirosaki, Y., Lee, J. M., et al. (2008). Use of PLGA 90:10 scaffolds enriched with in vitro-differentiated neural cells for repairing rat sciatic nerve defects. *Tissue engineering Part A*, 14(6), 979-93, 2008/05/02.

[59] Luis, A. L., Rodrigues, J. M., Geuna, S., Amado, S., Simoes, MJ, Fregnan, F., et al. (2008). Neural cell transplantation effects on sciatic nerve regeneration after a standardized crush injury in the rat. *Microsurgery*, 28(6), 458-70, 2008/07/16.

[60] Fu, Y.-S, Shih, Y.-T, Cheng, Y.-C, & Min, M.-Y. (2004). Transformation of Human Umbilical Mesenchymal Cells into Neurons in vitro. *Journal of Biomedical Science*, 11(5), 652-60.

[61] Weiss, M. L., Medicetty, S., Bledsoe, A. R., Rachakatla, R. S., Choi, M., Merchav, S., et al. (2006). Human Umbilical Cord Matrix Stem Cells: Preliminary Characterization and Effect of Transplantation in a Rodent Model of Parkinson's Disease. *Stem Cells*, 24(3), 781-92.

[62] Conconi, M. T., Burra, P., Di Liddo, R., Calore, C., Turetta, M., Bellini, S., et al. (2006). CD105(+) cells from Wharton's jelly show in vitro and in vivo myogenic differentiative potential. *Int J Mol Med*, 18(6), 1089-96, 2006/11/08.

[63] Fu, Y. -S, Cheng, Y. -C, Lin, M. -Y, Cheng, A., Chu, H. -MP., Chou, S. -C, et al. (2006). Conversion of Human Umbilical Cord Mesenchymal Stem Cells in Wharton's Jelly to Dopaminergic Neurons In Vitro: Potential Therapeutic Application for Parkinsonism. *Stem Cells*, 24(1), 115-24.

[64] Jomura, S., Uy, M., Mitchell, K., Dallasen, R., Bode, C. J., & Xu, Y. (2007). Potential treatment of cerebral global ischemia with Oct-4+ umbilical cord matrix cells. *Stem Cells*, 25(1), 98-106, 2006/09/09.

[65] Levy, Y. S., Bahat-Stroomza, M., Barzilay, R., Burshtein, A., Bulvik, S., Barhum, Y., et al. (2008). Regenerative effect of neural-induced human mesenchymal stromal cells in rat models of Parkinson's disease. *Cytotherapy*, 10(4), 340-52, 2008/06/25.

[66] Gong, Z., Calkins, G., Cheng, E. C., Krause, D., & Niklason, L. E. (2009). Influence of culture medium on smooth muscle cell differentiation from human bone marrow-derived mesenchymal stem cells. *Tissue Eng Part A*, 15(2), 319-30, 2009/01/01.

[67] Kirouac, D. C., & Zandstra, P. W. (2008). The systematic production of cells for cell therapies. *Cell Stem Cell*, 3(4), 369-81, 08/10/23.

[68] Semenov, O. V., Koestenbauer, S., Riegel, M., Zech, N., Zimmermann, R., Zisch, A. H., et al. (2010). Multipotent mesenchymal stem cells from human placenta: critical parameters for isolation and maintenance of stemness after isolation. *American journal of obstetrics and gynecology*, e1-e13, 2009/12/29.

[69] Caplan, A. I. (2009). Why are MSCs therapeutic? New data: new insight. *The Journal of Pathology*, 217(2), 318-24.

[70] Barry, F. P., & Murphy, J. M. (2004). Mesenchymal stem cells: clinical applications and biological characterization. *Int J Biochem Cell Biol*, 36(4), 568-84, 2004/03/11.

[71] Joyce, N., Annett, G., Wirthlin, L., Olson, S., Bauer, G., & Nolta, J. A. (2010). Mesenchymal stem cells for the treatment of neurodegenerative disease. *Regen Med*, 5(6), 933-46, 2010/11/19.

[72] Dimmeler, S., Burchfield, J., & Zeiher, A. M. (2008). Cell-based therapy of myocardial infarction. *Arteriosclerosisthrombosis, and vascular biology*, 28(2), 208-16, 2007/10/24.

[73] Strioga, M., Viswanathan, S., Darinskas, A., Slaby, O., & Michalek, J. (2012). Same or Not the Same? Comparison of Adipose Tissue-Derived Versus Bone Marrow-Derived Mesenchymal Stem and Stromal Cells. *Stem Cells Dev*, 2012/04/04.

[74] Health USNIo. (2012). ClinicalTrials.gov. http://clinicaltrials.gov/, [3 August].

[75] Battiston, B., Geuna, S., Ferrero, M., & Tos, P. (2005). Nerve repair by means of tubulization: Literature review and personal clinical experience comparing biological and synthetic conduits for sensory nerve repair. *Microsurgery*, 25(4), 258-67.

[76] Mackinnon, S. E., Doolabh, V. B., Novak, C. B., & Trulock, E. P. (2001). Clinical outcome following nerve allograft transplantation. *Plast Reconstr Surg*, 107(6), 1419-29, 2001/05/04.

[77] Kehoe, S., Zhang, X. F., & Boyd, D. (2012). FDA approved guidance conduits and wraps for peripheral nerve injury: a review of materials and efficacy. *Injury*, 43(5), 553-72, 2011/01/29.

[78] Siemionow, M., Bozkurt, M., & Zor, F. (2010). Regeneration and repair of peripheral nerves with different biomaterials: Review. *Microsurgery*, 30(7), 574-88.

[79] Mackinnon, S. E., Hudson, A. R., & Hunter, D. A. (1985). Histologic assessment of nerve regeneration in the rat. *Plast Reconstr Surg*, 75(3), 384-8, 1985/03/01.

[80] Ronchi, G., Nicolino, S., Raimondo, S., Tos, P., Battiston, B., Papalia, I., et al. (2009). Functional and morphological assessment of a standardized crush injury of the rat median nerve. *Journal of Neuroscience Methods*, 179(1), 51-7.

[81] de Ruiter, G. C., Malessy, M. J., Yaszemski, M. J., Windebank, A. J., & Spinner, R. J. (2009). Designing ideal conduits for peripheral nerve repair. *Neurosurg Focus*, 26(2), E5, 2009/05/14.

[82] Maurício, A. C., Gärtner, A., Armada-da-Silva, P., Amado, S., Pereira, T., Veloso, A. P., et al. (2011). Cellular Systems and Biomaterials for Nerve Regeneration in Neurotmesis Injuries. *Pignatello R, editor. Biomaterials Applications for Nanomedicine*, 978-9-53307-661-4, Available from: InTech.

[83] Wang, Y., Zhao, Z., Ren, Z., Zhao, B., Zhang, L., Chen, J., et al. (2012). Recellularized nerve allografts with differentiated mesenchymal stem cells promote peripheral nerve regeneration. *Neurosci Lett*, 514(1), 96-101, 2012/03/13.

[84] Madduri, S., & Gander, B. (2010). Schwann cell delivery of neurotrophic factors for peripheral nerve regeneration. *J Peripher Nerv Syst*, 15(2), 93-103, 2010/07/16.

[85] Hall, S. (2001). Nerve repair: a neurobiologist's view. *Journal of hand surgery (Edinburgh, Scotland)*, 26(2), 129-36, 2001/04/03.

[86] Torigoe, K., Tanaka, H. F., Takahashi, A., Awaya, A., & Hashimoto, K. (1996). Basic behavior of migratory Schwann cells in peripheral nerve regeneration. *Exp Neurol*, 137(2), 301-8, 1996/02/01.

[87] Schlosshauer, B., Muller, E., Schroder, B., Planck, H., & Muller, H. W. (2003). Rat Schwann cells in bioresorbable nerve guides to promote and accelerate axonal regeneration. *Brain Res*, 963(1-2), 321-6, 2003/02/01.

[88] Keilhoff, G., Goihl, A., Langnase, K., Fansa, H., & Wolf, G. (2006). Transdifferentiation of mesenchymal stem cells into Schwann cell-like myelinating cells. *European Journal of Cell Biology*, 85(1), 11-24.

[89] Matsuse, D., Kitada, M., Kohama, M., Nishikawa, K., Makinoshima, H., Wakao, S., et al. (2010). Human umbilical cord-derived mesenchymal stromal cells differentiate into functional Schwann cells that sustain peripheral nerve regeneration. *J Neuropathol Exp Neurol*, 69(9), 973-85, 2010/08/20.

[90] Luis, A. L., Rodrigues, J. M., Amado, S., Veloso, A. P., Armada-Da-Silva, P. A., Raimondo, S., et al. (2007). PLGA 90/10 and caprolactone biodegradable nerve guides for the reconstruction of the rat sciatic nerve. *Microsurgery*, 27(2), 125-37, 2007/02/10.

[91] Simoes, MJ, Amado, S., Gartner, A., Armada-Da-Silva, P. A., Raimondo, S., Vieira, M., et al. (2010). Use of chitosan scaffolds for repairing rat sciatic nerve defects. *Ital J Anat Embryol*, 115(3), 190-210, 2011/02/04.

[92] Luis, A. L., Amado, S., Geuna, S., Rodrigues, J. M., Simoes, MJ, Santos, JD, et al. (2007). Long-term functional and morphological assessment of a standardized rat sciatic nerve crush injury with a non-serrated clamp. *J Neurosci Methods*, 163(1), 92-104, 2007/04/03.

[93] Amado, S., Rodrigues, J. M., Luis, A. L., Armada-da-Silva, P. A., Vieira, M., Gartner, A., et al. (2010). Effects of collagen membranes enriched with in vitro-differentiated N1E-115 cells on rat sciatic nerve regeneration after end-to-end repair. *J Neuroeng Rehabil*, 7, 7, 2010/02/13.

[94] Luís, A. L., Amado, S., Geuna, S., Rodrigues, J. M., Simões, MJ, Santos, JD, et al. (2007). Long-term functional and morphological assessment of a standardized rat sciatic nerve crush injury with a non-serrated clamp. *Journal of Neuroscience Methods*, 163(1), 92-104.

[95] Luis, A. L., Rodrigues, J. M., Lobato, J. V., Lopes, MA, Amado, S., Veloso, A. P., et al. (2007). Evaluation of two biodegradable nerve guides for the reconstruction of the rat sciatic nerve. *Biomed Mater Eng*, 17(1), 39-52, 2007/02/01.

[96] Frattini, F., Pereira, Lopes. F. R., Almeida, F. M., Rodrigues, R. F., Boldrini, L. C., Tomaz, M., et al. (2012). Mesenchymal stem cells in a polycaprolactone conduit promote sciatic nerve regeneration and sensory neuron survival after nerve injury. *Tissue Eng Part A*, 2012/06/01.

[97] Spivey, E. C., Khaing, Z. Z., Shear, J. B., & Schmidt, C. E. (2012). The fundamental role of subcellular topography in peripheral nerve repair therapies. *Biomaterials*, 33(17), 4264-76, 2012/03/20.

[98] Dellon, A. L., & Mackinnon, S. E. (1989). Selection of the appropriate parameter to measure neural regeneration. *Ann Plast Surg*, 23(3), 197-202, 1989/09/01.

[99] Kanaya, F., Firrell, J. C., & Breidenbach, W. C. (1996). Sciatic function index, nerve conduction tests, muscle contraction, and axon morphometry as indicators of regeneration. *Plast Reconstr Surg*, 98(7), 1264-71, discussion 72-4., 1996/12/01.

[100] Shen, N., & Zhu, J. (1995). Application of sciatic functional index in nerve functional assessment. *Microsurgery*, 16(8), 552-5, 1995/01/01.

[101] Almquist, E., & Eeg-Olofsson, O. (1970). Sensory-nerve-conduction velocity and two-point discrimmination in sutured nerves. *J Bone Joint Surg Am*, 52(4), 791-6, 1970/06/01.

[102] de Medinaceli, L., Freed, W. J., & Wyatt, R. J. (1982). An index of the functional condition of rat sciatic nerve based on measurements made from walking tracks. *Exp Neurol*, 77(3), 634-43, 82/09/01.

[103] Varejao, AS, Cabrita, A. M., Meek, M. F., Bulas-Cruz, J., Melo-Pinto, P., Raimondo, S., et al. (2004). Functional and morphological assessment of a standardized rat sciatic nerve crush injury with a non-serrated clamp. *Journal of neurotrauma*, 21(11), 1652-70, 2005/02/03.

[104] Varejao, AS, Cabrita, A. M., Meek, M. F., Bulas-Cruz, J., Filipe, V. M., Gabriel, R. C., et al. (2003). Ankle kinematics to evaluate functional recovery in crushed rat sciatic nerve. *Muscle Nerve*, 27(6), 706-14, 2003/05/27.

[105] Varejao, AS, Cabrita, A. M., Meek, M. F., Fornaro, M., Geuna, S., & Giacobini-Robecchi, M. G. (2003). Morphology of nerve fiber regeneration along a biodegradable poly (DLLA-epsilon-CL) nerve guide filled with fresh skeletal muscle. *Microsurgery*, 23(4), 338-45, 2003/08/28.

[106] Koka, R., & Hadlock, T. A. (2001). Quantification of Functional Recovery Following Rat Sciatic Nerve Transection. *Experimental Neurology*, 168(1), 192-5.

[107] Masters, D. B., Berde, C. B., Dutta, S. K., Griggs, C. T., Hu, D., Kupsky, W., et al. (1993). Prolonged regional nerve blockade by controlled release of local anesthetic from a biodegradable polymer matrix. *Anesthesiology*, 79(2), 340-6, 1993/08/01.

[108] Bain, J. R., Mackinnon, S. E., & Hunter, D. A. (1989). Functional evaluation of complete sciatic, peroneal, and posterior tibial nerve lesions in the rat. *Plast Reconstr Surg*, 83(1), 129-38, 1989/01/01.

[109] Dijkstra, J. R., Meek, M. F., Robinson, P. H., & Gramsbergen, A. (2000). Methods to evaluate functional nerve recovery in adult rats: walking track analysis, video analysis and the withdrawal reflex. *J Neurosci Methods*, 96(2), 89-96, 2000/03/18.

[110] Cappozzo, A., Della Croce, U., Leardini, A., & Chiari, L. (2005). Human movement analysis using stereophotogrammetry. Part 1: theoretical background. *Gait & posture*, 21(2), 186-96, 2005/01/11.

[111] Goulermas, J. Y., Findlow, A. H., Nester, C. J., Howard, D., & Bowker, P. (2005). Automated design of robust discriminant analysis classifier for foot pressure lesions using kinematic data. *IEEE transactions on bio-medical engineering*, 52(9), 1549-62, 2005/09/30.

[112] Johnson, A., Aibinder, W., & Deland, J. T. (2008). Clinical tip: partial plantar plate release for correction of crossover second toe. *Foot & ankle international / American Orthopaedic Foot and Ankle Society [and] Swiss Foot and Ankle Society*, 29(11), 1145-7, 2008/11/26.

[113] Varejao, AS, Cabrita, A. M., Meek, M. F., Bulas-Cruz, J., Gabriel, R. C., Filipe, V. M., et al. (2002). Motion of the foot and ankle during the stance phase in rats. *Muscle Nerve*, 26(5), 630-5, 2002/10/29.

[114] Raimondo, S., Fornaro, M., Di Scipio, F., Ronchi, G., Giacobini-Robecchi, M. G., & Geuna, S. (2009). Chapter 5: Methods and protocols in peripheral nerve regeneration experimental research: part II-morphological techniques. *Int Rev Neurobiol*, 87, 81-103, 2009/08/18.

[115] Geuna, S., Gigo-Benato, D., & Rodrigues, Ade. C. (2004). On sampling and sampling errors in histomorphometry of peripheral nerve fibers. *Microsurgery*, 24(1), 72-6, 2004/01/30.

[116] Geuna, S., Tos, P., Battiston, B., & Guglielmone, R. (2000). Verification of the two-dimensional disector, a method for the unbiased estimation of density and number of myelinated nerve fibers in peripheral nerves. *Ann Anat*, 182(1), 23-34, 2000/02/11.

[117] Vander Heiden, M. G., Cantley, L. C., & Thompson, C. B. (2009). Understanding the Warburg effect: the metabolic requirements of cell proliferation. *Science*, 324(5930), 1029-33, 2009/05/23.

[118] Fehrer, C., Brunauer, R., Laschober, G., Unterluggauer, H., Reitinger, S., Kloss, F., et al. (2007). Reduced oxygen tension attenuates differentiation capacity of human mesenchymal stem cells and prolongs their lifespan. *Aging Cell*, 6(6), 745-57, 2007/10/11.

[119] Brushart, T. M., Hoffman, P. N., Royall, R. M., Murinson, B. B., Witzel, C., & Gordon, T. (2002). Electrical stimulation promotes motoneuron regeneration without increasing its speed or conditioning the neuron. *J Neurosci*, 22(15), 6631-8, 2002/08/02.

[120] Geuna, S., Raimondo, S., Ronchi, G., Di Scipio, F., Tos, P., Czaja, K., et al. (2009). Chapter 3: Histology of the peripheral nerve and changes occurring during nerve regeneration. Int Rev Neurobiol 2009/08/18., 87, 27-46.

Section 5

Special Applications of Biomaterials

Chapter 19

Hydroxylapatite (HA) Powder for Autovaccination Against Canine Non Hodgkin's Lymphoma

Michel Simonet, Nicole Rouquet and
Patrick Frayssinet

Additional information is available at the end of the chapter

http://dx.doi.org/10.5772/55129

1. Introduction

Lymphoma is the most common neoplasm of the canine hemolymphatic system. It represents about 15% of all malignant neoplasms. It has a very poor prognosis, the mean survival with high grade non Hodgkin's lymphomas being two months without treatment [1]. Incidence is increasing.

Lymphomas in dogs, as in humans, can be divided into numerous types depending on the cell line involved and their immunophenotype [2]. T lymphomas have a worse prognosis than B lymphomas and late clinical stages obviously have a very short survival period.

Lymphomas are known in human medicine to respond to chemotherapy, and some of them can even be cured by complex chemotherapy protocols, although severe side effects are noted.

Chemotherapy protocols have also been developed for dogs in the last 40 years. They have proved to be effective for the overall survival of the treated animal. Although the order of drug administration and duration of the maintenance part of the protocol vary considerably, most oncologists agree that a doxorubicin-based (eg CHOP) combination chemotherapy protocol provides the longest period of disease control and overall survival [3].

Even with chemotherapy, survival is relatively short and adjuvant therapies have been developed to improve prognosis. Immunotherapy protocols are of particular interest for this purpose as they may arm immune system cells against the abnormal proteins synthesized by cancer cells [4]. It is a selective way of destroying cancer cells and a treatment with much fewer side effects than chemo- or radiotherapy.

The immune system can destroy cancer cells by different methods:

- Synthesis of TNF-α (tumor necrosis factor alpha) and an oxygen intermediate such as nitric oxide by macrophages activated by IFN-γ (interferon gamma)
- Activation of NK (natural killer) cells by IL-2 (interleukin 2)
- Adsorption of an antibody against tumor cell antigens targeting macrophages and cytotoxic T lymphocytes.
- Activation of CD8+ T lymphocytes by an MHC(major histocompatibility complex) - mediated cell contact mechanism between antigen-presenting cells (APCs) and CD8 cells

Several kinds of immunotherapy protocol are available both in human and veterinary medicine [4, 5, 6, 7]. Heat shock proteins (HSPs) such as gp96 or HSPs70 which are synthesized by the cells submitted to stress are advantageous vaccination adjuvants due to their chaperone properties and their role in antigen presentation [8]. As chaperone molecules, almost all the cell peptides are associated with these proteins and HSP purification provides a fingerprint of the cell's protein synthesis [9]. This is particularly useful for cancer cells which synthesize numerous abnormal proteins during their natural progress [10]. These cells being genetically unstable, their abnormal protein synthesis differs from patient to patient and during the course of the disease. The HSPs and their associated peptides (AAPs) have special receptors (CD91) on dendritic cells which allow the internalization of the AAPs and their modification in order to be expressed at the surface of class I HLA proteins on the cell membrane, triggering activation of the CD8 T cells if they are abnormal [11].

Cancer cells are stressed by the mechanical and metabolic characteristics of the tumour. They synthesize many HSPs [12]. We thus isolated these proteins in order to make an autologous vaccine against the tumour. The HSPs were purified using a hydroxylapatite powder (HA) column. The powder carrying the HSPs was then injected subcutaneously to stimulate the immune system's response to the tumour.

Purification of HSPs using classical way is long and tedious. The use of hydroxylapatite powder allows a much faster purification process. Hydroxylapatite chromatography has been described by Tiselius in the early seventies. Hydroxylapatite chromatography is an adsorption chromatography. The adsorption mechanism is very poorly understood. The surface of the material is occupied bu Ca^{++} and PO_4^-. These ions are supposed to interact with the chemical groups of the proteins. However, post synthesis treatment such as sintering or spray- drying process modify the physico-chemical properties of the material surface. Furthermore, the interaction of the material surface with biological fluids triggers epitaxial growth of carbonated apatite at the surface of the material [13].

It was decided that the proteins purified will be injected carried by the particles for several reasons: the hydroxylapatite particles have been described as vaccine adjuvant, they are phagocytosed by the APCs and can deliver the proteins directly in the APCs, they trigger an afflux of APCs at the injection site [14]. Most of the adjuvants used in antiinfectious vaccines are nano or microparticular.

The aims of this study was to check the feasibility of this protocol using HA-particles with dogs suffering of high grade lymphoma and to know if secondary effects were detected.

2. Materials and methods

2.1. Dogs to be treated

Two dogs (Poodle, 6 years, Jagdterrier, 8 years) suffered from polyadenopathy without any sign of immune deficiency. One dog (mixed breed, 10 years) had a liver metastasis. The last one had a cutaneous form. Their general condition remained reasonable, with no real weight loss or fever. One dog (Poodle) had a splenomegaly. The blood numeration revealed that the white cells and calcaemia for both dogs were in the normal range. Node biopsies were performed. One was sent to the pathologist and the other was frozen (-20°C) and used to manufacture the vaccine. Disease staging was performed using the WHO-staging criteria for canine lymphoma.

A chemotherapy treatment was proposed by the physician but rejected by the dog's owners due to side effects and an informed consent was signed.

After vaccination, the dogs were subjected to a clinical examination whenever they came for the following injection and side effects noted. The volume of a control node was externally measured.

2.2. Toxicity and response assessment

The side effects were graded every week for the first month and every month for the next five months according to the National Cancer Institute's common Toxicity Criteria (version 2.0).

The dogs were also submitted to a physical examination at the same frequency to assess their clinical response. A complete response (CR) was defined as the disappearance of all the nodes and metastasis for at least 4 weeks. A partial response (PR) was defined as a decrease by at least 50% of the product of the 2 longest lengths of all the nodes without the appearance of new lesions for at least 4 weeks. A minor response (MR) was defined as a decrease by less than 50% using the same criteria. Stable disease (SD) was defined as the same criteria unchanged for at least 4 weeks. Progressive disease (PD) was defined as an increase by >25% of the product of the 2 longest lengths of all the nodes for at least 4 weeks or when new lesions appeared.

2.3. Hydroxylapatite powder characteristics

Hydroxylapatite powders (Fig. 1) can be used for adsorption chromatography under atmospheric or high pressure. In order that the protein solution did not fill in the column and to avoid the compaction, the powder was spray-dried then sintered at 1000°C.

The HA was transformed into a ceramic powder according to the following protocol. The synthesized calcium phosphate was suspended in a slurry which is liquid, spray-dried, then sintered. The spray-dried material was heated almost to fusion temperature which favors the migration of matter between the grains and the formation of bridges. As the surface energy was smaller for large than for small grains, their size increased and the distance between the grain centers and the particle surface area decreased.

Powder characteristics	
Nature of the charged groups	PO_4^{3-}, OH^-, Ca^{2+}
Electrocinetical potential (mV)	-35
Hydrophobicity	+
Surface pH	7,8
Granulometry (µm)	0-25
Surface area (m²/gr)	4
Shape	spherical

Table 1. Characteristics of the powder

The sintering considerably reduced the surface area making the amount of proteins adsorbed on the powder lower. It also stabilized the powder structure. The powder was submitted to dissolution/ precipitation processes when soaked inside a saline solution [13]. The modification occurring at the powder surface affected the adsorption properties of the powder. The reduced surface area decreased the interactions with the saline solution containing the proteins. There was a microporosity between the ceramic grains inside the same particle allowing the protein solution to diffuse inside the particles. The characteristics of the powder used in this experiment was given in table 1

As the chromatography was carried out under atmosphere, the granulometry of the powder was an important factor in order to avoid any plugging of the column. The HSPs could be eluted from the powder by 200-300 mM NaCl solutions. The powder solution in NaCl was not stable enough to be injected as it decanted too fast in the syringe. Thus to improve the injectability the powder was put in suspension in a 2% solution of carboxymethylcellulose in 20 mM NaCl.

2.4. Vaccine manufacturing protocol

The tumor tissue and all the materials used to prepare the vaccine were handled in sterile conditions under a laminar flow. The frozen tumor (200 mgr) tissue was homogenized using a bead tissue homogenizer. 1 ml of $NaHCO_3$ (30 mM, pH 7) was added for 1 ml of homogenate. The resulting homogenate was then centrifuged at 1000 g for 15 mn at 4°C to remove all cell fragments.

The supernatant containing the cytoplasmic proteins was used for protein purification by HA column chromatography as follows: a) two precipitations with ammonium sulphate (first at 50% and then at 70%) recovering the pellets. The last pellet was resuspended in 1 ml phosphate buffer (20 mM, pH 7). The column was filled (chromatography columns, Poly-prep, Cat. 731-1550, Bio Rad) with 0.2 gr of HA (0-25 µm), equilibrated with 10 volumes of phosphate buffer (20 mM pH7). The resuspended pellet was then added. The column was then washed with 3 ml of a 100 mM NaCl solution (fig.2).

http://dx.doi.org/10.5772/55129

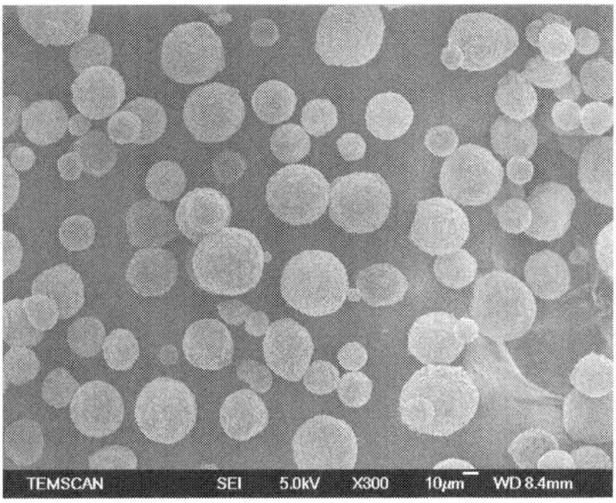

Figure 1. SEM of the HA-powder used for the vaccine.

The powder was then suspended in 5 ml carboxymethylcellulose (CMC) solution (2% in 20mM NaCl). 0.5 ml of this solution was used for each vaccine shot.

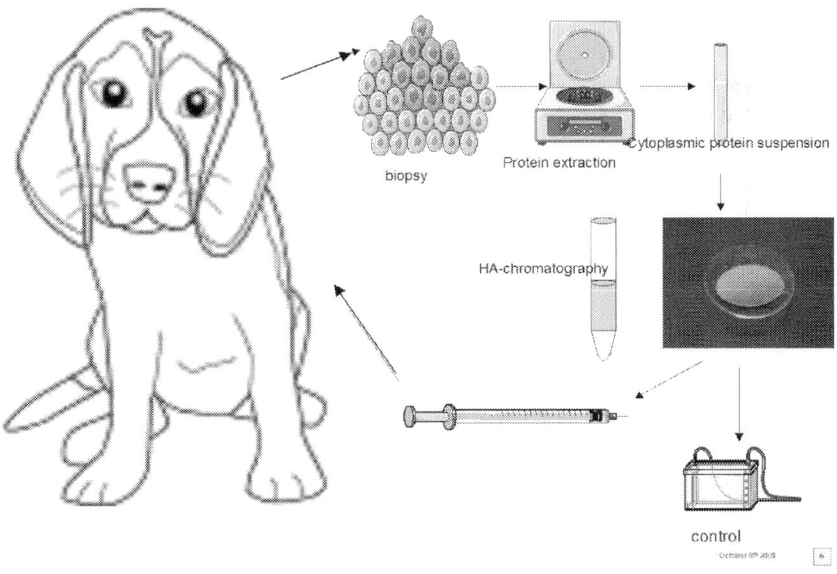

Figure 2. Autovaccine manufacturing scheme

0.2 ml of the previous solution was used to make electrophoretic control. The solution was centrifuged at 1000g 30 for seconds. The supernatant was discarded and the powder in the pellet washed with 0.1 ml of a 0.5 M NaCl solution. The solution was again centrifuged and the supernatant was used for a SDS-Page and for protein quantification using UV spectrometer. 10 μl of the solution was also used for dot blot with anti HSP70 and anti gp90 antibodies on a nitrocellulose membrane. The antibody labelling was evidenced using a westernbreeze™ (invitrogen) kit according to the manufacturer instructions.

2.5. Vaccination protocol

The dogs were injected with 0.5 ml (about 40 μg of proteins) of the vaccine subcutaneously on day 0. The dose schema consisted of 0.5 ml/dose every week for one month, followed by one dose every month for four months.

2.6. activation of TLR2 and 4

Some HSPs are TLR (Toll like receptor) agonist and the TLR activation was checked using two cell lines secreting an embryonic phosphatase alkaline when the TLR 2 or 4 are stimulated.

Two cell lines were used for TLR receptor stimulation. HEK-Blue™-hTLR2 and HEK-Blue™-hTLR4 cells (invivogen –France) were grown in DMEM supplemented with 10% fetal calf serum and 0,5% normocin (invivogen-France) to avoid mycoplasma contamination. 20000 cells were introduced in wells of 96-well plate, 4 hours later 20 μl of different concentrations of the vaccine was added and incubated at 37°C for 20 hours. Then the medium was removed and replaced by 180 μl of resuspended QUANTI- blueTM (Invivogen) and incubated for 1 hour and the positivity evaluated. The negative control was the HA- powder without any protein immobilized on its surface. The minimal concentration turning blue the medium was noted. It was estimated that the vaccine was functional when the minimal active concentration was less than the vaccine concentration.

3. Results

All the vaccine dilutions stimulated the TLR 4 but not the TLR2. The negative control (HA powder) did not stimulate any TLR. The SDS PAGE revealed two bands in the 95 kDa region which were previously demonstrated to be HSPs rich bands (fig. 3) The dot blots revealed that all the vaccine contained gp96 and HSP70 but the amount was different (fig.3 and 4).

Neither dog showed side effects after the injections, whether systemic or local. Two showed a decrease in nodule volume of less than 50% following the first month of injections and were rated MR. The other were rated SD. All dogs were rated PD about one month before they died.

No sign of infection such as fever was observed during the first month of vaccination. The Jagdterrier was diagnosed with an abscess of the collar by the fourth month, which was cured surgically. The lung radiographs did not reveal any lung metastases. The blood count did not show any anomaly in the white cell count up to the last month of survival.

http://dx.doi.org/10.5772/55129

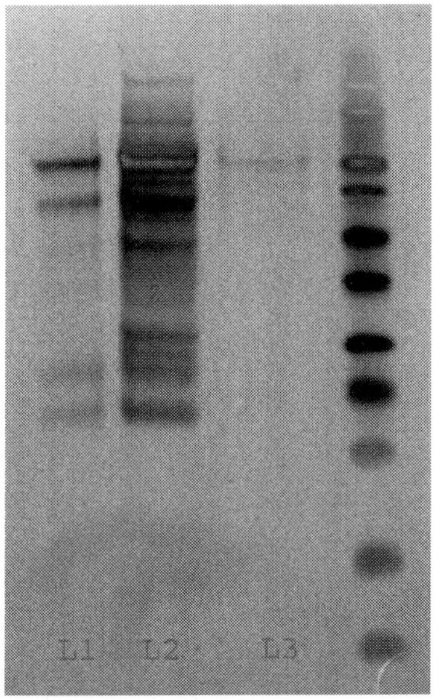

Figure 3. L1 was obtained after washing the powder with 0.3M NaCl, L2 was obtained after washing at 0.02M, L3 was obtained with 0.5M. At 0.02M almost all the contaminant proteins are eluted from the powder. At 0.3 M, the remaining proteins after the 0.02M elution are located in the 95 and 70 kDa zone. The 0.5 M column shows that after after 0.3 M there is almost no proteins remaining on the powder.

The poodle was euthanazised 11 months, and the Jagdterrier 6 months, after the disease was discovered. During their survival period, both dogs had a normal activity. The Jagdterrier was still hunting two weeks before to be euthanazied. The two other dogs were euthanazied at 155 and 173 days (table 2). The mean overall survival time is 210 days.

Dog number	Breed	Histology	Stage	OS	Immunophenotype
1	Jagd terrier	Centoblastic polymorphic	IIIb	330	B cell
2	Poodle	Centroblastic polymorphic	IIIb	180	B cell
3	mixed	Centroblastic polymorphic	IVb	173	B cell
4	cocker	cutaneous	IVb	155	T cell

Table 2. Characteristics of the dogs treated by the vaccination protocol and overall survival time (OS) in days.

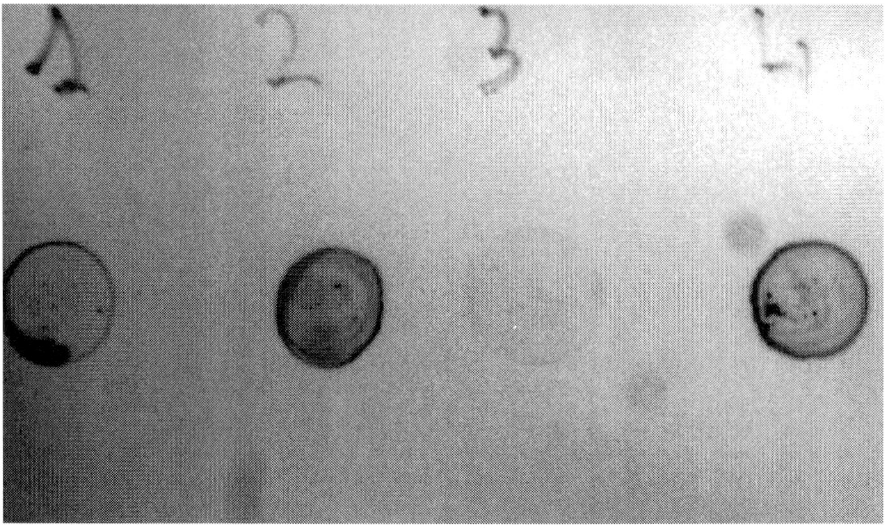

Figure 4. Dot blot after anti gp90 labelling

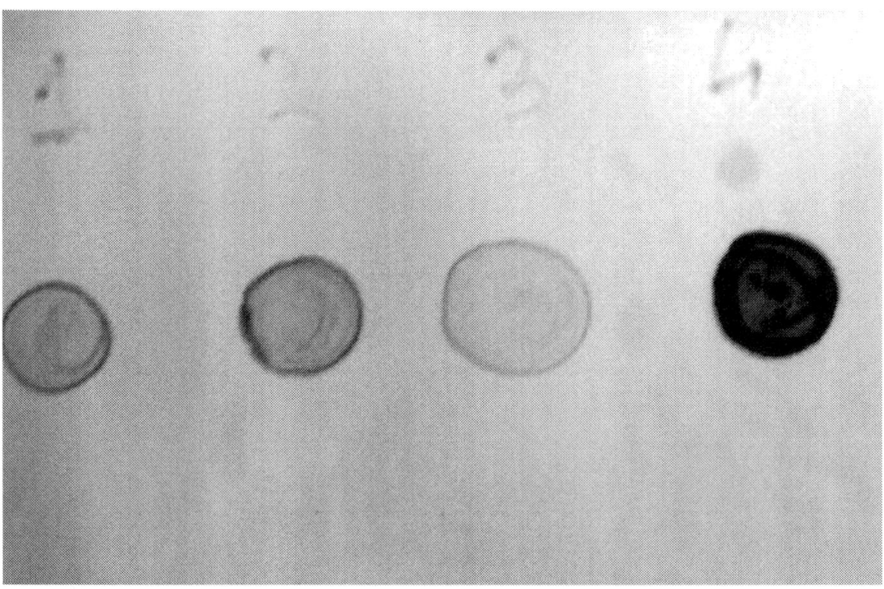

Figure 5. Dot blot after anti HSP70 labelling

4. Discussion

This experiment showed the feasibility of this protocol as an autologous vaccine for cancer in dogs in veterinary practice. We have also developed an *in vitro* test to assess the functionality of purified heat shock proteins. This test checks in a few minutes the vaccine's ability to stimulate antigen-presenting cells (APCs) through toll-like receptor (TLR) activation. Even if the sample size is small, the dogs' overall survival (210 days) was much higher than expected, as the average OS for this type of pathology is two months. In larger series of stage Vb lymphomas associating chemotherapy and the same vaccine protocol, we could demonstrate that after vaccination the dogs showed a delayed cutaneous hypersensibility after their own tumor antigen injection in the derm (Marconato, unpublished results).

High grade lymphomas in dogs are an interesting model for cancer vaccines because survival is very short, so the effect on OS is very easy to measure. The clinical efficiency of vaccine is not always related to the vaccine's effect on biological parameters. Models with a short life expectancy are thus of interest. Furthermore vaccination without the use of other drugs is of particular interest.

The amount of proteins used in this vaccine was justified by the different experiments published about autovaccines using HSPs in animals and humans. Furthermore we checked that the amount of proteins could activate 20000 APCs *in vitro.* The presence of the tumor proteins at the surface of the HA-particles allows the activation of the TLRs. It indicates that the proteins are immobilized at the surface of the particles and their adsorption does not denaturize these proteins.

Cancer vaccines are the focus of great interest. Although cancer cells synthesise abnormal proteins, they are are not recognised by the immune system. Different immunosuppression mechanisms by cancerous tumours have been described [15]. The use of multiantigen vaccines could reduce the cancer cell's "invisibility" to the immune system compared to monoproteic vaccines. Different kinds of immune therapy are under investigation. Cell engineering immunotherapy protocols have been tested, including activation of dendritic cells *in vitro* by tumour antigens before being reinjected into the patient [16]. Other trials concern amplification of the intratumoural lymphocytes (TILs) *in vitro* [17] before being reinjected into the donor. Recently, different types of antibodies were approved for the oncology field, in particular antibodies against VEGF (Vascular endothelial growth factor) to inhibit tumour vascularization [18, 19].

Heat Shock Proteins (HSPs) have proved to be of therapeutic interest in human medicine for some applications. In order to be useful as a cancer vaccination, these HSPs must be made available to APCs. HA-powder is a good material for use as a vector of HSPs to APCs. It has been shown that when injected in the dermis or subcutaneous tissue, it triggers a foreign body reaction and that the cells of this foreign body reaction could be transfected with a DNA molecule carried by the particles [20]. It suggests that, further to DNA vectorisation, the particles could help in the transfection of HSPs and their associated peptides in APCs.

Although gp96 has been described as able to stimulate TLR4, it is not sure that these proteins are the only proteins responsible for TLRs stimulation in this case. There are contaminating proteins in various bands of the SDS which could interfere with the TLRs. The non stimulation of the TLR2 indicates that the TLR4 is not activated by contaminating endotoxins.

Hydroxyapatite has been used as an adjuvant for various infectious vaccines such as diphtheria and tetanus [21]. The Hydroxyapatite lattice is a hexagonal structure which allows numerous substitutions. Ions and small amino acids can thus be trapped in the HA lattice. Consequently, HA powder has been used to purify DNA, proteins or even viruses from biological solutions [22]. In this case, the surface properties of the particles allowed both the purification of HSPs and their use as a vector to APCs and as an adjuvant.

The HA-powder characteristics seems well suited for its role of cancer vaccine adjuvant. The granulometry range allows the phagocytosis by the APCs as it was demonstrated previously [14]. Furthermore, the grain boundaries in each particle is degraded by the cells making the particles fragmented and thus decreasing the real granulometry a few days after injection. It was also demonstrated that the phagocytosis of these particles by the APCs induced the synthesis of various cytokines and lymphokines necessary for the cross-priming of the CD8. The HA-adjuvant effect does not seem to be due to TLR activation as HA-particles alone do not trigger TLR2 and 4 activation. Other mineral adjuvants such as aluminum oxide have been demonstrated to activate the inflamasome. The inflamasome is a multi-protein complex involved in the production of mature IL-1β. The alumn-induced release of IL-1β in macrophages is done under activation of the NLRP3 [23]. It is suggested from unpublished results that the adjuvancy effect of the HA-particles is also due to the inflamasome activation.

Tumor associated antigens show a very poor antigenicity. Thus the presence of an adjuvant like calcium phosphate particles is essential in order to increase the visibility of these antigens by the patient immune system.

This method was used in a short pilot study in humans and proved to be very safe, as only local effects (erythema) were reported in some patients (24). Although it was not the goal of the pilot study, some remission in the extent of the tumour was observed and constituted a good proof of concept. Gp96 has already been used in human medicine to treat a series of patients with indolent non Hodgkin's lymphoma [23]. The results are difficult to compare, as the patients could have been treated by chemo- or radiotherapy more than six weeks before the vaccination protocol. However, at three months most of the patients were rated SD, including those who were resistant to previous therapy. No patients suffered side effects.

5. Conclusions

The HA-particles are essential in this protocol both for their adsorption properties for proteins of interest and their adjuvant properties. They constitute a tool which allows to purify tumor associated antigens in a fast and reproducible way.

The overall survival of the four dogs was much better than expected. These preliminary results nonetheless show that the technique is feasible in private veterinary medicine. They are consistant with other results obtained with canine osteosarcomas or with a series of stage Vb lymphomas treated by chemotherapy and this protocol.

The association of the HA-particles to tumor associated antigens and HSPs did not trigger any adverse effects, confirming the safe results obtained for other applications in human medicine.

This technique could also be combined with conventional chemo- or radiotherapy to increase the animal's overall survival.

Author details

Michel Simonet[1], Nicole Rouquet[2] and Patrick Frayssinet[2*]

1 Veterinary hospital, Nuit Saint Georges, France

2 Urodelia, St Lys, France

References

[1] C. Fournel-Fleury, F. Ponce, P., Felman, A. Blavier, C. Bonnefont, L. Chabanne, T. Marchal, J.L. Cadore, I. Goy-Thollot, D. Ledieu, I. Ghernathi, J.P. Magnol, Canine T-cell lymphomas: a morphological, immunological, and clinical study of 46 new cases. Vet Pathol 39: 92-109 (2002).

[2] R. Chun, Lymphoma: which chemotherapy protocol and why? Top Companion Anim Med.;24: 157-62 (2007)

[3] L. Marconato, The staging and treatment of multicentric high-garde lymphoma in dogs: A review of recent developments and future prospects. The Veterinary Journal, 188: 34-38 (2011).

[4] T., F., Greten, E.M., Jaffee, Cancer vaccines, Journal of clinical oncology, 17: 1047-1060 (1999).

[5] C. J. Wheeler, K. L. Black, G. Liu, M. Mazer, X.-X. Zhang, S. Pepkowitz, D. Goldfinger, H. Ng, D. Irvin, and J. S. Yu. Vaccination elicits correlated immune and clinical responses in glioblastoma multiforme patients. Cancer Research 68 : 5955-5964, (2008).

[6] B. Klein, K. Tarte, A. M. Conge, L. Zhao-Yang, and J. F. Rossi. Stratégie de vaccination antitumorale. Biotech Médecine 23, (2002).

[7] Y. Akiyama, R. Tanosaki, N. Inoue, M. Shimada, Y. Hotate, A. Yamamoto, N. Yamazaki, I. Kawashima, I. Nukaya, K. Takesako, K. Maruyama, Y. Takaue, and K. Yama-

gushi. Clinical response in japanese metastatic melanoma patients treatted with peptide cocktail-pulsed dendritic cells. Journail of Translational Medicine 3: 4-14, (2005).

[8] A. Murshid, J. Gong, and S. K. Calderwood. Heat-shock proteins in cancer vaccines: agents of antigen cross presentation. Expert Rev Vaccines 7: 1019-1030, (2008).

[9] D. R. Ciocca and S. K. Calderwood. Heat shock proteins in cancer: diagnostic, prognostic, predictive, and treatment implications. Cell Stress & Chaperones 10: 263-270, (2005)

[10] J. Kaiser. A detailed genetic portrait of the deadliest human cancers. Science 321: 1280-1281, (2008).

[11] C. V. Nicchitta, D. M. Carrick, and J. C. Baker-LePain. The messenger and the message: gp96 (grp94)-peptide interactions in cellular immunity. Cell Stress & Chaperones 9 : 325-331, (2004).

[12] M. Romanucci, A. Marinelli, S. Giuseppe, and L. Della Salda. Heat shock protein expression in canine malignant mammary tumours. BMC Cancer 6: 171-183, (2006).

[13] M. Heughebaert, R.Z. LeGeros, M. Gineste ,A. Guilhem, G. Bonel, Physico chemical characterization of deposits associated with HA-ceramics implanted in osseous sites. J Biomed Mater Res. 22(3 Suppl): 257-68 (1988)

[14] P. Frayssinet, A. Guilhem, Cell transfection using HA-ceramics. Bioprocessing Journal, 3, 4 (2004)

[15] W. Zou. Immunosuppressive networks in the tumour environment and their therapeutic relevance. Nature Reviews 5: 263-274, (2005).

[16] L. M. Liau, R. M. Prins, S. M. Kiertscher, S. K. Odesa, T. J. Kremen, A. J. Giovannone, J.-W. Li, D. J. Chute, P. S. Mischel, T. F. Cloughesy, and M. D. Roth. Dendritic cell vaccination in glioblastoma patients induces systemic and intracranial T-cell responses modulated by the local central nervous system tumor microenvironment. Clinical Cancer Research 11: 5515-5525, (2005).

[17] G. P. Dunn, I. F. Dunn, and T. W. Curry. Focus on TILs: prognostic significance of tumor infiltrating lymphocytes in human glioma. Cancer Immunity 7: 12-27, (2007).

[18] K. Miller, M. Wang , J. Gralow, M. Dickler, M. Cobleigh, E.A. Perez, T. Shenkier, D. Cella, N.E. Davidson: Paclitaxel plus bevacizumab versus paclitaxel alone for metastatic breast cancer. N Engl J Med , 357: 2666-2676, (2007).

[19] N. Ferrara, K. J. Hillan, H.-P. Gerber, W. Novotny. Discovery and development of bevacizumab, an anti-VEGF antibody for treating cancer. Nature Reviews Drug Discovery, 3: 391-400, (2004)

[20] P. Frayssinet, N. Rouquet, M. Bausero, and P.O. Vidalain, Calcium Phosphates for Cell Transfection. In: P. Ducheyne, K.E. Healy, D.W. Hutmacher, D.W. Grainger, C.J. Kirkpatrick (eds.) Comprehensive Biomaterials, vol. 1, pp. 259-265 Elsevier (2011).

[21] Q. He, A.R. Mitchell, S.L. Johnson, C. Wagner-Bartak, T. Morcol, S. J. D., Bell, Calcium Phosphate Nanoparticle Adjuvant, Clinical and diagnostic laboratory immunology, 899–903 (2000)

[22] P. Gagnon, P., Ng,J., Zhen, C. Aberin, J. He, H. Mekosh, L. Cummings, , S. Zaidi, R. Richieri, A ceramic hydroxyapatite-based purification platform. Simultaneous removal of leached protein A, aggregates, DNA, and endotoxins from Mabs. Bioprocess International, 50-60 (2006).

[23] V. Hornung, F. Bauernfeind, A. Halle, E. Samstad, H. Kono, K.L. Rock, K. A. Fitzgerald, E. Latz, Silica crystals and aluminum salts mediate NALP-3 inflammasome activation via phagosomal destabilization. Nat Immunol.; 9(8): 847–856 (2008)

[24] D.R. Ciocca, P. Frayssinet, F.D. Cuello-Carion, A pilot study with a therapeutic vaccine based on hydroxyapatite ceramic particles and self-antigens in cancer patients. Cell Stress &Chaperones; 12: 33-43, 2007

[25] Y. Oki, P. McLaughlin, L. E. Fayad, B. Pro, P. F. Mansfield, G. L. Clayman, L. J. Medeiros, L. W. Kwak, P. K. Srivastava, and A. Younes. Experience with Heat shock protein-peptide complex 96 vaccine therapy in patients with indolent non-hodgkin lymphoma. Cancer 109: 77-83, (2007).

Chapter 20

Dental Materials

Junko Hieda, Mitsuo Niinomi, Masaaki Nakai and Ken Cho

Additional information is available at the end of the chapter

http://dx.doi.org/10.5772/55049

1. Introduction

Metallic biomaterials used for dental applications, which are called dental alloys, and such alloys require a high corrosion resistance because the pH and temperature vary widely in the oral environment where foods and beverages are taken in. These alloys also require biocompatibility in order to prevent an allergic reaction to the metals. Dental alloys are mainly used to make devices for filling cavities and as substitutes for teeth that are lost because of decay and periodontal disease. A variety of dental devices have been developed, which include metallic fillers, inlays, crowns, bridges, clasps, dentures, dental implants composed of a fixture and an abutment, and fixed braces (train tracks). These forms of dental restoration, custom-shaped for an individual, are made by casting; therefore, the castability of alloys is another requirement for dental applications.

Dental alloys are mainly classified into two groups: precious and nonprecious metals. Suitable alloys are employed according to the intended use. Alloys of precious metals such as gold (Au), palladium (Pd), and silver (Ag) are usually employed because of their high corrosion resistance, biocompatibility, and castability, as compared to those of nonprecious metals. Precious alloys are grouped into high-carat alloy (high-precious or -nobility alloy) and low-carat alloy (low-precious or –nobility alloy). The high-carat alloy contains more than 75 % precious metals. Non-precious metal alloys such as stainless steels, cobalt-chromium, nickel-chromium, and titanium alloys are also commonly used.

Among the dental alloys, precious alloys are widely used. Au alloys have been commonly used in dental applications from past to the present, and many commercial variations of alloy compositions have been developed, despite their high cost. American Dental Association classifies these Au alloys on the basis of their mechanical properties. Many studies have been carried out to improve the mechanical properties of the Au alloys containing copper (Cu) (Au–Cu–Pd [1, 2], Au–Ag–Pd–In [3], Au–Cu–Zn [4], Au–Cu–Zn–Ag [5], and Au–Ag–Cu–Pd [6] alloys), with the main focus on the microstructural changes produced by heat treatment. However, recent trends have shown that low-carat dental alloys (Ag and Pd alloys) are attracting much attention as alternatives to Au alloys because of their lower price. Thus, Ag alloys such as Ag–Pd–Cu–Au [7-17], and Ag–Cu–Pd–Au [18, 19] alloys have been developed for commercial applications. The hardness of these alloys increases with aging or solution treatment, and many studies have reported on the behaviors of these alloys in response to various heat treatments and the various mechanisms. Dental casting of Ag–20Pd–14.5Cu-12Au alloy (mass%) has been developed and used widely in Japan. In general, this alloy is subjected to aging treatment (AT) at around 673 K after solution treatment (ST) at 1023 K in order to enhance its mechanical strength. Recently, it has been reported that the mechanical strength of Ag–20Pd–14.5Cu–12Au alloy is significantly enhanced when the alloy is subjected to ST at temperatures higher than 1073 K and subsequently water quenched without any AT [9-16]. The Vickers hardness of this alloy increases with an increase in the cooling rate after ST [11]. This unique hardening behavior and the increase in mechanical strength induced by the high-temperature ST have been explained in terms of the precipitation hardening caused by the precipitation of an $L1_0$-type ordered β' phase.

This chapter describes with focusing on the relationship between the unique hardening behavior exhibited by the as-solutionized dental Ag–Pd–Cu–Au alloys and the corresponding microstructural changes. Other mechanical properties (fatigue, fretting-fatigue, and friction wear properties) and corrosion properties are also described.

2. Age hardening behavior of Ag–Pd–Cu–Au alloys

Ag-Pd alloys have complete miscibility in all composition ratios. Addition of Cu leads to age hardening in these alloys. In the early researches during 1960–70s, mechanisms for age hardening of Ag-Pd-Cu alloys were proposed [20-22]. According to them, the age hardening was caused by the formation of a CuPd ordered phase ($L2_0$-type) [20], precipitation of a Cu-rich α_1 phase [21], and phase separation of α solid solution to a CuPd ordered phase ($L2_0$-type) and Ag-rich α_2 phase [22]. More recently, Au has been added to Ag-Pd-Cu alloys to increase their corrosion resistance. Therefore, since 1980 to the present, many studies have focused on the age-hardening mechanism of the Ag-Pd-Cu-Au alloys (Table 1) [8, 19, 23, 24]. Ohta et al. [23] reported that the precipitation of a $L1_0$-type face-centered tetragonal CuPd ordered platelet (β') inside the grains and discontinuous precipitation of the Ag-rich α_2 phase and CuPd ordered phase (β) in the grain boundary regions enhance the hardening of Ag-Pd-Cu-Au alloys with a low Au content (Au = 10 mass%). According to their report, TEM images of the precipitates (β and α_2) along the grain boundaries in the alloys show strain contrast and moiré fringes, which indicates that β and α_2 phases are coherent with each other. Solution hardening behavior was also found in alloys subjected to ST at 1223 K followed by slow quenching (SQ) before aging. Researches since 2000 have focused on the age hardening behavior of the Ag-Pd-Cu-Au alloys as a function of the treatment duration [8, 19]. It was found that during the early stage of the aging, diffusion and aggregation of Cu atoms from the Ag-rich α phase occur, and in the later stage, the hardness of the alloy decreases because of the coarsening of the Cu-rich lamellar precipitates [8, 19]. At a Cu concentration of 20 mass%, the CuPd ordered phase (β) does not exhibit any change after aging and thus does not contribute to the age hardening [19].

The age-hardening mechanism of Ag-Pd-Cu-Au alloys with an Au content of 20 mass% was also investigated, and it was concluded that β′ and the discontinuous Ag-rich α_2 phase contribute to the age hardening of these alloys [24]. Fig.1 shows the hardness (Hv) of the precipitates measured independently inside the grains (inter grain) and along the grain boundaries (nodule) [24]. These curves indicate that there are two hardening stages of the alloys in terms of aging temperature: the formation of β′ phase in the grain interior at low aging temperatures and the precipitation of the Ag-rich α_2 phase along the grain boundaries at high aging temperatures. It was also found that the Cu concentration influences the formation of β′ phase in these alloys (Fig. 2) [24]. The hardness (H_v) of each alloy increases with an increase of the Cu concentration, i.e. the volume fraction of β′ phase, at any temperatures.

Compositions of Ag-Pd-Cu-Au alloys (mass%)	References
Ag-20.9Pd-8.5Cu-12.5Au-2.5Zn-0.5Sn-0.1Ir	[8]
Ag-20Pd-20Cu-12Au-2(Zn, Ir, and In)	[19]
Ag-25.4Pd-21.8Cu-10Au	[23]
Ag-25.20Pd-9.88Cu-20Au, Ag-30.32Pd-9.66Cu-5.04Au, Ag-25.4Pd-12.82Cu-9.96Au, Ag-28Pd-9.12Cu-12.04Au	[24]

Table 1. Compositions of Ag-Pd-Cu-Au alloys (mass%).

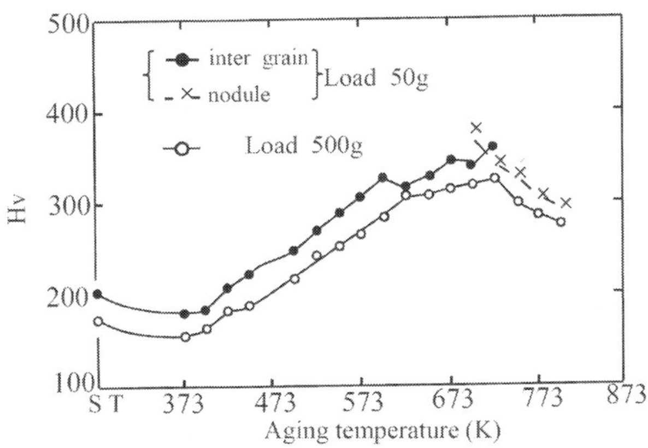

Figure 1. Anisothermal age hardening curve of Ag-25.2Pd-9.88Cu-20Au alloy.

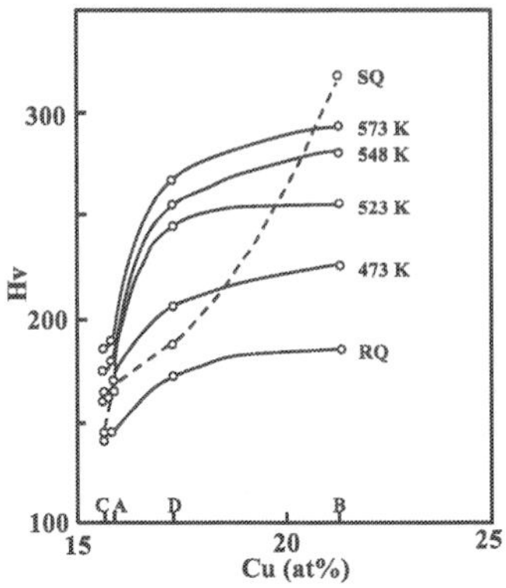

Figure 2. Effect of Cu concentration (at%) on hardness (Hv) of Ag-30.32Pd-9.66Cu-5.04Au (A), Ag-25.4Pd-12.82Cu-9.96Au (B), Ag-28Pd-9.12Cu-12.04Au (C), and Ag-25.2Pd-9.88Cu-20Au (D) alloys. SQ and RQ indicate specimens subjected to slow quenching and rapid quenching after ST, respectively.

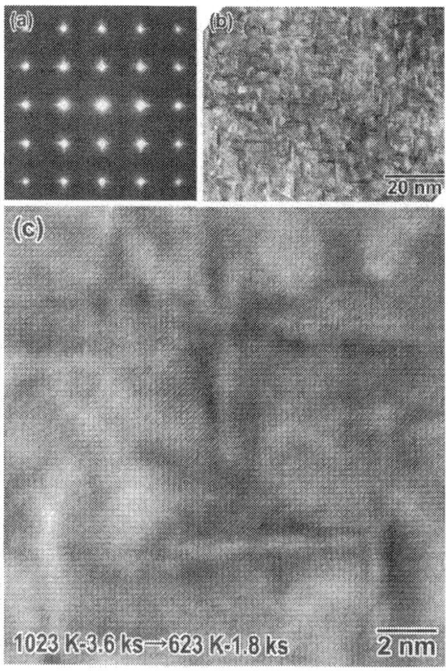

Figure 3. (a) Diffraction pattern, (b) TEM image, and (c) HRTEM image of Ag-Pd-Cu-Au alloy subjected to ST at 1023 K for 3.6 ks followed by aging treatment at 623 K for 1.8 ks.

Figure 3 shows a diffraction pattern, a transmission electron microscopy (TEM) image, and a high resolution TEM (HRTEM) image of β′ phase formed in the Ag–20Pd–14.5Cu–12Au alloy subjected to ST at 1023 K for 3.6 ks followed by AT at 623 K for 1.8 ks [25]. β′ phase is platelet shape with the size of about 10 nm long, which precipitates parallel to {200} crystal plane of matrix. A lattice constant of a-axis matches with that of the matrix, which exhibits coherent with the matrix [25].

3. Mechanical properties of Ag–20Pd–14.5Cu–12Au alloy

3.1. Unique hardening behavior

3.1.1. Precipitation of β′ phase after high temperature solution treatment

Commercial Ag–20Pd–14.5Cu–12Au dental alloy (as-received) is fabricated by a rolling process. The commercial Ag–20Pd–14.5Cu–12Au alloy has a multiphase microstructure (α_1, α_2, β). It is well known that the Ag–Pd–Cu–Au alloy exhibits age hardening behavior as described in the section 2, but the drastic increase in the hardness of Ag–20Pd–14.5Cu–12Au alloy through ST at temperatures over 1073 K subjected to water quenching has been newly reported [10]. This unique hardening behavior has been explained in terms of two hardening mechanisms: (1) solid solution hardening mechanism in which the alloying elements are dissolved into the matrix (α phase) during ST, and (2) precipitation hardening mechanism, in which $L1_0$-type ordered phases are precipitated during the quenching process after ST. Recent studies on the hardening mechanism of Ag–20Pd–14.5Cu–12Au alloy have revealed that the precipitation hardening mechanism is the probable mechanism for the unique hardening behavior exhibited [11, 26].

Conventionally, dentists have employed AT for the hardening of Ag–20Pd–14.5Cu–12Au alloy as mentioned above. Figure 4 [10] shows the effects of heat treatment (AT and ST) temperatures on the mechanical properties (tensile strength, elongation, and hardness (H_{RA})) of this alloy. The tensile strength and hardness increase until the temperature reaches 673 K, and then decrease for up to 923 K due to the removal of strain from the alloy. At temperatures higher than 923 K, this alloy exhibits a unique hardening behavior. Under AT, the tensile strength and hardness of this alloy drastically increase and the elongation decreases after treatment at 673 K. On the other hand, under ST, the tensile strength and hardness still increase but the elongation does not decrease after treatment at 1073 K. Since high temperature ST is very useful for the hardening of this alloy, this treatment will be widely adopted in the future.

The relationship between the microstructural changes in the $L1_0$-type ordered β′ phase and the hardening behavior in the solutionized alloys was investigated by changing the cooling rate. Figure 5 shows a schematic drawing of various cooling rates employed after ST [11]. The Vickers hardness of the as-received alloy and of the alloys subjected to ST followed by water quenching (WQ), air cooling (AC), and cooling in a furnace (FC) are shown in Fig. 6 [11]. ST subjected to WQ and AC leads to a significant and slight increase in the hardness of the alloy,

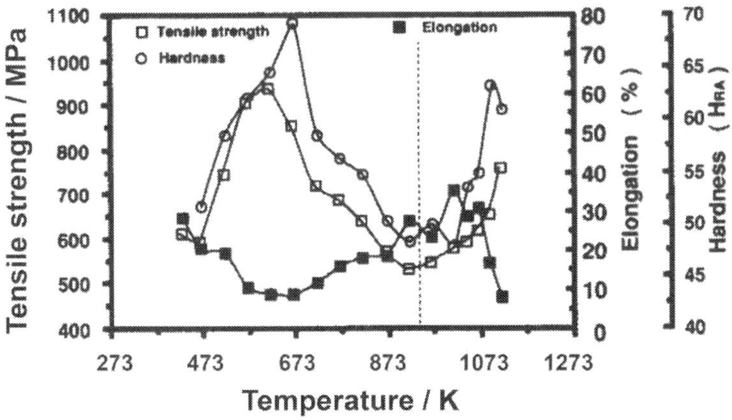

Figure 4. Effect of heat treatment temperature on mechanical properties of Ag–20Pd–14.5Cu–12Au alloy.

respectively, while ST subjected to FC decreases the hardness of the alloy. Thus, the hardness tends to decrease with a decrease in the cooling rate after ST.

Figures 7 and 8 show the microstructures of the as-received and solutionized alloy, respectively, as measured by backscattered electron (BSE) and energy dispersive X-ray spectroscopy (EDX) analysis [13]. The as-received alloy composed of a Cu-rich α_1 phase, an Ag-rich α_2 phase, and Cu–Pd intermetallic β phase, while the solutionized alloy composed of α_2 and β phases. After ST, the α_1 phase dissolved into the Ag-rich α_2 phase and the β phase remained in the matrix. The β′ phase precipitated in the matrix could not be observed by BSE, but could be observed by TEM.

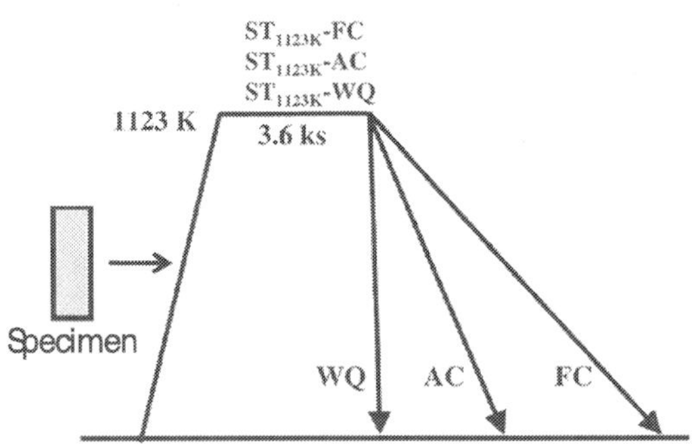

Figure 5. Schematic drawing of heat treatments with various cooling rates after ST. ST_{1123K}-WQ, ST_{1123K}-AC, and ST_{1123K}-FC indicate specimens subjected to ST at 1123 K followed by water, air, and furnace cooling, respectively.

http://dx.doi.org/10.5772/55049

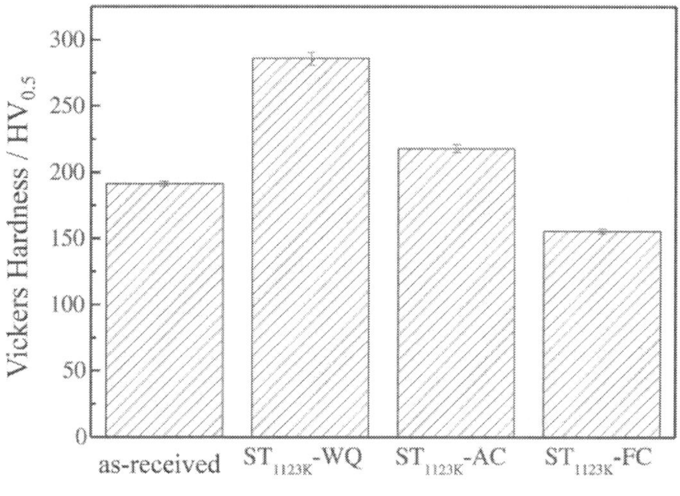

Figure 6. Vickers hardness of as-received, ST_{1123K}-WQ, ST_{1123K}-AC, and ST_{1123K}-FC.

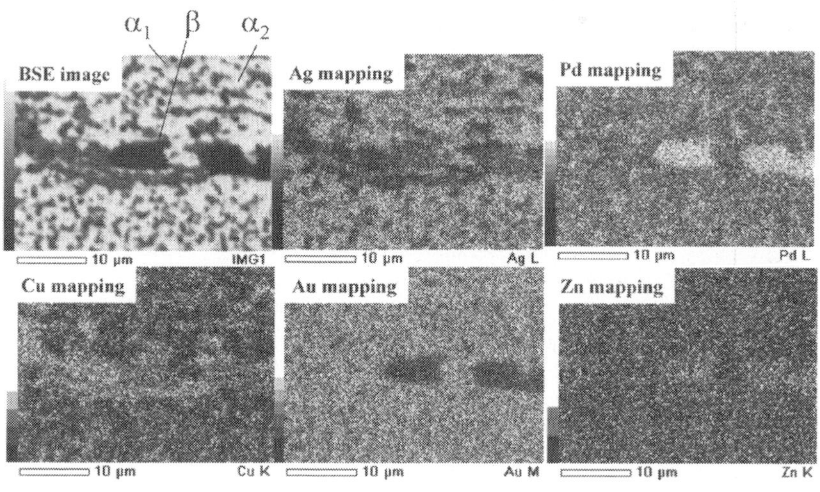

Figure 7. BSE image and elemental mappings obtained by EDX of as-received Ag–20Pd–14.5Cu–12Au alloy.

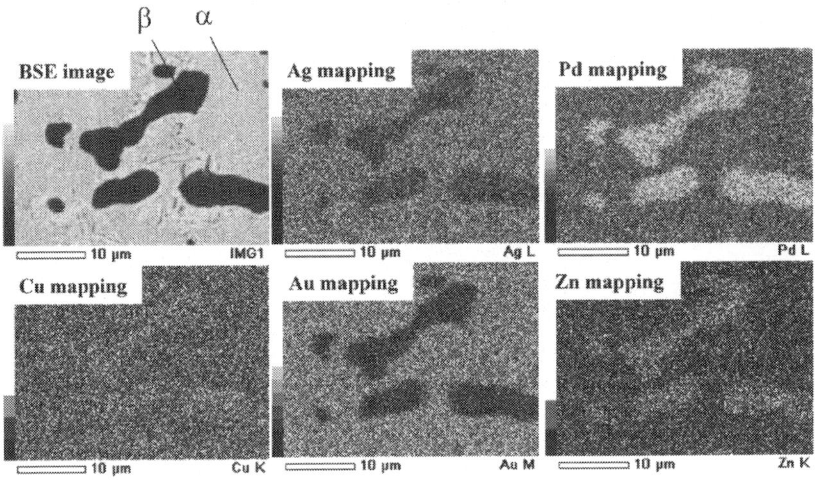

Figure 8. BSE image and elemental mappings obtained by EDX of Ag–20Pd–14.5Cu–12Au alloy subjected to ST at 1123 K for 3.6 ks.

The selected area diffraction patterns (SADP) and key diagrams obtained from TEM for the alloy subjected to ST and WQ, given in Fig. 9, show that three variants of the $L1_0$-type ordered β′ phase are present when the beam direction is parallel to [100] crystal direction: one variant (P) has a c-axis parallel to the electron beam; the other two variants (N_1, N_2) have c-axes normal to the electron beam [11]. The SADPs indicate that an $L1_0$-type ordered β′ phase is densely precipitated in the matrix. The dark field images using a (001)N_1 and (001)N_2 reflections (Fig. 9 (d) and (e)) show that the β′ phase is 2–6 nm wide and 20–60 nm long. In the alloy subjected to ST and AC, the β′ phase is 3–13 nm wide and 50–400 nm long, and in the alloy subjected to ST and FC the β′ phase is 3–25 nm wide and 70–700 nm long, as confirmed by the dark field images using a (001)N_1 reflection. In the alloys subjected to ST followed by AC and FC, the β ′ phase is less densely precipitated than the alloy subjected to ST and WQ. These dark field images indicate that the size and number of the β′ phase vary with the cooling rates after ST. In the solutionized alloy subjected to WQ, the size of the β′ phase is small and the amount of the β′ phase is large. The size of the β′ phase decreased and the amount of the β′ phase increased with an increase in the cooling rate after ST. The driving force behind the nucleation of the β′ phase in the solutionized alloy subjected to WQ is stronger than that in the solutionized alloy subjected to AC and FC, because the degree of undercooling in the solutionized alloy subjected to WQ is larger than that in the solutionized alloy subjected to AC and FC. The extent of nucleation of the β′ phase in the solutionized alloy subjected to WQ is larger than that in the solutionized alloy subjected to AC and FC. The β′ phase in the solutionized alloys subjected to AC and FC grew more coarsely than that in the solutionized alloy subjected to WQ, because the diffusion occurred more easily owing to the slow cooling process in the solutionized alloys subjected to AC and FC. Therefore, it is likely that the β′ phase is formed during the cooling process and grows by diffusion.

http://dx.doi.org/10.5772/55049

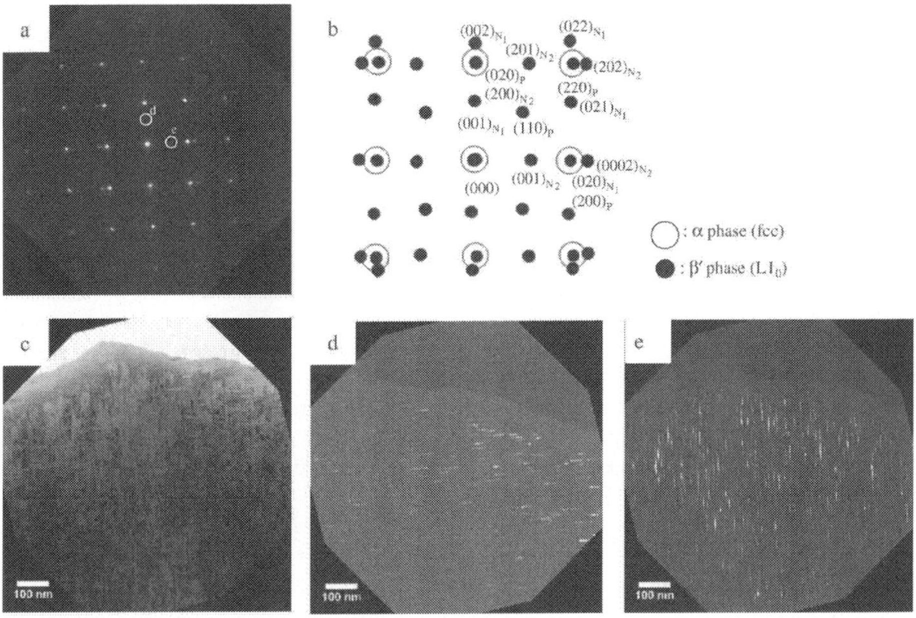

Figure 9. TEM micrographs of ST_{1123K}-WQ: (a) a selected area diffraction pattern, (b) a key diagram, (c) a bright-field image, and (d) and (e) dark-field images using (001)N_1 and (001)N_2 reflections, respectively. The beam direction is parallel to [100] crystal direction.

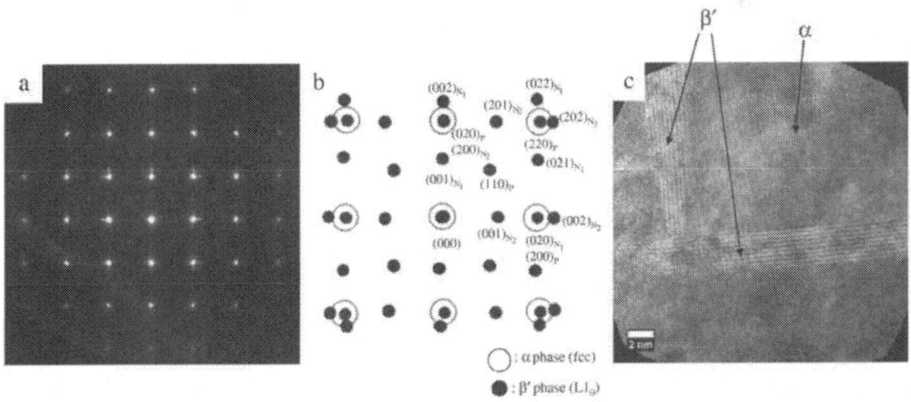

Figure 10. HRTEM micrographs of ST_{1123K}-WQ: (a) a selected area diffraction pattern, (b) a key diagram, and (c) a bright-field image. The beam direction is parallel to [100].

Figure 10 shows the SADP, key diagram, and HRTEM bright field image of ST_{1123K}-WQ [11]. The SADP and the key diagram in Fig. 10 (a) and (b) indicate that the FCC α phase and the three variants of the $L1_0$-type ordered β' phase are superimposed. The $L1_0$-type ordered β' phase is precipitated in the FCC matrix after ST. The streaks on the reflection spots of the $L1_0$-type ordered β' phase indicate that the shape of the β' phase is similar to a thin plate. The HRTEM bright field image in Fig. 10 (c) shows that several nanometer-sized thick plate-shaped β' phase with two variants (N_1, N_2), whose c-axes are normal to the electron beam, are precipitated in the matrix.

It is well known that it is difficult to make a single phase by quenching after high temperature ST; the supersaturated vacancies help the solutes to diffuse more easily and also help to form more clusters, G–P zones, and metastable phases during quenching after high temperature ST. The precipitated β' phase of ST_{1123K}-WQ shown in Fig.9 (d), (e) and Fig. 10 (c) is like a thin plate with a nanometer-scale thickness. These images also suggest that the formation of the β' phase is diffusion controlled. Generally, the formation of precipitates can be considered to be order–disorder transition, diffusionless transformation (martensitic transformation), or diffusional transformation. In the as-solutionized Ag–Pd–Cu–Au alloy used in this case, the dependence of microstructural changes in the precipitated β' phase on both the cooling rate after ST and the ST temperature show that the precipitated $L1_0$-type ordered β' phase is formed during the cooling process and that the growth of the β' phase is influenced by the diffusion process. The hardness increases with an increase in the cooling rate after ST, and consequently, the hardness of ST_{1123K}-WQ increases significantly by quenching after ST (Fig. 6). The fine β' phase in ST_{1123K}-WQ is densely precipitated in the matrix. The hardness of ST_{1123K}-AC increased only slightly, while the hardness of ST_{1123K}-FC decreases, as only coarse β' phases are precipitated in the matrix of ST_{1123K}-AC and ST_{1123K}-FC. Thus, the increase in hardness may be strongly affected by the presence of finely precipitated β' phase. The coherent precipitation of β' phases with long and short axes of around 100 nm and 10 nm, respectively, also occured during ST, although the amount of β' phase decreases with an increase in the ST time. The effect of solid solution hardening in the α, α_1, and α_2 phases is lower than that exerted by the precipitation hardening due to β' phases.

3.1.2. Hardening behavior of Ag–20Pd–12Au–14.5Cu alloy fabricated by liquid rapid solidification

An Ag–20Pd–14.5Cu–12Au alloy with a single α phase can be fabricated using a liquid rapid solidification (LRS) method that employs a melting mechanism, as shown schematically in Fig. 11 [12]. The critical temperature for the order–disorder transformation in the Cu–Pd binary phase diagram is below 1023 K, and hence, at 1023 K, the Cu-rich phase α_1 and Ag-rich phase α_2 decompose, as shown in the Ag–Cu binary phase diagram. Figure 12 shows TEM micrographs of an Ag–20Pd–14.5Cu–12Au alloy fabricated by the LRS method [13]. No precipitation is observed in the matrix.

As shown in Fig. 12 (b) and (c), the LRS alloy consists of a single α phase with face centered cubic structure (FCC). The tensile properties of the as-received Ag–20Pd–14.5Cu–12Au alloy (AS), AS subjected to ST at 1123 K for 3.6 ks in vacuum ($ST_{AS/3.6\,ks}$), LRS alloy (LRS), and LRS alloy subjected to ST at 1123 K for 3.6 ks in vacuum ($ST_{LRS/3.6\,ks}$) are shown in Fig. 13 [13]. The

http://dx.doi.org/10.5772/55049

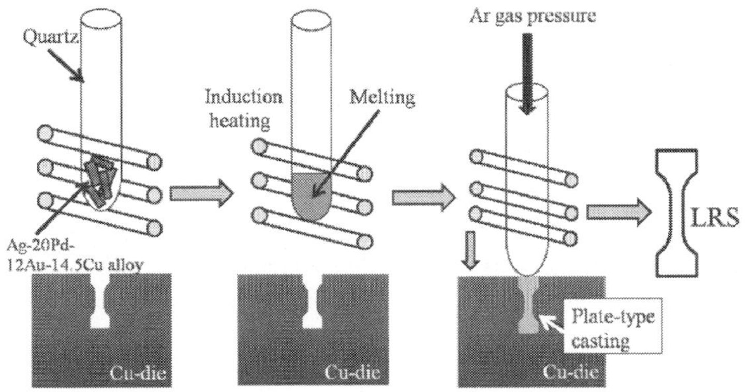

Figure 11. Schematic drawing of LRS method.

tensile strength and 0.2% proof stress of the AS alloy drastically increased after ST. The elongation of AS subjected to ST is smaller than that of the AS alloy. On the other hand, the tensile strength and 0.2% proof stress of the LRS alloy decrease after ST. The reduction in strain and the coarsening of the α phase during ST result in the decrease in the tensile strength and 0.2% proof stress and the increase in elongation. The tensile strength of the LRS alloy and solutionized LRS are relatively smaller than those of the AS alloy and the AS alloy subjected to the ST.

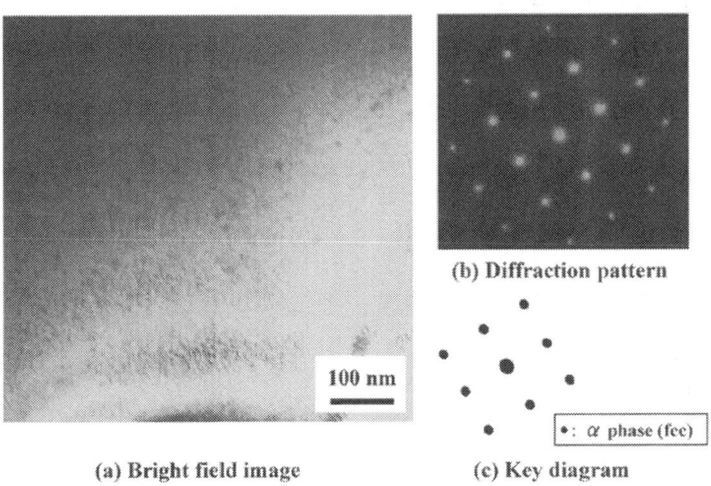

Figure 12. TEM micrographs of alloy fabricated by LRS: (a) bright field image, (b) diffraction pattern and (c) key diagram. Beam direction is parallel to [100].

Figure 14 shows the XRD profiles of the LRS alloy solutionized at 1173 K for 3.6 ks ($1173WQ_{LRS}$/3.6 ks) and the LRS alloy subjected to AT at 673 K for 1.8–28.8 ks following the ST ($673WQ_{LRS}$/1.8 ks-28.8 ks) [12]. A single α phase is identified in the solutionized LRS alloy, whereas, in the LRS alloy subjected to AT after the ST, α_2 and β phases are observed. The Vickers hardness of the solutionized LRS alloy and the LRS alloy subjected to AT are shown in Fig. 15. The hardness of the LRS alloy subjected to AT increases greatly as compared to that of the solutionized LRS alloy with a single α phase owing to the precipitation of β phase.

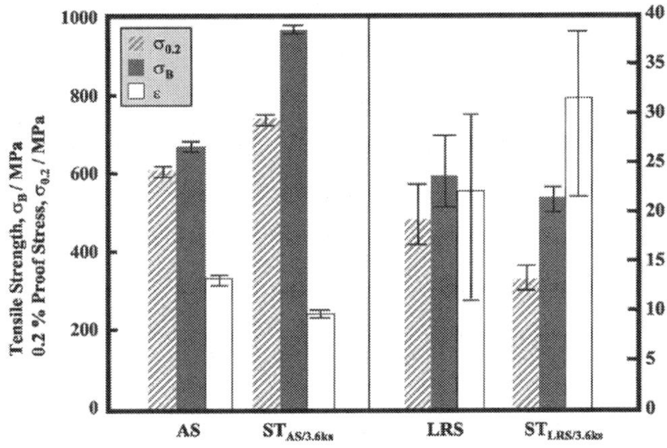

Figure 13. Tensile properties of AS, $ST_{AS/3.6ks}$, LRS, and $ST_{LRS/3.6ks}$.

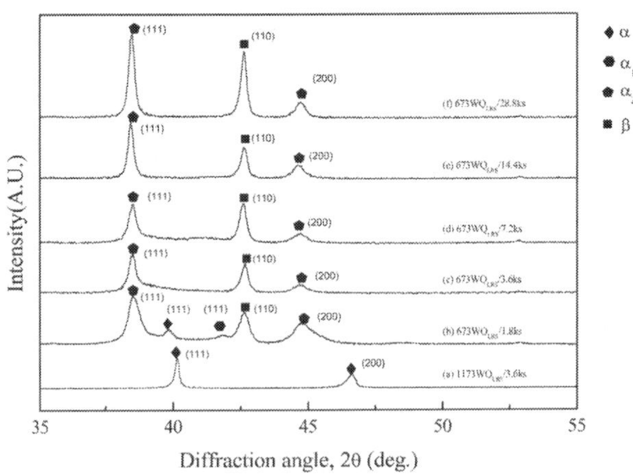

Figure 14. XRD profiles of (a) $1173WQ_{LRS}$/3.6 ks, (b) $673WQ_{LRS}$/1.8 ks, (c) $673WQ_{LRS}$/3.6 ks, (d) $673WQ_{LRS}$/7.2 ks, (e) $673WQ_{LRS}$/14.4 ks, and (f) $673WQ_{LRS}$/28.8 ks.

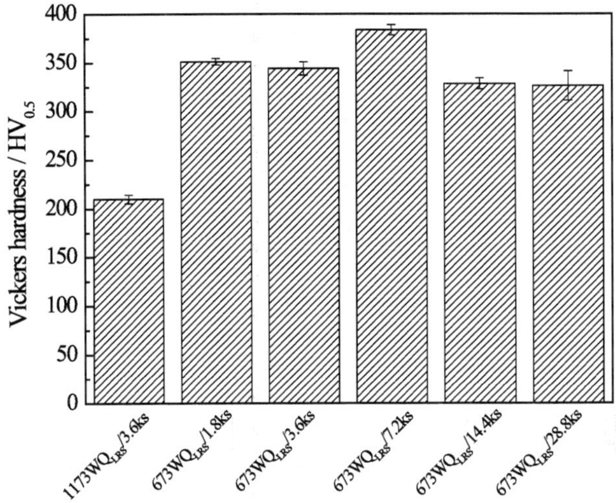

Figure 15. Vickers hardness of 1173WQ$_{LRS}$/3.6 ks, 673WQ$_{LRS}$/1.8 ks, 673WQ$_{LRS}$/3.6 ks, 673WQ$_{LRS}$/7.2 ks, 673WQ$_{LRS}$/14.4 ks, and 673WQ$_{LRS}$/28.8 ks.

3.2. Fatigue properties

Dental prosthetic products produced by dental casting method are subjected to cyclic stress, i.e., fatigue, because of mastication. Dental castings contain a number of casting defects such as microshrinkages, pores, and surface roughness. The effects of these casting defects on the fatigue properties of cast Ag–20Pd–14.5Cu–12Au alloy were investigated in comparison with the fatigue properties of a drawn Ag–20Pd–14.5Cu–12Au alloy. The tensile properties of the cast specimens and drawn specimens are shown in Fig. 16 [14]. The cast specimens were prepared using a lost wax method. The drawn alloy bars were solutionized at 1073K for 3.6 ks in vacuum and then cooled in air (D-1073AC (drawn)). The heat treatment produced a microstructure with a matrix similar to that in an as-cast alloy. For the mechanical testing, the following specimen dimension were used: gauge diameter 3 mm and gauge length 20 mm for cast specimens, gauge diameter 2 mm and gauge length 20 mm for drawn specimens. The surfaces of the cast specimens for tensile tests were sand blasted (non-polished). The tensile strength and 0.2% proof stress of the cast specimens were higher than those of the drawn specimens. The distribution of the elongation of the cast specimens was larger than that of the elongation of the drawn specimens. The amount of intermetallic β phase that leads to the high strength and low ductility is greater in the cast specimens than in the drawn specimens. The microstructure of the cast specimens may be coarser than that of the drawn specimens. The relationship between the volume fraction and the number of the microshrinkage as measured on the fracture surface of the non-polished cast specimen and the elongation is shown in Fig. 17 [14]. Although no correlation was obtained between the volume fractions of the micro-shrinkage and elongation, the elongation decreases with an increase in the number of the microshrinkage. The surfaces of some fatigue test specimens were finished by buff-polishing (polished). Drawn specimens with a gauge diameter of 5 mm and a gauge length of 20 mm were used for the fatigue tests. Fatigue tests were carried out in order to obtain an S–N curve for each specimen at a stress ratio, R, of 0.1 and a frequency of 10 Hz with a sine waveform in air at room temperature (295 K). As can be seen in Fig.18 [14], the fatigue strength of the non-polished specimen is similar to that of the polished specimen in both the low-cycle fatigue life region and the high-cycle fatigue life region. The distribution of the fatigue strength of both the cast specimens is greater than that of the fatigue strength of the drawn specimens. The fatigue strength of the non-polished cast specimens was 352–492 MPa ($\Delta\sigma_{max}$ (range of σ_{max}) = 140 MPa) at 10^5 cycles and 209–284 MPa ($\Delta\sigma_{max}$ = 75 MPa) at 10^6 cycles. The fatigue strength of the polished cast specimens was 347–565 MPa ($\Delta\sigma_{max}$ = 218 MPa) at 10^5 cycles and 233–251 MPa ($\Delta\sigma_{max}$ = 18 MPa) at 10^6 cycles. The fatigue strength of both cast specimens was lower than that of the drawn specimens, and is extremely low in the high-cycle fatigue life region, where N_f exceeds 10^5 cycles. The fatigue limits, which represent the fatigue strength for which the number of cycles to failure is over 10^7 cycles, are around 210 MPa. On the other hand, although the fatigue limit of the drawn specimen was not obtained, it is expected to be near 400 MPa. Figure 19 [14] shows SEM fractographs taken near the site at which fatigue crack initiated in the cast specimens and the drawn specimen, which then broke in the high-cycle fatigue life region. In the case of the drawn specimen, the fatigue crack initiates at the slip band on the specimen surface, but in the case of the cast specimens, the fatigue crack initiates from the microshrinkage near the specimen surface. In general, slip damage accumulates on the specimen surface, after which extrusion and intrusion take place. Next, the stress concentration occurs and a fatigue crack initiates along the slip plane. If there are polishing scars, defects, etc. on the specimen surface, stress concentration occurs, and the fatigue strength becomes lower. The sizes of the microshrinkage areas and pores whose size exceeds 10 μm were measured, because the microshrinkage areas and pores whose size is less than 10 μm were difficult to distinguish from dimples on the fracture surface in the measurement on the fractographs. The number and size of the microshrinkage, whose size exceeds 10 μm as measured on the fatigue fracture surface, are greater than those measured on the cross- section near the fatigue fracture surface. Therefore, the fatigue crack propagates by preferentially linking the areas of microshrinkage. The number and size of the pores measured on the fatigue fracture surface is nearly equal to those measured on the cross-section near the fatigue fracture surface. Therefore, the effect of the pores on the fatigue properties is much smaller than that of the microshrinkage.

In general, dental prosthetic materials sustain a stress of 20–230MPa during mastication. Moreover, they must be able to sustain such cyclic stress over 10,000,000 times (10^7 cycles), which is the calculated number of cycles that is equivalent to the number of times food will be chewed in a span of ten years. Therefore, the target value of the fatigue limit of the cast specimen is considered to be 230MPa, which is the greatest mastication stress. Since the fatigue strength of this cast specimen is strongly dependent on the size of the microshrinkage that acts as the fatigue crack initiation site, it is prudent to estimate the size of the microshrinkage. It is also beneficial to know the size that can be tolerated, in order to achieve reliability in casting that is subjected to fatigue fracture.

http://dx.doi.org/10.5772/55049

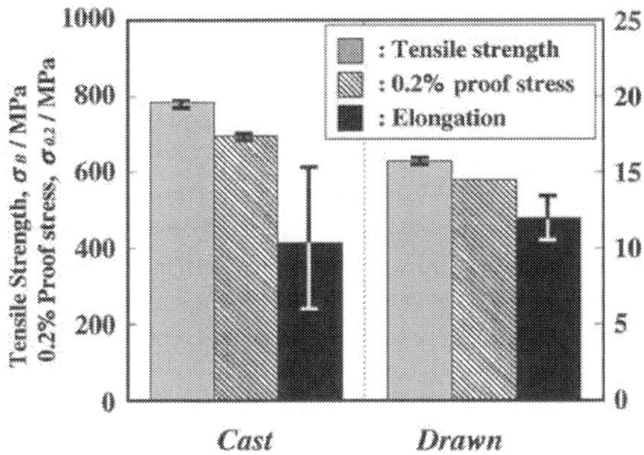

Figure 16. Tensile properties of non-polished cast specimens (Cast) and drawn specimens (Drawn: D-1073AC (drawn)).

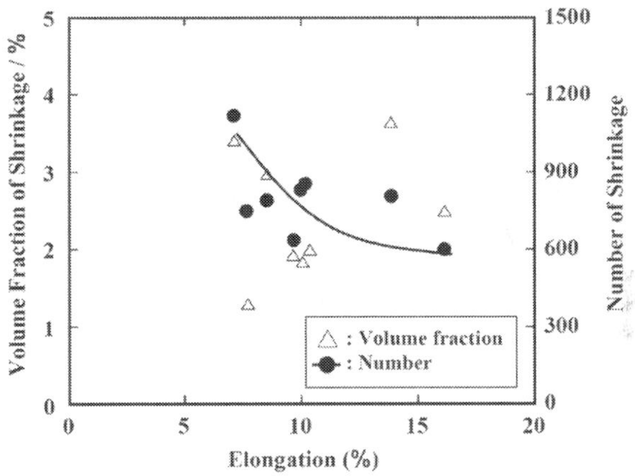

Figure 17. Relationships between volume fraction or number of shrinkage on fractograph and elongation of non-polished cast specimen.

3.3. Fretting–fatigue properties

Fretting–fatigue properties are also important for alloys that are used for dental applications, because during mastication, fretting occurs between the alloys and the teeth opposite them. Fig. 20 [15] shows the S–N curves of AS alloys subjected to ST at 1123 K for 3.6 ks followed by WQ and AT at 673 K for 1.8 ks followed by WQ that were obtained from plain fatigue and fretting–fatigue tests. The fretting–fatigue strength of the Ag–20Pd–14.5Cu–12Au alloy subjected to ST and AT decreases significantly as compared to the fatigue strength without fretting (plain–fatigue strength). The fretting–fatigue strength after the ST decreases by

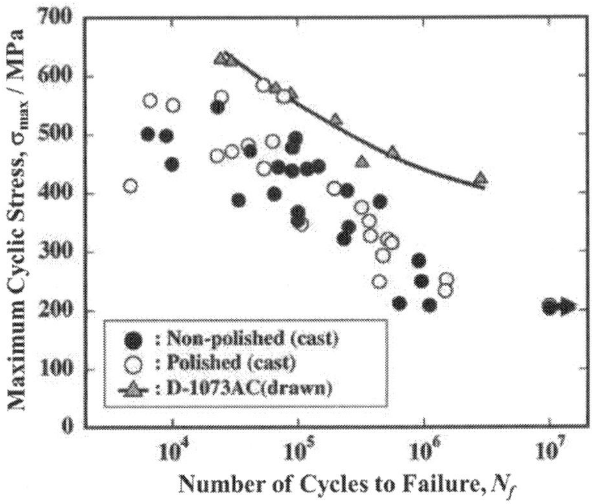

Figure 18. S-N_f curves of non-polished cast specimen (Non-polished (cast)), polished cast specimen (Polished (cast)), and drawn specimen (D-1073AC(drawn)).

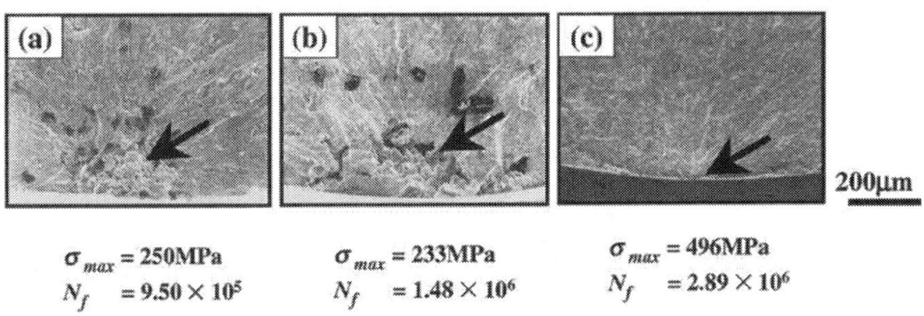

Figure 19. SEM fractographs in high-cycle fatigue life region: (a) non-polished cast specimen, (b) polished cast specimen, and (c) drawn specimen. Arrows indicate crack initiation sites. σ_{max} and N_f are the maximum cyclic stress and number of cycles to failure, respectively.

approximately 13% in the low-cycle fatigue life region and by approximately 40% in the high-cycle fatigue life region, as compared to the fatigue strength of the solutionized alloy. Moreover, the fretting–fatigue strength after the AT decreases by approximately 60% as compared to that after the ST, especially in the high-cycle fatigue life region. A schematic drawing of crack initiation from fretting damage region is shown in Fig.21 [15]. Although a slip between the specimen and a fretting pat does not occur in the stick region, a microslip between the specimen and the fretting pat occurs in the slip region during the deformation of the specimen. A significant stress concentration is generated by the damage to the specimen surface caused

http://dx.doi.org/10.5772/55049

by the microslip. Fracture morphologies caused by crack initiation and propagation then appear. It can be observed that several traces of fretting wear are distributed in the slip region of both materials. These wear traces are generated by the accumulation of wear debris on the fretting pad or on the fretting–fatigue specimen. These traces of fretting wear are distributed more closely in the slip region of the material that was subjected to AT. Therefore, the fatigue life decreases significantly because the fretting–fatigue crack initiation life and the propagation life decrease in the material that was subjected to AT.

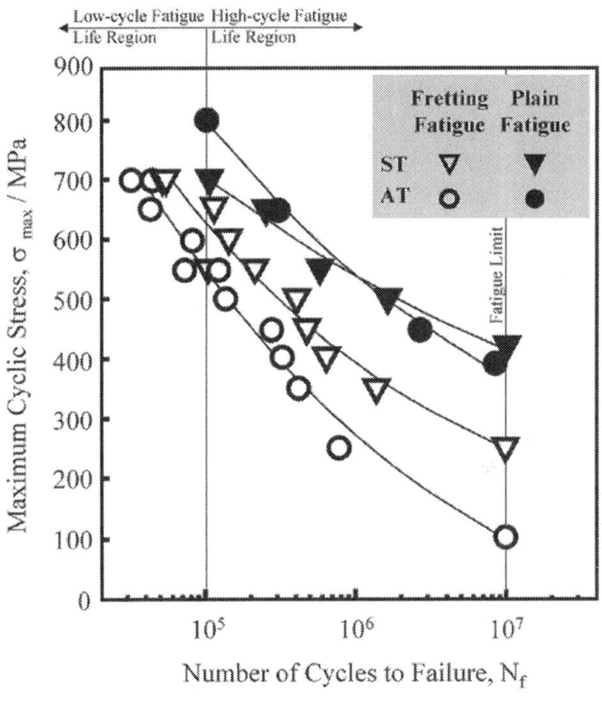

Figure 20. S–N curves of Ag–20Pd–14.5Cu–12Au alloy subjected to ST and AT obtained from plain fatigue and fretting fatigue tests.

3.4. Friction wear properties of Ag–20Pd–12Au–14.5Cu alloy in corrosive environments

Mastication also leads to friction wear in dental alloys and in teeth. As the friction wear progresses, this eventually causes problems with mastication. Therefore, the evaluation of the friction wear properties of dental alloys is quite important to the health of the teeth and oral cavity. The friction wear property of Ag–20Pd–14.5Cu–12Au alloy that was subjected to various heat treatments was evaluated in three corrosive environments: distilled water, 0.9% NaCl solution, and 3% NaCl solution. In the friction wear testing, the alloys were subjected to

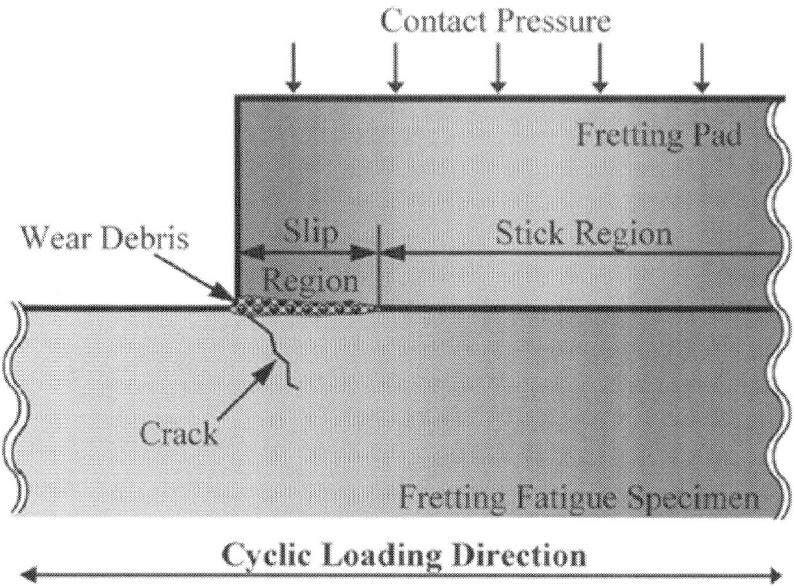

Figure 21. Schematic drawing of crack initiation from fretting damage region.

STs at 1073 and 1123 K subjected to WQ, AT at 673 K, and ST at 1073 K subjected to AC. It was found that the friction wear properties are influenced by the microstructures of this alloy as well as by the corrosive environments. The friction wear tests were performed using a pin-on-disk-type friction wear tester. The applied load, sliding diameter, sliding velocity, sliding distance, and test duration were 9.8 N, 3 mm, 31.4 mm/s (100 rpm), 4.71×10^5 mm, and 115 ks, respectively. The temperature of all the solutions was 310 K. The weight loss was calculated by subtracting the combined weight of a specimen and the mating material it after the friction wear test, from their combined weight before the test. Figure 22 [16] shows the total weight loss and wear surface roughness of specimens that were subjected to each heat treatment in each corrosive solution. The total weight loss (the sum of the weight losses of the specimen and the material opposite it) of the specimens was largest in distilled water, 0.9% NaCl and then 3% NaCl solution. A pin with a diameter of 1 mm, which was made from the as-received alloy, was used as the mating material. The weight losses of the specimens and the mating materials increased with a decrease in the kinematic viscosity of the solutions, due to the increase in the average friction coefficient during the friction wear tests. In every environment, the total weight loss of the specimen subjected to AT at 673 K, which gives a large amount of precipitated β phase, was small compared with that of the other heat-treated specimens, due to the higher degree of hardness near the contact surface, as also shown in Fig.4 [10]. The total weight loss increased with the surface roughness of the contact surface. The hardness near the contact surface changed the surface roughness of the contact surface. The mode of wear of these alloys in a corrosive environment is adhesive wear. No tarnish, which is corrosion caused by chloride ion and the production of silver chloride coatings were found on the wear marks or wear particles after the friction wear tests in each environment.

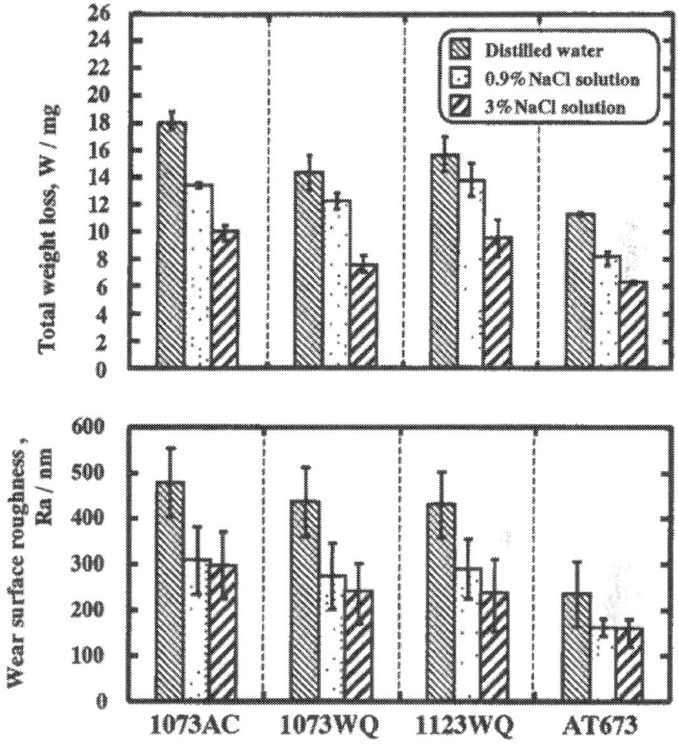

Figure 22. Total weight loss and wear surface roughness of specimen subjected to each heat treatment in each corrosive solution. 1073AC, 1073WQ, 1123WQ, and AT673 indicate specimens subjected to ST at 1073 K for 3.6 ks followed by AC or WQ, ST at 1123 K for 3.6 ks followed by WQ and AT at 673 K for 1.8 ks.

3.5. Corrosion properties

The study of the corrosion properties of dental alloys is very important due to their usage in severe oral environments. The formation of corrosion compounds on the surface causes the alloy to tarnish. The corrosion behaviors of commercial Ag-22.41Pd-15.64Cu-12.1Au alloys in various solutions were investigated by Ichinose [17]. The alloy was casted into a plate with 2 mm in thickness by heating at 973 K for 30 min. Table 2 [17] shows the rest-potential of the as-cast alloy in 1% NaCl solution, artificial saliva, human saliva, and 1% $C_3H_6O_3$ solution. The rest-potentials in 1% NaCl solution and 1% $C_3H_6O_3$ solution show obviously higher values than those for artificial and human saliva. The organic compounds in the artificial and human saliva inhibit the redox reaction by adsorbing on the surface of the alloy or by the formation of complexes with the metal. AgCl was formed on the surface of the alloy after anode polarization sweeping up to 1800 mV in 1% NaCl solution, artificial saliva, and human saliva, which contain Cl ion. 90% of the released ions of the as-cast alloy immersed in human saliva for 24 h were Cu ions (Fig.23 [17]). Cu contributes to the enhancement the hardness of the alloy. However, a large amount of Cu in the alloy decreases the corrosion resistance. The as-cast alloy was subjected to a softening treatment at 1073 K for 3 min followed by WQ (softened). An ingot of the commercial Ag-22.41Pd-15.64Cu-12.1Au alloy was subjected to ST at 1123 K for 2 h and then heated at 723 K for 1 h or 20 h (723 K 1h aged and 723 K 20 h aged, respectively). As shown in Fig. 24 [17], the amounts of released Cu and Ag ions depend on the heat treatment that the alloy is subjected to, because the corrosion behavior is influenced by the microstructure of the alloy. The amount of Cu ions released from the as-cast alloy is higher than that released from heat-treated alloys. Cavities formed by shrinkage during casting in the as-cast alloy are the reason for this phenomenon.

Solution	Rest–potential
1% NaCl solution	112.7 ± 22.4 mV
Artificial saliva	24.7 ± 17.2 mV
Human saliva	28.0 ± 20.4 V
1% $C_3H_6O_3$ solution	113.6 ± 37.7 mV

Table 2. Rest-potential of as-cast Ag-22.41Pd-15.64Cu-12.10Au alloy in various corrosive solutions.

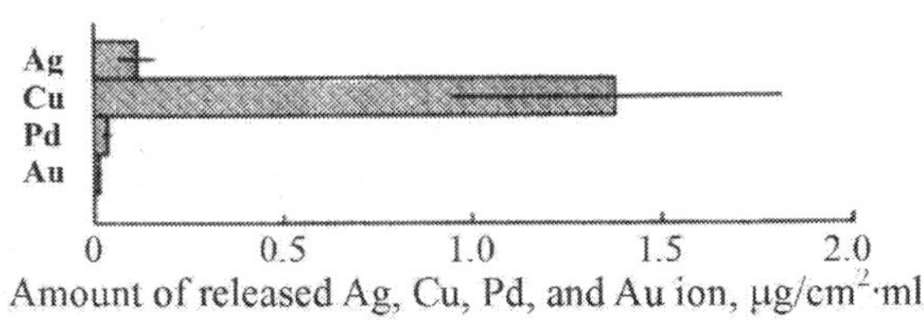

Figure 23. Amounts of ions released from as-cast Ag-22.41Pd-15.64Cu-12.10Au alloy immersed in human saliva for 24 h.

4. Summary

This chapter details the microstructure, the mechanical properties (hardness, fatigue, fretting-fatigue, and friction-wear), and the corrosion properties of the Ag–Pd–Cu–Au alloys, especially the Au-20Pd-14.5Cu-12Au alloy. Most studies on these alloys have been carried out in Japan and Korea. Although other dental materials such as high carat gold alloys, amalgam,

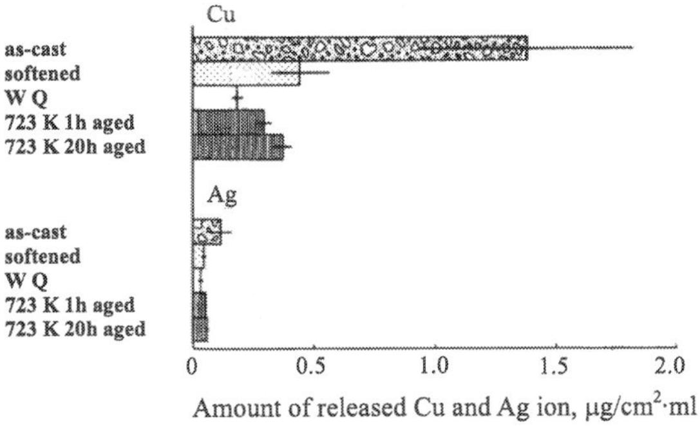

Figure 24. Cu and Ag release amounts from as-cast, softened, WQ, 723 K 1h aged, and 723 K 20 h aged Ag-22.41Pd-15.64Cu-12.10Au alloys immersed in human saliva for 24 h.

and non-precious alloys such as titanium and its alloys, cobalt-chromium alloys, and nickel–chromium alloys are also important, the authors intended to introduce present results of studies on the Ag–Pd–Cu–Au alloys to researchers who study dental materials all over the world through this chapter.

A variety of test results showed that the hardness of the alloy can be drastically increased through ST at a temperature over 1073 K subjected to WQ. The $L1_0$-type ordered phase (β′ phase) precipitated during the quenching process leads to this unique hardening behavior. The hardness of the alloy increased with an increase in the cooling rate following ST, because the size of the β′ phase decreased and the number of the β′ phase increased with an increase in the cooling rate following ST.

It was also found that the fatigue strength of a cast alloy was considerably less than that of a drawn alloy. The fatigue crack of a cast alloy initiated preferentially at the shrinkage near the specimen surface. The deviation of the fatigue strength of the cast alloy became small by relating the fatigue life to the maximum stress intensity factor that was calculated, assuming that the shrinkage that begins as a fatigue crack initiation site becomes an initial crack. This means that the size of the shrinkage strongly affects the fatigue strength of this cast alloy. A tolerable shrinkage size that satisfies the target value of the fatigue limit (230 MPa) of this cast alloy was calculated to be below 80 μm, using a derived equation that describes the relationship between the maximum stress intensity factor and the number of the cycles to failure.

The fretting–fatigue strength of the Ag–20Pd–12Au–14.5Cu alloy subjected to ST and aging treatment decreased significantly as compared to the fatigue strength without fretting. The fretting–fatigue strength after the aging treatment decreased by approximately 60% as compared to that following the ST, especially in the high-cycle fatigue life region.

The total weight losses (the sum of the weight losses of a specimen and of the mating material) of specimens were largest in distilled water, subjected to 0.9% NaCl and then 3% NaCl solution. In every environment, the total weight loss of a specimen subjected to aging treatment at 673

K, which gives a large amount of precipitated β phase, was small compared with that of the other heat treated specimens, due to the higher degree of hardness near the contact surface.

The Ag-Pd-Cu-Au alloys have excellent mechanical and corrosion properties, but the relatively high cost and supplied amounts of these alloys are issues to be used widely. The authors hope that the Ag-Pd-Cu-Au alloys is further investigated by many researchers for its wide usage.

Acknowledgements

This study was supported in part by ISHIFUKU Metal Industry Co., Ltd. for their support during this study. This study was financially supported in part by a Grant-in-aid for Scientific Research from the Japan Society for the Promotion of Science (Grant Number 21360332), the Global COE program "Materials Integration International Center of Education and Research, Tohoku University," Ministry of Education, Culture, Sports, Science and Technology (MEXT) of Japan, and Tohoku Leading Women's Jump Up Project, Tohoku University. The authors are very grateful to Dr. YH. Kim (Institute for Materials Research, Tohoku University) and Prof. H. Fukui (Aichi-Gakuin University) for their experimental work and fruitful discussion.

Author details

Junko Hieda, Mitsuo Niinomi, Masaaki Nakai and Ken Cho

Department of Biomaterials Science, Institute for Materials Research, Tohoku University, Katahira, Aoba-ku, Sendai, Japan

References

[1] Ohta M, Shiraishi T, Yamane M. Phase transformation and age-hardening of Au-Cu-Pd ternary alloys. Journal of Materials Science 1986;21(2) 529-535.

[2] Winn H, Tanaka Y, Shiraishi T, Udoh K, Miura E, Hernandez RI, Takuma Y, Hisatsune K. Two types of checkerboard-like microstructures in Au–Cu–Pd ternary alloys. Journal of Alloys and Compounds 2000;306(1-2) 262-269.

[3] Seol HJ, Son KH, Yu CH, Kwon YH, Kim HI. Precipitation hardening of a Cu-free Au–Ag–Pd–In dental alloy. Journal of Alloys and Compounds 2005;402(1-2) 130-135.

[4] Seol HJ, Shiraishi T, Tanaka Y, Miura E, Hisatsune K, Kim HI. Ordering behaviors and age-hardening in experimental AuCu–Zn pseudobinary alloys for dental applications Biomaterials 2002;23(24) 4873-4879.

[5] Hisatsune K, Sakrana A, Hamasaki K, Hernandez R, Salonga JP. Phase transformation in a dental gold alloy for soldering. Journal of Alloys and Compounds 1997;261(1-2) 308-312.

[6] Lee JH, Yi SJ, Seol HJ, Kwon YH, Lee JB, Kim HI. Age-hardening by metastable phases in an experimental Au–Ag–Cu–Pd alloy. Journal of Alloys and Compounds 2006;425(1-2) 210-215.

[7] Santos ML, Acciari HA, Vercik LCO, Guastaldi AC. Laser weld: microstructure and corrosion study of Ag–Pd–Au–Cu alloy of the dental application. Materials Letters 2003;57(13-14) 1888-1893.

[8] Seol HJ, Kim GC, Son KH, Kwon YH, Kim HI. Hardening mechanism of an Ag–Pd–Cu–Au dental casting alloy, Journal of Alloys and Compounds 2005;387(1-2) 139-146.

[9] Fukui H, Shinoda S, Mukai M, Yasue K, Hasegawa J. Effect of heat treatment on mechanical properties of type IV gold and 12wt% Au-Pd-Ag alloys. The Japanese Society for Dental Materials and Devices 1992;11(1) 141-148.

[10] Fukui H, Mukai M, Shinoda S, Hasegawa J. Strengthening mechanism of Au-Pd-Ag-Cu system. The Japanese Society for Dental Materials and Devices 1993;12(6) 685-690.

[11] Kim YH, Niinomi M, Hieda J, Nakai M, Fukui H. Formation of L10-type ordered β′ phase in as-solutionized dental Ag–Pd–Au–Cu alloys and hardening behavior. Materials Science and Engineering: C 2012;32(3) 503–509.

[12] Kim YH, Niinomi M, Nakai M, Akahori T, Kanno T, Fukui H. Mechanism of unique hardening of dental Ag–Pd–Au–Cu alloys in relation with constitutional phases. Journal of Alloys and Compounds 2012;519 15–24.

[13] Akahori T, Niinomi M, Nakai M, Tsutsumi H, Kanno T, Kim YH, Fukui H. Relationship between unique hardening behavior and microstructure of dental silver alloy subjected to solution treatment. Journal of the Japan Institute of Metals 2010;74(6) 337-344.

[14] Mizumoto T, Niinomi M, Nakano Y, Akahori T, Fukui H. Fatigue properties of cast Ag-Pd-Cu-Au-Zn alloy for dental applications in the relation with casting defects. Materials Transactions 2002;43(12) 3160-3166.

[15] Akahori T, Niinomi M, Nakai M, Kawagishi W, Fukui H, Fretting-fatigue properties and fracture mechanism of semi-precious alloy for dental applications. Journal of The Japan Institute of Metals 2008;72(1) 63-71.

[16] Mizumoto T, Niinomi M, Akahori T, Katou K, Fukui H. Friction wear properties of dental Ag-Pd-Cu-Au alloy in corrosive environments. The Japanese Society for Dental Materials and Devices 2003; 22(6) 459-468.

[17] Ichinose S. Corrosion behavior of dental Au-Pd-Cu-Au alloy in various solutions. The Japanese Society for Dental Materials and Devices 1992;11(1) 149-168.

[18] Seol HJ, Lee DH, Lee HK, Takada Y, Okuno O, Kwon YH, Kim HI. Age-hardening and related phase transformation in an experimental Ag–Cu–Pd–Au alloy. Journal of Alloys and Compounds 2006;407(1-2) 182-187.

[19] Yu CH, Park MG, Kwon YH, Seol HJ, Kim HI. Phase transformation and microstructural changes during ageing process of an Ag-Pd-Cu-Au Alloy. Journal of Alloys and Compounds 2008;460(1-2) 331-336.

[20] Kanzawa Y, Uzuka T, Kondo, Shoji M. Study on the Ag-Pd alloy. Journal of the Japan Society for Dental Apparatus and Materials 1963;4 157-160.

[21] Yasuda K. Study on the age-hardeability of dental precious metal alloys. Journal of the Japan Society for Dental Apparatus and Materials 1969;10 156-166.

[22] Ohta M, Hisatsune K, Yamane M. Study on the age-hardenable silver alloy. Journal of the Japan Society for Dental Apparatus and Materials 1975;16 144-149.

[23] Ohta M, Hisatsune K, Yamane M. Age Hardening of Ag-Pd-Cu dental alloy. Journal of the Less-Common Metals 1979;65(1) 11-21.

[24] Ohta M, Shiraishi T, Hisatsune K, Yamane M. Age-hardening of dental Ag-Pd-Cu-Au alloys. Journal of Dental Research 1980;59(11) 1966-1971.

[25] Tanaka Y, Miura E, Shiraishi T, Hisatsune K. Nano-precipitates generated in the dental silver–palladium–gold alloy. Materia Japan 2004;12 1036.

[26] Tanaka Y, Seol HJ, Ogata T, Miura E, Shiraishi T, Hisatsune K. Hardening mechanism of dental casting gold-silver-palladium alloy by higher-temperature heat-treatment. The Japanese Society for Dental Materials and Devices 2003;22 (2) 69.

Chapter 21

Ceramic-On-Ceramic Joints: A Suitable Alternative Material Combination?

Susan C. Scholes and Thomas J. Joyce

Additional information is available at the end of the chapter

http://dx.doi.org/10.5772/53237

1. Introduction

Total hip replacement (THR) surgery has been available for several decades and is now a relatively common procedure. Since the introduction of the Charnley metal-on-ultra-high molecular weight polyethylene (UHMWPE) hip prosthesis, THR is seen as one of the most successful orthopaedic operations available today. There are currently over 80,000 hip replacement procedures carried out in England and Wales [1] each year and THR has now become more popular with the younger, more active patient. This is shown in the statistics reported in the National Joint Registry; 12% of the patients who undergo THR are under the age of 55 and 85% of these are recorded as being either fit and healthy (16%) or with mild disease that is not incapacitating (69%) [1]. However, failure of these artificial joints does occur, leading to the need for revision surgery; approximately 10% of the THR procedures reported are revision operations [1].

Failure, in many cases, is due to aseptic loosening [1, 2]. With the conventional metal-on-UHMWPE joint this has been shown to be due to wear particle induced osteolyisis [3]. Although 90% of these joints are operating well 15 years after implantation [2] this wear particle induced osteolysis may lead to repetitive revision requirements for the younger, more active patient.

An alternative to the conventional metal-on-UHMWPE type of hip joint is to use ceramic-on-ceramic joints. Ceramic-on-ceramic joints were first introduced in the early 1970s but often resulted in poor performance due to fixation problems, poor quality alumina as a result of inadequately controlled grain size and other material properties (leading to catastrophic wear), and sub optimal design parameters such as too large a clearance. The work of many people over the years, including material scientists and engineers, has improved the quality of these ceramics. Therefore, since the introduction of the standard for ceramic production

(ISO 6474:1981 (second-generation ceramics), which was replaced with ISO 6474:1994 (third-generation ceramics) and has now been replaced by ISO 6474-1:2010 (fourth-generation ceramics)) the performance of these all ceramic bearings has been greatly improved [4, 5]. The third and fourth-generation alumina ceramics are manufactured using hot isostatic pressing. This produces a material that is highly pure with a small grain size (≤ 2.5 μm and many manufacturers produce ceramics with even smaller grain sizes) that provides material strength and minimises the risk of fracture. The majority of the ceramic-on-ceramic joints discussed in this chapter were produced using third-generation ceramics.

Ceramic-on-ceramic hip prosthesis performance will be reviewed in this chapter (*in vitro* and *in vivo*) along with a discussion of the concerns with ceramic-on-ceramic joints that happen in a minority of cases such as joint "squeaking" and component fracture. The majority of articles reviewed discuss the performance of alumina-on-alumina joints. There are, however, other ceramic materials available on the market today for use in orthopaedic surgery, for example BIOLOX® delta (an alumina matrix composite containing 72.5% alumina, 25.5% zirconia, and 2% mixed oxides). The reader will be made aware when joints made from this material are discussed.

2. Ceramic-on-ceramic hip prosthesis performance

The performance of ceramic-on-ceramic hip joints has been evaluated using data obtained from joints operating well within the body, prostheses retrieved due to joint failure and also tests performed within the laboratory. Firstly, the *in vitro* laboratory tests results will be discussed, then the *in vivo* data will be given. In addition to this, "squeaking", one of the most common concerns relating to ceramic-on-ceramic joints will be discussed.

2.1. *In vitro*

The majority of recent papers discussing the tribology (lubrication, friction and wear) of ceramic-on-ceramic hip joints detail tests done under 'severe' conditions such as malpositioning, edge-loading or microseparation. There are, however, some earlier studies that describe how these joints operate under 'standard' conditions.

A well positioned ceramic-on-ceramic hip, tested under the loads and motions expected during the standard walking cycle, performs exceptionally well in terms of friction, lubrication and wear [6-24]. Friction tests using different viscosities of carboxy-methyl cellulose (CMC) solution show that these joints operate close to full-fluid film lubrication with very low friction factors (0.002 at physiological viscosities) [6, 12]. These ceramic-on-ceramic joints have been shown to have very low surface roughness values that play a part in this low friction [6]. Tests were also performed using different viscosities of bovine serum [6, 12]. Bovine serum is often used in the laboratory as a replacement for the body's natural lubricating fluid, synovial fluid, as it contains proteins that act in a similar manner to those present in synovial fluid and although CMC fluids replicate the shear-thinning behaviour of synovial fluid, they do not contain any proteins. The introduction of these proteins into the lubricating fluid resulted in higher friction (0.03 at physiological viscosities) and mixed lubrication. These results are shown as Stribeck plots in Figure 1. A Stribeck plot shows the measured friction factor plotted against Sommerfeld number (a dimensionless parameter dependent on the lubricant viscosity, the entraining velocity of the bearing surfaces, the joint radius and the load applied). A rising trend of friction factor with increasing Sommerfeld number is indicative of a full-fluid film lubrication regime, whereas a falling trend is normally indicative of mixed lubrication. In full-fluid film lubrication, the surfaces are completely separated by the lubricant and the friction generated is due solely to the shearing of the lubricant film. In mixed lubrication, the load is carried in part by the contact between the asperities of the bearing surfaces and also by the pressure generated within the lubricant. Although the bovine serum tests suggested that these ceramic-on-ceramic joints were operating in mixed lubrication with some asperity contact, there was no surface damage evident on the joints after testing. It was speculated that the proteins that adhere to the ceramic surfaces when using protein-based lubricants produce a sufficiently thick layer to penetrate the fluid film and result in protein-to-protein contact and shearing. It is likely that the subsequent friction developed by the protein-to-protein contact is greater than that due to the shearing of the lubricant film alone. Therefore, although higher friction and mixed lubrication is encountered when testing under more physiological lubricating conditions, there is still little asperity contact during the normal walking cycle [12].

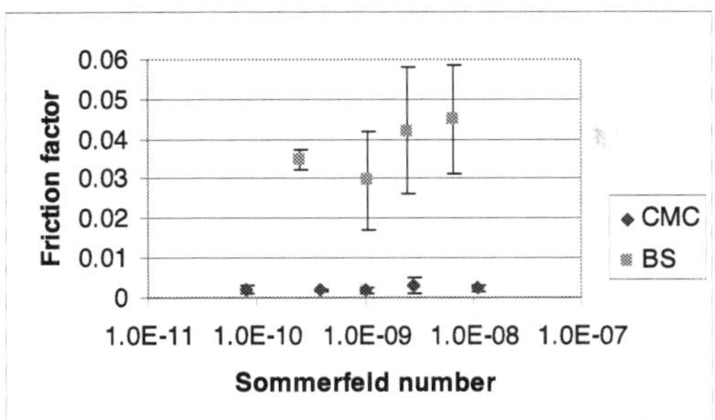

Figure 1. Stribeck plot for ceramic-on-ceramic joints (Scholes *et al* (2006) [12]). CMC: CMC fluids; BS: bovine serum.

With this good lubrication and low surface roughness the wear volumes produced under 'standard' conditions in the laboratory have, inevitably, been shown to be very low (less than 0.4 mg/million cycles cf. approximately 35 mg/million cycles for conventional metal or ceramic-on-UHMWPE joints, see Table 1), and sometimes almost immeasurable. As these ceramic-on-ceramic joints are working close to full-fluid film lubrication there is little or no contacting of the asperities on the joint surfaces leading to this low wear and friction [6, 12]. Also, laboratory studies have shown that cup malpositioning and elevated swing phase load testing have not significantly affected the wear of these joints [21, 22, 25, 26], see Table 2. This combination should, therefore, lead to a very successful artificial joint.

Reference	Material combination	Components	Wear rate (mg/million cycles)
[27]	Metal-on-UHMWPE	UHMWPE cup	47.4
[27]	Ceramic (zirconia)-on-UHMWPE	UHMWPE cup	37.9
[28]	Metal-on-UHMWPE	UHMWPE cup	~33.7
[28]	Ceramic (zirconia)-on-UHMWPE	UHMWPE cup	~29.0
[29]	Metal-on-UHMWPE	UHMWPE cup	~40.9
[29]	Ceramic (zirconia)-on-UHMWPE	UHMWPE cup	~29.5
[13]	Ceramic-on-ceramic	Head and cup	~0.25
[16]	Ceramic-on-ceramic	Head and cup	~0.16
[17]	Ceramic-on-ceramic	Head and cup	< 0.04
[18]	Ceramic-on-ceramic	Head and cup	0.09
[24]	Ceramic-on-ceramic	Head and cup	~0.20

Table 1. Wear rates found for well-positioned conventional metal or ceramic-on-UHMWPE and ceramic-on-ceramic joints under standard testing conditions

More severe loading conditions have, however, given slightly different results. Microseparation was first introduced during the swing phase of the walking cycle on the Leeds hip wear simulator [30] to replicate the stripe wear sometimes seen on retrieved ceramic-on-ceramic joints. Microseparation was incorporated in the simulator studies at Leeds because a year earlier, it was suggested by Mallory *et al* (1999) [31] that this separation of the femoral head and acetabular cup can occur in conventional metal-on-UHMWPE joints. Nevelos *et al* (2000) [30] hypothesised that a similar mechanism may take place in ceramic-on-ceramic joints resulting in the visible stripe wear found on some retrievals. Microseparation was, therefore, set up in the simulator and involved separation of the femoral head and acetabular cup during the swing phase of walking leading to relocation (with rim contact and edge-loading) during heel strike before the head then relocated in the cup in the stance phase (see Figure 2). The surface damage caused by this rim contact resulted in what is known as stripe wear.

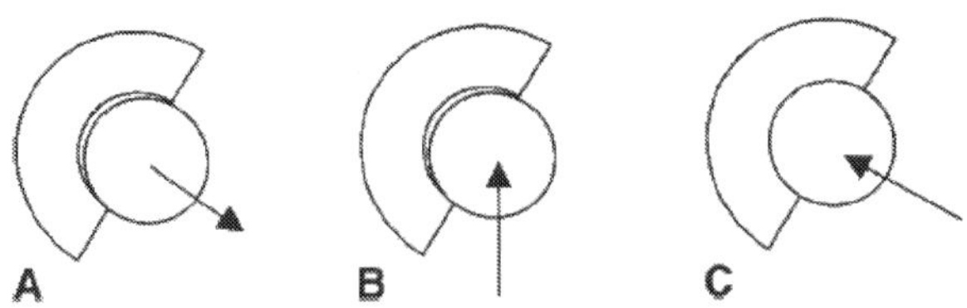

Figure 2. Microseparation as applied in the Leeds simulator by Nevelos *et al* (2000) [30]. (A) Swing phase: microseparation. (B) Heel-strike: rim contact. (C) Stance phase: relocation.

Using this microseparation technique, Nevelos *et al* (2000) [30] found slightly higher wear rates with ceramic-on-ceramic joints than found under 'standard' conditions (see Table 2). This study, however, was relatively short-term (800,000 cycles). Longer-term tests performed by Stewart *et al* (2001) [32] on the same simulator gave similar results to those found by Nevelos *et al* (2000) [30]. Other workers have also found an increase in wear rate with microseparation [33]. More recent work performed by Al-Hajjar *et al* (2010) [25] studied BIOLOX® delta couplings. These joints gave lower wear than that found for alumina-on-alumina joints. Microseparation of the BIOLOX® delta joint surfaces, again, resulted in higher wear. The wear rates found using microseparation are, however, still extremely low in comparison with conventional joints (see Tables 1 and 2).

Reference	Wear conditions	Components	Average wear rate (mg/million cycles)
[11]	Standard	Cup only	< 0.02
[13]	Standard	Not stated	~0.25
[16]	Standard	Head and cup	~0.16
[17]	Standard	Head and cup	< 0.04
[18]	Standard	Head and cup	0.09
[19]	Standard	Cup only	~0.40
[21]	Standard	Not stated	~0.24
[22]	Standard	Cup only	~0.32
[24]	Standard	Head and cup	~0.20
[21]	Elevated cup angle	Not stated	~0.20
[25]	Elevated cup angle	Not stated	0.20
[22]	Elevated swing phase load	Cup only	0.36
[25]	Microseparation	Not stated	0.52
[30]	Microseparation	Not stated	4.78 (after 800,000 cycles)
[32]	Microseparation	Not stated	(mild) ~0.40 (severe) ~5.17
[33]	Microseparation	Head and cup	1.55

Table 2. Wear rates found for alumina-on-alumina joints under standard testing conditions and microseparation

It is still unknown if it is this microseparation that causes the stripe wear that is observed on some retrievals or whether this type of wear is due simply to edge-loading of the head on the cup through a different mechanism. Higher rates of wear including the appearance of stripe wear may also be due to a steep acetabular cup implantation angle or repeated dislocations [34]. Microseparation, or edge-loading, may occur during various different physical activities such as stair climbing, standing from squat position and deep flexion. It may, however, also occur during walking. These simulator studies incorporating microseparation of the head and cup into the walking cycle are therefore a severe testing method. The resulting rim contact and stripe wear does, however, replicate the more severe conditions that these joints may encounter thus producing the same effect that is seen on some retrievals but not necessarily through the correct corresponding actions.

If, and when, wear of these ceramic components occurs it is important to understand how the body is likely to react to these wear particles. It is well known that wear particle induced osteolysis is a major concern for conventional metal-on-UHMWPE joints [3] but is this the case for ceramic-on-ceramic prostheses? Promisingly, several workers have shown that the cellular response to ceramic particles is less severe than that due to polyethylene particles [35, 15]. An important point to note with the laboratory tests discussed above is that, even under extreme loading and motion conditions, these joints provide low wear with a minimal adverse tissue reaction to these wear particles. As stated by Fisher *et al* (2006) [15], the combination of low wear and low reactivity means that '...ceramic-on-ceramic bearings address the tribological lifetime demand of highly active patients.'.

2.2. *In vivo*

As shown above, the laboratory test results for ceramic-on-ceramic joints are very promising. Is this mirrored by the *in vivo* performance? There are many publications detailing the performance of ceramic-on-ceramic hip joints in the body. A selection of these papers is discussed below to give a general overview of joint performance using this material combination.

The short-term (mean follow-up of 50.4 months) performance of ceramic-on-ceramic joints was compared to that of metal-on-highly cross linked polyethylene (XLPE) joints in a study reported by Bascarevic *et al* (2010) [36]. Seventy-five metal-on-highly XLPE hips (72 patients) and 82 ceramic-on-ceramic hips (78 patients) were assessed. Both were found to work well with no revisions performed on the ceramic-on-ceramic joints compared with 2 revisions necessary for the metal-on-highly XLPE group. The authors commented that ceramic-on-ceramic components manufactured using third-generation ceramics were "especially suited for hip arthroplasty in young and active persons".

A comparative study was also performed by Amanatullah *et al* (2011) [37]. The short-term performance (60 months) of 125 ceramic-on-ceramic hip joints was assessed against 95 ceramic-on-UHMWPE prostheses. In this study both the material combinations preformed well and no statistically significant difference was found between the clinical outcome scores for the two types of prosthesis, however, one important point to note is that audible noise such as "squeaking" did occur in 3.1%. of the ceramic-on-ceramic hips. "Squeaking" is a phenomenon reported by others and this will be discussed later. It was noted that this is a short-term study (five years) and, as the metal-on-UHMWPE hips did not produce as low wear rates in the radiographic analysis, osteolysis may occur at a later stage.

Another short-term study comparing the results of 525 hips (421 ceramic-on-ceramic and 104 metal-on-UHMWPE) was reported by Johansson *et al* (2011) [38]. These joints had been implanted for an average of 59 months. The survival rates were 98% and 92% for the ceramic-on-ceramic and metal-on-UHMWPE hip joints respectively.

http://dx.doi.org/10.5772/53237

A short-term study (60 month follow-up) reported by Nikolaou *et al* (2012) [39] assessed and compared the performance of 36 metal-on-UHMWPE, 32 metal-on-highly XLPE and 34 third-generation ceramic-on-ceramic hip joints. The apparent wear of the bearing surfaces was assessed radiologically. The ceramic-on-ceramic joints showed the lowest wear (mean linear wear 0.035 mm) and the metal-on-highly XLPE gave nearly 3 times lower wear than the metal-on-UHMWPE joints (0.329 mm cf. 0.869 mm). At 60 months, no difference in clinical outcome was found for these three different material combinations. "Squeaking" was observed in 3 (8.8%) of the ceramic-on-ceramic joints; no revision procedures were necessary as a results of this squeaking though.

Early to mid-term results were also reported by Stafford *et al* (2012) [40]. At a mean follow-up of 59 months six of these 250 ceramic-on-ceramic hips were revised. Two were revised for recurrent dislocation secondary to impingement, two for deep infection, one for recurrent dislocation and one due to fracture of the femoral head. Although no patients experienced "squeaking", six described a grinding or crunching noise that was experienced mainly during deep flexion. One of these "squeaking joints" was a BIOLOX® delta ceramic prosthesis. Again, no revision surgery was performed on these noisy joints.

Mesko *et al* (2011) [41] assessed the outcome of 930 ceramic-on-ceramic hips over 10 years that were implanted by nine different surgeons. The survivorship at 10 years (96.8%) was referred to as 'excellent' with 0.9% fracture, 2.3% dislocation and 2.5% reported incidents of noise such as clicking, squeaking, popping, or creaking.

Another 10 year follow-up was reported by Yeung *et al* (2012) [42]. The clinical information was available for 244 hips (227 patients) and the radiographic information was available for 184 hips (172 patients). The success of these joints led to an overall survival rate of 98% (again, with revision for any reason as the end point).

In a multicentre study performed by Capello *et al* (2008) [43], 452 patients (475 hips) were assessed. The majority of the hips implanted were ceramic-on-ceramic (380) whilst the remainder (95) were conventional metal-on-UHMWPE and used as a comparison. The average patient age was 53 years and there was an average 8 year follow-up period. The authors found the clinical results to be excellent. The ten-year Kaplan-Meier survivorship (with revision of either component for any reason given as the end-point) was stated as 95.9% for the ceramic-on-ceramic hip joints and 91.3% for metal-on-UHMWPE. The ceramic joints, therefore, performed better than the conventional joints and because one of the major causes of failure for conventional metal-on-UHMWPE joints is late aseptic loosening due to osteolysis, a longer term study is likely to give a greater difference in survivorship ratings. Of the ceramic-on-ceramic failures requiring revision, only 2/380 (0.5%) were due to ceramic fractures. Therefore, overall it was concluded that the third-generation ceramic-on-ceramic hip joints performed well clinically and radiographically in this young patient group. Although these joints performed relatively well "squeaking" was reported in 3/380 (0.8%) cases for the ceramic-on-ceramic joints.

In the study reported by Lee *et al* (2010) [44], the ten year survival rate of 88 ceramic-on-ceramic hip joints, with revision of either the head or the cup for any reason being as the end point, was 99.0%. These results were taken after a minimum of 10 years postoperatively and are, indeed, very promising.

A long-term study on earlier generation ceramic joints was performed by Hernigou *et al* (2009) [45]. This was a comparative study which investigated the wear and osteolysis of 28 bilateral arthroplasties (one ceramic-on-ceramic and the contralateral ceramic-on-UHMWPE). All of these joints (ceramic-on-ceramic and ceramic-on-UHMWPE) had lasted 20 years without the need for revision surgery. Computer tomography (CT scan) was used to assess the number and volume of osteolytic lesions present. Fewer osteolytic lesions were found on the side with the ceramic-on-ceramic joint; there were 13 cases of pelvic osteolysis found and 15 joints showed evidence of femoral osteolysis (cf. 24 joints with pelvic osteolysis and 23 with femoral osteolysis for the ceramic-on-UHMWPE joints). Also, in each patient, the diameter, surface and volume of osteolysis was substantially lower on the side with the ceramic-on-ceramic hip implant than the ceramic-on-UHMWPE joint. It must also be remembered that these joints were manufactured using first-generation ceramics and the ceramics produced to the standard expected for third and fourth-generation ceramics are expected to perform even better.

Another longer-term study (mean 20.8 years), again with earlier generation ceramics, performed by Petsatodis *et al* (2010) [46] also showed good results. There were 100 patients in this study, all with at least one ceramic-on-ceramic hip joint (109 hips) and, again, a young patient population (average age 46 years). The cumulative rate of survival was quoted as being 84.4% at 20.8 years.

Although these results are promising, in the majority of cases, the follow-up period particularly for the third generation ceramics, was only short-term. It will, therefore be very interesting to evaluate the performance of these third and fourth-generation ceramic-on-ceramic joints, along with the metal and ceramic-on-highly XLPE, on a longer-term basis. These results are eagerly anticipated.

The papers discussed so far have shown exceptional performance of ceramic-on-ceramic joints and suggests that these joints may perform better than metal or ceramic-on UHMWPE. This, however, is not reflected in the data described in the National Joint Registry (NJR) of England and Wales (2011) [1]. Promisingly though, the NJR states that there is 'little substantive difference' in the risk of revision for ceramic-on-ceramic, ceramic-on-polyethylene or metal-on-polyethylene joints. It is, however, not stated whether this polyethylene is UHMWPE or XLPE and the differences in performance between these two materials is not listed.

Although, as discussed above, many of these ceramic-on-ceramic hip joints perform exceptionally well, early dislocation is seen as a possible concern due to limited modular neck length and other factors. In the majority of the published literature referred to in this text [47, 36, 48, 43, 49, 50, 46, 51-57, 41, 40], the occurrence of dislocation is 0% - 2.3%. The number of dislocations was higher (6%) in a study reported by Chevillotte *et al* (2011) [58], however, no reoperation was required for any of these cases. Colwell *et al* (2007) [53] summarised other published literature detailing the number of dislocations in ceramic-on-ceramic hip joints in comparison with metal-on-polyethylene and found no difference between the two material combinations. In addition to this Amanatulla *et al* (2011) [37] and Bascarevic *et al* (2010) [36] showed no difference in the number of dislocations for ceramic-on-ceramic and ceramic-on-UHMWPE or metal-on-highly XLPE hip prostheses. Often there is no need for revision surgery after dislocation, unless it is recurrent dislocation.

Component fracture in ceramic-on-ceramic hips is another cause for concern for many surgeons and patients. The fracture rates of ceramic joints have been dramatically reduced since the introduction of third and fourth-generation ceramics with the new material processing methods. Ceramic-on-ceramic joints have strict regulations that must be abided by with regard to material properties such as burst strength; as discussed in work reported by Salih *et al* (2009) [59]. However, fracture of these components does still occur, leading to catastrophic joint failure and revision surgery. In the majority of cases though, this fracture rate is extremely low (0% - 0.5%) [60, 61, 43, 49, 45, 50, 46, 51, 48, 55, 57, 40]. Some reports do, however, give a higher rate of post-operative fracture (1-2.3%) [62, 44, 37, 42, 47, 39]. Fracture is often associated with trauma, however a multicentre review reported by Park *et al* (2006) [63] discussed the performance of 357 third-generation ceramic-on-ceramic THRs and fracture occurred under normal activities in six of these hips (1.7%). This design used a polyethylene-ceramic composite liner within a titanium alloy shell and the high rate of fracture led to the authors discontinuing its use. It must be recognised though that, in general, the rate of fracture for ceramic-on-ceramic hip joints is very low. In fact, an article describing the fracture of ceramic joints (Hannouche *et al* (2003) [64]) stated that during a 25 year period (1977 – 2001) 11 (less than 0.004%) of the 3300 ceramic-on-ceramic joints reported on failed due to fracture. This low rate of fracture is from ceramics produced using earlier versions of the ISO standard and, so, this is expected to reduce even further and many studies using third-generation ceramics have reported no failures due to fracture [51, 36, 65, 66], Fracture must, however, still be recognised as a risk (albeit very low) with ceramic-on-ceramic joints.

For what other reasons does failure occur? Savarino *et al* (2009) [67] analysed the clinical, radiographic, laboratory and microbiological data from 30 retrieved ceramic-on-ceramic hip components. They concluded that failure was due to malpositioning of the joint during surgery leading to mechanical instability, or trauma or infection. Loosening in this selection of joints was not due to wear debris induced osteolysis. They indicated that the wear debris produced by these joints and the osteolysis present were the effect of the loosening, rather than the cause. It is stated that correct positioning of the implant is crucial. This has also been stated by other workers [54].

A case study was reported by Nam *et al* (2007) [68] discussing a failed ceramic-on-ceramic hip joint where failure was stated as being caused by alumina debris-induced osteolysis. A sixty-three year old woman underwent bilateral THR in 1998. Eight years later the patient returned to the clinic for routine follow-up and was experiencing no discomfort or worrying symptoms. However, the radiographs taken showed expansive osteolytic lesions. After revision surgery a histologic analysis of the retrieved tissues showed alumina particles within the cytoplasm of macrophages and in intercellular tissue suggesting wear particle induced osteolysis. Alumina wear particle-induced osteolysis is, however, a very rare phenomenon.

Chang *et al* (2009) [69] reported on the clinical and radiographic outcomes when using third-generation ceramic-on-ceramic joints in revision THR of 42 failed metal-on-UHMWPE hips. This was an interesting study as most published literature discuss the choice of bearing material for primary surgery and, as stated by the authors, '...few studies have focussed on the choice of bearing surface in revision...'. This was a young patient group (mean age: 48.8 years, range: 32 – 59 years) and the mean length of time between primary and revision surgery was 9.5 years (range: 3.3 – 16.1 years). The mean duration of follow-up after this revision surgery was 64 months (range: 38 – 96 months). At the time of publication of this article, no hips needed additional revision surgery and no hips showed radiolucent lines, acetabular cup migration or osteolysis. This study gives very favourable results for the use of ceramic-on-ceramic hip joints in revision surgery, especially for the younger patient as the likelihood of the need for further revision is greater.

3. The "squeaking" hip

Another concern with hard-on-hard bearing couples such as ceramic-on-ceramic and metal-on-metal is the incidence of noise or "squeaking" in these joints. Audible sounds such as squeaking, clicking, snapping, cracking, grinding, rustling, crunching and tinkling are all referred to in this text as "squeaking". "Squeaking" can be present during different kinds of activity including stair climbing, bending forward, squatting, standing from a chair and walking. The occurrence of this "squeaking" has been reported by many workers to different degrees.

Jarrett *et al* (2009) [52] described a group of 131 patients from which 14 (10.7%) suffered an audible "squeak" during normal activities (however, only 4 of these patients were able to reproduce the "squeak" during the clinical review session). They stated that none of the patients had undergone revision surgery specifically because of "squeaking", however, longer-term follow-up was needed to monitor this noise.

Mai *et al* (2010) [70] reported noise or "squeaking" in 17% of 320 ceramic-on-ceramic hips and Keurentjes *et al* (2008) [71] found "squeaking" in 20.9% of cases. From their study they concluded that short neck length of the femoral component was a potential risk factor for "squeaking".

A study performed by Cogan *et al* (2010) [72] reported on the occurrence of "squeaking" in a patient population with at least one ceramic-on-ceramic hip; 10.6% of 265 hips demonstrated "squeaking". Two other published studies report on "squeaking" in 15% and 8.8% of the patient population [44, 39]. Other published literature suggests that "squeaking" occurs in 0% - 6% of patients [73-75, 40, 65, 50, 61, 56, 76, 43, 49, 51, 37, 58, 41, 77]. The "squeaking" phenomenon tends to appear at an average of 28.8 months (range of averages 5.7 - 66 months) post-operatively.

After 10 years of follow-up Chevillotte *et al* (2012) [73] discussed the performance of 100, third-generation ceramic-on-ceramic joints. By use of a questionnaire, 5% of these patients reported

http://dx.doi.org/10.5772/53237

the occurrence of "squeaking". All of these patients were active, sporty and heavy men. "Squeaking" was not related to any malpositioning, wear or loosening of the joint and none of the 100 patient group suffered from any component fracture. In this paper it was stated that "squeaking" noise seems to be an isolated phenomenon with no consequences for the patient regarding functional results and on the implant longevity at 10 years of follow-up'.

The largest study to date (Sexton *et al* (2011)) [75] reported on the occurrence of "squeaking" and the role of patient factors and implant positioning in 2406 ceramic-on-ceramic hips at a mean follow-up of 10.6 years. Seventy-four hips (73 patients, 3.1%) made "squeaking" sounds at a mean time post-operatively of 40 months. Taller, heavier and younger patients with higher activity levels were found to be more at risk of developing a "squeaking" hip. However, there was no relationship between BMI and the prevalence of "squeaking". Interestingly, it was found that at a mean follow-up of 9.5 years, 11 of these hips (15%) had stopped "squeaking". Therefore, "squeaking" is not necessarily a permanent complication.

There is great debate over the cause of "squeaking" in ceramic-on-ceramic THRs. Several possible causes of "squeaking" are edge-loading [78], component malpositioning [78-80] or component or stem design [74, 81, 70, 77]. Other workers have also related "squeaking" to patient weight, height or age [70, 78, 79]. It is possible that a number of factors need to be present to initiate the "squeaking" phenomenon. The occurrence of "squeaking" has not been found to compromise the results of ceramic-on-ceramic hip joints; however, some patients do request revision surgery in order to solve the "squeaking" issue [82].

Recently, laboratory studies have been performed in an attempt to re-create the conditions required to generate this "squeaking". Some authors have observed "squeaking" with ceramic-on-ceramic hip joints lubricated under dry conditions [83, 84, 73]. However, such an extreme lubrication regime is not expected to occur *in vivo* [85]. Work performed by Currie *et al* (2010) [86] suggested that the rolling/sliding mechanism of the bearing surfaces can induce vibration, of an audible frequency, resulting in "squeaking". Under lubricated conditions, Sanders *et al* (2012) [85] were able to reproduce the "squeak" using edge-loading during short-term wear tests. The "squeaking" was found to occur with a high contact force centred above or near the margin of the wear patch on a previously edge-worn femoral head. The authors state 'the results reveal key conditions that yield recurrent squeaking *in vitro* in various scenarios without resorting to implausible dry conditions'. In support of this, Walter *et al* (2011) [87] analysed 12 ceramic-on-ceramic components retrieved from "squeaking" joints and compared these with 33 'silent' ceramic-on-ceramic hip retrievals. All 12 "squeaking" hips showed evidence of edge-loading with up to 45 times greater wear than that reported for then the 33 'silent' hips [54]. The authors suggested that although the causes of "squeaking" are unknown, the high contact pressures experienced during edge-loading may result in a breakdown of the fluid film lubrication leading to some asperity contact and an increase in friction. Also, any surface damage, which may have been caused by the edge-loading, will result in an increased roughness of the bearing surfaces thus leading to possible destruction of the fluid film leading to asperity contact.

Although there are a few reports on poor ceramic-on-ceramic hip prosthesis performance, the majority of authors give good and optimistic results. There have been little or no fractures, dislocation, infection or osteolysis. Also, the few patients suffering from "squeaking" hips have, in the majority of cases, had no need for revision surgery. These are, therefore, excellent results, but Lee *et al* (2010) [44] did suggest that these small risks should be a concern to surgeons and that patients should also be made aware of these before surgery. In addition to this, they mentioned that longer-term follow-up is needed to assess the effects of these small risks on the prosthesis performance.

4. The younger patient

As these artificial hip joints have been found to perform well in the younger patient (45 to 55 years), some surgeons have chosen to replace the diseased joints of even younger patients with this material combination.

A case study was reported by Capello and Feinberg (2009) [88] where ceramic-on-ceramic joints were implanted in a 13 year-old child with bilateral end-stage arthritis of the hip. Seven and eight years post-operatively the patient had no pain, no limp, and was able to walk long distances. The radiographs showed no implant loosening, osteolysis or wear. This is a very encouraging result, however, it was stated that the patient is still very young (20 years of age at the time of report) and, therefore, the need for revision surgery will be more than likely.

Other studies on younger patients have not had as good results as those reported by Capello though. Nizard *et al* (2008) [89] reported on ceramic-on-ceramic hips that had been implanted in a group of 101 patients (132 hips) younger than 30 years old (mean age: 23.4 years, range: 13 – 30 years). These joints were implanted from 1977 to 2004 and, because of this, different implant designs and modes of fixation were used. Of these 132 joints, 17 were revised for aseptic loosening leading to a survivorship of 82.1% at 10 years and 72.4% at 15 years. These survivorship rates are quite low and may create cause for concern. It was, however, found that the higher rate of failures of joints replaced for treatment of slipped capital epiphysis or trauma influenced the survivorships greatly. Also, these artificial hip joints were implanted over a period of 26.5 years, during which time the ceramic materials have been improved and the fixation methods have changed. It is hoped that ceramic of the new generation with improved prosthesis design and mode of fixation will perform better and provide improved longer-term results.

5. Overview

From the results reported here it is clear that ceramic-on-ceramic hip joints have good tribological results: low friction, good lubrication and very low wear *in vitro* and *in vivo*. In addition to this, ceramic particles are biologically inert. Also, the fracture risk is relatively low. With good implant positioning these joints have the potential to perform incredibly well. These bearings, therefore, deserve to be high on the list for both primary and revision implants, especially for

http://dx.doi.org/10.5772/53237

the younger, more active patient. However, for the best results, the choice of bearing combination/design should be patient-specific; as one design does not suit all.

Author details

Susan C. Scholes and Thomas J. Joyce

*Address all correspondence to: susan.scholes@newcastle.ac.uk

School of Mechanical and Systems Engineering, Newcastle University, Newcastle-upon-Tyne, UK

References

[1] National Joint Registry for England and Wales. 8th Annual Report. 2011.

[2] The Swedish Hip Arthroplasty Register. Annual Report 2008. 2008.

[3] Harris WH. The Problem Is Osteolysis. Clin Orthop Rel Res. 1995(311):46-53.

[4] Sedel L, Kerboull L, Christel P, Meunier A, Witvoet J. Alumina-on-Alumina Hip-Replacement - Results and Survivorship in Young-Patients. J Bone Joint Surg-Br Vol. 1990;72(4):658-63.

[5] Walter A. On the Material and the Tribology of Alumina Alumina Couplings for Hip-Joint Prostheses. Clin Orthop Rel Res. 1992(282):31-46.

[6] Scholes SC, Unsworth A. Comparison of friction and lubrication of different hip prostheses. Proc Inst Mech Eng Part H-J Eng Med. 2000;214(H1):49-57.

[7] Scholes SC, Unsworth A, Hall RM, Scott R. The effects of material combination and lubricant on the friction of total hip prostheses. Wear. 2000;241(2):209-13.

[8] Scholes SC, Unsworth A, Goldsmith AAJ. A frictional study of total hip joint replacements. Phys Med Biol. 2000;45(12):3721-35.

[9] Vassiliou K, Scholes SC, Unsworth A. Laboratory studies on the tribology of hard bearing hip prostheses: ceramic on ceramic and metal on metal. Proc Inst Mech Eng Part H-J Eng Med. 2007;221(H1 Special Issue):11-20.

[10] Scholes SC, Green SM, Unsworth A. The friction and lubrication of alumina-on-alumina total hip prostheses - The effect of radial clearance and wear testing. Bioceramics 14. Key Engineering Materials, 2002. p. 535-9.

[11] Scholes SC, Green SM, Unsworth A. Nanotribological characterisation of alumina femoral heads using an atomic force microscope. Bioceramics 14. Key Engineering Materials, 2002. p. 543-7.

[12] Scholes SC, Unsworth A. The effect of proteins on the friction and lubrication of artificial joints. Proc Inst Mech Eng Part H-J Eng Med. 2006;220(H6):687-93.

[13] Spinelli M, Affatat S, Corvil A, Viceconti M. Ceramic-on-ceramic vs. metal-on-metal in total hip arthroplasty (THA): do 36-mm diameters exhibit comparable wear performance? Materialwissenschaft Und Werkstofftechnik. 2009;40(1-2):94-7. doi: 10.1002/mawe.200800381.

[14] Brockett C, Williams S, Jin Z-M, Isaac G, Fisher J. Friction of total hip replacements with different bearings and loading conditions. Journal of Biomedical Materials Research Part B-Applied Biomaterials. 2007;81B:508-15.

[15] Fisher J, Jin ZM, Tipper J, Stone M, Ingham E. Presidential guest lecture - Tribology of alternative beatings. Clin Orthop Rel Res. 2006(453):25-34. doi:10.1097/01.blo.0000238871.07604.49.

[16] Essner A, Sutton K, Wang A. Hip simulator wear comparison of metal-on-metal, ceramic-on-ceramic and crosslinked UHMWPE bearings. Wear. 2005;259:992-5. doi: 10.1016/j.wear.2005.02.104.

[17] Richardson HA, Clarke IC, Williams P, Donaldson T, Oonishi H. Precision and accuracy in ceramic-on-ceramic wear analyses: influence of simulator test duration. Proceedings of the Institution of Mechanical Engineers Part H-Journal of Engineering in Medicine. 2005;219(H6):401-5. doi:10.1243/095441105x34428.

[18] Oonishi H, Clarke IC, Good V, Amino H, Ueno M. Alumina hip joints characterized by run-in wear and steady-state wear to 14 million cycles in hip-simulator model. Journal of Biomedical Materials Research Part A. 2004;70A(4):523-32. doi:10.1002/jbm.a.30021.

[19] Smith SL, Unsworth A. An in vitro wear study of alumina-alumina total hip prostheses. Proceedings of the Institution of Mechanical Engineers Part H-Journal of Engineering in Medicine. 2001;215(H5):443-6.

[20] Clarke IC, Good V, Williams P, Schroeder D, Anissian L, Stark A et al. Ultra-low wear rates for rigid-on-rigid bearings in total hip replacements. Proc Inst Mech Eng Part H-J Eng Med. 2000;214(H4):331-47.

[21] Nevelos JE, Ingham E, Doyle C, Nevelos AB, Fisher J. The influence of acetabular cup angle on the wear of "BIOLOX Forte" alumina ceramic bearing couples in a hip joint simulator. J Mater Sci-Mater Med. 2001;12(2):141-4.

[22] Nevelos JE, Ingham E, Doyle C, Nevelos AB, Fisher J. Wear of HIPed and non-HIPed alumina-alumina hip joints under standard and severe simulator testing conditions. Biomaterials. 2001;22(16):2191-7.

[23] Shishido T, Clarke IC, Williams P, Boehler M, Asano T, Shoji H et al. Clinical and simulator wear study of alumina ceramic THR to 17 years and beyond. Journal of Biomedical Materials Research Part B-Applied Biomaterials. 2003;67B(1):638-47. doi: 10.1002/jbm.b.10048.

[24] Tipper JL, Firkins PJ, Besong AA, Barbour PSM, Nevelos J, Stone MH et al. Characterisation of wear debris from UHMWPE on zirconia ceramic, metal-on-metal and alumina ceramic-on-ceramic hip prostheses generated in a physiological anatomical hip joint simulator. Wear. 2001;250:120-8.

[25] Al-Hajjar M, Leslie IJ, Tipper J, Williams S, Fisher J, Jennings LM. Effect of cup inclination angle during microseparation and rim loading on the wear of BIOLOX (R) delta ceramic-on-ceramic total hip replacement. Journal of Biomedical Materials Research Part B-Applied Biomaterials. 2010;95B(2):263-8. doi:10.1002/jbm.b.31708.

[26] Affatato S, Traina F, Toni A. Microseparation and stripe wear in alumina-on-alumina hip implants. International Journal of Artificial Organs. 2011;34(6):506-12. doi: 10.5301/ijao.2011.8457.

[27] Smith SL, Unsworth A. A comparison between gravimetric and volumetric techniques of wear measurement of UHMWPE acetabular cups against zirconia and cobalt-chromium-molybdenum femoral heads in a hip simulator. Proc Inst Mech Eng Part H-J Eng Med. 1999;213(H6):475-83.

[28] Bigsby RJA, Hardaker CS, Fisher J. Wear of ultra-high molecular weight polyethylene acetabular cups in a physiological hip joint simulator in the anatomical position using bovine serum as a lubricant. Proc Inst Mech Eng Part H-J Eng Med. 1997;211(3):265-9.

[29] Barbour PSM, Stone MH, Fisher J. A hip joint simulator study using simplified loading and motion cycles generating physiological wear paths and rates. Proc Inst Mech Eng Part H-J Eng Med. 1999;213(H6):455-67.

[30] Nevelos J, Ingham E, Doyle C, Streicher R, Nevelos A, Walter W et al. Microseparation of the centers of alumina-alumina artificial hip joints during simulator testing produces clinically relevant wear rates and patterns. J Arthroplast. 2000;15(6):793-5.

[31] Mallory TH, Lombardi AV, Dennis DA, Komistek RD, Fada RA, Northcut E, editors. Do total hip arthroplasties piston during leg lift manoeuvres and gait. An in vivo determination of total hip arthroplasty during abduction/adduction leg lift and gait. 66th Annual Meeting of the American Academy of Orthopaedic Surgeons; 1999; Anaheim, California.

[32] Stewart T, Tipper J, Streicher R, Ingham E, Fisher J. Long-term wear of HIPed alumina on alumina bearings for THR under microseparation conditions. J Mater Sci-Mater Med. 2001;12(10-12):1053-6.

[33] Manaka M, Clarke IC, Yamamoto K, Shishido T, Gustafson A, Imakiire A. Stripe wear rates in alumina THR - Comparison of microseparation simulator study with retrieved implants. Journal of Biomedical Materials Research Part B-Applied Biomaterials. 2004;69B(2):149-57. doi:10.1002/jbm.b.20033.

[34] Nevelos JE, Ingham E, Doyle C, Fisher J, Nevelos AB. Analysis of retrieved alumina ceramic components from Mittelmeier total hip prostheses. Biomaterials. 1999;20(19): 1833-40.

[35] Catelas I, Huk OL, Petit A, Zukor DJ, Marchand R, Yahia LH. Flow cytometric analysis of macrophage response to ceramic and polyethylene particles: Effects of size, concentration, and composition. J Biomed Mater Res. 1998;41(4):600-7.

[36] Bascarevic Z, Vukasinovic Z, Slavkovic N, Dulic B, Trajkovic G, Bascarevic V et al. Alumina-on-alumina ceramic versus metal-on-highly cross-linked polyethylene bearings in total hip arthroplasty: a comparative study. International Orthopaedics. 2010;34(8):1129-35. doi:10.1007/s00264-009-0899-6.

[37] Amanatullah DF, Landa J, Strauss EJ, Garino JP, Kim SH, Di Cesare PE. Comparison of Surgical Outcomes and Implant Wear Between Ceramic-Ceramic and Ceramic-Polyethylene Articulations in Total Hip Arthroplasty. J Arthroplast. 2011;26(6):72-7. doi:10.1016/j.arth.2011.04.032.

[38] Johansson HR, Johnson AJ, Zywiel MG, Naughton M, Mont MA, Bonutti PM. Does Acetabular Inclination Angle Affect Survivorship of Alumina-ceramic Articulations? Clin Orthop Rel Res. 2011;469(6):1560-6. doi:10.1007/s11999-010-1623-y.

[39] Nikolaou VS, Edwards MR, Bogoch E, Schemitsch EH, Waddell JP. A prospective randomised controlled trial comparing three alternative bearing surfaces in primary total hip replacement. J Bone Joint Surg-Br Vol. 2012;94B(4):459-65. doi: 10.1302/0301-620x.94b4.27735.

[40] Stafford GH, Ul Islam S, Witt JD. Early to mid-term results of ceramic-on-ceramic total hip replacement ANALYSIS OF BEARING-SURFACE-RELATED COMPLICATIONS. J Bone Joint Surg-Br Vol. 2011;93B(8):1017-20. doi:10.1302/0301-620x. 93b8.26505.

[41] Mesko JW, D'Antonio JA, Capello WN, Bierbaum BE, Naughton M. Ceramic-on-Ceramic Hip Outcome at a 5- to 10-Year Interval. J Arthroplast. 2011;26(2):172-7. doi: 10.1016/j.arth.2010.04.029.

[42] Yeung E, Bott PT, Chana R, Jackson MP, Holloway I, Walter WL et al. Mid-Term Results of Third-Generation Alumina-on-Alumina Ceramic Bearings in Cement less Total Hip Arthroplasty A Ten-Year Minimum Follow-up. J Bone Joint Surg-Am Vol. 2012;94A(2):138-44. doi:10.2106/jbjsj.00331.

[43] Capello WN, D'Antonio JA, Feinberg JR, Manley MT, Naughton M. Ceramic-on-Ceramic Total Hip Arthroplasty: Update. J Arthroplast. 2008;23(7):39-43. doi:10.1016/ j.arth.2008.06.003.

[44] Lee YK, Ha YC, Yoo JJ, Koo KH, Yoon KS, Kim HJ. Alumina-on-Alumina Total Hip Arthroplasty A Concise Follow-up, at a Minimum of Ten Years, of a Previous Report. J Bone Joint Surg-Am Vol. 2010;92A(8):1715-9. doi:10.2106/jbjs.i.01019.

[45] Hernigou P, Zilber S, Filippini P, Poignard A. Ceramic-Ceramic Bearing Decreases Osteolysis: A 20-year Study versus Ceramic-Polyethylene on the Contralateral Hip. Clin Orthop Rel Res. 2009;467(9):2274-80. doi:10.1007/s11999-009-0773-2.

[46] Petsatodis GE, Papadopoulos PP, Papavasiliou KA, Hatzokos IG, Agathangelidis FG, Christodoulou AG. Primary Cementless Total Hip Arthroplasty with an Alumina Ceramic-on-Ceramic Bearing Results After a Minimum of Twenty Years of Follow-up. J Bone Joint Surg-Am Vol. 2010;92A(3):639-44. doi:10.2106/jbjs.h.01829.

[47] Sugano N, Takao M, Sakai T, Nishii T, Miki H, Ohzono K. Eleven- to 14-year Follow-up Results of Cementless Total Hip Arthroplasty Using a Third-generation Alumina Ceramic-on-ceramic Bearing. J Arthroplast. 2012;27(5):736-41. doi:10.1016/j.arth.2011.08.017.

[48] Mai K, Hardwick ME, Walker RH, Copp SN, Ezzet KA, Colwell CW, Jr. Early dislocation rate in ceramic-on-ceramic total hip arthroplasty. Hss J. 2008;4(1):10-3.

[49] Garcia-Cimbrelo E, Garcia-Rey E, Murcia-Mazon A, Blanco-Pozo A, Marti E. Alumina-on-alumina in THA - A multicenter prospective study. Clin Orthop Rel Res. 2008;466(2):309-16. doi:10.1007/s11999-007-0042-1.

[50] Greene JW, Malkani AL, Kolisek FR, Jessup NM, Baker DL. Ceramic-on-Ceramic Total Hip Arthroplasty. J Arthroplast. 2009;24(6):15-8. doi:10.1016/j.arth.2009.04.029.

[51] Kim YH, Choi Y, Kim JS. Cementless total hip arthroplasty with ceramic-on-ceramic bearing in patients younger than 45 years with femoral-head osteonecrosis. International Orthopaedics. 2010;34(8):1123-7. doi:10.1007/s00264-009-0878-y.

[52] Jarrett CA, Ranawat AS, Bruzzone M, Blum YC, Rodriguez JA, Ranawat CS. The Squeaking Hip: A Phenomenon of Ceramic-on-Ceramic Total Hip Arthroplasty. J Bone Joint Surg-Am Vol. 2009;91A(6):1344-9. doi:10.2106/jbjs.f.00970.

[53] Colwell CW, Jr., Hozack WJ, Mesko JW, D'Antonio JA, Bierbaum BE, Capello WN et al. Ceramic-on-ceramic total hip arthroplasty early dislocation rate. Clin Orthop Relat Res. 2007;465:155-8.

[54] Lusty PJ, Watson A, Tuke MA, Walter WL, Walter WK, Zicat B. Orientation and wear of the acetabular component in third generation alumina-on-alumina ceramic bearings - An analysis of 33 retrievals. J Bone Joint Surg-Br Vol. 2007;89B(9):1158-64.

[55] D'Antonio J, Capello W, Manley M, Naughton M, Sutton K. Alumina ceramic bearings for total hip arthroplasty - Five-year results of a prospective randomized study. Clin Orthop Rel Res. 2005(436):164-71. doi:10.1097/01.blo.0000162995.50971.39.

[56] Lusty PJ, Tai CC, Sew-Hoy RP, Walter WL, Walter WK, Zicat BA. Third-generation alumina-on-alumina ceramic bearings in cementless total hip arthroplasty. J Bone Joint Surg-Am Vol. 2007;89A(12):2676-83. doi:10.2106/jbjs.f01466.

[57] Murphy SB, Ecker TM, Tannast M. Two- to 9-year clinical results of alumina ceramic-on-ceramic THA. Clin Orthop Rel Res. 2006(453):97-102. doi:10.1097/01.blo.0000246532.59876.73.

[58] Chevillotte C, Pibarot V, Carret JP, Bejui-Hugues J, Guyen O. Nine years follow-up of 100 ceramic-on-ceramic total hip arthroplasty. International Orthopaedics. 2011;35(11):1599-604. doi:10.1007/s00264-010-1185-3.

[59] Salih S, Currall VA, Ward AJ, Chesser TJS. Survival of ceramic bearings in total hip replacement after high-energy trauma and periprosthetic acetabular fracture. J Bone Joint Surg-Br Vol. 2009;91B(11):1533-5. doi:10.1302/0301-620x.91b11.22737.

[60] Lewis PM, Al-Belooshi A, Olsen M, Schemitch EH, Waddell JP. Prospective Randomized Trial Comparing Alumina Ceramic-on-Ceramic With Ceramic-On-Conventional Polyethylene Bearings in Total Hip Arthroplasty. J Arthroplast. 2010;25(3):392-7. doi:10.1016/j.arth.2009.01.013.

[61] Baek SH, Kim SY. Cementless total hip arthroplasty with alumina bearings in patients younger than fifty with femoral head osteonecrosis. J Bone Joint Surg-Am Vol. 2008;90A(6):1314-20. doi:10.2106/jbjs.g.00755.

[62] Yoo JJ, Kim YM, Yoon KS, Koo KH, Song WS, Kim HJ. Alumina-on-alumina total hip arthroplasty - A five-year minimum follow-up study. J Bone Joint Surg-Am Vol. 2005;87A(3):530-5. doi:10.2106/jbjs.d01753.

[63] Park YS, Hwang SK, Choy WS, Kim YS, Moon YW, Lim SJ. Ceramic failure after total hip arthroplasty with an alumina-on-alumina bearing. J Bone Joint Surg-Am Vol. 2006;88A(4):780-7. doi:10.2106/jbjs.e.00618.

[64] Hannouche D, Nich C, Bizot P, Meunier A, Nizard RM, Sedel L. Fractures of ceramic bearings - History and present status. Clin Orthop Rel Res. 2003(417):19-26. doi:10.1097/01.blo.0000096806.78689.50.

[65] Byun J-W, Yoon T-R, Park K-S, Seon J-K. Third-generation ceramic-on-ceramic total hip arthroplasty in patients younger than 30 years with osteonecrosis of femoral head. The Journal of Arthroplasty. 2012;27(7):1337-43.

[66] Solarino G, Piazzolla A, Notarnicola A, Moretti L, Tafuri S, De Giorgi S et al. Long-term results of 32-mm alumina-on-alumina THA for avascular necrosis of the femoral head. Journal of orthopaedics and traumatology : official journal of the Italian Society of Orthopaedics and Traumatology. 2012;13(1):21-7.

[67] Savarino L, Baldini N, Ciapetti G, Pellacani A, Giunti A. Is wear debris responsible for failure in alumina-on-alumina implants? Acta Orthopaedica. 2009;80(2):162-7. doi:10.3109/17453670902876730.

[68] Nam KW, Yoo JJ, Kim YL, Kim YM, Lee MH, Kim HJ. Alumina-debris-induced osteolysis in contemporary alumina-on-alumina total hip arthroplasty - A case report. J Bone Joint Surg-Am Vol. 2007;89A(11):2499-503. doi:10.2106/jbjs/g.00130.

[69] Chang JD, Kamdar R, Yoo JH, Hur M, Lee SS. Third-Generation Ceramic-on-Ceramic Bearing Surfaces in Revision Total Hip Arthroplasty. J Arthroplast. 2009;24(8):1231-5. doi:10.1016/j.arth.2009.04.016.

[70] Mai K, Verioti C, Ezzet KA, Copp SN, Walker RH, Colwell CW, Jr. Incidence of 'squeaking' after ceramic-on-ceramic total hip arthroplasty. Clin Orthop Relat Res. 2010;468(2):413-7.

[71] Keurentjes JC, Kuipers RM, Wever DJ, Schreurs BW. High incidence of squeaking in THAs with alumina ceramic-on-ceramic bearings. Clin Orthop Rel Res. 2008;466(6): 1438-43. doi:10.1007/s11999-008-0177-8.

[72] Cogan A, Nizard R, Sedel L. Occurrence of noise in alumina-on-alumina total hip arthroplasty. A survey on 284 consecutive hips. Orthopaedics & Traumatology-Surgery & Research. 2011;97(2):206-10. doi:10.1016/j.otsr.2010.11.008.

[73] Chevillotte C, Pibarot V, Carret JP, Bejui-Hugues J, Guyen O. Hip Squeaking A 10-Year Follow-Up Study. J Arthroplast. 2012;27(6):1008-13. doi:10.1016/j.arth.2011.11.024.

[74] Stanat SJC, Capozzi JD. Squeaking in Third- and Fourth-Generation Ceramic-on-Ceramic Total Hip Arthroplasty Meta-Analysis and Systematic Review. J Arthroplast. 2012;27(3):445-53. doi:10.1016/j.arth.2011.04.031.

[75] Sexton SA, Yeung E, Jackson MP, Rajaratnam S, Martell JM, Walter WL et al. The role of patient factors and implant position in squeaking of ceramic-on-ceramic total hip replacements. J Bone Joint Surg-Br Vol. 2011;93B(4):439-42. doi:10.1302/0301-620x.93b4.25707.

[76] Restrepo C, Matar WY, Parvizi J, Rothman RH, Hozack WJ. Natural History of Squeaking after Total Hip Arthroplasty. Clin Orthop Rel Res. 2010;468(9):2340-5. doi:10.1007/s11999-009-1223-x.

[77] Parvizi J, Adeli B, Wong JC, Restrepo C, Rothman RH. A Squeaky Reputation: The Problem May Be Design-dependent. Clin Orthop Rel Res. 2011;469(6):1598-605. doi:10.1007/s11999-011-1777-2.

[78] Walter WL, Waters TS, Gillies M, Donohoo S, Kurtz SM, Ranawat AS et al. Squeaking Hips. J Bone Joint Surg-Am Vol. 2008;90A:102-11. doi:10.2106/jbjs.h.00867.

[79] Walter WL, O'Toole GC, Walter WK, Ellis A, Zicat BA. Squeaking in ceramic-on-ceramic hips - The importance of acetabular component orientation. J Arthroplast. 2007;22(4):496-503. doi:10.1016/j.arth.2006.06.018.

[80] Savarino L, Padovanni G, Ferretti M, Greco M, Cenni E, Perrone G et al. Serum Ion Levels after Ceramic-on-Ceramic and Metal-on-Metal Total Hip Arthroplasty: 8-Year Minimum Follow-up. J Orthop Res. 2008;26(12):1569-76. doi:10.1002/jor.20701.

[81] Swanson TV, Peterson DJ, Seethala R, Bliss RL, Spellmon CA. Influence of Prosthetic Design on Squeaking After Ceramic-on-Ceramic Total Hip Arthroplasty. J Arthroplast. 2010;25(6):36-42. doi:10.1016/j.arth.2010.04.032.

[82] Matar WY, Restrepo C, Parvizi J, Kurtz SM, Hozack WJ. Revision Hip Arthroplasty for Ceramic-on-Ceramic Squeaking Hips Does Not Compromise the Results. J Arthroplast. 2010;25(6):81-6. doi:10.1016/j.arth.2010.05.002.

[83] Taylor S, Manley MT, Sutton K. The role of stripe wear in causing acoustic emissions from alumina ceramic-on-ceramic bearings. J Arthroplast. 2007;22(7):47-51. doi: 10.1016/j.arth.2007.05.038.

[84] Hothan A, Huber G, Weiss C, Hoffmann N, Morlock M. The influence of component design, bearing clearance and axial load on the squeaking characteristics of ceramic hip articulations. J Biomech. 2011;44(5):837-41. doi:10.1016/j.jbiomech.2010.12.012.

[85] Sanders A, Tibbitts I, Brannon R. Concomitant evolution of wear and squeaking in dual-severity, lubricated wear testing of ceramic-on-ceramic hip prostheses. J Orthop Res. 2012;30(9):1377-83. doi:10.1002/jor.22080.

[86] Currier JH, Anderson DE, Van Citters DW. A proposed mechanism for squeaking of ceramic-on-ceramic hips. Wear. 2010;269(11-12):782-9. doi:10.1016/j.wear.2010.08.006.

[87] Walter WL, Kurtz SM, Esposito C, Hozack W, Holley KG, Garino JP et al. Retrieval analysis of squeaking alumina ceramic-on-ceramic bearings. J Bone Joint Surg-Br Vol. 2011;93B(12):1597-601. doi:10.1302/0301-620x.93b12.27529.

[88] Capello WN, Feinberg JR. Use of an alumina ceramic-on-alumina ceramic bearing surface in THA in a 13 year old with JIA--a single case study. Bull NYU Hosp Jt Dis. 2009;67(4):384-6.

[89] Nizard R, Pourreyron D, Raould A, Hannouche D, Sedel L. Alumina-on-alumina hip arthroplasty in patients younger than 30 years old. Clin Orthop Rel Res. 2008;466(2): 317-23. doi:10.1007/s11999-007-0068-4.